BIOINFORMATICS AND MOL

Bioinformatics and Molecular Evolution

Paul G. Higgs and *Teresa K. Attwood*

Blackwell
Publishing

© 2005 by Blackwell Science Ltd
a Blackwell Publishing company

BLACKWELL PUBLISHING
350 Main Street, Malden, MA 02148-5020, USA
9600 Garsington Road, Oxford OX4 2DQ, UK
550 Swanston Street, Carlton, Victoria 3053, Australia

The right of Paul G. Higgs and Teresa K. Attwood to be identified as the Authors of this Work has
been asserted in accordance with the UK Copyright, Designs, and Patents Act 1988.

First published 2005 by Blackwell Publishing Ltd

7   2010

Library of Congress Cataloging-in-Publication Data

Higgs, Paul G.
     Bioinformatics and molecular evolution / Paul G. Higgs and Teresa K. Attwood.
          p. cm.
     Includes bibliographical references and index.
     ISBN 978-1-4051-0683-2 (pbk. : alk. paper)
     1. Molecular evolution—Mathematical models.   2. Bioinformatics.
     [DNLM: 1. Computational Biology.   2. Evolution, Molecular.   3. Genetics.   QU 26.5 H637b 2005]
I. Attwood, Teresa K.   II. Title.

     QH371.3.M37H54 2005
     572.8—dc22

                                                                            2004021066

A catalogue record for this title is available from the British Library.

Set in 9½/12pt Photina
by Graphicraft Ltd, Hong Kong
Printed and bound in Singapore
by Fabulous Printers Pte Ltd

The publisher's policy is to use permanent paper from mills that operate a sustainable forestry policy,
and which has been manufactured from pulp processed using acid-free and elementary chlorine-free
practices. Furthermore, the publisher ensures that the text paper and cover board used have met
acceptable environmental accreditation standards.

For further information on
Blackwell Publishing, visit our website:
www.blackwellpublishing.com

Short Contents

Preface		x
Chapter plan		xiii
1	Introduction: The revolution in biological information	1
2	Nucleic acids, proteins, and amino acids	12
3	Molecular evolution and population genetics	37
4	Models of sequence evolution	58
5	Information resources for genes and proteins	81
6	Sequence alignment algorithms	119
7	Searching sequence databases	139
8	Phylogenetic methods	158
9	Patterns in protein families	195
10	Probabilistic methods and machine learning	227
11	Further topics in molecular evolution and phylogenetics	257
12	Genome evolution	283
13	DNA Microarrays and the 'omes	313
Mathematical appendix		343
List of Web addresses		355
Glossary		357
Index		363

Color plates fall between pages 194 and 195

Full Contents

Preface   x

1 INTRODUCTION: THE REVOLUTION IN BIOLOGICAL INFORMATION   1
1.1   Data explosions   1
1.2   Genomics and high-throughput
      techniques   5
1.3   What is bioinformatics?   6
1.4   The relationship between population
      genetics, molecular evolution, and
      bioinformatics   7
Summary ● References ● Problems   10

2 NUCLEIC ACIDS, PROTEINS, AND AMINO ACIDS   12
2.1   Nucleic acid structure   12
2.2   Protein structure   14
2.3   The central dogma   16
2.4   Physico-chemical properties of the
      amino acids and their importance in
      protein folding   22
Box   2.1 Polymerase chain reaction
           (PCR)   23
2.5   Visualization of amino acid properties
      using principal component
      analysis   25
2.6   Clustering amino acids according to
      their properties   28
Box   2.2 Principal component analysis
           in more detail   29
Summary ● References ● Self-test
Biological background   34

3 MOLECULAR EVOLUTION AND POPULATION GENETICS   37
3.1   What is evolution?   37
3.2   Mutations   39
3.3   Sequence variation within and
      between species   40
3.4   Genealogical trees and coalescence   44
3.5   The spread of new mutations   46
3.6   Neutral evolution and adaptation   49
Box   3.1 The influence of selection on
           the fixation probability   50
Box   3.2 A deterministic theory for
           the spread of mutations   51
Summary ● References ● Problems   54

4 MODELS OF SEQUENCE EVOLUTION   58
4.1   Models of nucleic acid sequence
      evolution   58
Box   4.1 Solution of the Jukes–Cantor
           model   61
4.2   The PAM model of protein sequence
      evolution   65
Box   4.2 PAM distances   70
4.3   Log-odds scoring matrices for amino
      acids   71
Summary ● References ● Problems ●
Self-test Molecular evolution   76

5 INFORMATION RESOURCES FOR GENES AND PROTEINS   81
5.1   Why build a database?   81
5.2   Database file formats   82

5.3	Nucleic acid sequence databases	83
5.4	Protein sequence databases	89
5.5	Protein family databases	95
5.6	Composite protein pattern databases	108
5.7	Protein structure databases	111
5.8	Other types of biological database	113
Summary ● References		115

6 SEQUENCE ALIGNMENT ALGORITHMS — 119

6.1	What is an algorithm?	119
6.2	Pairwise sequence alignment – The problem	121
6.3	Pairwise sequence alignment – Dynamic programming methods	123
6.4	The effect of scoring parameters on the alignment	127
6.5	Multiple sequence alignment	130
Summary ● References ● Problems		136

7 SEARCHING SEQUENCE DATABASES — 139

7.1	Similarity search tools	139
7.2	Alignment statistics (in theory)	147
Box 7.1	Extreme value distributions	151
Box 7.2	Derivation of the extreme value distribution in the word-matching example	152
7.3	Alignment statistics (in practice)	153
Summary ● References ● Problems ● Self-test Alignments and database searching		155

8 PHYLOGENETIC METHODS — 158

8.1	Understanding phylogenetic trees	158
8.2	Choosing sequences	161
8.3	Distance matrices and clustering methods	162
Box 8.1	Calculation of distances in the neighbor-joining method	167
8.4	Bootstrapping	169
8.5	Tree optimization criteria and tree search methods	171

8.6	The maximum-likelihood criterion	173
Box 8.2	Calculating the likelihood of the data on a given tree	174
8.7	The parsimony criterion	177
8.8	Other methods related to maximum likelihood	179
Box 8.3	Calculating posterior probabilities	182
Summary ● References ● Problems ● Self-test Phylogenetic methods		185

9 PATTERNS IN PROTEIN FAMILIES — 195

9.1	Going beyond pairwise alignment methods for database searches	195
9.2	Regular expressions	197
9.3	Fingerprints	200
9.4	Profiles and PSSMs	205
9.5	Biological applications – G protein-coupled receptors	208
Summary ● References ● Problems ● Self-test Protein families and databases		216

10 PROBABILISTIC METHODS AND MACHINE LEARNING — 227

10.1	Using machine learning for pattern recognition in bioinformatics	227
10.2	Probabilistic models of sequences – Basic ingredients	228
Box 10.1	Dirichlet prior distributions	232
10.3	Introducing hidden Markov models	234
Box 10.2	The Viterbi algorithm	238
Box 10.3	The forward and backward algorithms	239
10.4	Profile hidden Markov models	241
10.5	Neural networks	244
Box 10.4	The back-propagation algorithm	249
10.6	Neural networks and protein secondary structure prediction	250
Summary ● References ● Problems		253

11   FURTHER TOPICS IN MOLECULAR EVOLUTION AND PHYLOGENETICS   257
11.1   RNA structure and evolution   257
11.2   Fitting evolutionary models to
         sequence data   266
11.3   Applications of molecular
         phylogenetics   272
Summary ● References   279

12   GENOME EVOLUTION   283
12.1   Prokaryotic genomes   283
Box 12.1 Web resources for bacterial
           genomes   284
12.2   Organellar genomes   298
Summary ● References   309

13   DNA MICROARRAYS AND THE 'OMES   313
13.1   'Omes and 'omics   313
13.2   How do microarrays work?   314
13.3   Normalization of microarray data   316
13.4   Patterns in microarray data   319
13.5   Proteomics   325
13.6   Information management for
         the 'omes   330

Box 13.1 Examples from the Gene
           Ontology   335
Summary ● References ● Self-test   337

MATHEMATICAL APPENDIX   343
M.1   Exponentials and logarithms   343
M.2   Factorials   344
M.3   Summations   344
M.4   Products   345
M.5   Permutations and combinations   345
M.6   Differentiation   346
M.7   Integration   347
M.8   Differential equations   347
M.9   Binomial distributions   348
M.10 Normal distributions   348
M.11 Poisson distributions   350
M.12 Chi-squared distributions   351
M.13 Gamma functions and gamma
         distributions   352
Problems ● Self-test   353

List of Web addresses   355
Glossary   357
Index   363

Preface

RATIONALE

Degree programs in bioinformatics at Masters or undergraduate level are becoming increasingly common and single courses in bioinformatics and computational biology are finding their way into many types of degrees in biological sciences. This book satisfies a need for a textbook that explains the scientific concepts and computational methods of bioinformatics and how they relate to current research in biological sciences. The content is based on material taught by both authors on the MSc program in bioinformatics at the University of Manchester, UK, and on an upper level undergraduate course in computational biology taught by P. Higgs at McMaster University, Ontario.

Many fundamental concepts in bioinformatics, such as sequence alignments, searching for homologous sequences, and building phylogenetic trees, are linked to evolution. Also, the availability of complete genome sequences provides a wealth of new data for evolutionary studies at the whole-genome level, and new bioinformatics methods are being developed that operate at this level. This book emphasizes the evolutionary aspects of bioinformatics, and includes a lot of material that will be of use in courses on molecular evolution, and which up to now has not been found in bioinformatics textbooks.

Bioinformatics chapters of this book explain the need for computational methods in the current era of complete genome sequences and high-throughput experiments, introduce the principal biological databases, and discuss methods used to create and search them. Algorithms for sequence alignment, identification of conserved motifs in protein families, and pattern-recognition methods using hidden Markov models and neural networks are discussed in detail. A full chapter on data analysis for microarrays and proteomics is included.

Evolutionary chapters of the book begin with a brief introduction to population genetics and the study of sequence variation within and between populations, and move on to the description of the evolution of DNA and protein sequences. Phylogenetic methods are covered in detail, and examples are given of application of these methods to biological questions. Factors influencing evolution at the level of whole genomes are also discussed, and methods for the comparison of gene content and gene order between species are presented.

The twin themes of bioinformatics and molecular evolution are woven throughout the book – see the Chapter Plan diagram below. We have considered several possible orders of presenting this material, and reviewers of this book have also suggested their own alternatives. There is no single right way to do it, and we found that no matter in which order we presented the chapters, there was occasionally need to forward-reference material in a later chapter. This order has been chosen so as to emphasize the links between the two themes, and to proceed from background material through standard methods to more advanced topics. Roughly speaking, we would consider everything up to the end of Chapter 9 as fundamental material, and Chapters 10–13 as more advanced methods or more recent applications.

Individual instructors are free to use any combination of chapters in whichever order suits them best.

This book is for people who want to understand bioinformatics methods and who may want to go on to develop methods for themselves. Intelligent use of bioinformatics software requires a proper understanding of the mathematical and statistical methods underlying the programs. We expect that many of our readers will be biological science students who are not confident of their mathematical ability. We therefore try to present mathematical material carefully at an accessible level. However, we do not avoid the use of equations, since we consider the theoretical parts of the book to be an essential aspect. More detailed mathematical sections are placed in boxes aside from the main text, where appropriate. The book contains an appendix summarizing the background mathematics that we would hope bioinformatics students should know. In our experience, students need to be encouraged to remember and practice mathematical techniques that they have been taught in their early undergraduate years but have not had occasion to use.

We discuss computational algorithms in detail but we do not cover programming languages or programming methods, since there are many other books on computing. Although we give some pointers to available software and useful Web sites, this book is not simply a list of programs and URLs. Such material becomes out of date extremely quickly, whereas the underlying methods and principles that we focus on here retain their relevance.

Features

• Comprehensive coverage of bioinformatics in a single text, including sequence analysis, biological databases, pattern recognition, and applications to genomics, microarrays, and proteomics.
• Places bioinformatics in the context of evolutionary biology, giving detailed treatments of molecular evolution and molecular phylogenetics and discussing evolution at the whole-genome level.
• Emphasizes the theoretical and statistical methods used in bioinformatics programs in a way that is accessible to biological science students.
• Extended problem questions provide reinforcement of concepts and application of chapter material.
• Periodic cumulative "self-tests" challenge the students to synthesize and apply their overall understanding of the material up to that point.
• Accompanied by a dedicated Web site at www.blackwellpublishing.com/higgs including the following:
  – all art in downloadable JPEG format (also available to instructors on CD-ROM)
  – all answers to self-tests
  – downloadable sequences
  – links to Web resources.

ACKNOWLEDGMENTS

We wish to thank Andy Brass for his tremendous input to the Manchester Bioinformatics MSc program over many years, and for his "nose" for which subjects are the important ones to teach. We thank Magnus Rattray for his help on the theory and algorithms module and for coming up with several of the problem questions in this book. Thanks are also due to several members of our research groups whose work has helped with some of the topics covered here: Vivek Gowri-Shankar, Daniel Jameson, Howsun Jow, Bin Tang, and Anna Gaulton.

At Blackwell publishers, we would like to thank Liz Marchant, from the UK office, who originally commissioned this book, and Elizabeth Wald and Nancy Whilton at the US office, who have given advice and support in the later stages.

Paul Higgs would also like to thank the enlightened administrative authorities at the University of Manchester for employing a physicist in the biology department and the equally enlightened representatives of McMaster University for employing a bioinformatician in the physics department. The jump between disciplines and between institutions is not too much of a stretch:

```
M--CMASTER
MANCHESTER
```

Paul Higgs
Terri Attwood
May 2004

**Bioinformatics and Molecular Evolution
Chapter Plan**

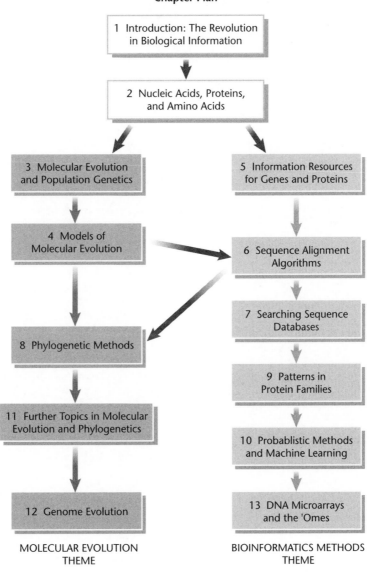

1 Introduction: The Revolution
in Biological Information

2 Nucleic Acids, Proteins,
and Amino Acids

3 Molecular Evolution
and Population Genetics

5 Information Resources
for Genes and Proteins

4 Models of
Molecular Evolution

6 Sequence Alignment
Algorithms

7 Searching Sequence
Databases

8 Phylogenetic Methods

9 Patterns in
Protein Families

11 Further Topics in Molecular
Evolution and Phylogenetics

10 Probablistic Methods
and Machine Learning

12 Genome Evolution

13 DNA Microarrays
and the 'Omes

MOLECULAR EVOLUTION
THEME

BIOINFORMATICS METHODS
THEME

Introduction: The revolution in biological information

CHAPTER

1

CHAPTER PREVIEW

Here we consider the rapid expansion in the amount of biological sequence data available and compare this to the exponential growth in computer speed and memory size that has occurred in the same period. The reader should appreciate why bioinformatics is now essential for understanding the information contained in the sequences, and for efficient storage and retrieval of the information. We also consider some of the history of bioinformatics, and show that many of its foundations are related to molecular evolution and population genetics. Thus, the reader should understand what is meant by the term "bioinformatics" and the role of bioinformatics in relation to other disciplines.

1.1 DATA EXPLOSIONS

In the past decade there has been an explosion in the amount of DNA sequence data available, due to the very rapid progress of genome sequencing projects. There are three principal comprehensive databases of nucleic acid sequences in the world today.

• The EMBL (European Molecular Biology Laboratory) database is maintained at the European Bioinformatics Institute in Cambridge, UK (Stoesser *et al.* 2003).

• GenBank is maintained at the National Center for Biotechnology Information in Maryland, USA (Benson *et al.* 2003).

• The DDBJ (DNA Databank of Japan) is maintained at the National Institute of Genetics in Mishima, Japan (Miyazaki *et al.* 2003).

These three databases share information and hence contain almost identical sets of sequences. The objective of these databases is to ensure that DNA sequence information is stored in a way that is publicly, and freely, accessible and that it can be retrieved and used by other researchers in the future. Most scientific journals require submission of newly sequenced DNA to one of the public databases before a publication can be made that relies on the sequence. This policy has proved tremendously successful for the progress of science, and has led to a rapid increase in the size and usage of sequence databases.

As a measure of the rapid increase in the total available amount of sequence data, Fig. 1.1 and Table 1.1 show the total length of all sequences in GenBank, and the total number of sequences in GenBank as a function of time. Note that the vertical scale is logarithmic and the curves appear approximately as straight lines. This means that the size of GenBank is increasing exponentially with time (see Problem 1.1). The dotted line in the figure is a straight-line fit to the data for the total sequence length (the 1982 point seemed to be an outlier and was excluded). From this we can estimate that the yearly multiplication factor (i.e., the factor by which the amount of data goes up each year) is about 1.6, and that the database doubles in size every 1.4 years. All those sequencing machines are working hard! Interestingly, the curve for the number of sequences almost exactly parallels the curve for the total length. This means that the typical length of one sequence entry in GenBank has remained at

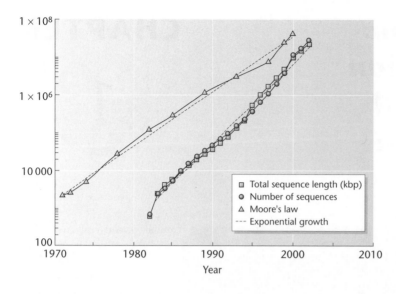

Fig. 1.1 Comparison of the rate of growth of the GenBank sequence (data from Table 1.1) with the rate of growth of the number of transistors in personal computer chips (Moore's law: data from Table 1.2). Dashed lines are fits to an exponential growth law.

Table 1.1 The growth of GenBank.

Year	Base pairs	Sequences
1982	680,338	606
1983	2,274,029	2,427
1984	3,368,765	4,175
1985	5,204,420	5,700
1986	9,615,371	9,978
1987	15,514,776	14,584
1988	23,800,000	20,579
1989	34,762,585	28,791
1990	49,179,285	39,533
1991	71,947,426	55,627
1992	101,008,486	78,608
1993	157,152,442	143,492
1994	217,102,462	215,273
1995	384,939,485	555,694
1996	651,972,984	1,021,211
1997	1,160,300,687	1,765,847
1998	2,008,761,784	2,837,897
1999	3,841,163,011	4,864,570
2000	11,101,066,288	10,106,023
2001	15,849,921,438	14,976,310
2002	28,507,990,166	22,318,883

Data obtained from http://www.ncbi.nih.gov/Genbank/genbankstats.html.

close to 1000. There are, of course, enormous variations in length between different sequence entries.

There is another famous exponentially increasing curve that goes by the name of Moore's law. Moore (1965) noticed that the number of transistors in integrated circuits appeared to be roughly doubling every year over the period 1959–65. Data on the size of Intel PC chips (Table 1.2) show that this exponential increase is still continuing. Looking at the data more carefully, however, we see that the estimate of doubling every year is rather overoptimistic. The chip size is actually doubling every **two** years and the yearly multiplication factor is 1.4. Although extremely impressive, this is substantially slower than the rate of increase of GenBank (see Fig. 1.1 and Table 1.3).

What about the world's fastest supercomputers? Jack Dongarra and colleagues from the University of Tennessee introduced the LINPACK benchmark, which measures the speed of computers at solving a complex set of linear equations. A list of the top 500 supercomputers according to this benchmark is published twice yearly (http://www.top500.org). Figure 1.2 shows the performance benchmark rate of the top computer at each release of the list. Once again, this is approximately an exponential (with large fluctuations). The best-fit straight line has a

Table 1.2 The growth of the number of transistors in personal computer processors.

Type of processor	Year of introduction	Transistors
4004	1971	2,250
8008	1972	2,500
8080	1974	5,000
8086	1978	29,000
286	1982	120,000
386™ processor	1985	275,000
486™ DX processor	1989	1,180,000
Pentium® processor	1993	3,100,000
Pentium II processor	1997	7,500,000
Pentium III processor	1999	24,000,000
Pentium 4 processor	2000	42,000,000

Data obtained from Intel
(http://www.intel.com/research/silicon/mooreslaw.htm).

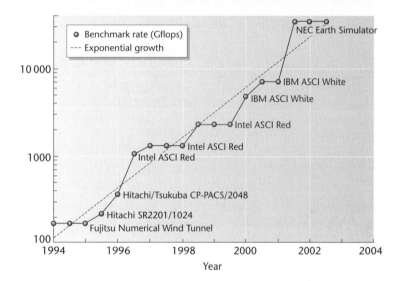

Fig. 1.2 The performance of the world's top supercomputers using the LINPACK benchmark (Gflops). Data from http://www.top500.org.

doubling time of 1.04 years. So supercomputers seem to be beating GenBank for the moment. However, most of us do not have access to a supercomputer. The PC chip size may be a better measure of the amount of computing power available to anyone using a desktop.

Clearly, we have reached a point where computers are essential for the storage, retrieval, and analysis of biological sequence data. However, we cannot simply rely on computers and stop thinking. If we stick with our same old computing methods, then we will be limited by the hardware. We still need people, because only people can think of better and faster algorithms for data analysis. That is what this book is about. We will discuss the methods and algorithms used in bioinformatics, so that hopefully you will understand enough to be able to improve those methods yourself.

Another important type of biological data that is exponentially increasing is protein structures. PDB is a database of protein structures obtained from X-ray crystallography and NMR experiments. From the number of entries in PDB in successive releases, we calculated that the doubling time for the number

Table 1.3 Comparison of rates of increase of several different data explosion curves.

Type of data	Growth rate, r	Doubling time, T (years)	Yearly multiplication factor, R
GenBank (total sequence length)	0.480	1.44	1.62
PC chips (number of transistors)	0.332	2.09	1.39
Supercomputer speed (LINPACK benchmark)	0.664	1.04	1.94
Protein structures (number of PDB entries)	0.209	3.31	1.23
Number of complete prokaryotic genomes	0.518	1.34	1.68
Abstracts: bioinformatics	0.587	1.18	1.80
Abstracts: genomics	0.569	1.22	1.77
Abstracts: proteomics	0.996	0.70	2.71
Abstracts: phylogenetic(s)	0.188	3.68	1.21
Abstracts: total	0.071	9.80	1.07

Table 1.4 The history of genome-sequencing projects.

Year	Archaea	Bacteria	Eukaryotes	Landmarks
1995	0	2	0	First bacterial genome: *Haemophilus influenzae*
1996	1	2	0	First archaeal genome: *Methanococcus jannaschii*
1997	2	4	1	First unicellular eukaryote: *Saccharomyces cerevisiae*
1998	1	5	1	First multicellular eukaryote: *Caenorhabditis elegans*
1999	1	4	1	—
2000	3	13	2	First plant genome: *Arabidopsis thaliana*
2001	2	24	3	First release of the human genome
2002	6	32	9	—
2003 (to July)	0	25	2	—
Total	16	111	19	—

Data from the Genomes OnLine Database (http://wit.integratedgenomics.com/GOLD/).

of available protein structures is 3.31 years (Table 1.3), which is considerably slower than the number of sequences. Since the number of experimentally determined structures is lagging further and further behind the number of sequences, computational methods for structure prediction are important. Many of these methods work by looking for similarities in sequence between a protein of unknown structure and a protein of known structure, and use this to make predictions about the unknown structure. These techniques will become increasingly useful as our knowledge of real examples increases.

In 1995, the bacterium *Haemophilus influenzae* entered history as the first organism to have its genome completely sequenced. Sequencing technology has advanced rapidly and has become increasingly automated. The sequencing of a new prokaryotic genome has now become almost commonplace. Table 1.4 shows the progress of complete genome projects with some historical landmarks. With the publication of the human genome in 2001, we can now truly say that we are in the "post-genome age". The number of complete prokaryotic genomes (total of archaea plus bacteria from Table 1.4) is going through its own data explosion. The doubling time is about 1.3 years and the yearly multiplication factor is about 1.7. For the present, complete eukaryotic genomes are still rather few, so that the publication of each individual genome still retains its status as a landmark event. It seems only a matter of time, however, before we shall be able to draw a data explosion curve for the number of eukaryotic genomes too.

This book emphasizes the relationship between bioinformatics and molecular evolution. The availability of complete genomes is tremendously important for evolutionary studies. For the first time we can begin to compare whole sets of genes between organisms, not just single genes. For the first time we can begin to study the processes that govern the evolution of whole genomes. This is therefore an exciting time to be in the bioinformatics area.

1.2 GENOMICS AND HIGH-THROUGHPUT TECHNIQUES

The availability of complete genomes has opened up a whole research discipline known as **genomics**. Genomics refers to scientific studies dealing with whole sets of genes rather than single genes. The advances made in sequencing technology have come at the same time as the appearance of new **high-throughput** experimental techniques. One of the most important of these is **microarray** technology, which allows measurement of the expression level (i.e., mRNA concentration) of thousands of genes in a cell simultaneously. For example, in the case of the yeast, *Saccharomyces cerevisiae*, where the complete genome is available, we can put probes for **all** the genes onto one microarray chip. We can then study the way the expression levels of all the genes respond to changes in external conditions or the way that they vary during the cell cycle. Complete genomes therefore change the way that experimental science is carried out, and allow us to address questions that were not possible before.

Another important field where high-throughput techniques are used is **proteomics**. Proteomics is the study of the proteome, i.e., the complete set of proteins in a cell. The experimental techniques used are principally two-dimensional gel electrophoresis for the separation of the many different proteins in a cell extract, and mass spectrometry for identifying proteins by their molecular masses. Once again, the availability of complete genomes is tremendously important, because the masses of the proteins determined by mass spectrometry can be compared directly to the masses of proteins expected from the predicted position of open reading frames in the genome.

High-throughput experiments produce large amounts of quantitative data. This poses challenges for bioinformaticians. How do we store information from a microarray experiment in such a way that it can be compared with results from other groups? How do we best extract meaningful information from the vast array of numbers generated? New statistical methods are needed to spot significant trends and patterns in the data. This is a new area of biological sciences where computational methods are essential for the progress of the experimental science, and where algorithms and experimental techniques are being developed side by side.

As a measure of the interest of the scientific community in genomics and related areas, let us look at the number of scientific papers published in these areas over the past few years. The ISI Science Citation Index allows searches for articles published in specific years that use specified words in their title, keywords, or abstract. Figure 1.3 shows the numbers of published articles (cumulative since 1981) for several important terms relevant to this book. Papers using the words "genomics" and "bioinformatics" increase at almost exactly the same rate, both having yearly multiplication factors of 1.8 and doubling times of 1.2 years. "Proteomics" is a very young field, with no articles found prior to 1998. The doubling time is 0.7 years: the fastest growth of any of the quantities considered in Table 1.3. References to "microarray" also increase rapidly. This curve appears significantly nonlinear because there are several different meanings for the term. Almost all the references prior to about 1996 refer to microarray electrodes, whereas in later years, almost all refer to DNA microarrays for gene expression. The rate of increase of the use of DNA microarrays is therefore steeper than it appears in the figure.

The number of papers using both "sequence" and "database" is much larger than those using any of the terms considered above (although it is increasing less rapidly). This shows how important biological databases and the algorithms for searching them have

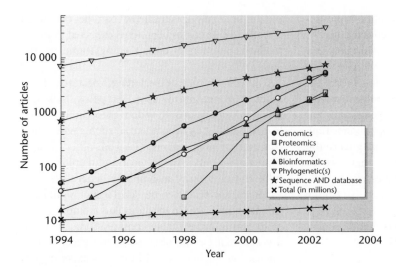

Fig. 1.3 Cumulative number of scientific articles published from 1981 to the date shown that use specific terms in the title, keywords, or abstract. Data from the Science Citation Index (SCI-EXPANDED) available at http://wos.mimas.ac.uk/ or http://isi6.isiknowledge.com.

become to the biological science community in the past decade. The number of papers using the term "phylogenetic" or "phylogenetics" dwarfs those using all the other terms considered here by at least an order of magnitude. This curve is a remarkably good exponential, although the doubling time is fairly long (3.7 years). Phylogenetics is a relatively old area, where morphological studies predate the availability of molecular sequences by several decades. The high level of interest in the field in recent years is largely a result of the availability of sequence data and of new methods for tree construction. Very large sequence data sets are now being used, and we are beginning to resolve some of the controversial aspects of evolutionary trees that have been argued over for decades.

As a comparison, for all the curves in Fig. 1.3 that refer to specific scientific terms, the figure also shows the total number of articles in the Science Citation Index (this curve is in millions of articles, whereas the others are in individual articles). The total level of scientific activity (or at least, scientific writing) has also been increasing exponentially, and hence we all have to read more and more in order to keep up. This curve is an almost perfect exponential, with a doubling time of 9.8 years. Thus, all the curves related to the individual subjects are increasing far more rapidly than the total accumulation of scientific knowledge.

At this point you will be suitably impressed by the importance of the subject matter of this book and will be eager to read the rest of it!

1.3 WHAT IS BIOINFORMATICS?

Since bioinformatics is still a fairly new field, people have a tendency to ask, "What is bioinformatics?" Often, people seem to worry that it is not very well defined, and tend to have a suspicious look in their eyes when they ask. These people would never trouble to ask "What is biology?" or "What is genetics?" In fact, bioinformatics is no more difficult or more easy to define than these other fields. Here is our short and simple definition.

Bioinformatics is the use of computational methods to study biological data.

In case this is too short and too general for you, here is a longer one.

Bioinformatics is:
(i) the development of computational methods for studying the structure, function, and evolution of genes, proteins, and whole genomes;
(ii) the development of methods for the management and analysis of biological information arising from genomics and high-throughput experiments.

If that is still too short, have another look at the contents list of this book to see what we think are the most important topics that make up the field of bioinformatics.

1.4 THE RELATIONSHIP BETWEEN POPULATION GENETICS, MOLECULAR EVOLUTION, AND BIOINFORMATICS

1.4.1 A little history . . .

The field of population genetics is concerned with the variation of genes within a population. The issues of natural selection, mutation, and random drift are fundamental to population genetics. Alternative versions of a gene are known as alleles. A large body of population genetics theory is used to interpret experimental data on allele frequency distributions and to ask questions about the behavior of the organisms being studied (e.g., effective population size, pattern of migration, degree of inbreeding). Population genetics is a well-established discipline with foundations dating back to Ronald Fisher and Sewall Wright in the first half of the twentieth century. These foundations predate the era of molecular sequences. It is possible to discuss the theory of the spread of a new allele, for example, without knowing anything about its sequence.

Molecular evolution is a more recent discipline that has arisen since DNA and protein sequence information has become available. Molecular techniques provide many types of data that are of great use to population geneticists, e.g., allozymes, microsatellites, restriction fragment length polymorphisms, single nucleotide polymorphisms, human mitochondrial haplotypes. Population geneticists are interested in what these molecular markers tell us about the organisms (see the many examples in the book by Avise 1994). In contrast, the focus of molecular evolution is on the molecules themselves, and understanding the processes of mutation and selection that act on the sequences. There are many genes that have now been sequenced in a large number of different species. This usually means that we have a representative example of a single gene sequence from each species. There are only a few species for which a significant amount of information about within-species sequence variation is available (e.g., humans and *Drosophila*). The emphasis in molecular evolution therefore tends to be on comparative molecular studies **between** species, while population genetics usually considers variation **within** a species.

The article by Zuckerkandl and Pauling (1965) is sometimes credited with inventing the field of molecular evolution. This was the first time that protein sequences were used to construct a molecular phylogeny and it set many people thinking about biological sequences in a **quantitative** way. 1965 was the same year in which Moore invented his law and in which computers were beginning to play a significant role in science. Indeed, molecular biology has risen to prominence in the biological sciences in the same time frame that computers have risen to prominence in society in general.

We might also argue that bioinformatics was beginning in 1965. The first edition of the *Atlas of Protein Sequence and Structure*, compiled by Margaret Dayhoff, appeared in printed form in 1965. The *Atlas* later became the basis for the PIR protein sequence database (Wu *et al.* 2002). However, this is stretching the point a little. The term bioinformatics was not around in 1965, and barring a few pioneers, bioinformatics was not an active research area at that time. As a discipline, bioinformatics is more recent. It arose from the recognition that efficient computational techniques were needed to study the huge amount of biological sequence information that was becoming available. If molecular biology arose at the same time as scientific computing, then we may also say that bioinformatics arose at the same time as the Internet. It is **possible** to imagine the existence of biological sequence databases without the Internet, but they would be a whole lot less useful. Database use would be restricted to those who subscribed to postal deliveries of database releases. Think of that cardboard box arriving each month and getting exponentially bigger each time. Amos Bairoch of the Swiss Institute of Bioinformatics comments (Bairoch 2000) that in 1988, the version of their PC/Gene database and software was shipped as 53 floppy disks! For that matter, think how difficult it would be to submit sequences to a database if it were not for email and the Internet.

At this point, the first author of this book starts to feel old. Coincidentally, I also first saw the light of day in 1965. Shortly afterwards, in 1985, I was happily writing programs with DO-loops in them for mainframes (students who are too young to know what mainframe computers are probably do not need to know). In 1989, someone first showed me how to use a mouse. I remember this clearly because I used the mouse for the first time when I began to write my Ph.D. thesis. It is scary to think almost all my education is pre-mouse. Possibly even more frightening is that I remember – it must have been in 1994 – someone explaining to our academic department how the World-Wide Web worked and what was meant by the terms URL and Netscape. A year or so after that, use of the Internet had become a daily affair for me. Now, of course, if the network is down for a day, it is impossible to do anything at all!

1.4.2 Evolutionary foundations for bioinformatics

Let's get back to the plot. Bioinformatics is a new discipline. Since this is a bioinformatics book, why do we need to know about the older subjects of molecular evolution and population genetics? There is a famous remark by the evolutionary biologist Theodosius Dobzhansky that, "Nothing in biology makes sense except in the light of evolution". You will find this quoted in almost every evolutionary textbook, but we will not apologize for quoting it once again. In fact, we would like to update it to, "Nothing in bioinformatics makes sense except in the light of evolution". Let's consider some examples to see why this is so.

The most fundamental and most frequently used procedure in bioinformatics is pairwise sequence alignment. When amino acid sequences are aligned, we use a scoring system, such as a PAM matrix, to determine the score for aligning any amino acid with any other. These scoring systems are based on evolutionary models. High scores are assigned to pairs of amino acids that frequently substitute for one another during protein sequence evolution. Low, or negative, scores are assigned to pairs of amino acids that interchange very rarely. When RNA sequences are aligned, we often use the fact that the secondary structure tends to be conserved, and that pairs of compensatory substitutions occur in the base-paired regions of the structure. Thus, creating accurate sequence alignments of both proteins and RNAs relies on an understanding of molecular evolution.

If we want to know something about a particular biological sequence, the first thing we do is search the database to find sequences that are similar to it. In particular, we are often interested in sequence motifs that are well conserved and that are present in a whole family of proteins. The logic is that important parts of a sequence will tend to be conserved during evolution. Protein family databases like PROSITE, PRINTS, and InterPro (see Chapter 5) identify important conserved motifs in protein alignments and use them to assign sequences to families. An important concept here is **homology**. Sequences are homologous if they descend from a common ancestor, i.e., if they are related by the evolutionary process of divergence. If a group of proteins all share a conserved motif, it will often be because all these proteins are homologous. If a motif is very short, however, there is some chance that it will have evolved more than once independently (**convergent evolution**). It is therefore important to try to distinguish chance similarities arising from convergent evolution from similarities arising from divergent evolution. The thrust of protein family databases is therefore to facilitate the identification of true homologs, by making the distinction between chance and real matches clearer.

Similar considerations apply in protein structural databases. It is often observed that distantly related proteins have relatively conserved structures. For example, the number and relative positions of α helices and β strands might be the same in two proteins that have very different sequences. Occasionally, the sequences are so different that it would be very difficult to establish a relationship between them if the structure were not known. When similar (or identical) structures are found in different proteins, it probably indicates homology, but the possibility of small structural motifs arising more than once still needs to be considered. Another important aspect of protein structure that is strongly linked to evolution is **domain shuffling**. Many large proteins are composed of smaller domains that are continuous sections of the sequence that fold into fairly well-defined three-dimensional structures; these assemble to form the overall protein structure. Particularly in eukaryotes, it is found that certain domains occur in many different proteins in different combinations and different orders. See the ProDom database (Corpet *et al.* 2000), for example. Although bioinformaticians will argue about what constitutes a domain and where the boundaries between domains lie, it is clear that the duplication and reshuffling of domains is a very useful way of evolving new complex proteins. The main message is that in order to create reliable information resources for protein sequences, structures, and domains, we need to have a good understanding of protein evolution.

In recent years, evolutionary studies have also become possible at the whole genome level. If we want to compare the genomes of two species, it is natural to ask which genes are shared by both species. This question can be surprisingly hard to answer. For each gene in the first species, we need to decide if there is a gene in the second species that is homologous to it. It may be difficult to detect similarity between sequences from different species simply because of the large amount of evolutionary change that has gone on since the divergence of the species. Most genomes contain many open reading frames that are thought to be genes, but for which no similar sequence can be found in other species. This is evidence for the limitations of our current methods as much as for the diversity generated by molecular evolution. In cases where we **are** able to detect similarity, then it can still be tough to decide which genes are homologous. Many genomes contain families of duplicated genes that often have slightly different functions or different sites of expression within the organism. Sequences from one species that are evolutionarily related and that diverged from one another due to a gene duplication event are called paralogous sequences, in contrast to orthologous sequences, which are sequences in different organisms that diverged from one another due to the split

between the species. Duplications can occur in different lineages independently, so that a single gene in one species might be homologous to a whole family in the other species. Alternatively, if duplications occurred in a common ancestor, then both species should contain a copy of each member of the gene family – unless, of course, some genes have been deleted in one or other species. Another factor to consider, particularly for bacteria, is that genomes can acquire new genes by horizontal transfer of DNA from unrelated species. This sequence comparison can show up genes that are apparently homologous to sequences in organisms that are otherwise thought to be extremely distantly related. A major task for bioinformatics is to establish sets of homologous genes between groups of species, and to understand how those genes got to be where they are. The flip side of this is to be able to say which genes are **not** present in an organism, and how the organism manages to get by without them.

The above examples show that many of the questions addressed in bioinformatics have foundations in questions of molecular evolution. A fair amount of this book is therefore devoted to molecular evolution. What about population genetics? There are many other books on population genetics and hence this book does not try to be a textbook of this area. However, there are some key points that are usually considered in population genetics courses that we need to consider if we are to properly understand molecular evolution and bioinformatics. These questions concern the way in which sequence diversity is generated in populations and the way in which new variant sequences spread through populations. If we run a molecular phylogeny program, for example, we might be asking whether "the" sequence from humans is more similar to "the" sequence from chimpanzees or gorillas. It is important to remember that these sequences have diverged as a result of the fixation of new sequence variants in the populations. We should also not forget that the sequences we have are just samples from the variations that exist in each of the populations.

There are some bioinformatics areas that have a direct link to the genetics of human populations. We are accumulating large amounts of information about variant gene sequences in human populations, particularly where these are linked to hereditary diseases. Some of these can be major changes, like deletions of all or part of a gene or a chromosome region. Some are single nucleotide polymorphisms, or SNPs, where just a single base varies at a particular site in a gene. Databases of SNPs potentially contain information of great relevance to medicine and to the pharmaceutical industry. The area of **pharmacogenomics** attempts to understand the way that different patients respond more or less well to drug treatments according to which alleles they have for certain genes. The hope is that drug treatments can be tailored to suit the genetic profile of the patient. However, many important diseases are not caused by a single gene. Understanding the way that variations at many different loci combine to affect the susceptibility of individuals to different medical problems is an important goal, and developing computational techniques to handle data such as SNPs, and to extract information from the data, is an important application of bioinformatics.

SUMMARY

The amount of biological sequence information is increasing very rapidly and seems to be following an exponential growth law. Computational methods are playing an increasing role in biological sciences. New algorithms will be required to analyze this information and to understand what it means. Genome sequencing projects have been remarkably successful, and comparative analysis of whole genomes is now possible. This provides challenges and opportunities for new types of study in bioinformatics. At the same time, several types of experimental methods are being developed currently that may be classed as "high-throughput". These include microarrays, proteomics, and structural genomics. The philosophy behind these methods is to study large numbers of genes or proteins simultaneously, rather than to specialize in individual cases. Bioinformatics therefore has a role in developing statistical methods for analysis of large data sets, and in developing methods of information management for the new types of data being generated.

Evolutionary ideas underlie many of the methods used in bioinformatics, such as sequence alignments, identifying families of genes and proteins, and establishing homology between genes in different organisms. Evolutionary tree construction (i.e., molecular phylogenetics) is itself a very large field within computational biology. Since we now have many complete genomes, particularly in bacteria, we can also begin to look at evolutionary questions at the whole-genome level. This book will therefore pay particular attention to the evolutionary aspects of bioinformatics.

REFERENCES

Avise, J.C. 1994. *Molecular Markers, Natural History, and Evolution.* New York: Chapman and Hall.

Bairoch, A. 2000. Serendipity in bioinformatics: The tribulations of a Swiss bioinformatician through exciting times. *Bioinformatics*, **16**: 48–64.

Benson, D.A., Karsch-Mizrachi, I., Lipman, D.J., Ostell, J., and Wheeler, D.L. 2003. GenBank. *Nucleic Acids Research*, **31**: 23–7.

Corpet, F., Sernat, F., Gouzy, J., and Kahn, D. 2000. ProDom and ProDom-CG: Tools for protein domain analysis and whole genome comparisons. *Nucleic Acids Research*, **28**: 267–9. http://prodes.toulouse.inra.fr/prodom/2002.1/html/home.php

Miyazaki, S., Sugawara, H., Gojobori, T., and Tateno, Y. 2003. DNA Databank of Japan (DDBJ) in XML. *Nucleic Acids Research*, **31**: 13–16.

Moore, G.E. 1965. Cramming more components onto integrated circuits. *Electronics*, **38**(8): 114–17.

Stoesser, G., Baker, W., van den Broek, A., Garcia-Pastor, M., Kanz, C., Kulikova, T., Leinonen, R., Lin, Q., Lombard, V., Lopez, R., Mancuso, R., Nardone, R., Stoehr, P., Tuli, M.A., Tzouvara, K., and Vaughan, R. 2003. The EMBL Nucleotide Sequence Database: Major new developments. *Nucleic Acids Research*, **31**: 17–22.

Wu, C.H., Huang, H., Arminski, L., Castro-Alvear, J., Chen, Y., Hu, Z.Z., Ledley, R.S., Lewis, K.C., Mewes, H.W., Orcutt, B., Suzek, B.E., Tsugita, A., Vinayaka, C.R., Yeh, L.L., Zhang, J., and Barker, W.C. 2002. The Protein Information Resource: An integrated public resource of functional annotation of proteins. *Nucleic Acids Research*, **30**: 35–7. http://pir.georgetown.edu/.

Zuckerkandl, E. and Pauling, L. 1965. Evolutionary divergence and convergence in proteins. In V. Bryson, and H.J. Vogel (eds.), *Evolving Genes and Proteins*, pp. 97–166. New York: Academic Press.

PROBLEMS

1.1 The data explosion curves provide us with a good way of revising some fundamental points in mathematics that will come in handy later in the book. Now would also be a good time to check the Mathematical appendix and maybe have a go at the Self-test that goes with it.

For each type of data considered in this chapter, we have a quantity $N(t)$ that is increasing with time t, and we are assuming that it follows the law:

$$N(t) = N_0 \exp(rt)$$

Here, N_0 is the value at the initial time point, and r is the growth rate. In the cases we considered, time was measured in years. We defined the yearly multiplication factor as the factor by which N increases each year, i.e.

$$R = \frac{N(t + 1)}{N(t)} = \exp(r)$$

If the data really follow an exponential law, then this ratio is the same at whichever time we measure it. Another way of writing the growth law is therefore:

$$N(t) = N_0 R^t$$

The other useful quantity that we measured was the doubling time T. To calculate T we require that the number after a time T is twice as large as its initial value.

$$\frac{N(T)}{N_0} = \exp(rT) = 2$$

Hence $rT = \ln(2)$ or $T = \ln(2)/r$.

If any of these steps is not obvious, then you should revise your knowledge of exponentials and logarithms. There are some helpful pointers in the Mathematical appendix of this book, Section M.1.

1.2 Use the data from Tables 1.1 and 1.2 and plot your own graphs. The figures in this chapter plot N directly against t and use a logarithmic scale on the vertical axis. This comes out to be a straight line because:

$$\ln(N) = \ln(N_0) + rt$$

so the slope of the line is r. The other way to do it is to calculate $\ln(N)$ at each time point with a calculator, and then to plot $\ln(N)$ against t using a linear scale on both axes. Plot the graphs both ways and make sure they look the same.

My graph-plotting package will do a best fit of an exponential growth law to a set of data points. This is how the values of r were obtained in Table 1.3. However, if your package will not do that, then you can also estimate r from the $\ln(N)$ versus t graph by using a straight-line fit. Try doing the fit to the data in both ways and make sure that you get the same answer.

1.3 The exponential growth law arises from the assumption that the rate of increase of N is proportional to its current value. Thus the growth law is the solution of the differential equation

$$\frac{dN}{dt} = rN$$

Now would be a good time to make sure you understand what this equation means (see Sections M.6 and M.8 for some help).

1.4 While the assumption in 1.3 might have some plausibility for the increase in the size of a rabbit population (if they have a limitless food supply), there does not seem to be a theoretical reason why the size of GenBank or the size of a PC chip should increase exponentially. It is just an empirical observation that it works that way. Presumably, sooner or later all these curves will hit a limit.

There are several other types of curve we might imagine to describe an increasing function of time.

Linear increase: $\qquad N(t) = A + Bt$

Power law increase: $\qquad N(t) = At^k$ (for some value of k not equal to 1)

Logarithmic increase: $N(t) = A + B\ln(t)$

In each case, A, B, and k are arbitrary constants that could be obtained by fitting the curve to the data. Try to fit the data in Tables 1.1 and 1.2 to these other growth laws. Is it true that the exponential growth law fits better than the alternatives?

If you had some kind of measurements that you believed followed one of the other growth laws, how would you plot the graph so that the points would lie on a straight line?

Nucleic acids, proteins, and amino acids

CHAPTER

2

CHAPTER PREVIEW

This chapter begins with a basic introduction to the chemical and physical structure of nucleic acids and proteins for those who do not have a background in biochemistry or biology. We also give a reminder of the processes of transcription, RNA processing, translation and protein synthesis, and DNA replication. We then give a detailed discussion of the physico-chemical properties of the amino acids and their relevance in protein folding. We use the amino acid property data as an example when introducing two statistical techniques that are useful in bioinformatics: principal component analysis and clustering algorithms.

2.1 NUCLEIC ACID STRUCTURE

There are two types of nucleic acid that are of key importance in cells: deoxyribonucleic acid (DNA) and ribonucleic acid (RNA). The chemical structure of a single strand of RNA is shown in Fig. 2.1. The backbone of the molecule is composed of ribose units (five-carbon sugars) linked by phosphate groups in a repeating polymer chain. Two repeat units are shown in the figure. The carbons in the ribose are conventionally numbered from 1 to 5, and the phosphate groups are linked to carbons 3 and 5. At one end, called the 5′ ("five prime") end, the last carbon in the chain is a number 5 carbon, whereas at the other end, called the 3′ end, the last carbon is a number 3. We often think of a strand as beginning at the 5′ end and ending at the 3′ end, because this is the direction in which genetic information is read. The backbone of DNA differs in that deoxyribose sugars are used instead of ribose. The OH group on carbon number 2 in ribose is simply an H in deoxyri-bose, but the molecules are otherwise the same.

Each sugar is linked to a molecule known as a base. In DNA, there are four types of base, called adenine, thymine, guanine, and cytosine, usually referred to simply as A, T, G, and C. The structures of these molecules are shown in Fig. 2.2. In RNA, the base uracil (U; Fig. 2.2) occurs instead of T. The structure of U is similar to that of T but lacks the CH_3 group linked to the ring of the T molecule. In addition, a variety of bases of slightly different structures, called modified bases, can also be found in some types of RNA molecule. A and G are known as purines. They both have a double ring in their chemical structure. C, T, and U are known as pyrimidines. They have a single ring in their chemical structure. The fundamental building block of nucleic acid chains is called a nucleotide: this is a unit of one base plus one sugar plus one phosphate. We usually think of the "length" of a nucleic acid sequence as the number of nucleotides in the chain. Nucleotides are also found as separate molecules in the cell, as well as being part of nucleic acid polymers. In this case, there are usually two or three phosphate groups attached to the same nucleotide. For example, ATP (adenosine triphosphate) is an important molecule in cellular metabolism, and it has three phosphates attached in a chain.

DNA is usually found as a double strand. The two strands are held together by hydrogen bonding

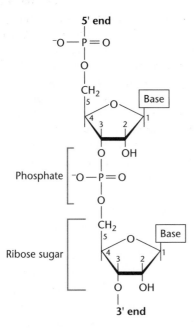

Fig. 2.1 Chemical structure of the RNA backbone showing ribose units linked by phosphate groups.

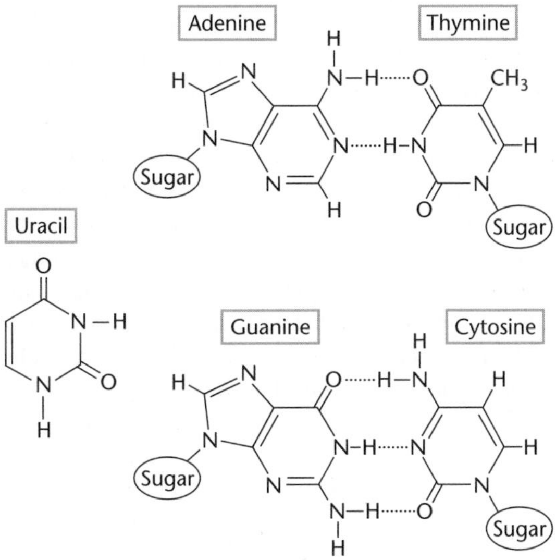

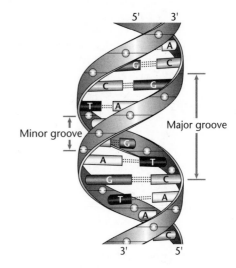

Fig. 2.3 Schematic diagram of the DNA double helical structure.

Fig. 2.2 The chemical structure of the four bases of DNA showing the formation of hydrogen-bonded AT and GC base pairs. Uracil is also shown.

between A and T and between C and G bases (Fig. 2.2). The two strands run in opposite directions and are exactly complementary in sequence, so that where one has A, the other has T and where one has C the other has G. For example:

$$5'-C-C-T-G-A-T-3'$$
$$3'-G-G-A-C-T-A-5'$$

The two strands are coiled around one another in the famous double helical structure elucidated by Watson and Crick 50 years ago. This is shown schematically in Fig. 2.3.

In contrast, RNA molecules are usually single stranded, and can form a variety of structures by base pairing between short regions of complementary sequences within the same strand. An example of this is the cloverleaf structure of transfer RNA (tRNA), which has four base-paired regions (stems) and three hairpin loops (Fig. 2.4). The base-pairing rules in RNA are more flexible than DNA. The principal pairs are GC and AU (which is equivalent to AT in DNA), but GU pairs are also relatively frequent, and a variety of unusual, so-called "non-canonical", pairs are also found in some RNA structures (e.g., GA pairs). A two-dimensional drawing of the base-pairing pattern is called a secondary structure

Nucleic acids, proteins, and amino acids ● **13**

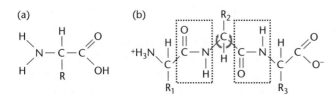

Fig. 2.4 Secondary structure of tRNA-Ala from *Escherichia coli* showing the anticodon position and the site of amino acid attachment.

diagram. In Chapter 11, we will discuss RNA secondary structure in more detail.

2.2 PROTEIN STRUCTURE

The fundamental building block of proteins is the amino acid. An amino acid has the chemical structure shown in Fig. 2.5(a), with an amine group on one side and a carboxylic acid group on the other. In solution, these groups are often ionized to give NH_3^+ and COO^-. There are 20 types of amino acid found in proteins. These are distinguished by the nature of the side-chain group, labeled R in Fig. 2.5(a). The central carbon to which the R group is attached is known as the α carbon. Proteins are linear polymers composed of chains of amino acids. The links are formed by removal of an OH from one amino acid and an H from the next to give a water molecule. The

resultant linkage is called a peptide bond. These are shown in boxes in Fig. 2.5(b), which illustrates a tripeptide, i.e., a chain composed of three amino acids. Proteins, or "polypeptides", are typically composed of several hundred amino acids.

The chemical structures of the side chains are given in Fig. 2.6. Each amino acid has a standard three-letter abbreviation and a standard one-letter code, as shown in the figure. A protein can be represented simply by a sequence of these letters, which is very convenient for storage on a computer, for example:

```
MADIQLSKYHVSKDIGFLLEPLQDVLPDYFAPWNR
LAKSLPDLVASHKFRDAVKEMPLLDSSKLAGYRQK
```

is the first part of a real protein. The two ends of a protein are called the N terminus and the C terminus because one has an unlinked NH_3^+ group and the other has an unlinked COO^- group. Protein sequences are traditionally written from the N to the C terminus, which corresponds to the direction in which they are synthesized.

The four atoms involved in the peptide bond lie in a plane and are not free to rotate with respect to one another. This is due to the electrons in the chemical bonds, which are partly delocalized. The flexibility of the protein backbone comes mostly from rotation about the two bonds on either side of each α carbon. Many proteins form globular three-dimensional structures due to this flexibility of the backbone. Each protein has a structure that is specific to its sequence. The formation of this three-dimensional structure is called "protein folding". The amino acids vary considerably in their properties. The combination of repulsive and attractive interactions between the different amino acids, and between the amino acids and water, determines the way in which a protein folds. An important role of proteins is as catalysts of

Fig. 2.5 Chemical structure of an amino acid (a) and the protein backbone (b). The peptide bond units (boxed) are planar and inflexible. Flexibility of the backbone comes from rotation about the bonds next to the α carbons (indicated by arrows).

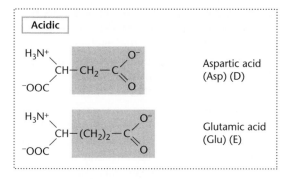

Acidic

Aspartic acid
(Asp) (D)

Glutamic acid
(Glu) (E)

Basic

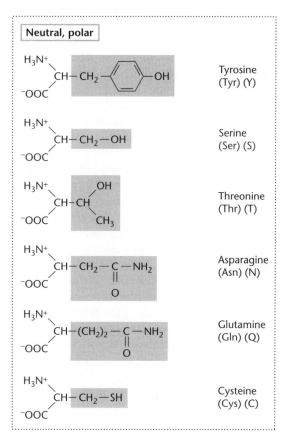

Lysine
(Lys) (K)

Arginine
(Arg) (R)

Histidine
(His) (H)

Neutral, nonpolar

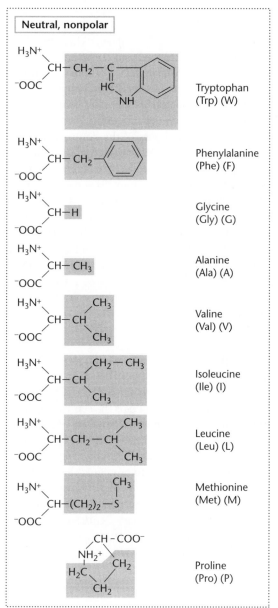

Tryptophan
(Trp) (W)

Phenylalanine
(Phe) (F)

Glycine
(Gly) (G)

Alanine
(Ala) (A)

Valine
(Val) (V)

Isoleucine
(Ile) (I)

Leucine
(Leu) (L)

Methionine
(Met) (M)

Proline
(Pro) (P)

Neutral, polar

Tyrosine
(Tyr) (Y)

Serine
(Ser) (S)

Threonine
(Thr) (T)

Asparagine
(Asn) (N)

Glutamine
(Gln) (Q)

Cysteine
(Cys) (C)

Fig. 2.6 Chemical structures of the 20 amino acid side chains.

biochemical reactions – protein catalysts are called enzymes. Proteins are able to catalyze a huge variety of chemical reactions due to the wide range of different side groups of the amino acids and the vast number of sequences that can be assembled by combining the 20 building blocks in specific orders.

Many protein structures are known from X-ray crystallography. Plate 2.1(a) illustrates the interaction between a glutaminyl-tRNA molecule and a protein called glutaminyl-tRNA synthetase. We will discuss the functions of these molecules below when we describe the process of translation and protein synthesis. At this point, the figure is a good example of the three-dimensional structure of both RNAs and proteins. Rather than drawing all the atomic positions, the figure is drawn at a coarse-grained level so that the most important features of the structure are visible. The backbone of the RNA is shown in yellow, and the bases in the RNA are shown as colored rectangular blocks. The tRNA has the cloverleaf structure with the four stems shown in Fig. 2.4. In three dimensions, the molecule is folded into an L-shaped globule. The anticodon loop (see Section 2.3.4) at the bottom of Fig. 2.4 is the top loop in Plate 2.1(a). The stem regions in the secondary structure diagram are the short sections of the double helix in the three-dimensional structure.

The backbone of the protein structure is shown as a purple ribbon in Plate 2.1(a). Two characteristic aspects of protein structures are visible in this example. In many places, the protein backbone forms a helical structure called an α helix. The helix is stabilized by hydrogen bonds between the CO group in the peptide link and the NH in another peptide link further down the chain. These hydrogen bonds are roughly parallel to the axis of the helix. The side groups of the amino acids are pointing out perpendicular to the helix axis. For more details, see a textbook on protein structure, e.g., Creighton (1993). The other features visible in the purple ribbon diagram of the protein are β sheets. The strands composing the sheets are indicated by arrows in the ribbon diagram that point along the backbone from the N to the C terminus. A β sheet consists of two or more strands that are folded to run more or less side by side with one another in either parallel or anti-

parallel orientations. The strands are held together by hydrogen bonds between CO groups on one strand and NH groups on the neighboring strand. The α helices and β sheets in a protein are called elements of secondary structure (whereas the secondary structure of an RNA sequence refers to the base-paired stem regions).

Plate 2.1(b) gives a second example of a molecular structure. This shows a dimer of the lac repressor protein bound to DNA. The *lac* operon in *Escherichia coli* is a set of genes whose products are responsible for lactose metabolism. When the repressor is bound to the DNA, the genes in the operon are turned off (i.e., not transcribed). This happens when there is no lactose in the growth medium. This is an example of a protein that can recognize and bind to a specific sequence of DNA bases. It does this without separating the two strands of the DNA double helix, as it is able to bind to the "sides" of the DNA bases that are accessible in the grooves of the double helix. The binding of proteins to DNA is an important way of controlling the expression of many genes, allowing genes to be turned on in some cells and not others.

Plates 2.1(a) and (b) give an idea of the relative sizes of proteins and nucleic acids. The protein in Plate 2.1(a) has 548 amino acids, which is probably slightly larger than the average globular protein. The tRNA has 72 nucleotides. Most RNAs are much longer than this. The diameter of an α helix is roughly 0.5 nm, whereas the diameter of a nucleic acid double helix is roughly 2 nm.

2.3 THE CENTRAL DOGMA

2.3.1 Transcription

There is a principle, known as the central dogma of molecular biology, that information passes from DNA to RNA to proteins. The process of going from DNA to RNA is called transcription, while the process of going from RNA to protein is called translation. Synthesis of RNA involves simply rewriting (or transcribing) the DNA sequence in the same language of nucleotides, whereas synthesis of proteins involves translating from the language of nucleotides to the language of amino acids.

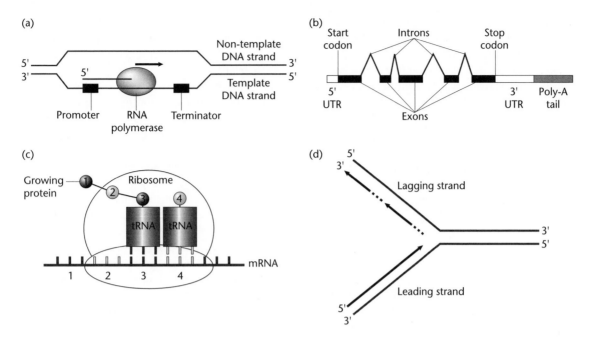

Fig. 2.7 Four important mechanisms. (a) Transcription. (b) Structure and processing of prokaryotic mRNA. (c) Translation. (d) DNA replication.

DNA is usually found in cells in very long pieces. For example, the genomes of most bacteria are circular loops of DNA a few million base pairs (Mbp) long, while humans have 23 pairs of linear chromosomes, with lengths varying from 19 to 240 Mbp. Genes are regions of DNA that contain the information necessary to make RNA and protein molecules. Transcription is the process of synthesis of RNA using DNA as a template. Typically sections of DNA a few thousand base pairs long are transcribed that correspond to single genes (or sometimes a small number of sequential genes). Transcription is carried out by an enzyme called RNA polymerase. The RNA polymerase binds to one of the two strands of DNA that are temporarily separated from one another during the transcription process (see Fig. 2.7(a)). This strand is called the template strand. The polymerase catalyzes the assembly of individual ribonucleotides into an RNA strand that is complementary to the template DNA strand. Base pairing occurs between the template strand and the growing RNA strand initially, but as the polymerase moves along, the RNA separates from the template and the two DNA strands close up again. When the template is a C, G, or T, the base added to RNA is a G, C, or A, as usual. If the template has an A, then a U base is added to the RNA rather than a T. Since the RNA is complementary to the template strand, it is actually **the same** as the **non-template** strand, with the exception that Ts are converted to Us. When people talk about the DNA sequence of a gene, they usually mean the non-template strand, because it is this sequence that is the same as the RNA, and this sequence that is subsequently translated into the protein sequence.

RNA polymerase moves along the template from the 3′ to the 5′ end, hence the RNA is synthesized from its 5′ end to its 3′ end. The polymerase needs to know where to stop and start. This information is contained in the DNA sequence. A promoter is a short sequence of DNA bases that is recognized as a start signal by RNA polymerase. For example, in *E. coli*, most promoters have a sequence TATAAT about 10 nucleotides before the start point of transcription and a sequence TTGACA about 35 nucleotides before the start. However, these sequences are not fixed, and there is considerable variation between genes. Since promoters are relatively short and relatively variable in sequence, it is actually quite a hard problem to write a computer program to reliably locate them.

The stop signal, or terminator, for RNA polymerase is often in the form of a specific sequence that has the ability to form a hairpin loop structure in the RNA sequence. This structure delays the progression of the polymerase along the template and causes it to dissociate. We still have a lot more to learn about these signals.

2.3.2 RNA processing

An RNA strand that is transcribed from a protein-coding region of DNA is called a messenger RNA (mRNA). The mRNA is used as a template for protein synthesis in the translation process discussed below. In prokaryotes, mRNAs consist of a central coding sequence that contains the information for making the protein and short untranslated regions (UTRs) at the 5′ and 3′ ends. The UTRs are parts of the sequence that were transcribed but will not be translated.

In eukaryotes, the RNA transcript has a more complicated structure (Fig. 2.7(b)). When the RNA is newly synthesized, it is called a pre-mRNA. It must be processed in several ways before it becomes a functional mRNA. At the 5′ end, a structure known as a cap is added, which consists of a modified G nucleotide and a protein complex called the cap-binding complex. At the 3′ end, a poly-A tail is added, i.e., a string of roughly 200 A nucleotides. Proteins called poly-A binding proteins bind to the poly-A tail. Many mRNAs have a rather short lifetime (a few minutes) in the cell because they are broken down by nuclease enzymes. These are proteins that break down RNA strands into individual nucleotides, either by chopping them in the middle (endonucleases), or by eating them up from the end one nucleotide at a time (exonucleases). Having proteins associated with the mRNA, particularly at the ends, slows down the nucleases. Variation in the types of binding protein on different mRNAs is an

important way of controlling mRNA lifetimes, and hence controlling the amount of protein synthesized by limiting the number of times an mRNA can be used in translation.

Probably the most important type of RNA processing occurs in the middle of the pre-mRNA rather than at the ends. Eukaryotic gene sequences are broken up into alternating sections called exons and introns. Exons are the pieces of the sequence that contain the information for protein coding. These pieces will be translated. Introns do not contain protein-coding information. The introns, indicated by the inverted Vs in Figure 2.7(b), are cut out of the pre-mRNA and are not present in the mRNA after processing. When an intron is removed, the ends of the exons on either side of it are linked together to form a continuous strand. This is known as splicing.

Splicing is carried out by the spliceosome, a complex of several types of RNA and proteins bound together and acting as a molecular machine. The spliceosome is able to recognize signals in the pre-mRNA sequence that tell it where the intron–exon boundaries are and hence which bits of the sequence to remove. As with promoter sequences, the signals for the splice sites are fairly short and somewhat variable, so that reliable identification of the intron–exon structure of a gene is a difficult problem in bioinformatics. Nevertheless, the spliceosome manages to do it.

Introns that are spliced out by the spliceosome are called spliceosomal introns. This is the majority of introns in most organisms. In addition, there are some interesting, but fairly rare, self-splicing introns, which have the ability to cut themselves out of an RNA strand without the action of the spliceosome. There are surprisingly large numbers of introns in many eukaryotic genes – 10 or 20 in one gene is not uncommon. In contrast, most prokaryotic genes do not contain introns. It is still rather controversial where and when introns appeared, and what is the use, if any, of having them.

In eukaryotes, the DNA is contained in the nucleus, and transcription and RNA processing occur in the nucleus. The mRNA is then transported out of the nucleus through pores in the nuclear membrane, and translation occurs in the cytoplasm.

Table 2.1 The standard genetic code. This is used in most prokaryotic genomes and in the nuclear genomes of most eukaryotes.

First position	Second position				Third position
	U	**C**	**A**	**G**	
U	Phe (F)	Ser (S)	Tyr (Y)	Cys (C)	U
	Phe (F)	Ser (S)	Tyr (Y)	Cys (C)	C
	Leu (L)	Ser (S)	STOP	STOP	A
	Leu (L)	Ser (S)	STOP	Trp (W)	G
C	Leu (L)	Pro (P)	His (H)	Arg (R)	U
	Leu (L)	Pro (P)	His (H)	Arg (R)	C
	Leu (L)	Pro (P)	Gln (Q)	Arg (R)	A
	Leu (L)	Pro (P)	Gln (Q)	Arg (R)	G
A	Ile (I)	Thr (T)	Asn (N)	Ser (S)	U
	Ile (I)	Thr (T)	Asn (N)	Ser (S)	C
	Ile (I)	Thr (T)	Lys (K)	Arg (R)	A
	Met (M)	Thr (T)	Lys (K)	Arg (R)	G
G	Val (V)	Ala (A)	Asp (D)	Gly (G)	U
	Val (V)	Ala (A)	Asp (D)	Gly (G)	C
	Val (V)	Ala (A)	Glu (E)	Gly (G)	A
	Val (V)	Ala (A)	Glu (E)	Gly (G)	G

2.3.3 The genetic code

We now need to consider the way information in the form of sequences of four types of base is turned into information in the form of sequences of 20 types of amino acid. The mRNA sequence is read in groups of three bases called codons. There are $4^3 = 64$ codons that can be made with four bases. Each of these codons codes for one type of amino acid, and since 64 is greater than 20, most amino acids have more than one codon that codes for them. The set of assignments of codons to amino acids is known as the genetic code, and is given in Table 2.1.

The table is divided into blocks that have the same bases in the first two positions. For example, codons of the form UCN (where N is any of the four bases) all code for Ser. There are many groups of four codons where all four code for the same amino acid and the base at the third position does not make any difference. There are several groups where there are two pairs of two amino acids in a block, e.g., CAY codes

for His and CAR codes for Gln (Y indicates a pyrimidine, C or U; and R indicates a purine, A or G). There are only two amino acids that have a single codon: UGG = Trp, and AUG = Met. Ile is unusual in having three codons, while Leu, Ser, and Arg all have six codons, consisting of a block of four and a block of two. There are three codons that act as stop signals rather than coding for amino acids. These denote the end of the coding region of a gene.

When the genetic code was first worked out in the 1960s, it was thought to be universal, i.e., identical in all species. Now we realize that it is extremely widespread but not completely universal. The standard code shown in Table 2.1 applies to almost all prokaryotic genomes (including both bacteria and archaea) and to the nuclear genomes of almost all eukaryotes. In mitochondrial genomes, there are several different genetic codes, all differing from the standard code in small respects (e.g., the reassignment of the stop codon UGA to Trp, or the reassignment of the Ile codon AUA to Met). There are also some changes in the nuclear genome codes for specific groups of organisms, such as the ciliates (a group of unicellular eukaryotes including *Tetrahymena* and *Paramecium*), and *Mycoplasma* bacteria use a slightly different code from most bacteria. These changes are all quite small, and presumably they occurred at a relatively late stage in evolution. The main message is that the code is shared between all three domains of life (archaea, bacteria, and eukaryotes) and hence must have evolved before the divergence of these groups. Thus the last universal common ancestor of all current life must have used this genetic code.

Here we will pause to write a letter of complaint to the BBC. When the release of the human genome sequence was announced in 2001, there were many current affairs broadcasters who commented on how exciting it is that we now know the complete "human genetic code". We have known the genetic code for 40 years! What is new is that we now have the complete genome sequence. Please do not confuse the genetic code with the genome. We now have the complete book, whereas 40 years ago we only knew the words in which the book is written. It will probably take us another 40 years to understand what the book means.

2.3.4 Translation and protein synthesis

Translation is the process of synthesis of a protein sequence using mRNA as a template. A key molecule in the process is transfer RNA (tRNA). The structure of tRNA was already shown in Fig. 2.4 and Plate 2.1(a). The three bases in the middle of the central hairpin loop in the cloverleaf are called the anticodon. The sequence shown in Fig. 2.4 is a tRNA-Ala, i.e., a tRNA for the amino acid alanine. The anticodon of this molecule is UGC (reading from 5' to 3') in the tRNA. This can form complementary base pairs with the codon sequence GCA (reading from 5' to 3' in the mRNA) like this:

```
    |       |
  C – G – U        tRNA
 –G – C – A–       mRNA
```

Note that GCA is an alanine codon in the genetic code. Organisms possess sets of tRNAs capable of

base pairing with all 61 codons that denote amino acids. These tRNAs have different anticodons, and are also different from one another in many other parts of the sequence, but they all have the same cloverleaf secondary structure.

It is not true, however, that there is one tRNA that exactly matches every codon. Many tRNAs can pair with more than one codon due to the flexibility of the pairing rules that occurs at the third position in the codon – this is known as wobble. For example, most bacteria have two types of tRNA-Ala. One type, with anticodon UGC, decodes the codons GCA and GCG, while the other type, with anticodon GGC, decodes the codons GCU and GCC. The actual number of tRNAs varies considerably between organisms. For example, the *E. coli* K12 genome has 86 tRNA genes, of which three have UGC and two have GGC anticodons. In contrast *Rickettsia prowazeckii*, another member of the proteobacteria group, has a much smaller genome with only 32 tRNAs, and only one of each type of tRNA-Ala. These figures are all taken from the genomic tRNA database (Lowe and Eddy, 1997). In eukaryotes, the wobble rules tend to be less flexible, so that a greater number of distinct tRNA types are required. Also the total number of tRNA genes can be much larger, due to the presence of duplicate copies. Thus, in humans, there are about 496 tRNAs in total, and for tRNA-Ala there are 10 with UGC anticodon, five with CGC, and 25 with AGC. In contrast, in most mitochondrial genomes, there are only 22 tRNAs capable of decoding the complete set of codons. In this case, whenever there is a box of four codons, only one tRNA is required. Pairing at the third position is extremely flexible (sometimes known as hyperwobble). For example the tRNA-Ala, with anticodon UGC, decodes all codons of the form GCN.

Transfer RNA acts as an adaptor molecule. The anticodon end connects to the mRNA, and the other end connects to the growing protein chain. Each tRNA has an associated enzyme, known as an amino acyl-tRNA synthetase, whose function is to attach an amino acid of the correct type to the 3' end of the tRNA. The enzyme and the tRNA recognize one another specifically, due to their particular shape and intermolecular interactions. The interaction between glutaminyl-tRNA synthetase and tRNA-glutamine is shown in Plate 2.1(a).

Protein synthesis is carried out by another molecular machine called a ribosome. The ribosome is composed of a large and a small subunit (represented by the two large ellipses in the cartoon in Fig. 2.7(c)). In bacteria, the small subunit contains the small subunit ribosomal RNA (SSU rRNA), which is typically 1500 nucleotides long, together with about 20 ribosomal proteins. The large subunit contains large subunit ribosomal RNA (LSU rRNA), which is typically 3000 nucleotides long, together with about 30 proteins and another smaller ribosomal RNA known as 5S rRNA. The ribosomes of eukaryotes are larger – the two major rRNA molecules are significantly longer and the number of proteins in each subunit is greater.

Figure 2.7(c) illustrates the mechanism of protein synthesis. The ribosome binds to the mRNA and moves along it one codon at a time. tRNAs, charged with their appropriate amino acid, are able to bind to the mRNA at a site inside the ribosome. The amino acid is then removed from the tRNA and attached to the end of a growing protein chain. The old tRNA then leaves and can be recharged with another molecule of the same type of amino acid and used again. The tRNA corresponding to the next codon then binds to the mRNA and the ribosome moves along one codon.

Just as with transcription, translation also requires signals to tell it where to start and stop. We already mentioned stop codons. These are codons that do not have a matching tRNA. When the ribosome reaches a stop codon, a protein known as a release factor enters the appropriate site in the ribosome instead of a tRNA. The release factor triggers the release of the completed protein from the ribosome.

There is also a specific start codon, AUG, which codes for methionine. The ribosome begins protein synthesis at the first AUG codon it finds, which will be slightly downstream of the place where it initially binds to the mRNA. In bacteria, mRNAs contain a conserved sequence of about eight nucleotides, called the Shine–Dalgarno sequence, close to their 5′ end. This sequence is complementary to part of SSU rRNA in the small subunit of the ribosome. This interaction triggers the binding of the ribosome to the mRNA. The first tRNA involved is known as an fMet initiator tRNA. This is a special type of tRNA-Met, where a formyl group has been added to the methionine on the charged tRNA. The fMet is only used when an AUG is a start codon. Other AUG codons occurring in the middle of a gene sequence lead to the usual form of Met being added to the protein sequence.

In the last few years, we have been able to obtain three-dimensional crystal structures of the ribosome (e.g., Yusupov *et al.* 2001), and we are getting closer to understanding the mechanism by which the ribosome actually works. The ribosome is acting as a catalyst for the process of peptide bond formation. "Ribozyme" is the term used for a catalytic RNA molecule, by analogy with "enzyme", which is a catalytic protein. It had previously been thought that rRNA was simply a scaffold onto which the ribosomal proteins attached themselves, and that it was the catalytic action of the proteins that achieved protein synthesis. Recent experiments are making it clear that rRNA plays an essential role in the catalysis, and hence that rRNA is a type of ribozyme.

2.3.5 *Closing the loop: DNA replication*

As stated above, the central dogma is the principle that information is stored in DNA, is transferred from DNA to RNA, and then from RNA to proteins. We have now briefly explained the mechanisms by which this occurs. In order to close the loop in our explanation of the synthesis of nucleic acids and proteins, we still need to explain how DNA is formed.

DNA needs to be replicated every time a cell divides. In a multicellular organism, each cell contains a full copy of the genome of the organism to which it belongs (with the exception of certain cells without nuclei, such as red blood cells). The DNA is needed in every cell in order that transcription and translation can proceed in those cells. DNA replication is also essential for reproduction, because DNA contains the genetic information that ensures heredity.

DNA replication is semi-conservative. This means that the original double strand is replicated to give two double strands, each of which contains one of the original strands and one newly synthesized strand that is complementary to it. Clearly, both strands of DNA contain the full information necessary to recreate the other strand. The key processes of DNA replication occur at a replication fork (Fig. 2.7(d)). At this point, the two old strands are separated from one another and the new strands are synthesized. The main enzyme that does this job is DNA polymerase III. This enzyme catalyzes the addition of nucleotides to the 3′ ends of the growing strands (at the heads of the arrows in Fig. 2.7(d)). The new strand is therefore synthesized in the 5′ to 3′ direction (as with mRNA synthesis during transcription). On one strand, called the leading strand, synthesis is possible in a continuous unbroken fashion. However, on the lagging strand on the opposite side, continuous synthesis is not possible and it is necessary to initiate synthesis independently many times. The new strand is therefore formed in pieces, which are known as Okazaki fragments.

DNA polymerase III is able to carry out the addition of new nucleotides to a strand but it cannot initiate a new strand. This is in contrast to RNA polymerase, which is able to perform both initiation and addition. DNA polymerase therefore needs a short sequence, called a primer, from which to begin. Primers are short sequences of RNA (indicated by dotted lines in Fig. 2.7(d)) that are synthesized by a form of RNA polymerase called primase. The processes of DNA synthesis initiated by primers has been harnessed to become an important laboratory tool, the polymerase chain reaction or PCR (see Box 2.1).

Once the fragments on the lagging strand have been synthesized, it is necessary to connect them together. This is done by two more enzymes. DNA polymerase I removes the RNA nucleotides of the primers and replaces them with DNA nucleotides. DNA ligase makes the final connection between the fragments. Both DNA polymerase I and III have the ability to excise nucleotides from the 3′ end if they do not match the template strand. This process of error correction is called proof-reading. This means that the fidelity of replication of DNA polymerase is increased by several orders of magnitude with respect to RNA polymerases. Errors in DNA replication cause heritable point mutations, whereas errors

in RNA replication merely lead to mistakes in a single short-lived mRNA. Hence accurate DNA replication is very important.

We called this section "closing the loop" because, in the order that we presented things here, DNA replication is the last link in the cycle of mechanisms for synthesis of the major biological macromolecules. There is, however, a more fundamental sense in which this whole process is a loop. Clearly proteins cannot be synthesized without DNA because proteins do not store genetic information. DNA **can** store this information, but it cannot carry out the catalytic roles necessary for metabolism in a cell, and it cannot replicate itself without the aid of proteins. There is thus a chicken and egg situation: "Which came first, DNA or proteins?" Many people now believe that RNA preceded both DNA and proteins, and that there was a period in the Earth's history when RNA played both the genetic and catalytic roles. This is a tempting hypothesis, because several types of catalytic RNA are known (both naturally occurring and artificially synthesized sequences), and because many viruses use RNA as their genetic material today. As with all conjectures related to the origin of life and very early evolution, however, it is difficult to prove that an RNA world once existed.

2.4 PHYSICO-CHEMICAL PROPERTIES OF THE AMINO ACIDS AND THEIR IMPORTANCE IN PROTEIN FOLDING

As we mentioned in Section 1.1, we have many protein sequences for which experimentally determined three-dimensional structures are unavailable. A long-standing goal of bioinformatics has been to predict protein structure from sequence. Some methods for doing this will be discussed in Chapter 10 on pattern recognition. In this section, we will introduce some of the physico-chemical properties that are thought to be important for determining the way a protein folds.

One property that obviously matters for amino acids is size. Proteins are quite compact in structure, and the different residues pack together in a way

BOX 2.1
Polymerase chain reaction (PCR)

The object of PCR is to create many copies of a specified sequence of DNA that is initially present in a very small number of copies. The amplified section can then be used in further experiments or for DNA sequencing. To carry out PCR, it is not necessary to know the sequence to be amplified, but it is necessary to know the sequence of two short sequences at either end of the region to be amplified. These will be used as primers and are indicated by white boxes below.

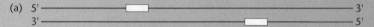

The long sequence of DNA containing the region of interest is denatured by heating, and mixed with oligonucleotides of the two primer sequences, indicated by black boxes below. The complementary strands are synthesized by a DNA polymerase called *Taq* polymerase from the thermophilic bacterium *Thermus aquaticus*. This enzyme is able to withstand the high temperatures used in the denaturing cycles used in PCR. The primers determine the position where the polymerase begins. The situation now looks like this:

These molecules are again denatured and the complementary strands are synthesized, and the cycle of denaturation and DNA synthesis is carried out many times. Included in the mixture of products are some pieces of DNA that are bounded by the primers, like this:

(c) 5' ▰▬▬▬▬▬▬▬▬▬▬ 3'
3' ▬▬▬▬▬▬▬▬▰ 5'

These strands can multiply exponentially because the products of the DNA synthesis can be used as templates at the next cycle. After many cycles, the specified sequence dominates the population of DNA sequences, with only negligible fractions of the longer DNA sequences being present.

that is almost space filling. The volume occupied by the side groups is important for protein folding, and also for molecular evolution. It would be difficult to substitute a very large amino acid for a small one because this would disrupt the structure. It is more difficult than we might think at first to define the volume of an amino acid. We have a tendency to think of molecules as "balls and sticks", but really molecules contain atomic nuclei held together by electrons in molecular orbitals. However, if you push atoms together too much, they repel and hence it is possible to define a radius of an atom, known as a van der Waals radius, on the basis of these repulsions. A useful measure of amino acid volume is to sum the volumes of the spheres defined by the van der Waals radii of its constituent atoms. These figures are given in Table 2.2 (in units of $\mathring{A}^3$). There is a significant variation in volume between the amino acids. The largest amino acid, tryptophan, has roughly 3.4 times the volume of the smallest amino acid, glycine. Creighton (1993) gives more information on van der Waals interactions and on amino acid volumes. Since protein folding occurs in water, another way to define the amino acid volume

Table 2.2 Physico-chemical properties of the amino acids.

			Vol.	Bulk.	Polarity	pl	Hyd.1	Hyd.2	Surface area	Fract. area
Alanine	Ala	A	67	11.50	0.00	6.00	1.8	1.6	113	0.74
Arginine	Arg	R	148	14.28	52.00	10.76	−4.5	−12.3	241	0.64
Asparagine	Asn	N	96	12.28	3.38	5.41	−3.5	−4.8	158	0.63
Aspartic acid	Asp	D	91	11.68	49.70	2.77	−3.5	−9.2	151	0.62
Cysteine	Cys	C	86	13.46	1.48	5.05	2.5	2.0	140	0.91
Glutamine	Gln	Q	114	14.45	3.53	5.65	−3.5	−4.1	189	0.62
Glutamic acid	Glu	E	109	13.57	49.90	3.22	−3.5	−8.2	183	0.62
Glycine	Gly	G	48	3.40	0.00	5.97	−0.4	1.0	85	0.72
Histidine	His	H	118	13.69	51.60	7.59	−3.2	−3.0	194	0.78
Isoleucine	Ile	I	124	21.40	0.13	6.02	4.5	3.1	182	0.88
Leucine	Leu	L	124	21.40	0.13	5.98	3.8	2.8	180	0.85
Lysine	Lys	K	135	15.71	49.50	9.74	−3.9	−8.8	211	0.52
Methionine	Met	M	124	16.25	1.43	5.74	1.9	3.4	204	0.85
Phenylalanine	Phe	F	135	19.80	0.35	5.48	2.8	3.7	218	0.88
Proline	Pro	P	90	17.43	1.58	6.30	−1.6	−0.2	143	0.64
Serine	Ser	S	73	9.47	1.67	5.68	−0.8	0.6	122	0.66
Threonine	Thr	T	93	15.77	1.66	5.66	−0.7	1.2	146	0.70
Tryptophan	Trp	W	163	21.67	2.10	5.89	−0.9	1.9	259	0.85
Tyrosine	Tyr	Y	141	18.03	1.61	5.66	−1.3	−0.7	229	0.76
Valine	Val	V	105	21.57	0.13	5.96	4.2	2.6	160	0.86
Mean			109	15.35	13.59	6.03	−0.5	−1.4	175	0.74
Std. dev.			28	4.53	21.36	1.72	2.9	4.8	44	0.11

Vol., volume calculated from van der Waals radii (Creighton 1993); Bulk., bulkiness index (Zimmerman, Eliezer, and Simha 1968); Polarity, polarity index (Zimmerman, Eliezer, and Simha 1968); pl, pH of the isoelectric point (Zimmerman, Eliezer, and Simha 1968); Hyd.1, hydrophobicity scale (Kyte and Doolittle 1982); Hyd.2, hydrophobicity scale (Engelman, Steitz, and Goldman 1986); Surface area, surface area accessible to water in unfolded peptide (Miller *et al.* 1987); Fract. area, fraction of accessible area lost when a protein folds (Rose *et al.* 1985).

is to consider the increase in volume of a solution when an amino acid is dissolved in it. This is known as the partial volume. Partial volumes are closely correlated with the volumes calculated from the van der Waals radii, and we do not show them in the table.

Zimmerman, Eliezer, and Simha (1968) presented data on several amino acid properties that are relevant in the context of protein folding. Rather than simply considering the volume, they defined the "bulkiness" of an amino acid as the ratio of the side chain volume to its length, which provides a measure of the average cross-sectional area of the side chain. These figures are shown in Table 2.2 (in Å^2). Zimmerman, Eliezer, and Simha (1968) also introduced a measure of the polarity of the amino acids. They calculated the electrostatic force of the amino acid acting on its surroundings at a distance of 10 Å. This is composed of the force from the electric charge (for the amino acids that have a charged side group) plus the force from the dipole moment (due to the non-uniformity of electronic charge across the amino acid). The total force (in units scaled for convenience) was used as a polarity index, and this is shown in Table 2.2. The electrostatic charge term, where it exists, is much larger than the dipole term. Hence, this measure clearly distinguishes between the charged and uncharged amino acids.

The polarity index does not distinguish between the positively and negatively charged amino acids, however, since both have high polarity. A quantity that does this is the pI, which is defined as the pH of the isoelectric point of the amino acid. Acidic amino acids (Asp and Glu) have pI in the range 2–3. This means that these amino acids would be negatively charged at neutral pH due to ionization of the COOH group to COO⁻. We need to put them in an acid solution in order to shift the equilibrium and balance this charge. The basic amino acids (Arg, Lys, and His) have pI greater than 7. All the others usually have uncharged side chains in real proteins. They have pI in the range 5–6. Thus, pI is a useful measure of acidity of amino acids that distinguishes clearly between positive, negative, and uncharged side chains.

A key factor in protein folding is the "hydrophobic effect", which arises as a result of the unusual characteristics of water as a solvent. Liquid water has quite a lot of structure due to the formation of chains and networks of molecules interacting via hydrogen bonds. When other molecules are dissolved in water, the hydrogen-bonded structure is disrupted. Polar amino acid residues are also able to form hydrogen bonds with water. They therefore disrupt the structure less than non-polar amino acids that are unable to form hydrogen bonds. We say that the non-polar amino acids are hydrophobic, because they do not "want" to be in contact with water, whereas the polar amino acids are hydrophilic, because they "like" water. It is generally observed that hydrophobic residues in a protein are in the interior of the structure and are not in contact with water, whereas hydrophilic residues are on the surface and are in contact with water. In this way the free energy of the folded molecule is minimized.

Kyte and Doolittle (1982) defined a hydrophobicity (or hydropathy) scale that is an estimate of the difference in free energy (in kcal/mol) of the amino acid when it is buried in the hydrophobic environment of the interior of a protein and when it is in solution in water. Positive values on the scale mean that the residue is hydrophobic: it costs free energy to take the residue out of the protein and put it in water.

Another version of the hydrophobicity scale was developed by Engelman, Steitz, and Goldman (1986), who were particularly interested in membrane proteins. The interior of a lipid bilayer is hydrophobic, because it mostly consists of the hydrocarbon tails of the lipids. They estimated the free energy cost for removal of an amino acid from the bilayer to water. These two scales are similar but not identical; therefore both scales are shown in the table.

Another property that is thought to be relevant for protein folding is the surface area of the amino acid that is exposed (accessible) to water in an unfolded peptide chain and that becomes buried when the chain folds. Table 2.2 shows the accessible surface areas of the residues when they occur in a Gly–X–Gly tripeptide (Miller *et al.* 1987, Creighton 1993). Rose *et al.* (1985) calculated the average fraction of the accessible surface area that is buried in the interior in a set of known crystal structures. They showed that hydrophobic residues have a larger fraction of the surface area buried, which supports the argument that the "hydrophobic effect" is important in determining protein structure.

2.5 VISUALIZATION OF AMINO ACID PROPERTIES USING PRINCIPAL COMPONENT ANALYSIS

So far, this chapter has summarized some of the fundamental aspects of molecular biology that we think every bioinformatician should know. In the rest of the chapter, we want to introduce some simple methods for data analysis that are useful in bioinformatics. We will use the data on amino acid properties.

Table 2.2 shows eight properties of each amino acid (and we could easily have included several more columns using data from additional sources). It would be useful to plot some kind of diagram that lets us visualize the information in this table. It is straightforward to take any two of the properties and use these as the coordinates for the points in a two-dimensional graph. Figure 2.8 shows a plot of volume against pI. This clearly shows the acidic amino acids at low pI, the basic amino acids at high pI, and all the rest in the middle. It also shows the

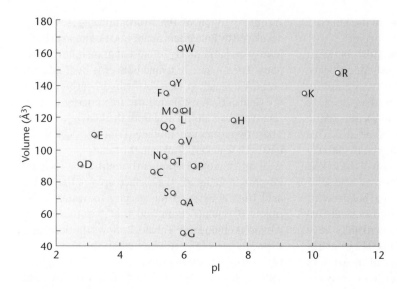

Fig. 2.8 Plot of amino acid volume against pI – two properties thought to be important in protein folding.

large spread of the middle group along the volume axis. However, the figure does not distinguish between the hydrophilic and hydrophobic amino acids in the middle group: N and Q appear very close to M and V, for example. We could separate these by using one of the hydrophobicity scales on the axis instead of pI, but then the acidic and basic groups would appear close together because both are hydrophilic (negative on the hydrophobicity scale). What we need is a way of combining the information from all eight properties into a two-dimensional graph. This can be done with principal component analysis (PCA).

In general with PCA, we begin with the data in the form of an $N \times P$ matrix, like Table 2.2. The number of rows, N, is the number of objects in our data set (in this case $N = 20$ amino acids), and the number of columns, P, is the number of properties of those objects (in this case $P = 8$). Each row in the data matrix can be thought of as the coordinates of a point in P-dimensional space. The whole data set is a cloud of these points. The PCA method transforms this cloud of points first by scaling them and shifting them to the origin, and then by rotating them in such a way that the points are spread out as much as possible, and the structure in the data is made easier to see.

Let the original data matrix be X_{ij} (i.e., X_{ij} is the value of the jth property of object i). The mean and standard deviation of the properties are

$$\mu_j = \frac{1}{N} \sum_i X_{ij}$$

and

$$\sigma_j = \left(\frac{1}{N} \sum_i (X_{ij} - \mu_j)^2 \right)^{1/2}$$

The mean and standard deviation are listed at the foot of Table 2.2. Since the properties all have different scales and different mean values, the first step of PCA is to define scaled data values by

$$z_{ij} = (X_{ij} - \mu_j)/\sigma_j$$

The z_{ij} matrix measures the deviation of the values from the mean values for each property. By definition, the mean value of each column in the z_{ij} matrix is 0 and the standard deviation is 1. Scaling the data in this way means that all the input properties are placed on an equal footing, and all the properties will contribute equally to the data analysis.

We now choose a set of vectors $v_j = (v_{j1}, v_{j2}, v_{j3}, \ldots v_{jP})$ that define the directions of the principal

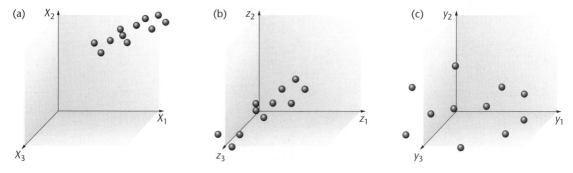

Fig. 2.9 Schematic illustration of principal component analysis. (a) Original data. (b) Scaled and centered on the origin. (c) Rotated onto principal components.

components. These vectors are of unit length, i.e., $\sum_k v_{jk}^2 = 1$ for each vector, and they are all orthogonal to one another, i.e., $\sum_k v_{ik} v_{jk} = 0$, when i and j are not equal. Each vector represents a new coordinate axis that is a linear combination of the old coordinates. The positions of the points in the new coordinate system are given by

$$y_{ij} = \sum_k v_{jk} z_{ik}$$

The new y coordinate system is a rotation of the z coordinate system – see Fig. 2.9.

There are still P coordinates, so we can only use two of them if we plot a two-dimensional graph. However, we can define the y coordinates so that as much of the variation between the points as possible is visible in the first few coordinates. We therefore choose the v_{1k} values so that the variance of the points along the first principal component axis, $\frac{1}{N} \sum_i y_{i1}^2$ is as large as possible. (Note that the means of the y's are all zero because the means of the z's were zero.) We then choose the v_{2k} for the second component by maximizing the variance $\frac{1}{N} \sum_i y_{i2}^2$, with the constraint that the second axis is orthogonal to the first, i.e., $\sum_k v_{1k} v_{2k} = 0$. If we wish, we can define further components by maximizing the vari-

ance with the constraint that each component is orthogonal to the previous ones. Calculation of the v_{jk} is discussed in more detail in Box 2.2.

The results of PCA for the amino acid data in Table 2.2 are shown in Fig. 2.10. The first two principal component vectors are shown in the matrix on p. 28. For component 1, the largest contributions in the vector are the negative contributions from the hydrophobicity scales. Thus hydrophobic amino acids appear on the left side and hydrophilic ones on the right. For component 2, the largest contributions are positive ones from volume, bulkiness, and surface area. Thus large amino acids appear near the top of the figure and small ones near the bottom. However, all the properties contribute to some extent to each of the components; therefore, the resulting figure is not the same as we would have got by simply plotting hydrophobicity against volume.

Figure 2.10 illustrates several points about the data that seem intuitive. There is a cluster of medium-sized hydrophobic residues, I, L, V, M, and F. The two acids, D and E, are close, and so are the two amides, Q and N. Two of the basic residues, R and K, are very close, and H is fairly close to these. The two largest residues, W and Y, are quite close to one another. The PCA diagram manages to do a fairly good job at illustrating all these similarities at the same time.

The PCA calculation in this section was done using the program pca.c by F. Murtagh (http://astro.u-strasbg.fr/~fmurtagh/mda-sw/).

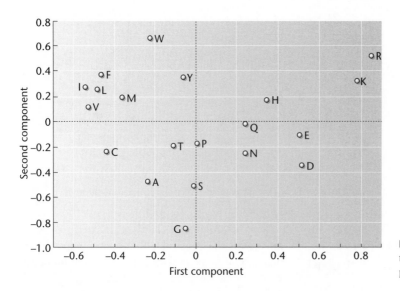

Fig. 2.10 Plot of the amino acids on the first two components of the principal component analysis.

	Vol	Bulk.	Pol.	pI	Hyd.1	Hyd.2	S.A.	Fr.A.
Comp. 1	(0.06,	−0.22,	0.44,	0.19,	−0.49,	−0.51,	0.10,	−0.45)
Comp. 2	(0.58,	0.48,	0.10,	0.25,	0.03,	−0.03,	0.56,	0.17)

2.6 CLUSTERING AMINO ACIDS ACCORDING TO THEIR PROPERTIES

2.6.1 Handmade clusters

When we look at a figure like 2.10, it is natural to try to group the points into "clusters" of similar objects. We already remarked above that I, L, V, M, and F look like a cluster. So, where would you put clusters? Before going any further, make a few photocopies of Fig. 2.10. Now take one of the copies and draw rings around the groups of points that you think should be clustered. You can decide how many clusters you think there should be – somewhere between four and seven is probably about right. You can also decide how big the clusters should be – you can put lots of points together in one cluster if you like, or you can leave single points on their own in a cluster of size one. OK, go ahead!

When we presented the chemical structures of the amino acids in Fig. 2.6, we chose four groups:

Neutral, nonpolar	W, F, G, A, V, I, L, M, P
Neutral, polar	Y, S, T, N, Q, C
Acidic	D, E
Basic	K, R, H

This is one possible clustering. We chose these four clusters because this is the way the amino acids are presented in most molecular biology textbooks. Try drawing rings round these four clusters on another copy of Fig. 2.10. The acidic and basic groups work quite well. The neutral polar group forms a rather spread-out cluster in the middle of the figure, but unfortunately it has P in the middle of it. The nonpolar group can hardly be called a cluster, as it takes up about half the diagram, and contains points that are very far from one another, like G and W. You probably think that you did a better job when you made up your own clusters a few minutes ago.

We now want to consider ways of clustering data that are more systematic than drawing rings on paper.

BOX 2.2
Principal component analysis in more detail

From the $N \times P$ data matrix, we can define a $P \times P$ matrix of correlation coefficients, C_{jk}, between the properties:

$$C_{jk} = \frac{1}{N\sigma_j\sigma_K} \sum_i (X_{ij} - \mu_j)(X_{ik} - \mu_k) = \frac{1}{N} \sum_i z_{ij} z_{ik}$$

The coefficients are always in the range –1 to 1. If $C_{jk} > 0$, the two properties are positively correlated, i.e., they both tend to be large at the same time and small at the same time. If $C_{jk} < 0$, the properties are negatively correlated, i.e., one tends to be large when the other is small. The correlation matrix for the amino acid data looks like this.

The matrix is symmetric ($C_{jk} = C_{kj}$) and all the diagonal elements are 1.00 by definition. The values illustrate features of the data that are not easy to see in the original matrix. For example, volume has a strong positive correlation with surface area and bulkiness, and a fairly weak correlation with the other properties. The two hydrophobicity scales have strong positive correlation with each other and also with the fractional area property, but they have a significant negative correlation with the polarity scale.

It can be shown that the vectors $\mathbf{v}_j$ that define the principal component axes are the eigenvectors of the correlation matrix, i.e., they satisfy the equation:

$$\sum_j v_{nj} C_{jk} = \lambda_n v_{nk}$$

where the λ_n are constants called eigenvalues. The first principal component (PC) vector is the eigenvector with the largest eigenvalue. Subsequent PCs can be listed in order of decreasing size of eigenvalue. The first two eigenvalues in this case are $\lambda_1 = 3.57$ and $\lambda_2 = 2.81$.

The variance along the nth PC axis is equal to the corresponding eigenvalue:

$$\frac{1}{N} \sum_i y_{in}^2 = \frac{1}{N} \sum_i \sum_j \sum_k v_{nj} z_{ij} v_{nk} z_{ik} = \sum_j \sum_k v_{nj} C_{jk} v_{nk} = \sum_k \lambda_n v_{nk}^2 = \lambda_n$$

We know that the variance of each of the z coordinates is 1, hence the total variance of all the coordinates is P. When we change the coordinates to the principal components, we just rotate the points in space, so the total variance in the PC space is still P. The fraction of the total variance represented by the first two PCs is therefore $(\lambda_1 + \lambda_2)/P$, which in our case is $(3.57 + 2.81)/8 = 0.797$. This is why it is useful to look at the data on the PC plot (as in Fig. 2.10). Roughly 80% of the variation in the positioning of the points in the original coordinates can be seen with just two PCs. When points appear close in the two-dimensional plot of the first two PCs, they really are close in the eight-dimensional space, because the remaining six dimensions that we can't see do not contribute much to the distance between the points. This means that if we spot patterns in the data in the PC plot, such as clusters of closely spaced points, then these are likely to give a true impression of the patterns in the full data.

	Vol	Bulk.	Pol.	pI	Hyd.1	Hyd.2	S.A.	Fr.A.
Vol.	1.00	0.73	0.24	0.37	−0.08	−0.16	0.99	0.18
Bulk.	0.73	1.00	−0.20	0.08	0.44	0.32	0.64	0.49
Pol.	0.24	−0.20	1.00	0.27	−0.69	−0.85	0.29	−0.53
pI	0.37	0.08	0.27	1.00	−0.20	−0.27	0.36	−0.18
Hyd.1	−0.08	0.44	−0.67	−0.20	1.00	0.85	−0.18	0.84
Hyd.2	−0.16	0.32	−0.85	−0.27	0.85	1.00	−0.23	0.79
S.A.	0.99	0.64	0.29	0.36	−0.18	−0.23	1.00	0.12
Fr.A.	0.18	0.49	−0.53	−0.18	0.84	0.79	0.12	1.00

In fact, there is a **huge** number of different clustering methods. This testifies to the fact that there are a lot of different people from a lot of different disciplines who find clustering useful for describing the patterns in their data. Unfortunately, it also means that there is not one single clustering method that everyone agrees is best. Different methods will give different answers when applied to the same data; therefore, there has to be some degree of subjectivity in deciding which method to use for any particular data set.

In the context of the amino acids, clustering according to physico-chemical properties is actually quite helpful when we come to do protein sequence alignments. We usually want to align residues with similar properties with one another, even if the residues are not identical. There are several sequence alignment editors that ascribe colors to residues, assigning the same color to clusters of similar amino acids. In well-aligned parts of protein sequences, we often find that all the residues in a column have the same color. The coloring scheme can thus help with constructing alignments and spotting important conserved motifs. When we look at protein sequence evolution (Chapter 4) it turns out that substitutions are more frequent between amino acids with similar properties. So, clustering according to properties is also relevant for evolution. In the broader context, however, clustering algorithms are very general and can be used for almost any type of data. In this book, they will come up again in two places: in Chapter 8 we discuss distance matrix methods for molecular phylogenetics, which are a form of hierarchical clustering; and in Chapter 13 we discuss applications of clustering algorithms on microarray data. It is therefore worth spending some time on these methods now, even if you are getting a bit bored with amino acid properties.

2.6.2 Hierarchical clustering methods

In a hierarchical clustering method, we need to choose a measure of similarity between the data points, then we need to choose a rule for measuring the similarity of clusters.

We will use the scaled coordinates z as in the previous section. There is a vector $\mathbf{z}_i$ from the origin to

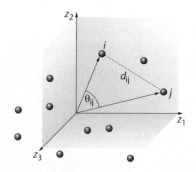

Fig. 2.11 Illustration of the data points as vectors in multidimensional space.

each point i in the data set (see Fig. 2.11). The length of the vector is:

$$|\mathbf{z}_i| = \left(\sum_k \mathbf{z}_{ik}^2 \right)^{1/2}$$

We want to measure how similar the vectors are for two points i and j. A simple way to do this is to use the cosine of the angle θ_{ij} between the vectors. If the two vectors are pointing in almost the same direction, θ_{ij} will be small and $\cos \theta_{ij}$ will be close to 1. Vectors with no correlation will have θ_{ij} close to 90° and $\cos \theta_{ij}$ close to 0. Vectors with negative correlation will have $\theta_{ij} > 90°$ and $\cos \theta_{ij} < 0$.

From standard geometry,

$$\cos \theta_{ij} = \frac{\sum_k z_{ik} z_{jk}}{|\mathbf{z}_i| |\mathbf{z}_j|}$$

Another possible similarity measure is the correlation coefficient between the $\mathbf{z}$ vectors:

$$R_{ij} = \frac{1}{P s_i s_j} \sum_k (z_{ik} - m_i)(z_{jk} - m_j)$$

where m_i and s_i are the mean and standard deviation of the elements in the ith row (see also Box 2.2, where we define the correlation between the columns). R_{ij} is in the range -1 to 1.

In what follows, we shall assume that we have calculated an $N \times N$ matrix of similarities between the data points that could be $\cos \theta_{ij}$ or R_{ij}, or any other measure of similarity that appears appropriate for the data in question. We will call the similarity matrix S_{ij} from now on, to emphasize that the method is general and works the same way, whichever measure we use for similarity.

During the process of hierarchical clustering, points are combined into clusters, and small clusters are combined to give progressively larger clusters. To decide in what order these clusters will be connected, we will need a definition of similarity between clusters. Suppose we already have two clusters A and B. We want to define the similarity S_{AB} of these clusters. There are (at least) three ways of doing this:
• Group average. S_{AB} = the mean of the similarities S_{ij} between the individual data points, averaged over all pairs of points, where i is in cluster A and j is in cluster B.
• Single-link rule. S_{AB} = maximum similarity S_{ij} for any i in A and j in B.
• Complete-link rule. S_{AB} = minimum similarity S_{ij} for any i in A and j in B.
The reasons for the terms "single link" and "complete link" will be made more clear in Section 2.6.3.

An algorithm is a computational recipe that specifies how to solve a problem. Algorithms come up throughout this book, and we will discuss some general points about algorithms in Chapter 6. For the moment, we will present a very simple algorithm for hierarchical clustering. This works in the same way, whatever the definitions of similarity between data points and between clusters. We begin with each point in a separate cluster of its own.

1 Join the two clusters with the highest similarity to form a single larger cluster.

2 Recalculate similarities between all the clusters using one of the three definitions above.

3 Repeat steps 1 and 2 until all points have been connected to a single cluster.

This procedure is called "hierarchical" because it generates a set of clusters within clusters within clusters. For this reason, the results of a hierarchical clustering procedure can be represented as a tree. Each branching point on the tree is a point where two smaller clusters were joined to form a larger one. Reading backwards from the twigs of the tree to the root tells us the order in which the clusters were connected.

Plate 2.2(a) shows a hierarchical clustering of the amino acid data. This was performed using the CLUTO package (Karypis 2002). The similarity measure used was $\cos \theta$ and the group-average rule was used for the similarity between clusters. The tree on the left of Plate 2.2(a) shows the order in which the amino acids were clustered. For example, L and I are very similar, and are clustered at the beginning. The LI cluster is later combined with V. In the meantime M and F are clustered, and then the MF cluster is combined with VLI, and so on. The tree indicates what happens if the clustering is continued to the point where there is only one cluster left. In practice, we want to stop the clustering at some stage where there is a moderate number of clusters left. The right side of Plate 2.2(a) shows the clusters we get if we stop when there are six clusters. These can be summarized as follows.

Cluster 1:	Basic residues	K, R, H
Cluster 2:	Acid and amide residues	E, D, Q, N
Cluster 3:	Small residues	P, T, S, G, A
Cluster 4:	Cysteine	C
Cluster 5:	Hydrophobic residues	V, L, I, M, F
Cluster 6:	Large, aromatic residues	W, Y

The central part of Plate 2.2(a) is a representation of the scaled data matrix z_{ij}. Red/green squares indicate that the value is significantly higher/lower than average; dark colors indicate values close to the average. This color scheme makes sense in the context of microarrays, as we shall see in Chapter 13. We have named the clusters above according to what seemed to be the most important feature linking members of the cluster. The basic cluster contains all the residues that are red on both the pI and polarity scales. The acid and amide cluster contains all the residues that are green on the hydrophobicity scales and also on the pI scale. Note that if we had stopped the clustering with a larger number of clusters, the acids and the amides would have been in separate clusters. We called cluster 3 "small" because the most noticeable thing is that these residues are all green on the volume and surface area scales. These residues are quite mixed in terms of hydrophobicities. Cluster 4 contains only cysteine. Cysteine has an unusual role in protein structure because of its potential to form disulfide bonds between pairs of cysteine residues. For this reason, cysteines tend to be important when they occur and it is difficult to interchange them for other residues. Cysteine does not appear to be particularly extreme in any of the eight properties used here, and none of the eight properties captures the important factor of disulfide bonding. Nevertheless, it is interesting that this cluster analysis manages to spot some of its uniqueness. Cluster 5 is clearly hydrophobic, and cluster 6 contains the two largest amino acids, which both happen to be aromatic. It is worth noting, however, that the other aromatic residue, phenylalanine (F), is in cluster 5. Phenylalanine has a simple hydrocarbon ring as a side group and therefore is hydrophobic. In contrast, tryptophan and tyrosine are only moderate on the hydrophobicity scales used here.

At the top of Plate 2.2(a), there is another tree indicating a clustering of the eight properties. This is done so that the properties can be ordered in a way that illustrates groups of properties that are correlated. The tree shows very similar information to the correlation matrix given in Box 2.2, i.e., volume and surface area are correlated, the two hydrophobicity scales are correlated with the fractional area scale, etc.

2.6.3 Variants on hierarchical clustering

Take another copy of Fig. 2.10 and draw rings around the six clusters specified by the hierarchical method. These clusters seem to make sense, and they are probably as good as we are likely to get with these data as input. They are not the only sensible set of clusters, however, and the details of the clusters we get depend on the details of the method.

First, the decision to stop at six clusters is subjective. If we use the same method ($\cos \theta$ and group average) and stop at seven, the difference is that the acids are separated from the amides. If we stop at five, cysteine is joined with the hydrophobic cluster.

A second point to consider is the rule for similarity between clusters. In hierarchical clustering methods, we could in principle plot the similarity of the pair of clusters that we connect at each step of the process as a function of the number of steps made. This level begins at one, and gradually descends and the clusters get bigger and the similarity between the clusters gets lower. In the group-average method, the similarity of the clusters is the mean of the similarities of the pairs of points in the cluster. Therefore, roughly half of the pairs of points will have similarities greater than or equal to the similarity level at which the connection is made. When the single-link rule is used, the level at which the connection is made is the similarity of the most similar pair of points in the two clusters connected. This means that clusters can be very spread out. Two points in the same cluster may be very different from one another as long as there is a chain of points between them, such that each link in the chain corresponds to a high similarity pair. In contrast, the complete-link rule will only connect a pair of clusters when all the pairs of points in the two clusters have similarity greater than the current connection level. Thus each point is completely linked to all other points in the cluster. In our case, using $\cos \theta$, the single-link rule and stopping at six clusters yields the same six clusters as with the group-average rule, except that WY is linked with VLIMF and QN is split from DE. Using $\cos \theta$ with the complete-link rule gives the same as the group-average method, with the exception that C is linked with VLIMF and TP is split from SGA.

These are relatively minor changes. We also tried using the correlation coefficient as the similarity measure instead of $\cos \theta$, and this gave a more significant change in the result. With the group-average rule we obtained: EDH; QNKR; YW; VLIMF; PT; SGAC. These clusters seem less intuitive than those obtained with the $\cos \theta$ measure, and also appear less well defined in the PCA plot. The correlation coefficient therefore seems to work less well on this particular data set. The general message is that it is worth considering several different methods on any real data, because differences will arise.

So far we have been treating the data in terms of similarities. It is also possible to measure distances between data points that measure how "far apart" the points are, rather than how similar they are. We already have points in our P-dimensional space defined by the z coordinates (Fig. 2.11). Therefore we can straightforwardly measure the Euclidean distance between these points:

$$d_{ij} = \left(\sum_k (z_{ik} - z_{jk})^2 \right)^{1/2}$$

We can use the matrix of distances between points instead of the matrix of similarities. The only difference in the hierarchical clustering procedure is always to connect the pair of clusters with the smallest distance, rather than the pair with the highest similarity. Group-average, single-link, and complete-link methods can still be used with distances. Even though the clustering rule is basically the same, clustering based on distances and similarities will give different results because the data are input to the method in a different way – the distances are not simple linear transformations of the similarities.

One of the first applications of clustering techniques, including the ideas of single-link, complete-link, and group-average clusters, was for construction of phylogenetic trees using morphological characters (Sokal and Sneath 1963). Distance-matrix clustering methods are still important in molecular phylogenetics. In that case, the data consist of sequences, rather than points in Euclidean space. There are many ways of defining distances between sequences (Chapter 4), but once a distance matrix has been defined, the clustering procedure is the same. In the phylogenetic context, the group-average method starting with a distance matrix is usually called UPGMA (see Section 8.3).

2.6.4 Non-hierarchical clustering methods

All the variants discussed above give rise to a nested set of clusters within clusters that can be represented by a tree. There are other types of clustering method, sometimes called "direct" clustering methods, where we simply specify the number, K, of clusters required and we try to separate the objects into K groups without any notion of a hierarchy between the groups. Direct clustering methods require us to define a function that measures how good a set of clusters is. One function that does this is

$$I_2 = \sum_A \sqrt{\sum_{i,j \in A} S_{ij}}$$

Here, A labels the cluster, and we are summing over all clusters $A = 1, 2 \ldots K$. The notation $i, j \in A$ means

that we are summing over all pairs of objects i and j that are in cluster A. We called this function I_2, following the notation in the manual for the CLUTO software (Karypis 2002). Given any proposed division of the objects into clusters, we can evaluate I_2. We can then choose the set of clusters that maximizes I_2.

There are many other optimization functions that we might think of to evaluate the clusters. Basically, we want to maximize some function of the similarities of objects within clusters or minimize some function of the similarities of objects in different clusters. I_2 is the default option in CLUTO, but several other functions can be specified as alternatives. Note that if a cluster has n objects, there are n^2 pairs of points in the cluster. The square root in I_2 provides a way of balancing the contributions of large and small clusters to the optimization function. Using the I_2 optimization function on the amino acid data with $K = 6$ gives the clusters: KRH; EDQN; PT; CAGS; VLIMF; WY. This is another slight variant on the one shown in Plate 2.2(a), but one that also seems to make sense intuitively and when drawn on the principal components plot.

Another well-known form of direct clustering, known as K-means (Hartigan 1975), treats the data in the form of distances instead of similarities. In this case, we define an error function E and choose the set of clusters that minimizes E. Let μ_{Aj} be the mean value of z_{ij} for all objects i assigned to cluster A. The square of the distance of object i from the mean point of the cluster to which it belongs is

$$d_{iA}^2 = \sum_j (z_{ij} - \mu_{Aj})^2$$

and the error function is

$$E = \sum_A \sum_{i \in A} d_{iA}^2$$

In direct clustering methods, we have a well-defined function that is being optimized. However, we do not have a well-defined algorithm for finding the set of clusters. It is necessary to write a computer program that tries out very many possible solutions and saves the best one that it finds. Typically, we might begin with some random partition of the data into K clusters and then try moving one object at a time into a different cluster in such a way as to make the best possible improvement in the optimization function. If there is no movement of an object that would improve the optimization function, then we have found at least a local optimum solution. If the process is repeated several times from different starting positions, we have a good chance of finding the global optimum solution.

For the hierarchical methods in the previous section, the algorithm tells us exactly how to form the clusters, so there is no trial and error involved. However, there is no function that is being optimized. Exactly the same distinction will be made when we discuss phylogenetic methods in Chapter 8: distance matrix methods have a straightforward algorithm but no optimization criterion, whereas maximum-parsimony and maximum-likelihood methods have well-defined optimization criteria, but require a trial-and-error search procedure to locate the optimal solution.

There are many issues related to clustering that we have not covered here. Some methods do not fit into either the hierarchical or the direct clustering categories. For example, we can also do top-down clustering where we make successive partitions of the data, rather than successive amalgamations, as in hierarchical methods. It is worth stating an obvious point about all the clustering methods discussed in this chapter: clusters are defined to be non-overlapping. An object cannot be in more than one cluster at once. When we run a clustering algorithm, we are forcing the data into non-overlapping groups. Sometimes the structure of the data may not warrant this, in which case we should be wary of using clustering methods or of reading too much into the clusters produced. Statistical tests for the significance of clusters are available, and these would be important if we were in doubt whether a clustering method was appropriate for our data.

To illustrate the limitations of non-overlapping clusters, we tried to plot a Venn diagram illustrating as many relevant properties of amino acids as possible: see Plate 2.2(b). These properties **do** overlap. For example, several amino acids are not strongly polar or nonpolar, and are positioned in the overlap area. There are aromatic amino acids on both the polar and nonpolar sides, so the aromatic ring overlaps the others. This diagram is surprisingly hard to draw (this is at least the fourth version we tried!). There were some things in earlier versions that got left out of this one, for example tyrosine (Y) is sometimes weakly acidic (so should it be in a ring with D and E?) and histidine is only weakly basic (so should we move it into the polar neutral area?). The general message is that clusters are useful, but they have limitations, and we should keep this in mind when clustering more complex data sets, such as the microarray data discussed in Chapter 13.

SUMMARY

DNA is composed of sequences of four types of nucleotide building blocks known as A, C, G, and T. It is the molecule that stores the genetic information of the cell. It usually exists as a double helix composed of two exactly complementary strands. RNA is also composed of four nucleotide building blocks, but U is used instead of T. RNA molecules are usually single stranded and fold to form complex stem-loop secondary structures by base pairing between short sections of the same strand. Proteins are polymers composed of sequences of 20 types of amino acid linked by peptide bonds.

The process of synthesis of RNA using a DNA strand as a template is called transcription. It is carried out by RNA polymerase. The process of protein synthesis using mRNA as a template is called translation. It is carried out by the ribosome. Protein-coding DNA sequences store information in the form of groups of three bases called codons. Each codon codes for either an amino acid or a stop signal. The mapping from codons to amino acids is known as the genetic code. During translation, the anti-codon sequences in tRNA molecules pair with the codon sequences in the mRNA. Each tRNA is charged with a specific amino acid, and this amino acid gets transferred from the tRNA to the growing protein chain due to the catalytic activity of the ribosome.

Amino acids vary greatly in size, charge, hydrophobicity, and other physical properties. Principal component analysis (PCA) is a way of visualizing the important features of multidimensional data sets, such as tables of amino acid properties. PCA chooses coordinates that are linear combinations of the original variables in such a way that the maximum variability between the data points is explained by the first few coordinates. Clustering analysis is another way of looking for patterns in complex data sets. A large variety of clustering methods is possible, including hierarchical and direct clustering. These methods give slightly different answers; hence some thought is required in order to interpret the resulting clusters and to decide which method is most appropriate for a given data set. Clustering and PCA are applied to the amino acid data in this chapter because they reveal interesting properties of the amino acids and also because the methods are general and are useful in many areas, such as microarray data analysis, as we shall discuss in Chapter 13.

REFERENCES

Bell, C.E. and Lewis, M. 2001. Crystallographic analysis of lac repressor bound to natural operator O1. *Journal of Molecular Biology*, **312**: 921–6.

Berman, H.M., Olson, W.K., Beveridge, D.L., Westbrook, J., Gelbin, A., Demeny, T., Hsieh, S.H., Srinivasan, A.R., and Schneider, B. 1992. The nucleic acid database: A comprehensive relational database of three-dimensional structures of nucleic acids. *Journal of Biophysics*, **63**: 751–9. (http://ndbserver.rutgers.edu/index.html)

Creighton, T.E. 1993. *Proteins: Structures and Molecular Properties*. New York: W.H. Freeman.

Engelman, D.A., Steitz, T.A., and Goldman, A. 1986. Identifying non-polar transbilayer helices in amino acid sequences of membrane proteins. *Annual Review of Biophysics and Biophysical Chemistry*, **15**: 321–53.

Hartigan, J.A. 1975. *Clustering Algorithms*. New York: Wiley.

Karypis, G. 2002. CLUTO – a clustering toolkit. University of Minnesota technical report #02–017 (http://www-users.cs.umn.edu/~karypis/cluto/)

Kyte, J. and Doolittle, R.F. 1982. A simple method for displaying the hydropathic character of a protein. *Journal of Molecular Biology*, **157**: 105–32.

Lowe, T.M. and Eddy, S.R. 1997. tRNA-scan-SE: A program for improved detection of transfer RNA genes in genomic sequences. *Nucleic Acids Research*, **25**: 955–64. (http://rna.wustl.edu/tRNAdb/)

Miller, S., Janin, J., Lesk, A.M., and Chothia, C. 1987. Interior and surface of monomeric proteins. *Journal of Molecular Biology*, **196**: 641–57.

Parry-Smith, D.J., Payne, A.W.R., Michie, A.D., and Attwood, T.K. 1998. CINEMA: A novel colour interactive editor for multiple alignments. *Gene*, **221**: GC57–GC63.

Rose, G.D., Geselowitz, A.R., Lesser, G.J., Lee, R.H., and Zehfus, M.H. 1985. Hydrophobicity of amino acid residues in globular proteins. *Science*, **228**: 834–8.

Sherlin, L.D., Bullock, T.L., Newberry, K.J., Lipman, R.S.A., Hou, Y.M., Beijer, B., Sproat, B.S., and Perona, J.J. 2000. Influence of transfer RNA tertiary structure on aminoacylation efficiency by glutaminyl- and cysteinyl-tRNA synthetases. *Journal of Molecular Biology*, **299**: 431–46.

Sokal, R.R. and Sneath, P.H.A. 1963. *Principles of Numerical Taxonomy*. San Francisco: W.H. Freeman.

Yusupov, M.M., Yusupova, G.Z., Baucom, A., Lieberman, K., Earnest, T.N., Cate, J.H.D., and Noller, H.F. 2001. Crystal structure of the ribosome at 5.5 angstrom resolution. *Science*, **292**: 883–96.

Zimmerman, J.M., Eliezer, N., and Simha, R. 1968. The characterization of amino acid sequences in proteins by statistical methods. *Journal of Theoretical Biology*, **21**: 170–201.

SELF-TEST
Biological background

This is a quick test to check that you have remembered some of the fundamental biological material in this chapter.

1 Which of the following are pyrimidines?
A: Adenine and Thymine
B: Cytosine and Thymine
C: Cysteine and Guanine
D: Cytosine and Guanine

2 Which of the following contain phosphorous atoms?
A: Proteins, DNA, and RNA
B: DNA and RNA
C: DNA only
D: Proteins only

3 Which of the following contain sulfur atoms?
A: Histidine
B: Lysine
C: Methionine
D: All of the above

4 Which of the following is not a valid amino acid sequence?
A: EINSTEIN
B: CRICK
C: FARADAY
D: WATSON

5 Which of the following "one-letter" amino acid sequences corresponds to the sequence Tyr-Phe-Lys-Thr-Glu-Gly ?
A: YFKTEG
B: WPKTEG
C: WFLTGY
D: YFLTDG

6 Which of the following peptides would have the largest positive charge in a solution at neutral pH?
A: LYAIRT
B: CTKPLH
C: VEMDAS
D: PHRYLD

7 Consider the following DNA oligomers. Which two are complementary to one another? All are written in the 5′ to 3′ direction.
(i) TTAGGC (ii) CGGATT
(iii) AATCCG (iv) CCGAAT
A: (i) and (ii)
B: (ii) and (iii)
C: (i) and (iii)
D: (ii) and (iv)

8 Which of the following statements about transcription is correct?
A: Transcription is initiated at a start codon.
B: Transcription is carried out by aminoacyl-tRNA synthetases.
C: RNA sequences must be spliced prior to transcription.
D: Transcription involves the complementary pairing of a DNA strand and an RNA strand.

9 Which of the following components of a cell does not contain RNA?
A: The nucleus
B: The ribosome
C: The spliceosome
D: The cell membrane

10 Which of the following statements about ribosomes is correct?
A: During translation, a ribosome binds to a messenger RNA near its 5′ end.
B: Ribosomes are essential for DNA replication.
C: Ribosomes can be synthesized by *in vitro* selection techniques.
D: Inside a ribosome there is one tRNA for each type of amino acid.

11 Which of the following statements about the genetic code is correct?
A: Cysteine and tryptophan have only one codon in the standard genetic code.
B: Serine and arginine both have six codons in the standard genetic code.
C: The fMet tRNA is a special tRNA that binds to the UGA stop codon.
D: All of the above.

12 Which of the following statements about polymerases is correct?
A: RNA polymerase has a much lower error rate than DNA polymerase due to proof-reading.
B: DNA polymerases move along an existing DNA strand in the 5′ to 3′ direction.
C: DNA polymerase I makes a primer onto which further nucleotides are added by DNA polymerase III.
D: None of the above.

Molecular evolution and population genetics

CHAPTER

3

CHAPTER PREVIEW

This chapter covers some basic ideas in evolution and population genetics that provide a foundation for work in bioinformatics. We discuss the role of mutation, natural selection, and random drift in molecular evolution. We present the basics of the coalescence theory that describes the descent of individuals from common ancestors, and discuss the relevance of this theory to human evolution. We consider the theory of fixation of new mutations in a population, and describe the way the fixation probability depends on strength of selection for or against the mutation. We compare the type of variation seen within a population with the differences seen between species, and compare the adaptationist and neutralist explanations for sequence variation both within and between populations.

3.1 WHAT IS EVOLUTION?

We argued in the Introduction (Section 1.4.2) that evolution is essential to understanding biology and bioinformatics. Biology is the study of life. While it is usually obvious to us whether something is alive, it is not so easy to come up with a satisfactory definition of what we mean by life. One definition that we like is that used by scientists in the NASA astrobiology program, who are interested in searching for signs of life on other planets. If you are going to search for something, it is important to know what you are looking for!

Life is a self-sustained chemical system capable of undergoing Darwinian evolution.

You may know of other definitions of life, or be able to come up with one of your own. One reason we like this one is that it underlines the importance of evolution in biology by including it in the very definition of life. Of course, this begs the question of how to define evolution.

Most people have an intuitive idea of what is meant by evolution, and are familiar with phrases like "survival of the fittest". Everyone also knows that modern ideas of evolution can be traced back to Charles Darwin, whose famous book *On the Origin of Species by Means of Natural Selection* appeared in 1859. Take a few moments now to think about what the word evolution means to you. Try to write down a definition of evolution in one or two sentences, in a form that would be suitable for use in a glossary of a scientific textbook. When you have done this, read on.

A definition that is often found in genetics textbooks is that evolution is a "change in frequency of genes in a population". While it is true that when a new variant of a gene spreads through a population, the population will have evolved, this definition does not really seem to get to the heart of the matter. Another one we came across is that evolution is "heritable changes in a population over many generations". This seems closer to the mark, but we are not quite satisfied with it. We would like to suggest that the following two points are the essential factors that define evolution: (i) error-prone self-replication; (ii) variation in success at self-replication.

By "self-replication", we mean that whatever is evolving must have the ability to make copies of itself. Evolution is not simply change. For example, changes that occur during the lifetime of an organism, such as the development of an embryo, aging processes, or the occurrence of mutations in non-germline cells, do not count as evolution. Dawkins (1976) uses the term "replicator" to describe a thing that can self-replicate. Clearly genes can self-replicate, at least in the context of the cell in which they occur. Asexual organisms like bacteria can self-replicate – when a bacterial cell divides, two cells are produced having all the same essential components as the original cell in roughly the same quantities. Sexual organisms like people can also self-replicate in a sense – the offspring of two people is recognizably another person. However, the situation is complicated by the mixing of genes from both parents into one offspring. For this reason, Dawkins (1976) makes a good case for considering genes rather than organisms as the fundamental replicators in biology.

By "error-prone", we mean that the copies are not always identical to the originals. If DNA replication were perfect, there would be no further evolution, because there would be no variations upon which natural selection could act. Also, modern-day genes have arisen by gradual changes that occurred when previous versions of these genes were replicated slightly incorrectly. Errors are thus essential for evolution, provided they do not occur too frequently. If there are too many errors, there will be no heredity, i.e., genetic information will not be passed on to the next generation.

Point (ii) is variation in success at self-replication. Most organisms could continue to increase exponentially if they were not limited by something (usually food or space or predators). If all organisms survived and multiplied at the same rate, then there would be no change in frequency of the variants, and there could be no evolution. When the population size is limited, not everyone survives, and there is the possibility of natural selection occurring. Some variants will have a greater probability than others of surviving and successfully replicating again. In evolutionary biology, the fitness of an organism is the expected number of offspring that it will have (or some quantity proportional to this). The principle of natural selection is that those variants with higher fitness will increase in number relative to those with lower fitness. The properties of the population therefore change in time, as is required by both the first two definitions of evolution that we mentioned. However, there is another important aspect of evolution that is covered within the "variation in success" in our definition. When population sizes are limited to finite numbers, chance effects can determine which variants survive. If we have a population composed initially of equal numbers of two different variants, these numbers will not stay exactly equal, even if the fitnesses of the variants are equal. Chance fluctuations in numbers will occur, known as random drift, so that sooner or later one variant will take over the whole population. This variation in success due to chance means that evolution can occur even without natural selection. This is referred to as neutral evolution, and it is an important part of the way evolution occurs at the molecular level, as we shall discuss below.

The evolutionary mechanism is surprisingly simple, to the extent of seeming almost inevitable, which is scientifically pleasing, as all we need is some kind of replicator and the notion of fallibility. It goes without saying that nothing is perfect, and hence that some errors are bound to occur in the replication. It also goes without saying that resources are limited and the replicators will have varying success, whether due to chance or to natural selection. Hence replicators are bound to evolve. Interestingly, it is not only biological replicators that can evolve. Dawkins (1976) introduced the term "meme" to describe an idea or fashion that replicates and evolves in human society. Blackmore (1999) has given a very stimulating discussion of this topic.

Given the beautiful simplicity and inevitability of evolution, it is rather surprising that it should have taken mankind until the middle of the nineteenth century to come up with the idea. It is even more surprising that some educated people today should still steadfastly refuse to accept that evolution occurs. In the context of a book on bioinformatics, there can be no better example of evolution than looking at a sequence alignment. The fact that similar gene sequences can be found in different organisms, even when these organisms look as different from one another as bacteria, mushrooms, and humans, is striking evidence that they evolved from a common ancestor. The other nice thing about comparing sequences is that it actually allows us to see the **mechanism** of evolution in a way that comparing morphological features does not. When we look at a picture of a chimpanzee and a human, we can say qualitatively that they look pretty similar but with some changes, although it is difficult to know exactly what the changes have been. When we look at an alignment of a chimpanzee gene and a human gene, however, we can say that they are different precisely because of a couple of amino acid substitutions at this point and that point. We will now go on to look at how gene sequences evolve.

3.2 MUTATIONS

At this point, it will be handy to introduce some genetic terminology that we will need in this chapter. A **locus** is the genetic name for a particular position on a chromosome. Usually a locus is a gene, but it might also be a molecular marker, i.e., a sequence that is detectable experimentally, but which is not necessarily a coding sequence. Alternative sequence variants that occur at the same locus are called **alleles**. If sequence information is available from many different individuals in a population, then the frequencies of the alleles at a given locus can be measured. A **polymorphic** locus (or a **polymorphism**) is one where there is more than one allele present in the population at a significant frequency. Sometimes there are rare alleles in a population, with frequencies of only a few percent or a fraction of a percent. It will be difficult to detect such rare alleles unless extremely large samples are available. For this reason, geneticists often count a locus as polymorphic only if the most common allele has a frequency less than a cutoff (usually 99%). This definition essentially ignores the presence of extremely rare alleles.

Most prokaryotic organisms are **haploid**, i.e., they have a single copy of each locus in the genome.

Many eukaryotic organisms are **diploid**, i.e., they have paired chromosomes, and hence two copies of each locus. If the two sequences at a given locus in an individual are copies of the same allele, then the individual is said to be **homozygous** at this locus, whereas if they are different alleles, the individual is **heterozygous**.

A **mutation** is any change in a gene sequence that can be passed on to offspring. Mutations occur either as a result of damage occuring to the DNA molecule (e.g., from radiation) or as a result of errors in the replication process. The simplest type of mutation is a point mutation, where a single DNA base is replaced by another base of a different type. As we discussed in Chapter 2, A and G bases are called purines, and C and T/U bases are called pyrimidines. A mutation from one purine to another or from one pyrimidine to another is known as a **transition**, whereas a mutation from a purine to a pyrimidine or vice versa is known as a **transversion**. When a mutation occurs, it will often create a new allele that is not already present in the population. However, in principle, the same mutation could occur more than once at different times in different individuals, in which case the mutation would create another copy of an existing allele.

Point mutations occur all over DNA sequences. The effect of these mutations will depend on where the mutation occurs. In some genomes, there are large regions of non-coding DNA between genes. In eukaryotic genomes, these often consist of repetitive sequence elements that are not transcribed and do not contribute to the functioning of the organism. A mutation occurring in such a region will have no effect on any protein sequences and is quite likely to be a neutral mutation, i.e., a mutation that does not affect the fitness of the organism. Not all non-coding regions are non-functional, however. Non-coding regions will contain signals that control initiation and termination of transcription and translation (see Chapter 2). A mutation that affects one of these regions could have a significant effect on fitness, because it would affect the level of expression of a gene.

When a mutation occurs within a protein-coding region, it has the potential to change the amino acid sequence of the protein that will be synthesized from this gene. Due to the structure of the genetic code (Table 2.1), not every DNA mutation causes a change in the protein sequence. A mutation that does not lead to a change in amino acid is called a **synonymous** substitution, while one that does lead to an amino acid change is called a **non-synonymous** substitution. Many changes at the third codon position are synonymous: e.g., the CCG codon codes for proline, but the G could change to A, C, or T without affecting the amino acid. There are also a few synonymous changes that can occur in other positions of the codon: e.g., in the leucine codon CTG, the first position C can change to a T. However, the majority of changes in first and second positions are non-synonymous.

Non-synonymous changes are sometimes classified as either **missense** mutations, where an amino acid is replaced by another amino acid, or **nonsense** mutations, where a stop codon is introduced into the middle of the sequence (e.g., the tryptophan codon TGG could change to the stop codon TAG). A nonsense mutation will lead to termination of translation in the middle of the protein, i.e., half the protein will be missing. This will almost always have a strong negative effect on fitness. A missense mutation might have a positive or negative effect or no effect on fitness, depending on where the change is in the protein and how different the new amino acid is from the old one.

Not all mutations are simple substitutions of one base by another. Insertions and deletions can also occur – these are collectively referred to as **indels**. Small indels of a single base or just a few bases are frequent. One way in which this can occur is by slippage during DNA replication. If the DNA replicase slips back a couple of bases, it can lead to insertion of those bases twice into the new strand. Repeated sequences are particularly prone to this type of indel because the two strands of DNA can still pair with one another if they slip slightly: e.g., a sequence GCGCGCGCGC could have extra GC units inserted or removed from it. Repeated sequences of short motifs like this are called **microsatellites**. The length of microsatellite alleles (i.e., the number of copies of the repeated motif) changes rapidly, hence microsatel-

lites are often polymorphic. They are useful in experimental population genetics studies because they reveal information about the history and geographical subdivisions of populations (see Schlötterer 2000). Repeated sequences occurring in protein-coding sequences can have dramatic effects, and often lead to disease (Karlin *et al.* 2002). For example, the protein huntingtin contains a CAG repeat region that is prone to expansion via slippage, which results in Huntington's disease, a severe hereditary disorder in humans.

If an indel occurs in a protein sequence that is not a multiple of three nucleotides, this causes a **frame shift** that changes the sequence of amino acids specified for the whole of the region downstream of the deletion or insertion. This is likely to cause loss of function of the gene. Thus a single base insertion or deletion is a much more drastic mutation than a single base substitution.

Mutations can sometimes involve much larger scale changes in DNA. For example, deletions or insertions of tens or even thousands of bases can sometimes occur in genes. Changes can also occur at the chromosome level. Sections of DNA involving whole genes or several genes can be **inverted**, i.e., reversed in direction, or **translocated**, i.e., cut out from one part of a genome and inserted into another. All these types of change can be classed as mutations, as they are changes in DNA that can be passed on to future generations.

3.3 SEQUENCE VARIATION WITHIN AND BETWEEN SPECIES

In this section, we will consider the example of one well-studied gene that allows us to see the patterns of sequence variation that arise due to mutations. We will compare the variations we see within the human species with those we see between humans and other animals.

The gene *BRCA1* has been shown to be associated with hereditary breast and ovarian cancer in humans (Miki *et al.* 1994). Women with mutations in this gene are significantly more likely to develop cancer. It is thought that the normal role of the

Table 3.1 Numbers of documented cancer-associated mutations in the *BRCA1* gene.

Mutation	Number
Nucleotide substitutions (missense/nonsense)	126
Small deletions	163
Small insertions	55
Small indels	5
Gross deletions	20
Gross insertions and duplications	2
Complex rearrangements (including inversions)	3
Nucleotide substitutions (splicing)	34
Total	408

Data taken from the Human Gene Mutation Database (Stenson *et al.* 2003; http://www.hgmd.org).

BRCA1 protein is in repair of oxidative DNA damage (Gowen *et al.* 1998). If this protein does not function correctly, DNA damage can accumulate in individual somatic cells and occasionally this causes cancer. The Human Gene Mutation Database (Stenson *et al.* 2003) describes mutations in human genes that are known to be associated with diseases. For *BRCA1*, the database lists 408 mutations. The positions of these mutations are distributed along the entire length of the sequence. The numbers of documented mutations of different types are shown in Table 3.1.

Roughly a third of these mutations are single nucleotide substitutions. Slightly over a third of the mutations are small deletions of one or a few nucleotides from the coding region of the DNA. A few gross deletions and insertions are also observed in this gene, as are a few inversions. Like many human genes, *BRCA1* has a complex intron/exon structure – there are 24 exons in this case. The correct synthesis of the BRCA1 protein therefore requires correct splicing of the mRNA. Nucleotide substitutions occurring at the splice sites can lead to incorrect splicing, and hence to large-scale errors in the resulting protein. For example, a splice site might disappear due to a mutation, which might cause the omission of a whole exon from a protein, or a sequence defining a new splice site might arise in the wrong place (this is called activation of a cryptic splice site). Many splice-site mutations are seen in this protein.

This example illustrates how many different ways mutations can impair the function of a gene. Genes have arisen through many millions of years of natural selection, so we would expect them usually to be rather good at their jobs. Most mutations introduced into a functioning gene sequence will be **deleterious**, i.e., they will be of reduced fitness in comparison to the original sequence. Mutations contained in disease-linked databases are those that are sufficiently deleterious to result in an obvious disease phenotype. These mutations are usually very rare in the population at large because there is significant natural selection acting against them. Other mutations may have only a very small effect on the function of the gene that would have no obvious consequences for the individual. However, small selective effects are very important on the evolutionary time scale. Natural selection, acting over many generations, tends to keep the frequencies of slightly deleterious mutations to low levels.

Of course, not all mutations are deleterious. Some will be neutral, and there must be occasional **advantageous** mutations that increase the fitness of the sequence. If there were no advantageous mutations, then the gene sequence could never have evolved to be functional in the first place. Nevertheless, we would expect advantageous mutations to be much rarer than deleterious ones, so when we look at a set of mutations of a gene in a population like this, the majority will be deleterious.

The *BRCA1* gene has been sequenced in many other species as well as humans. Figure 3.1 shows an alignment of part of the *BRCA1* DNA sequence from 14 different mammals. Sites shaded in black are identical in all species, and sites shaded in gray show a noticeable degree of conservation across species. All these sequences must have descended from a common ancestor at some point in the past. They have gradually diverged from one another due to the fixation of different mutations in different species. Deleterious mutations will not often be fixed

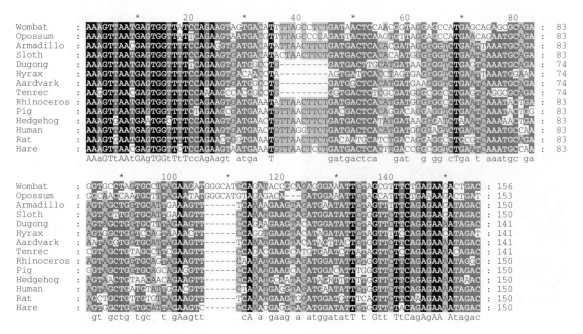

Fig. 3.1 Part of the alignment of the DNA sequences of the *BRCA1* gene.

in a population, thus the differences between species will usually arise due to the accumulation of either neutral or advantageous mutations.

Figure 3.2 shows the protein alignment for the same part of the *BRCA1* sequence as Fig. 3.1. The first few amino acids in this sequence are the same in all these species. In the DNA alignment, the only substitutions that have occurred in this region are synonymous changes. These occur in columns 3, 6, 9, and 12 of Fig. 3.1 (check these with the genetic code in Table 2.1). Since almost all synonymous changes are at the third position of codons, third-position sites tend to evolve more rapidly than first- and second-position sites. For each amino acid where there are just two synonymous codons in the genetic code, the alternative bases are either two purines or two pyrimidines. This means that many synonymous substitutions tend to be transitions, and this is one reason why transitions tend to be seen more frequently than transversions. For example, in Fig. 3.1 we have A ↔ G transitions at columns 3 and 12, and C ↔ T transitions in column

9 (the C seems to have arisen more than once in unrelated species). In column 6 there is a T ↔ G transversion. This is still synonymous because there are four synonymous codons for valine.

When examining protein alignments, we see that not all amino acid substitutions are equally likely. Those that occur between amino acids of similar chemical properties tend to be more frequent. These are often called **conservative** changes. As clear examples in Fig. 3.2, we see that in column 8 there is a substitution from an arginine to a lysine (two basic amino acids) and in column 12 there are several substitutions between isoleucine, leucine, valine, and methionine (all of which are hydrophobic, as we discussed in Chapter 2). In Chapter 4, we will describe methods for quantifying the differences in substitution rates between different amino acids.

The alignments in Figs. 3.1 and 3.2 also contain several gap regions. There appears to have been a nine-base (i.e., three-amino acid) deletion in the four species dugong, hyrax, aardvark, and tenrec, from positions 37–45. This deletion was pointed out by

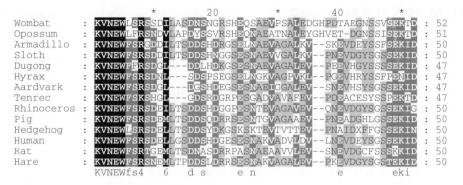

```
                      *         20          *          40          *
Wombat     : KVNEWLSRSSDILASDNSNGRSHEQSAEVPSALEDGHPDTAEGNSSVSEKTD : 52
Opossum    : KVNEWLFRSNDVLAPDYSSVRSHEQNAEATNALEYGHVET-DGNSSISEKTD : 51
Armadillo  : KVNEWFSRGDDILTSDDSHDRGSELNAEVAGALKV--SKEVDEYSSFSEKID : 50
Sloth      : KVNEWFSRSDDILTSDDSHNGGSESNAEVVGALKV--PNEVDGYSGSSEKID : 50
Dugong     : KVNEWFERSDGL---DDLHDRGSESNAEVAGALEV--PEEVHGYSSSSEKID : 47
Hyrax      : KVNEWFSRSDNL---SDSPSEGSELNGKVAGPVKL--PGEVHRYSSFPENID : 47
Aardvark   : KVNEWFSRSDGL---DGSHDEGSESNAEIGGALEV--SNEVHSYSGSSEKID : 47
Tenrec     : KVNEWFSKSHGL---GDSRDGRPESGADVAVAFEV--PDEACESYSSPEKTD : 47
Rhinoceros : KVNEWFSRSDEILTSDDSHDGGPESNTEVAGAVEV--QNEVDGYSGSSEKIG : 50
Pig        : KVNEWFSRSDEMLTSDDSQDRRSESNTGVAGAAEV--PNEADGHLGSSEKID : 50
Hedgehog   : KVNEWLSRSDELLTSDDSYDKGSKSKTEVTVTTEV--PNAIDXFFGSSEKIN : 50
Human      : KVNEWFSRSDELLGSDDSHDGESESNAKVADVLDV--LNEVDEYSGSSEKID : 50
Rat        : KVNEWFSRTGEMLTSDNASDRRPASNAEAAVVLEV--SNEVDGCFSSSKKID : 50
Hare       : KVNEWFSRSNEMLTPDDSLDRRSESNAKVAGALEV--PKEVDGYSGSTEKID : 50
             KVNEWfs4    6   d s    e  n                e       eki
```

Fig. 3.2 Alignment of the BRCA1 protein sequences for the same region of the gene as Fig. 3.1.

Madsen *et al.* (2001), who showed that the deletion was common to a group of species known as Afrotheria, which also includes golden moles, elephant shrews, and elephants. The Afrotheria group is extremely diverse morphologically, and it is only by using molecular evidence such as this that we have become aware that these species are relatives. Mammalian phylogeny is discussed in more detail in Chapter 11. For now, we note that there are two more deletions in this part of the *BRCA1* gene. Three bases have been deleted from positions 121–3 in the opossum, and six bases have been deleted in positions 106–11 in all species except the wombat and opossum. This example is also illustrative of mammalian phylogeny: the wombat and opossum are marsupials, whereas the rest are eutherian mammals. Strictly speaking, whenever there is a gap in an alignment, we have no way of knowing whether there has been a deletion in one group of species or an insertion in the opposite group. For the first two examples given above, it seems safe to assume that the deletion is a derived feature, and that the shorter sequences have evolved from longer ancestors. However, in the third example, the information in the alignment does not tell us whether there was a deletion in the common ancestor of the eutherian mammals or an insertion in the common ancestor of the marsupials.

One thing that **is** clear in all three indels is that they have occurred in multiples of three bases, i.e., whole codons. None causes a frame shift. An indel of

a small number of amino acids into a protein will somtimes not affect the protein much, especially if it does not disrupt a region of secondary structure. So, like some amino acid substitutions, some small indels can be nearly neutral. We know from studying human mutations (e.g., Table 3.1) that frame-shift-causing indels do occur. Presumably they occur in other species too, although data on sequence variants are not usually available for non-human species. The reason we do not see these frame shifts in between-species comparisons is that they are very likely to be deleterious mutations that do not become fixed in populations.

This section has emphasized that there is an important difference between the sequence variation that we see within a species and the variation that we see between species. Most sequence variants within a species will be deleterious mutations that will probably go on to be eliminated by selection. The differences that we see between species are as a result of fixation of mutations in the population of one species or the other. These changes are the ones that did not get eliminated by selection, i.e., they are usually either neutral or advantageous. The question of the relative importance of neutral versus advantageous mutations in driving molecular evolution is a key issue to which we will return at the end of this chapter. For the next two sections, we need to look at two fundamental aspects of population genetics: coalescence theory and the process of fixation of mutations in populations.

3.4 GENEALOGICAL TREES AND COALESCENCE

3.4.1 Adam and Eve

Each gene that we have is a copy of a gene in one of our parents, and this was itself a copy of a gene in one of our grandparents. It is possible to trace the lines of descent of a gene back through the generations. It is simplest to begin with genes that are inherited through only one sex. In humans, there are two good examples of this: mitochondrial DNA (mtDNA) is inherited through the maternal line (both boys and girls get it from their mother), and Y chromosome DNA is inherited through the paternal line (boys get it from their father). As far as we know, both types of DNA are inherited without recombination. Thus, in the absence of new mutations, each gene will be an exact copy of the parent's gene. Figure 3.3 shows the "family tree" of such a gene. Each circle represents a copy of a gene in a population. Each row represents a generation. Each gene is copied from one of the genes in the previous generation, indicated by a line. Suppose we consider all the individuals in the current generation (bottom of the figure) and trace back their genes. The lines of descent leading to the present generation are shown in bold. All these lines of descent eventually converge on a single copy of the gene if we go far enough back in time. This process is known as coalescence. The number of generations that we have to go back until all the lines of descent coalesce depends on the size of the population and other details, but it is inevitable that coalescence will occur eventually. The reason for this inevitability is simply that not all individuals in any one generation pass on copies of their genes. A woman with no daughters does not pass on her mitochondrial DNA, and a man with no sons does not pass on his Y chromosome. These are dead ends in the gene family tree. Equally there are some gene copies that are passed on more than once (e.g., mothers with two or more daughters). These are branch points in the tree when we move forwards in time, and coalescence points when we move back in time.

In the case of human mtDNA, it is believed that all existing sequence variants can be traced back to a single woman living in Africa roughly 200,000 years ago, who is usually known as mitochondrial Eve (Cavalli-Sforza, Menozzi, and Piazza 1996). This does **not** mean that there was only one woman in Africa at that time. There must have been a whole population, but Eve was the only one whose mtDNA happened to survive. There was probably nothing much to distinguish her from anyone else around at the time. We know that humans were found in many different areas of the world well before the estimated date of mitochondrial Eve. The "out of Africa" hypotheses, supported by the mtDNA evidence, is that modern humans are all descended from an African population that successfully spread throughout the world relatively recently and replaced previously existing groups.

Most studies of human mtDNA have used a highly variable non-coding part of the genome called the D-loop. More recently complete mitochondrial genome sequences have become available from many individuals and these have also been used to reconstruct the history of human populations. Ingman *et al.* (2000) observed that the mean number of sites

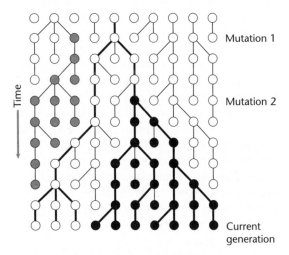

Fig. 3.3 Illustration of the coalescence process. Each circle represents one gene copy. Bold lines show the lines of descent of genes in the current generation. Thin lines show lines of descent that do not lead to the current generation. Shaded circles show the inheritance of two different mutations.

that differed between pairs of genomes taken from a worldwide sample was 61.1 out of a total length of 16,553 bases. When the sample was divided into Africans and non-Africans, the mean pairwise difference was 76.7 between Africans and only 38.5 between non-Africans. This suggests that there have been different divergent populations in Africa for much longer than in the rest of the world. The part of the genome excluding the D-loop almost entirely codes for proteins, rRNAs, and tRNAs. Although the D-loop has a substantially higher density of variable sites than the rest of the genome, Ingman *et al.* (2000) argue that the rest of the genome evolves in a more predictable fashion and is more reliable for estimating substitution rates and times of divergences between groups.

Plate 3.1 shows their results for complete genomes excluding the D-loop. This shows only sites that are informative for construction of phylogenetic trees. Sites that are identical in all individuals, or where a single individual differs from all the rest are not informative because they give no information about shared common ancestors. The sites shown are polymorphic, i.e., there is more than one base that occurs at this site with an appreciable frequency in the total human population. The sites are shown in order of decreasing frequency of the polymorphism, not according to their position on the genome. Sequence variants are shared by related individuals because the original substitution occurred in a common ancestor of the individuals. For example, all individuals in the blue group in Plate 3.1 possess A at the second site, in contrast to all the other sequences, which have a G at this point. Thus, there must have been a substitution from a G to an A in a common ancestor of the blue group. The "blockiness" of the data illustrates the fact that mtDNA is inherited without recombination. Thus, the individuals with an A at the second site also have a T at the third and fourth sites, showing that several substitutions occurred in the same line of descent. Recombination would tend to obscure such patterns of correlations between sites. The pattern of ancestry is also obscured when the same substitution occurs more than once at a site. For example, at the first site, there seems to have

been more than one independent substitution from A to G. Sites like this give conflicting phylogenetic information from other sites, but as long as there are sufficient segregating sites in the whole sequence, it is likely that the true relationship will be apparent.

The left side of Plate 3.1 shows the deduced phylogenetic tree. The early branching sequences (shaded pink) are all African. The group shaded green is also all African, and is the most closely related African group to sequences from the rest of the world. The yellow and blue groups include individuals from many other parts of the world (e.g., Europe, China, Australia) but no Africans. The most important branch points on this tree are strongly supported by bootstrapping (a method of judging the reliability of nodes on phylogenetic trees that we will discuss in Chapter 8). Thus, the data strongly suggest that a relatively small population left Africa and spread through the rest of the world, taking its mitochondrial DNA along with it. The estimate for the age of the most recent common ancestor of all the sequences in the study (i.e., Eve) is 170,000 ± 50,000 years ago, while the estimated date of the most recent common ancestor of the non-African sequences (i.e., the date of the migration out of Africa) is 52,000 ± 27,000 years ago.

A similar story is told by human Y chromosomes. Underhill *et al.* (2000) studied 167 polymorphisms in the non-recombining part of the Y chromosome in a sample of over 1,000 men. These consisted mostly of single nucleotide substitutions, together with a few small insertions and deletions, and one insertion of an *Alu* element. They also found that the most divergent sequences were African. Thompson *et al.* (2000) estimated the date for the most recent common male ancestor, "Y-chromosome Adam", to be around 59,000 years ago, and the date for the expansion out of Africa to be around 44,000 years ago (with wide confidence intervals on both these dates). There is no reason why Adam and Eve should have existed in the same time and the same place, because the lines of descent of the two different types of DNA are independent of one another. Patterns of migration of men and women over time may also have been different.

3.4.2 A model of the coalescence process

After this brief diversion into human evolution, let us consider a simple theory for the coalescence process. Suppose there are N women in the population, and the population size is constant throughout time. For simplicity, assume that all individuals are equally fit, and therefore that they all have the same expected number of offspring. This means that any individual in the present generation is equally likely to have had any of the individuals in the previous generation as a mother. If we choose two random individuals in the present generation, the probability that they had the same mother according to this model is $1/N$, and the probability that they had different mothers is $1 - 1/N$. The probability that their most recent common ancestor lived T generations ago is

$$P(T) = \left(1 - \frac{1}{N}\right)^{T-1} \frac{1}{N} \qquad (3.1)$$

This means that as we follow the two lines of descent backwards, they must have had different mothers for $T - 1$ generations and then the same mother on the Tth generation. To a good approximation (see Eq. 35 in the Mathematical Appendix) this can be written as

$$P(T) = \frac{1}{N} e^{-T/N} \qquad (3.2)$$

i.e., the distribution of times to coalescence of the lines of descent of any two individuals is exponential, with a mean time of N generations. It can also be shown that the mean time till coalescence of all N individuals is $2N$, i.e., for a constant non-expanding population, we expect that the most recent common ancestor lived roughly $2N$ generations ago.

We have so far spoken of genes passed down in one sex only. However, the majority of our genes are on the autosomes (i.e., non-sex chromosomes) and are passed down through both sexes. We have two copies of each autosomal gene, one of which is descended from each of our parents. Thus, for a population of size N, there are $2N$ copies of each gene. However, more importantly, recombination can occur in the autosomes. There are typically one

or two crossover events per generation on every pair of autosomes, i.e., there is a lot of reshuffling of gene combinations. We cannot draw a simple genealogical tree like Fig. 3.3 for a whole chromosome because different parts of the chromosome will be descended from different ancestors. If we are interested in just one gene on an autosome, then the tree picture still makes some degree of sense. What is relevant is the probability of crossover occurring **within** the gene, which is usually very small, because any one gene represents only a small proportion of the whole chromosome. There are some genes for which we have sequence data from many individuals and we can use coalescence theory to tell us something about the probability that recombination events have occurred at different points within the sequence. If there is more than one segregating site in the gene, recombination can create new sequence variants that have not been seen before. In the following section, however, we are only concerned with the spread of a single point mutation through a population. In this case, we can follow back the tree that describes the inheritance of this single site in the DNA sequence. For this calculation it does not really matter whether there is recombination in the gene or not.

3.5 THE SPREAD OF NEW MUTATIONS

3.5.1 Fixation of neutral mutations

Consider a point mutation that has just occurred in a gene. One individual in a population now contains a gene sequence that is different from all the other members of the population. What happens to this new sequence? In Fig. 3.3, the different shadings of the circles represent different mutations. Mutation number 1 (shaded gray) first arises at the second generation in the figure. This mutation is transmitted to several individuals and survives in the population for a number of generations before extinction occurs. This mutation does not survive to the present generation because it does not arise on the lines of descent of the gene copies in the present generation (the thicker black lines in the figure). In contrast, mutation 2 (shaded black) does arise on the line of descent of the gene copies in the present generation. This mutation has risen to a high frequency in the population. Mutation 2 stands a large chance of spreading to take over the whole population, i.e., of becoming fixed.

Very few new mutations become fixed – most disappear after only a few generations. If a mutation is neutral, it is easy to calculate the probability, p_{fix}, that it will become fixed by considering the tree diagram in Fig. 3.3. We know that we can trace all individuals of the present generation back to some single individual in an ancestral population. Suppose a mutation occurred in the ancestral population. There were N copies of the gene, and each one was equally likely to mutate. The probability that the mutation occurred in the gene copy that happened to become the ancestor of the present generation is therefore $1/N$. The same argument works if we think forwards from the present generation. If we go sufficiently far forwards in time, only one of the present generation will have surviving descendants. Only mutations that occur in this one individual will be fixed. The fraction of neutral mutations in the present generation that will become fixed is therefore

$$p_{fix} = 1/N \qquad (3.3)$$

Suppose that the probability of a new mutation arising at each site in a DNA sequence is u per generation in each copy of the gene. The mean number of new mutations arising at a given site per generation in the whole population is therefore Nu. The rate of fixation of new mutations is the rate at which mutations arise, multiplied by the probability that each mutation is fixed:

$$u_{fix} = Nu \times p_{fix} = u \qquad (3.4)$$

This says that the rate of fixation of neutral mutations is equal to the underlying mutation rate and is independent of the population size. This is one of the most fundamental results of neutral evolution theory (Kimura 1983). However, for non-neutral mutations, p_{fix} is not equal to $1/N$ (see Box 3.1), and therefore it is only true that $u_{fix} = u$ for neutral mutations.

3.5.2 Simulation of random drift and fixation

Neutral mutations spread through a population by **random drift**. This means that the number of copies of the mutation in the population changes only due to chance effects. Suppose there are m copies of a neutral mutant sequence in a population at one generation, and we would like to know the number n of copies that there will be at the next generation. On average, we expect that $n = m$; however, random drift means that n could be more or less than m. We will use a population genetics model known as the Wright–Fisher model to investigate this effect. In this model, it is assumed that each copy of the gene in the new generation is descended from one randomly chosen gene copy in the previous generation. Thus, each gene in the new generation possesses the mutation with probability $a = m/N$, and lacks the mutation with probability $1 - a$. The probability that there are n copies of the mutation in the new generation is therefore given by a binomial distribution (see Section M.9 in the Appendix if you need a reminder):

$$P(n) = C_n^N a^n (1 - a)^{N-n} \qquad (3.5)$$

From this distribution, the mean value of n is $Na = m$, as expected. However, there is fluctuation about this mean. This formula provides the basis for a computer simulation of the fixation process shown in Fig. 3.4. The simulation considers a population of $N = 200$ gene copies. At time zero, one copy of a new sequence arises by mutation. We then calculate the value of n at each subsequent generation by choosing a random value with the probability distribution given by Eq. (3.5). The simulation continues until either the mutation becomes extinct ($n = 0$) or it goes to fixation ($n = N$). The simulation was repeated 2,000 times to follow the progress of 2,000 independent mutations. Figure 3.4 shows n as a function of time for several of these 2,000 runs. The vast majority of runs become extinct after only a few generations, and hence are hardly visible on the graph. One run is shown where n rises to quite a high value but eventually disappears after about 75 generations. Another is shown in which fixation occurs after about 225 generations. Out of 2,000

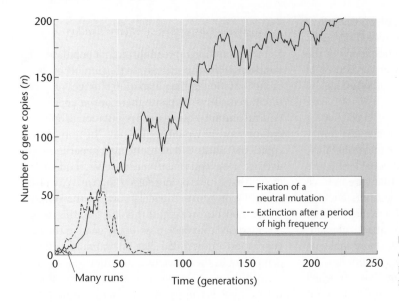

Fig. 3.4 Simulations of the spread of neutral mutations through a population under the influence of random drift.

runs, it was found that fixation occurred 11 times. According to the theory, the probability of fixation is $1/N$, hence the expected number of fixations is $2{,}000 \times 1/200 = 10$. The observation is therefore consistent with this expectation (see Problem 3.2).

3.5.3 Introducing selection

We can extend the simulation to deal with the case of an advantageous mutation that has a fitness $1 + s$, with respect to the original sequence that has fitness 1. The constant s is known as the selection coefficient. Remember that fitness is defined to be proportional to the expected number of offspring: an individual with the mutation is expected to have $1 + s$ times as many offspring on average as one without.

If there are m copies of the mutation at some point in time, then the mean fitness of the population is

$$\bar{w} = \frac{m(1 + s) + (N - m)}{N} \qquad (3.6)$$

According to the Wright–Fisher model, each gene copy is selected to be a parent of the next generation with a probability proportional to its fitness. The probability that a gene in the new generation possesses the advantageous mutation is

$$a = \frac{m(1 + s)}{N\bar{w}} \qquad (3.7)$$

The distribution of the number of copies in the next generation is still given by the binomial distribution in Eq. (3.5), but with the new value of a. The mean number of copies in the next generation is now greater than m. Figure 3.5(a) shows the results of simulations where the selection coefficient is $s = 0.05$. Even with this advantage, there are still many runs of the simulation where the mutation becomes extinct. One example where the mutation becomes fixed is also shown.

The way that p_{fix} depends on selection is discussed in more detail in Box 3.1. The most important result in the Box is that when s is small, such that $s < 1/N$, the spread of the mutation is more influenced by random drift than it is by selection, and the probability of fixation is virtually equal to $1/N$, as it is for neutral mutations. Mutations with small s in this regime are called "nearly neutral", and behave essentially the same as neutral mutations. This result means that the range of selective values that behave as nearly neutral depends on the population size. A mutation with a given s could thus behave as an advantageous mutation in a large

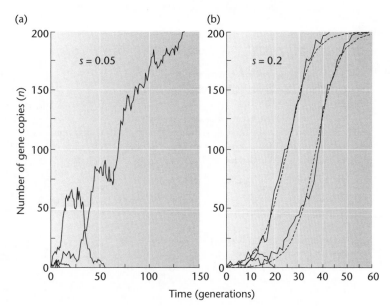

Fig. 3.5 Simulations of the spread of advantageous mutations through a population. (a) For selection coefficient $s = 0.05$ both selection and random drift are important. (b) For $s = 0.2$ selection dominates random drift. The dashed lines show the predictions of the deterministic theory in Box 3.2.

population and a nearly neutral mutation in a smaller population.

In Fig. 3.5(b), the selective advantage of the mutation is increased to $s = 0.2$. In this case, a greater fraction of the runs become fixed (but by no means all of them). In cases where a mutation goes to fixation, the time it takes to do so becomes shorter, the higher the value of the selection coefficient. This can be seen by comparing the times to fixation in Figs. 3.4, 3.5(a), and 3.5(b). If the selective advantage of the mutation is large, it spreads rapidly through the population in the form of a selective sweep. In this case, the spread can be approximated by a deterministic theory, as described in Box 3.2.

For a deleterious mutation, the fitness of the mutation is defined as $1 - s$ instead of $1 + s$ but otherwise things are the same. The fixation probability for a deleterious mutation is less than that for a neutral mutation, as we might expect, but it is not altogether zero. The way that p_{fix} depends on selection for deleterious mutations is also discussed in Box 3.1.

A word of caution may be necessary when interpreting the simulations shown here. We chose a rather small population size for convenience (to save computer time and memory). For this reason, we needed to use rather large selection coefficients in order for the selective effect to be apparent. The case

of $s = 0.05$ is behaving like a slightly advantageous mutation here when $N = 200$, with a lot of random drift evident in the curves in Fig. 3.5(a). However, 5% would be considered a rather big selective effect in real life, because mutations of this degree of advantage are probably rare, and because most population sizes would be much larger than 200. This is an example where it is useful to have both simulations that show visually what is going on, and analytical calculations that give us results for parameter values where it is difficult to do simulations. Simulations also allow us to check the mathematical results, and to make sure that any approximations that were made in the analytical calculations were valid.

3.6 NEUTRAL EVOLUTION AND ADAPTATION

Evolutionary biologists working at the whole-organism level are used to thinking of selection as the primary influence shaping the creatures that they study. It is easy to find examples of characters that must have adapted to their present forms from simpler, or more general, forms due to the action of natural selection (e.g., horses' hooves, human

BOX 3.1
The influence of selection on the fixation probability

It is possible to make an exact calculation of the probability of fixation of an advantageous mutation with fitness $1 + s$. The calculation uses what is known as diffusion theory (see Crow and Kimura 1970). The result is:

$$p_{fix} = \frac{(1 - e^{-2s})}{(1 - e^{-2Ns})} \qquad (3.8)$$

The shape of this function is shown in Fig. 3.6. If the mutation is **extremely advantageous**, i.e., $s \gg 1$, then both the exponentials in Eq. (3.8) are much less than 1, and so $p_{fix} \approx 1$. However, evolution usually proceeds

by very small steps, and it is highly unlikely that a mutation with such a large advantage would arise.

It is more likely that a mutation would occur in the regime of **slightly advantageous** mutations, where $s \ll 1$, but $Ns \gg 1$. In this case, the exponential in the denominator is negligible, but the exponential in the top line can be expanded as $e^{-2s} \approx 1 - 2s$. Thus we obtain $p_{fix} \approx 2s$ for slightly advantageous mutations. This is the sloping dashed line in Fig. 3.6. Since we assumed that $s \ll 1$, the fixation probability for slightly advantageous mutations is still very small, even though they are advantageous.

The final regime is that of **nearly neutral** mutations, where $Ns \ll 1$. In this case, both the exponentials can be expanded:

$$p_{fix} = \frac{1 - (1 - 2s \dots)}{1 - (1 - 2Ns \dots)} \approx \frac{1}{N} \qquad (3.9)$$

Thus, in the limit of very small s, the fixation probability becomes equal to the value that we already calculated for strictly neutral mutations. This now tells us more precisely what is meant by a "nearly neutral" mutation: it is one such that s is sufficiently small that $Ns \ll 1$.

A similar calculation can be done for deleterious mutations with a fitness $1 - s$. The result is:

$$p_{fix} = \frac{(e^{2s} - 1)}{(e^{2Ns} - 1)} \qquad (3.10)$$

This curve is also shown in Fig. 3.6. In the nearly neutral regime, where $Ns \ll 1$, p_{fix} again tends to $1/N$. If s increases, the fixation probability drops steeply, so that the fixation probability of significantly deleterious mutations becomes extremely small. (Note that in this section, N is the number of gene copies, since we are thinking of haploid genes. In many places, this formula is quoted specifically for diploid genes where there are $2N$ gene copies for a population of size N. This just changes the N to $2N$ in the formulae, if we assume that the fitness of a heterozygous individual is midway between the fitness of the two homozygotes.)

Fig. 3.6 Fixation probability in a population of $N = 200$ as a function of selection coefficient s, for both advantageous and deleterious mutations. When $Ns \ll 1$, both types of mutation behave as nearly neutral mutations.

hands, flounders lying on their sides, camouflage patterns on many animals). We can think of this type of selection as "positive" selection, because it acts to select new variants. When molecular sequences first became available, it was natural to assume

that positive selection would be important in causing adaptations at the sequence level too. **Adaptationists** are people who argue that positive selection is the major driving force in molecular evolution. In contrast, **neutralists** are people who argue that

BOX 3.2
A deterministic theory for the spread of mutations

The spread of mutations through populations is an inherently stochastic process, where chance plays a large role. However, if the population size is very large, then we can sometimes approximate the process using a deterministic equation. In this case, we forget the number of copies of the mutation n and instead we think of the frequency of the mutation in the population, $x = n/N$. If the frequency is x, at generation t, then the frequency at generation $t + 1$, is

$$x(t + 1) = \frac{x(1 + s)}{\bar{w}} = \frac{x(1 + s)}{1 + sx} \qquad (3.11)$$

This is analogous to Eq. (3.7) in the stochastic case. Now we use two approximations:

$$x(t + 1) \approx x + \frac{dx}{dt} + \ldots \qquad (3.12)$$

$$(1 + sx)^{-1} \approx 1 - sx + \ldots \qquad (3.13)$$

These are valid provided s is small and the expected change in frequency in a single generation is small. In this case, Eq. (3.11) becomes

$$\frac{dx}{dt} = sx(1 - x) \qquad (3.14)$$

This will be familiar to ecologists as the logistic equation for population growth. In the ecological case, x would be the population density and s would be the rate of multiplication of the population (usually written as r).

Equation (3.14) has the solution

$$x = \frac{\exp(s(t - T))}{1 + \exp(s(t - T))} \qquad (3.15)$$

This function is an S-shaped growth curve. The constant T can take any arbitrary value: in fact, T is the time at which the frequency rises to $1/2$. In Fig. 3.5(b), the dashed lines show the logistic growth curve for the case where $s = 0.2$. The values of T have been chosen to match the two simulated curves as closely as possible. The deterministic result works quite well in this case, as long as x is neither very small nor very close to 1. When x is close to one of the two extremes, stochastic effects are important. This is why the two runs of the simulation do not rise up at exactly the same time, even though they have the same parameters.

---

neutral evolution is responsible for most changes at the molecular level. Neutralists argue that mutation is the driving force in molecular evolution and that positive selection has a relatively minor role. In this section, we consider these two points of view more carefully.

An important technique developed in the 1960s and 70s is the study of **allozymes**. These are alternative alleles of proteins that can be separated by electrophoresis due to differing charges. These studies revealed that many protein loci are polymorphic. This was somewhat unexpected according to the prevailing adaptationist point of view, because selection should remove lower fitness alleles at a polymorphic locus relatively quickly, leaving only a single allele. The allozyme technique only detects a fraction of the sequence variation that is present,

because synonymous substitutions do not lead to changes in the protein, and because amino acid changes that do not alter the net charge on the protein are usually not detected either. It therefore became an important issue to explain why many allozyme loci are polymorphic.

Another important technique, which was developed in the 1970s and 80s, is the use of restriction fragment length polymorphisms (RFLPs). Restriction enzymes are nucleases that recognize particular nucleotide sequences in DNA, and chop the DNA at these points. Restriction enzymes will cut a long section of DNA into fragments that can be separated according to their lengths by electrophoresis. This technique has often been used with mitochondrial DNA. When differences in fragment lengths are observed between individuals, this reveals polymorphisms in the DNA bases at the restriction sites. Both allozymes and RFLPs reveal only certain types of sequence polymorphism. In recent years, it has become possible to do DNA sequencing on a large scale, and hence we are beginning to get large data sets of complete DNA sequences of genes for samples of individuals in populations. These studies reveal complete information about the pattern of single nucleotide polymorphisms in the genes studied. At both DNA and protein levels, it is important to ask why sites are polymorphic and what this reveals about the mechanism of molecular evolution.

In its simplest form, positive selection removes deleterious alleles, and hence reduces the number of polymorphic loci. Adaptationists therefore propose more complicated selective scenarios for maintenance of polymorphisms. One possibility is that there may be alleles that are advantageous in certain circumstances and disadvantageous in others. For example, a species may inhabit two different areas with different climates, and there may be different alleles that are optimal in the two areas. Selection would then lead to the maintenance of both alleles in the population. Another possibility is that a heterozygous individual may have a higher fitness than either of the homozygotes. This would lead to selection towards a state with equal frequencies of the two alleles.

Although scenarios like this are probably true for some genes, it seems unlikely that they are responsible for the majority of polymorphisms that we see. The neutralist explanation for polymorphisms is simply that there is a constant process of creation of new alleles by mutation and loss of old alleles when they disappear due to random drift. Thus, at any point in time, some loci will be polymorphic, without the need for any particular selective scenario. The neutralist explanation for variability within populations is simply that there is a relatively large mutation rate. A useful measure of variability at a locus is the heterozygosity. For a diploid population, the heterozygosity is the fraction of heterozygous individuals. A heterozygous individual obtains one allele from each parent. If we assume that mating is random with repect to the alleles at the locus, the two alleles possessed by any individual will be randomly

selected alleles from the population at the parents' generation. Therefore, it is also possible to define heterozygosity as the probability that two alleles selected randomly from the population are different from one another. The average heterozygosity maintained by mutation and random drift at a neutral locus can be shown to be a function of Nu, where N is the population size and u is the mutation rate. For $Nu \geq 1$, the mean heterozygosity will be large and most loci will be polymorphic. For $Nu \ll 1$, the mean heterozygosity will be very low, and there will be few polymorphic loci.

One advantage of the neutral theory is that many quantities can be calculated exactly. This means that the neutral model can be used as a null hypothesis against which to test real data. If it can be shown that the data differ significantly from expectations under the neutral model, then this demonstrates that selection has played a role. A feature of the neutral theory is that there are large fluctuations in quantities of interest. We already saw that the mean time to coalescence of two lineages is N generations, but the distribution of coalescence times is exponential (Eq. 3.2). The distribution of times is not sharply peaked around its mean value. The same effect happens with heterozygosity. The mean heterozygosity is easy to calculate, but the distribution of heterozygosities across loci has a complex shape that is not sharply peaked around the mean (see examples in Fuerst, Chakraborty, and Nei 1977 and Higgs 1995). This means that statistical tests for deviations from neutrality tend to be quite complex, and tend to require large data sets in order to be effective (Watterson 1978, Tajima 1989).

Just as for within-population variation, there are also adaptationist and neutralist explanations for the variation of sequences between species. Adaptationists would argue that most differences in protein sequences between species are the result of positively selected amino acid changes in one lineage or the other. The new sequence variant might function better in the context of one species and not the other. This context might include other molecules with which the protein interacts in the cell, as well as factors in the lifestyle and external environment of the species that might cause selection on the protein.

Proteins in different species would thus become selected for optimal function within those species, and selection would be the force driving divergence between species.

However, the picture that emerges when looking at sequence alignments, such as the example of BRCA1 in Section 3.3, is that the most obvious type of selection going on is not positive selection of new advantageous variants, but stabilizing selection acting to remove deleterious mutations and retain the current version of the gene. As we saw above, we tend to see a preponderance of synonymous to non-synonymous substitutions, a preponderance of conservative amino acid changes over changes between vastly different amino acids, and an avoidance of frame shifts. Another example of stabilizing selection is the frequent occurrence of compensatory pairs of substitutions in RNA genes, whereby the sequence changes but the secondary structure remains conserved (see Chapter 11). One thing to come to light from the genome projects is how many vertebrate proteins have recognizable homologs in organisms as apparently diverse as yeast and bacteria. In other words, it is the similarities of sequences between organisms that strikes one first, rather than the specializations. Thus, it seems reasonable to conclude that sequences diverge **despite** the action of stabilizing selection that is trying to keep them the same. Neutralists therefore argue that it is mutation that is the dominant force driving the divergence between sequences, and that most of the differences we see between sequences in different species are the result of fixation of nearly neutral mutations by chance in one or other lineage.

King and Jukes (1969) produced an influential paper arguing for the importance of neutrality, and the neutral theory has famously been developed and championed by Kimura (1983). There has since been a lot of criticism from adaptationists, who dislike the apparently large role of chance in neutral evolution. It is nevertheless important to understand exactly what the neutral hypothesis says. Neutralists do not deny that selection happens. They recognize the role of stabilizing selection in removing deleterious mutations, and recognize that occasionally advantageous mutations must occur.

However, they argue that (nearly) neutral mutations are much more frequent than advantageous ones, and hence that the majority of mutations that manage to become fixed in a population are neutral. Advantageous mutations are hard to spot because they spread rapidly through a population and it is difficult to catch one in the act. Once the mutation has gone to fixation, there will be no variability remaining in this part of the sequence. This is called a **selective sweep**. When the advantageous mutation goes to fixation, it may cause other mutations at closely linked points in the sequence to hitch-hike to fixation at the same time. The other mutations may be neutral or even slightly disadvantageous. If they are close to the original mutation, so that there is no recombination between them during the time in which the selective sweep is occuring, then they will be dragged along with the advantageous mutation.

In recent years, the argument for and against neutral evolution has died down, even if it has not been fully resolved. Neutral mutations are seen as part of a spectrum of possibilities, and the neutral theory has an established role as a null model. Even if the neutral hypothesis is true on a statistical basis, it is nevertheless of interest to look for particular examples in which gene sequences seem to have been under directional selection in particular organisms. One way of doing this is to look for genes in which there is a large ratio of non-synonymous to synonymous substitutions (more details in Section 11.2.3). Such genes are rare, although the *BRCA1* gene is actually an example where there seems to have been adaptive evolution in the lineage of humans and chimpanzees since their divergence from gorillas (Huntley *et al.* 2000). Kreitman and Comeron (1999) have also reviewed recent studies looking for evidence of selection acting in coding sequences.

We will conclude this section with a brief mention of the phenomenon known as **codon bias**. We might expect that synonymous substitutions would be a prime case where the neutral theory was likely to apply. However, there is considerable evidence that weak selection can also act between synonymous codons. As a result, synonymous codons are not all used with equal frequency in gene sequences, i.e., there is a bias in the usage of codons, such that some codons are apparently preferred over others. One reason for this is purely mutational. If mutation rates between the four bases are different, then the expected frequencies of the bases at the third codon position will be different from one another, even in the absence of selection. However, a key observation is that codon usage sometimes differs between genes in the same genome. It is often found that genes that are highly expressed are more biased in their codon usage (i.e., they use the preferred codons more frequently) than less strongly expressed genes. This is thought to be due to selection for increasing the efficiency of the translation process. It is known that certain tRNAs are present in the cell at higher concentration than others, and that the preferred codons seem to correspond to the anticodons of the tRNAs that are most frequent. Using the preferred codons therefore means there is less time spent during translation waiting for an appropriate tRNA to come along. Codon bias is quite a subtle problem because it arises as a result of weak selective effects that may not be the same in all situations. For interesting recent examples in this field, see Musto *et al.* (1999), Coghlan and Wolfe (2000), and Duret (2000).

SUMMARY

Evolution requires error-prone replication. As DNA replication is not perfect, errors are bound to occur, even if only rarely. Any change that occurs in a DNA sequence that can be passed on to the next generation is called a mutation. Mutations may be single-base substitutions, insertions, or deletions of one or more bases, or may involve large-scale insertions, deletions, and rearrangements of the sequence.

Mutations create new variant sequences in a population and increase genetic diversity. This diversity can be quantified by measuring the fraction of polymorphic loci, or by measuring the average level of heterozygosity of loci. Genetic diversity is reduced by natural selection, which tends to eliminate lower fitness alleles from a population. Random drift also reduces genetic diversity because some alleles can be lost by chance, even when there is no selection acting. The level of genetic diversity in a population is therefore determined by a balance between mutation, selection, and drift.

Gene sequences in a population are related to one another by descent from common ancestors. If the lines of descent of a gene from two individuals in a current population are traced back in time, they will coalesce, i.e., they will have a common ancestor at some point in the past. The typical time back to the coalescence point will be of the order N generations, where N is the population size. The coalescence process is easiest to interpret for genes that are inherited uniparentally, like mitochondrial DNA or the Y chromosome. Studies of human mitochondrial DNA indicate the presence of a common ancestor for all present-day humans who lived in Africa roughly 200,000 years ago.

When a new mutation occurs in a population, there will initially be only one copy. The number of copies of this mutation will increase and decrease due to the action of selection and drift. Most new mutations will be eliminated from the population within a few generations because there is a high probability of them being lost by chance when the copy number is small, even if they are selectively advantageous. Occasionally, a mutation will become fixed in the population, i.e., it will spread through the population and reach high frequency. The probability of fixation of a neutral mutation is $1/N$. An advantageous mutation, with fitness $1 + s$, has a significantly larger probability of becoming fixed if $s > 1/N$. Similarly, a deleterious mutation, with fitness $1 - s$, has a very small chance of fixation if $s > 1/N$. However, when $s < 1/N$, both advantageous and deleterious mutations may be classed as nearly neutral. This means that random drift is more important than selection in determining their fate, and their probability of fixation is very close to that of a neutral mutation.

Studies of human populations indicate the presence of many low-frequency mutant alleles. Information is available particularly for disease-linked genes. This suggests that most mutations are deleterious and will eventually be eliminated by selection. When we compare sequences between species, the differences we see are the result of fixation of mutations in one lineage or the other. The most frequent types of change are conservative ones, i.e., synonymous substitutions occur more rapidly than non-synonymous ones, and amino acid changes occur more rapidly when the amino acids have similar properties. This suggests that the major mode of selection acting is stabilizing, and that the changes that we do see are those that were nearly neutral and were thus not selected against.

REFERENCES

Blackmore, S. 1999. *The Meme Machine*. Oxford, UK: Oxford University Press.

Bulmer, M.G. 1987. Coevolution of codon usage and transfer RNA abundance. *Nature*, **325**: 728–30.

Cavalli-Sforza, L.L., Menozzi, P., and Piazza, A. 1996. *The History and Geography of Human Genes*. Princeton, NJ: Princeton University Press.

Coghlan, A. and Wolfe, K.H. 2000. Relationship of codon bias to mRNA concentration and protein length in *Saccharomyces cerevisiae*. *Yeast*, **16**: 1131–45.

Crow, J. and Kimura, M. 1970. *An Introduction to Population Genetics Theory*. New York: Harper & Row.

Dawkins, R. 1976. *The Selfish Gene*. Oxford, UK: Oxford University Press.

Duret, L. 2000. tRNA gene number and codon usage in the *C. elegans* genome are co-adapted for optimal translation of highly expressed genes. *Trends in Genetics*, **16**: 287–9.

Fuerst, P.A., Chakraborty, R., and Nei, M. 1977. Statistical studies on protein polymorphism in natural populations. I. Distribution of single locus heterozygosity. *Genetics*, **86**: 455–83.

Gowen, L.C., Avrutskaya, A.V., Latour, A.M., Koller, B.H., and Leadon, S.A. 1998. BRCA1 required for transcription-coupled repair of oxidative DNA damage. *Science*, **281**: 1009–12.

Higgs, P.G. 1995. Frequency distributions in population genetics parallel those in statistical physics. *Physical Review E*, **51**: 95–101.

Huntley, G.A., Easteal, S., Southey, M.C., Tesoriero, A., Giles, G.G., McCredie, M.R.E., Hopper, J.L., and Venter, D.J. 2000. Adaptive evolution of the tumour supressor *BRCA1* in humans and chimpanzees. *Nature Genetics*, **25**: 410–13.

Ikemura, T. 1985. Codon usage and tRNA content in unicellular and multicellular organisms. *Molecular Biology and Evolution*, **2**: 13–34.

Ingman, M., Kaessmann, H., Pääbo, S., and Gyllensten, U. 2000. Mitochondrial genome variation and the origin of modern humans. *Nature*, **408**: 708–13.

Karlin, S., Brocchieri, L., Bergman, A., Mrázek, J., and Gentles, A.J. 2002. Amino acid runs in eucaryotic proteomes and disease associations. *Proceedings of the National Academy of Sciences USA*, **99**: 333–8.

Kimura, M. 1983. *The Neutral Theory of Molecular Evolution*. Cambridge, UK: Cambridge University Press.

King, J.L. and Jukes, T.H. 1969. Non-Darwinian evolution. *Science*, **164**: 788–98.

Kreitman, M. and Comeron, J.M. 1999. Coding sequence evolution. *Current Opinion in Genetics and Development*, **9**: 637–41.

Madsen, O., Scally, M., Douady, C.J., Kao, D.J., DeBry, R.W., and Adkins, R. 2001. Parallel adaptive radiations in two major clades of placental mammals. *Nature*, **409**: 610–14.

Miki, Y. *et al.* 1994. A strong candidate for the breast and ovarian cancer susceptibility gene *BRCA1*. *Science*, **266**: 66–71.

Musto, H., Romero, H., Zavala, A., Jabbari, K., and Bernardi, G. 1999. Synonymous codon choices in the extremely GC-poor genome of *Plasmodium falciparum*. *Journal of Molecular Evolution*, **49**: 27–35.

Schlötterer, C. 2000. Evolutionary dynamics of microsatellite DNA. *Chromosoma*, **109**: 365–71.

Stenson, P.D., Ball, E.V., Mort, M., Phillips, A.D., Shiel, J.A., Thomas, N.S., Abeysinghe, S., Krawczak, M., and Cooper, D.N. 2003. Human Gene Mutation Database (HGMD): 2003 update. *Human Mutation*, **21**: 577–81. (http://www.hgmd.org)

Tajima, F. 1989. Statistical method for testing the neutral mutation hypothesis by DNA polymorphism. *Genetics*, **123**: 585–95.

Thompson, R., Pritchard, J.K., Shen, P., Oefner, P.J., and Feldman, M.W. 2000. Recent common ancestry of human Y chromosomes: Evidence from DNA sequence data. *Proceedings of the National Academy of Sciences USA*, **97**: 7360–5.

Underhill, P.A. *et al.* 2000. Y chromosome sequence variation and the history of human populations. *Nature Genetics*, **26**: 358–61.

Watterson, G.A. 1978. The homozygosity test of neutrality. *Genetics*, **88**: 405–17.

PROBLEMS

3.1 A rare allele is present at a locus at 5% frequency in a population. A sample of n gene copies is studied. What is the probability that the rare allele will be found at least once in the sample? Plot a graph of the way this probability depends on n. How large must the sample be before there is >95% probability of detecting the variant?

3.2 Simulations of the fixation process were carried out using the method described in Section 3.5 for a range of different selection coefficients, both advantageous and deleterious. 2000 runs were made for each case, and the number of observed fixations is given in Table 3.2. The table also gives the fixation probability p_{fix} calculated using the results of diffusion theory given in Box 3.1. Show that the observed numbers of fixations in the simulation are consistent with the expectations from the theory. Hint: you can either use a χ^2 test to check that observed and expected numbers are not significantly different (Section M.12), or you can use a z-score (Section M.10) to show that the observed number is not more than a couple of standard deviations away from the mean.

Table 3.2 Results of computer simulations of the fixation process.

s	Ns	p_{fix}	Observed number of fixations
Neutral mutations			
0.0	0.0	5.00×10^{-3}	11
Advantageous mutations			
0.002	0.4	7.25×10^{-3}	15
0.01	2.0	0.0202	40
0.02	4.0	0.0392	92
0.05	10.0	0.0952	190
0.2	40.0	0.3297	639
Deleterious mutations			
0.002	0.4	3.27×10^{-3}	7
0.01	2.0	3.77×10^{-4}	1
0.02	4.0	1.37×10^{-5}	0

3.3 For the case of a mutation spreading deterministically through a very large population (Box 3.2), the frequency x is given by Eq. (3.15). Prove that this solution satisfies the logistic equation by substituting the result back into Eq. (3.14).

3.4 This question investigates the way that biased usage of synonymous codons is related to tRNA concentration. The data and theory are taken from Ikemura (1985) and Bulmer (1987).

Iso-acceptor transfer RNAs are different types of tRNA from the same organism that bind to the same amino acid but recognize different codons. It is possible for one type of tRNA to recognize more than one codon due to wobble base pairing between the tRNA anticodon and the third codon position in the mRNA. Organisms expend a large amount of time and energy in protein synthesis; therefore, it is to be expected that there is a considerable selective pressure acting to improve the efficiency of translation. If a particular tRNA type is rare in the organism it will take longer for the corresponding amino acid to be incorporated into a protein under synthesis. One way in which the organism can improve translational efficiency is to use those codons in its genes more frequently for which the corresponding tRNAs are most common.

Table 3.3 shows the results of measurements made on *E. coli* to investigate the relationship between codon bias and tRNA concentration. For each amino acid, the types of iso-acceptor tRNA are listed with the codons they recognize. The concentration of these tRNAs is expressed in units such that the leucine tRNA number 1 has a concentration of one unit. For each amino acid, the tRNA with the highest concentration is listed first (this is referred to as the major tRNA), and the others (referred to as minor tRNAs) are listed subsequently.

The number of occurrences of different codons was counted in several *E. coli* gene sequences. The genes were divided into two groups – strongly expressed (many copies of each protein per cell) and weakly expressed (fewer copies of each protein per cell). The number of occurrences of each codon in the two groups is shown in Table 3.3. Where a tRNA type recognizes more than one codon, the number given is the total number of occurrences of all the codons recognized.

(a) For each minor tRNA, calculate the relative concentration C_2/C_1, where C_2 is the concentration of the minor tRNA and C_1 is the concentration of the corresponding major tRNA. Also calculate the degree of codon bias p_2/p_1 for both sets of genes, where p_2 is the frequency of usage of the minor tRNA codon (or codons) and p_1 is the frequency of usage of the major tRNA codon (or codons). Plot a graph of p_2/p_1 as a function of C_2/C_1. Use a scale of 0 to 1 on both axes, and plot results for both strongly and weakly expressed genes on the same graph.

Table 3.3 Number of occurrences of codons.

Amino acid	tRNA type (codons recognized)	Concentration of tRNA	Number of occurrences of codon in two groups of genes Strongly expressed	Weakly expressed
Leu	1. CUG	1.0	53	96
	2. CUU/CUC	0.3	3	32
	3. UUA/UUG	0.25	0	45
	4. CUA	0.05	0	9
Gly	1. GGU/GGC	1.1	79	87
	2. GGA/GGG	0.15	2	24
	3. GGG	0.1	2	14
Ile	1. AUU/AUC	1.0	58	84
	2. AUA	0.05	0	1
Gln	1. CAG	0.4	16	51
	2. CAA	0.3	0	26
Ser	1. UCU/UCA/UCC	0.25	20	39
	2. AGU/AGC	0.2	1	33
	3. UCG	0.05	0	18
Arg	1. CGU/CGC/CGA	0.9	46	87
	2. CGG	0.05	0	3
	3. AGA/AGG	0.05	0	3

(b) Do these results support the argument that codon bias is influenced by tRNA concentration? Would you say that relative codon frequency is correlated with relative tRNA concentration?

It has been suggested that the frequencies of codon usage are maintained by a balance between mutation and selection. Mutation makes random changes between synonymous codons and tends to eliminate codon bias. Selection penalizes gene sequences having disadvantageous codons and hence increases the codon bias. Consider two types of iso-acceptor tRNA with concentrations C_1 and C_2, and let p_1 and p_2 be the proportions of codons in the genome of the two types. There is a mutation rate u from type 1 to type 2 codons, and vice versa. It is proposed that there is a selective disadvantage s for use of type 2 codons, where

$$s = k\left(\frac{C_1}{C_2} - 1\right)$$

and k is an unknown constant. Note that s is zero if $C_2 = C_1$, and is very large if $C_2 \ll C_1$. The theory shows that at mutation–selection balance

$$p_2 = \frac{1}{2}\left(1 + \frac{2u}{s} - \sqrt{1 + \left(\frac{2u}{s}\right)^2}\right)$$

and $p_1 = 1 - p_2$.

(c) Use this theory to plot graphs of p_2/p_1 as a function of C_2/C_1 for three different values of u/k:
(i) $u/k = 1$; (ii) $u/k = 0.01$; (iii) $u/k = 100$.
(d) Comment qualitatively on the shape of the three theoretical curves. Why must all the curves pass through (0,0) and (1,1)? What is the relative importance of mutation and selection in each case? Does this theory satisfactorily explain the observed data?
(e) What other factors might influence codon usage, the speed of translation, and the translational error rate (i.e., incorporation of the wrong amino acid into a growing protein)? How might these be subject to evolution? You may wish to consider some of the following: the possibility of adjusting tRNA concentrations at the same time as codon frequencies; the rate of various reaction steps; the accuracy of molecular recognition processes; the possibility of random drift in codon frequencies; the G + C content in gene sequences; the way in which codons are assigned to amino acids in the genetic code.

Models of sequence evolution

CHAPTER PREVIEW

Models of the evolution of nucleic acids and proteins are used in phylogenetic methods as the basis of defining evolutionary distances and calculating the likelihood of a set of sequences evolving on a phylogenetic tree. In this chapter, we describe the details of these models. The scoring systems used in protein sequence alignment algorithms are also related to evolutionary models. Here we describe the derivation of PAM and BLOSUM substitution matrices in detail.

4.1 MODELS OF NUCLEIC ACID SEQUENCE EVOLUTION

4.1.1 Why do we need evolutionary models?

Whenever we have two similar-looking sequences, it is natural to ask just how different they are. We want to be able to ask questions like, "Is a human gene more similar to a chimpanzee gene than a gorilla gene?", or "What is the rate of substitutions that has occurred in the human and chimpanzee genes since the species diverged?". To answer these questions we need a quantitative model of evolution.

The simplest way to measure distance between sequences is to align them one below the other and just count what fraction of sites is different in the two sequences. In Fig. 4.1(a), two sequences, 1 and 2, have descended from a common ancestor 0. The fraction of sites that differ is $D = 3/10$ in this case. The difference will increase as a function of time because substitutions accumulate in one sequence or the other. If the time since divergence of the sequences is short, then there will be few differences between the sequences. It is very likely that each of the substitutions that occurs will happen at

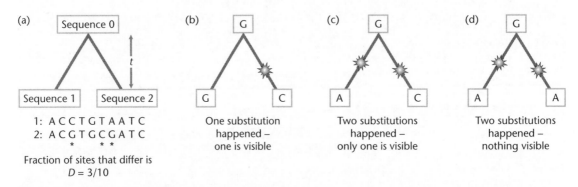

Fig. 4.1 The accumulation of substitutions in two sequences descending from a common ancestor.

a different site in the sequence. Thus, each of the substitutions is visible by comparing the two sequences. This is shown in Fig. 4.1(b), where a substitution from a G to a C has happened in one species. Substitutions are assumed to occur randomly with a constant rate. Hence, the average number of substitutions occurring in a given time will be proportional to the time, t, and D should increase linearly with t when t is small.

If the time since divergence is slightly larger, then it becomes possible that there has been more than one substitution occurring at the same site. In Fig. 4.1(c), there have been two substitutions. When we compare the two present-day species, we see that they are different, but we are only aware that one substitution has occurred. In Fig. 4.1(d), two substitutions have occurred from G to A, but there is no visible difference between the present-day species. The number of visible substitutions between two sequences is always less than or equal to the number of substitutions that have actually occurred. For this reason, at larger times, D increases more slowly than linearly with time. Now, if the time since divergence of the sequences is extremely large, there will have been many substitutions at every site. The sequences will therefore be completely randomized with respect to their common ancestor. If we align two random sequences containing equal frequencies of the four bases, on average $3/4$ of sites will differ between the two sequences, i.e., D tends to $3/4$ when the time since divergence is very large.

The problem with using D as a measure of distance between sequences is that it does not increase linearly with time: it would be better if we had a distance measure that was twice as big if the time since the divergence was twice as large. Another problem with D is that it is not additive. To understand this, imagine that we knew the sequence of the common ancestor 0 in Fig. 4.1(a). We could then measure the distances D_{01} and D_{02} from the ancestor to the two present-day sequences. It would be nice if the distance from 1 to 2 were simply the sum of the two distances to 0, i.e., if $D_{12} = D_{01} + D_{02}$. If this were true, these would be called additive distances. However, if D is simply the fraction of sites that differ, then the distances will not be additive.

Usually, D_{12} will be smaller than the sum of the other two distances.

It is therefore useful to define an evolutionary distance that is both additive and a linear function of time. The usual way to do this is to define a distance d to be the average number of substitutions that have occurred per site between the two sequences. If substitutions occur randomly at a constant rate, then the number of substitutions per site is directly proportional to time. Also the number of substitutions between 1 and 2 is by definition the sum of the number of substitutions from 0 to 1 and 0 to 2. Thus, d is a useful measure of evolutionary distance between sequences; it is a quantitative way of telling us how much evolution has happened between two sequences. However, d is not directly observable when we look at two sequences, unlike D. To calculate d, we need an evolutionary model, as we shall see below.

The next step from comparing pairs of sequences is to compare whole sets of related sequences. The natural way to represent the relationship between sequences is on a phylogenetic tree. There are several ways to draw trees, and we will leave the full discussion of this until the beginning of Chapter 8. However, it is worth noting here that we often want to draw trees where the branch lengths are proportional to evolutionary distance. In that way, we can see by looking at the tree where are the branches on which most evolutionary change has occurred. Quantitative measures of distance are therefore a key part of tree drawing.

Evolutionary models are also essential for phylogenetic methods based on likelihoods (which we will also discuss in Chapter 8). If we have a quantitative model, we can calculate the likelihood that our set of sequences would have evolved on any proposed phylogenetic tree. We can therefore distinguish between alternative possible trees according to their relative likelihoods.

The phylogenetic tree tells us the evolutionary relationships between the gene sequences used, but often we are making the assumption that the tree also tells us the evolutionary relationships between the species from which the genes came. When we do this, we are assuming that the differences between the species are larger than the differences between sequence variants within one species. For this reason it does not really matter which sequence we use as a representative of the species. The models of evolution that we use in this chapter are defined in terms of rates of substitution. These substitutions represent changes that have occurred in populations due to fixation of new sequence variants, rather than simply changes in individual sequences due to mutations. Thus the substitution rates are determined by mutation, selection, and random drift, as we discussed in Chapter 3, and not simply by mutation rates. When discussing phylogenetic trees we often do not need to think much about population genetics. Nevertheless, it is important to realize that the models used in phylogenetics are actually describing the outcome of fixation processes going on at the level of population genetics.

In addition to their use in molecular phylogenetics, another motivation for developing evolutionary models of protein sequences is that they can be used to define scoring systems for sequence alignment. We want to assign high scores to pairs of amino acids that we expect will substitute frequently for one another during evolution. In this way, the alignment with the optimal score should reflect the alignment that is evolutionarily most likely. The final section of this chapter is therefore devoted to scoring matrices for amino acids.

The material in this chapter merits its position relatively near the beginning of the book because it follows directly on from the introduction to molecular evolution in Chapter 3, and because it is a foundation for the basic bioinformatics methods of sequence alignment (Chapter 6) and phylogenetic methods (Chapter 8). However, this is one of the most mathematical chapters in the book. Some readers may wish to skip the more mathematical sections below and return to them again after seeing the applications of the models introduced in subsequent chapters.

4.1.2 The Jukes–Cantor model

The simplest possible model of sequence evolution is that of Jukes and Cantor (1969) – henceforward, the JC model. The model describes one single site in an alignment of DNA sequences. The base at this site can be either an A, C, G, or T. The model assumes that all four bases have equal frequency and that there is a rate of substitution α from any of the four DNA bases to any other base.

As described above, the fraction of sites that differ between two sequences, D, is directly observable by comparing the two sequences, but does not count all the changes that have happened, because there may have been more than one substitution per site. Therefore, we wish to calculate the evolutionary distance, d, defined as the estimated number of substitutions that have occurred per site. According to the model, the rate of a substitution from any one base to each of the other bases is α. The net rate of change of a base to any other base is therefore 3α, because there are three other bases. To calculate the mean number of changes we simply multiply the rate of change by the length of time. In Fig. 4.1(a) the length of time on the branch leading to each of the species is t, and hence the total amount of time available for changes to occur is $2t$. Thus, the mean number of substitutions occurring per site is

$$d = 2t \times 3\alpha = 6\alpha t \qquad (4.1)$$

However, in practice, we usually do not know t or α with any certainty. Therefore, it is not possible to calculate d directly from Eq. (4.1). Nevertheless, we can get round this problem in the following way. In Box 4.1 we calculate the way in which D depends on time:

$$D = \frac{3}{4} - \frac{3}{4}e^{-8\alpha t} \qquad (4.2)$$

If t is very small we can expand the exponential to first order in t, and hence $D \approx 6\alpha t$. This confirms our expectation that D increases linearly with t initially. When t is large, the exponential term dies away, and D tends to $^3/_4$, as we expected from the argument about random sequences given above. The dependences of D and d on time are compared to one another in Fig. 4.3(b).

We notice in Eqs. (4.1) and (4.2) that both d and D are functions of αt. It is therefore possible to eliminate αt from these two equations. By rearranging (4.2) we find

BOX 4.1
Solution of the Jukes–Cantor model

Consider a single site in a sequence. We define $P_{ij}(t)$ to be the probability that the site is in state j at time t given that the site in the ancestral sequence was in state i at time 0. Here i and j label the base (A, C, G, or T) that is present at that site. Figure 4.2 shows a site that was initially in state A, that is in some state k at time t, and that is in state A a short time interval δt after that. The probability of this situation is (by definition) $P_{AA}(t + \delta t)$. This can be written as a sum over the four possible bases that could exist at time t. If base k is a C, then there is a probability $P_{AC}(t)$ that an A changes to a C in time t multiplied by the probability $\alpha\delta t$ that a substitution from a C to an A happens in the interval δt. The terms for the cases where k is G or T are similar to this. The term for the case where k is an A is the probability $P_{AA}(t)$ that the base is an A after time t, multiplied by the probability that there is no substitution in the interval δt. The total rate of change is 3α; therefore the probability that there is no substitution is $1 - 3\alpha\delta t$. Adding these four terms, we obtain:

$$P_{AA}(t+\delta t) = \alpha\delta t\left(P_{AC}(t) + P_{AG}(t) + P_{AT}(t)\right) + (1 - 3\alpha\delta t)P_{AA}(t)$$

(4.5)

For small δt we know that $P_{AA}(t + \delta t) = P_{AA}(t) + \delta t\, \dfrac{dP_{AA}}{dt}$

(see Section M.6). After substituting this into the above equation, the term $P_{AA}(t)$ can be subtracted from both sides, and we can then cancel out the δt, which results in the following differential equation for P_{AA}:

$$\frac{dP_{AA}}{dt} = \alpha(P_{AC} + P_{AG} + P_{AT}) - 3\alpha P_{AA}$$

(4.6)

We also know that $P_{AC} + P_{AG} + P_{AT} = 1 - P_{AA}$: this just says that the sum of the probabilities that base k is a C, G, or T is equal to the probability that base k is **not** an A. Putting this into Eq. (4.6), we obtain:

$$\frac{dP_{AA}}{dt} = -4\alpha P_{AA} + \alpha$$

(4.7)

This is an equation that involves only the single unknown function P_{AA}. It is of a standard form that allows us to guess that the solution must be of the form $P_{AA}(t) = Ae^{-4\alpha t} + B$, where A and B are two constants to be determined. By substituting this trial solution back into Eq. (4.7), following the method described in Section M.8, we find that the equation is satisfied if $B = \frac{1}{4}$. By using the initial condition that $P_{AA}(0) = 1$, we find that $A = \frac{3}{4}$. Therefore the solution is:

$$P_{AA}(t) = \frac{3}{4}e^{-4\alpha t} + \frac{1}{4}$$

(4.8)

Due to the symmetry of the model, $P_{CC}(t)$, $P_{GG}(t)$ and $P_{TT}(t)$ are all equal to $P_{AA}(t)$. Also, $P_{AC}(t)$, $P_{AG}(t)$, $P_{AT}(t)$ are all equal to one another, hence:

$$P_{AC}(t) = \frac{1}{3}(1 - P_{AA}(t)) = \frac{1}{4} - \frac{1}{4}e^{-4\alpha t}$$

(4.9)

The functions $P_{AA}(t)$ and $P_{AC}(t)$ are plotted in Fig. 4.3(a). Both functions tend to $\frac{1}{4}$ for long times. This is because after a long time the site is equally likely to be any of the four bases, irrespective of which base it was at time 0.

The probability that the base at time t is different from the base at time 0 is equal to $P_{AC}(t) + P_{AG}(t) + P_{AT}(t) = 3P_{AC}(t)$. When we are comparing two species with a time t since the common ancestor, the total time separating the species is $2t$. Therefore, the probability that these two species have a different base is $D = 3P_{AC}(2t)$, which gives us an $8\alpha t$ in the exponential of Eq. (4.9). Thus we obtain the Eq. (4.2) for D in the main text.

$$\ln\left(1 - \frac{4}{3}D\right) = -8\alpha t$$

(4.3)

and by comparing with (4.1), we obtain:

$$d = -\frac{3}{4}\ln\left(1 - \frac{4}{3}D\right)$$

(4.4)

This is the most important equation in this section. We can use it to calculate the evolutionary distance that we want, d, in terms of a quantity that we can directly measure from the sequence, D, even though we do not know the values of t or α. The value of d is known as the Jukes–Cantor distance between the sequences. Figure 4.4 shows the way d depends

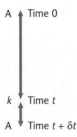

A   Time 0

k   Time t

A   Time $t + \delta t$

Fig. 4.2 Illustration of a single line of descent – used for the derivation in Box 4.1.

on D. If D is small, $d \approx D$, whereas if $D > 1/2$, d is substantially greater than D, because there is a high likelihood of multiple substitutions per site. The distance becomes infinite as D approaches $3/4$. Examples of calculations of the JC distance for real sequences are given in Fig. 8.5 when we discuss the use of evolutionary distances for phylogenetic methods.

4.1.3 More complex models of DNA sequence evolution

The JC model is the simplest of a range of models of sequence evolution that are often used in molecular phylogenetics. Substitution rate models are required to get evolutionary distances for input into distance

matrix methods (see Section 8.3), and they also form the basis for maximum-likelihood methods (Section 8.6). A substitution rate model is defined by a rate matrix, whose off-diagonal elements r_{ij} define the rate of substitution from state i (row) to state j (column). The matrix for the Jukes–Cantor model is:

$$
\begin{array}{ccccc}
 & A & G & C & T \\
A & -3\alpha & \alpha & \alpha & \alpha \\
G & \alpha & -3\alpha & \alpha & \alpha \\
C & \alpha & \alpha & -3\alpha & \alpha \\
T & \alpha & \alpha & \alpha & -3\alpha
\end{array}
\tag{4.10}
$$

The diagonal elements of the matrix, r_{ii}, are negative and are equal to the total rate of substitution away from state i to anything else. This is given by the sum of the elements on the same row. For any given rate matrix, r_{ij}, the substitution probabilities must satisfy the equation:

$$
\frac{dP_{ij}}{dt} = \sum_k P_{ik} r_{kj}
\tag{4.11}
$$

This can be derived using a similar argument to that in Eqs. (4.5) and (4.6) in Box 4.1. Solutions for the functions $P_{ij}(t)$ can be calculated for all the rate models proposed below.

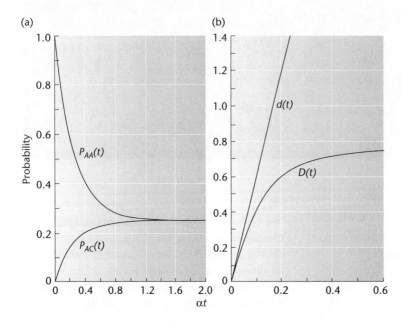

(a)

(b)

Fig. 4.3 The quantities $P_{AA}(t)$, $P_{AC}(t)$, $D(t)$, and $d(t)$ arising from the solution of the Jukes–Cantor model shown as a function of αt.

The next most simple model after JC is the Kimura two-parameter (K2P) model (Kimura 1983). As it is often observed in real data that transitions occur more frequently than transversions, the K2P model has a parameter α for the rate of transitions and a parameter β for the rate of transversions. In general these parameters are not equal, and usually $\alpha > \beta$. The rate matrix is:

$$
\begin{array}{c@{\quad}cccc}
 & A & G & C & T \\
A & -\alpha-2\beta & \alpha & \beta & \beta \\
G & \alpha & -\alpha-2\beta & \beta & \beta \\
C & \beta & \beta & -\alpha-2\beta & \alpha \\
T & \beta & \beta & \alpha & -\alpha-2\beta
\end{array}
\tag{4.12}
$$

An evolutionary distance formula can be calculated for this model. For a given pair of sequences, we can count the fraction of sites that differ by a transition, S, and the fraction that differ by a transversion, V. The total fraction of sites that differ is $D = S + V$. It can be shown that the estimated number of substitutions per site in this model is:

$$
d = -\frac{1}{2}\ln(1-2S-V) - \frac{1}{4}\ln(1-2V)
\tag{4.13}
$$

See Problem 4.1 at the end of the chapter for the derivation of this result.

For the two sequences in Fig. 4.1(a), we have $S = 0.2$, $V = 0.1$, and $D = 0.3$. The distance given by Eq. (4.13) is 0.402. The Jukes–Cantor distance for the same sequences, from Eq. (4.4) is 0.383. Thus the estimate of the number of substitutions per site depends on the model of evolution that we use. One of the main uses of evolutionary distances is for construction of phylogenetic trees, and tree-building methods can be sensitive to small changes in input distances. We are therefore more likely to obtain reliable results if we use a model of evolution that is appropriate for the sequences being studied.

An important property of real sequences that is not accounted for by either of the above models is that the frequencies of the four bases are often not equal to one another. The model of Hasegawa, Kishino, and Yano (1985), referred to as the HKY model, introduces the four base frequencies π_A, π_C, π_G and π_T into the rate matrix:

$$
\begin{array}{c@{\quad}cccc}
 & A & G & C & T \\
A & -\alpha\pi_G-\beta(\pi_C+\pi_T) & \alpha\pi_G & \beta\pi_C & \beta\pi_T \\
G & \alpha\pi_A & -\alpha\pi_A-\beta(\pi_C+\pi_T) & \beta\pi_C & \beta\pi_T \\
C & \beta\pi_A & \beta\pi_G & -\alpha\pi_T-\beta(\pi_A+\pi_G) & \alpha\pi_T \\
T & \beta\pi_A & \beta\pi_G & \alpha\pi_C & -\alpha\pi_C-\beta(\pi_A+\pi_G)
\end{array}
\tag{4.14}
$$

The substitution probability functions for the HKY model have the property that $P_{ij}(t)$ tends to π_j for very large times. In other words, the probability of being in state j after a long time is equal to the equilibrium frequency of base j, irrespective of the starting state i. This model has parameters α and β to control rates of transitions and transversions. If the four bases all have frequency $1/4$, the HKY model becomes equivalent to the K2P model.

The observed number of substitutions per unit time from state i to state j is given by the probability that a site is in state i multiplied by the rate of substitution from i to j. The models in this section all have the property of **time reversibility**:

$$
\pi_i r_{ij} = \pi_j r_{ji}
\tag{4.15}
$$

This means that for any pair of states i and j, the number of substitutions per site per unit time in the forward direction is equal to the number of substitutions in the reverse direction. It follows that for any time t

$$
\pi_i P_{ij}(t) = \pi_j P_{ji}(t)
\tag{4.16}
$$

and also that the base frequencies remain constant in time on average. This property is important for phylogenetic methods, and time reversibility is almost always assumed. The most general rate matrix that satisfies time reversibility is the general reversible (GR) model, which has the following rate matrix:

$$
\begin{array}{c@{\quad}cccc}
 & A & G & C & T \\
A & * & \alpha_{AG}\pi_G & \alpha_{AC}\pi_C & \alpha_{AT}\pi_T \\
G & \alpha_{AG}\pi_A & * & \alpha_{GC}\pi_C & \alpha_{GT}\pi_T \\
C & \alpha_{AC}\pi_A & \alpha_{GC}\pi_G & * & \alpha_{CT}\pi_T \\
T & \alpha_{AT}\pi_A & \alpha_{GT}\pi_G & \alpha_{CT}\pi_C & *
\end{array}
\tag{4.17}
$$

The diagonal elements (denoted * to save space) are equal to minus the sum of the other elements on the same row, as for all the other models. This model has four frequency parameters, as for HKY, and six parameters that control substitution rates, α_{AG}, α_{AC} etc. Whichever pair of bases is chosen, the elements of this matrix satisfy Eq. (4.15). Since this model has more free parameters than the others, it is in principle able to describe the evolution of real sequences more accurately than the simpler models. These models are defined in terms of parameters, and the values of these parameters can be determined by fitting the model to real sequence data, usually using the maximum-likelihood criterion. This will be described in Chapter 8.

4.1.4 Variability of rates between sites

So far we assumed that all the sites in the sequence were evolving at the same rate, i.e., that the probability of a substitution occurring is the same at each site. In fact, this is hardly ever true for real sequences. Sites that are structurally or functionally important tend to evolve more slowly than less important sites. In the case of very strong selection it may be that a site is invariant, i.e., unable to change at all. Suppose there is a fraction f of sites that are invariant, but the remaining sites evolve at a constant rate according to the JC model. It can be shown that the evolutionary distance becomes

$$d = -\frac{3}{4}(1 - f)\ln\left(1 - \frac{4D}{3(1 - f)}\right) \qquad (4.18)$$

which reduces to the usual form for the JC distance (4.4) when f tends to zero. A relatively small fraction of invariant sites makes a fairly large difference to the distance. Figure 4.4 shows this relationship when $f = 0.25$. Note that in Eq. (4.18), D is still the observed fraction of sites that differ across the whole molecule, and d is still defined as the average number of substitutions that have occurred per site across the whole molecule. Since a fraction f of sites have no substitutions, the average number of substitutions at the sites which **are** variable is $d/(1 - f)$.

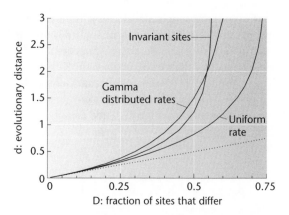

Fig. 4.4 Evolutionary distances d as a function of observed fraction of differences D according to the Jukes–Cantor model with: (i) uniform rate of evolution at all sites; (ii) a fraction $f = 0.25$ of invariant sites; (iii) with rate variation across sites described by a gamma distribution with $a = 1$. These are shown in comparison to the uncorrected distance $d = D$ (dashed line).

In real sequences we find a range of rates at different sites, and it is not sufficient to simply class sites as invariant or variable. Therefore it is desirable to have a model to describe this distribution. The most frequently used model is the gamma distribution, defined in Section M.13.

$$f(r,a) = const \times r^{a-1}e^{-ar} \qquad (4.19)$$

In this case, r is the rate of substitution at a given site relative to the average rate of substitution in the whole molecule, $f(r,a)$ is the probability that a site has a relative rate r, and a is a parameter that controls the shape of the distribution. The shapes of the gamma distributions for different values of a are shown in Fig. M.5 in the Appendix. When $a > 1$, the distribution is peaked with a maximum fairly close to 1. The larger a, the smaller the degree of variabililty of rates there is between sites. If a is very large, it becomes a single spike at $r = 1$, i.e., all sites evolve at a constant rate. When $a = 1$, the distribution is a simple exponential. When $a < 1$, the distribution has a high weight close to $r = 0$, meaning that there are a significant number of sites that are almost invariant. There is also a long tail at high values of r, meaning that there are a significant number of sites that evolve much more rapidly than average. The evolutionary distance for the JC model when a gamma distribution of rates is introduced is

$$d = \frac{3}{4}a((1 - 4D/3)^{-1/a} - 1) \qquad (4.20)$$

This function is also shown in Fig. 4.4 for the case where $a = 1$. It can be seen that both the invariant sites model and the gamma distribution model give higher estimates of the mean number of substitutions per site for any given observed difference D than the standard model of constant evolutionary rates. Thus, if we ignore rate variability we get a systematic underestimation of the evolutionary distance. The effect is small if the distance is small, but it becomes very large for higher distances. This is of practical importance, if we are interested in estimating dates of splits between groups on a phylogenetic tree. Yang (1996) gives an instructive example of the way the estimated divergence times between humans and various apes depend on rate variability.

The gamma function is used because by introducing only one extra parameter to the model, a, we are able to describe a variety of situations from very little to very large amounts of rate variability between sites. There is no reason why the distribution of rates in a real sequence should obey the gamma distribution exactly, but the function is sufficiently flexible that it can fit real data quite well. The optimal value of a to fit a real data set can be found by maximum likelihood. For more details, see Yang (1996) and references therein.

4.2 THE PAM MODEL OF PROTEIN SEQUENCE EVOLUTION

4.2.1 Counting amino acid substitutions

Whereas for a model of DNA sequence evolution we need a 4×4 matrix, for a model of protein sequences we need a 20×20 matrix to account for substitutions between all the amino acids. The original substitution rate matrices for amino acids were termed PAM matrices, where PAM stands for "point accepted mutation". An accepted mutation is one that spreads through the population and goes to fixation. Perhaps a better name for a PAM would be a single amino acid substitution; however, the term PAM is well established, and we stick to it here. The following derivation of the PAM model of amino acid substitutions follows that given by Dayhoff, Schwartz, and Orcutt (1978). At that time, relatively few protein sequences were available. Several groups have since used much larger data sets to derive improved PAM models, but the methods used have remained similar. Dayhoff, Schwartz, and Orcutt made alignments of 71 families of closely related proteins. Sequences in the same family were not more than 15% different from one another. They constructed an evolutionary tree for each family using the parsimony method. Parsimony is described in Section 8.7, but the principle is straight-forward to state here – the selected tree is the one that minimizes the total number of amino acid substitutions required.

As an example, consider the short protein alignment shown in Fig. 4.5. This is based on a real protein alignment, but in order to keep the example simple, sites have been chosen so that only six of the amino acids occur. The evolutionary tree is also shown. The sequences A–G on the tips of the tree are known, but the sequences 1–5 are chosen by a computer program in order to minimize the total number of substitutions over the whole tree. The required substitutions are labeled in Fig. 4.5 on the branches where they occur.

Having determined all the substitutions, we obtain a matrix whose elements, A_{ij}, are the number of times that amino acid i is substituted by amino acid j. The parsimony method does not tell us in which direction a change occurred (because it produces unrooted trees – see Section 8.1). Therefore when a change between i and j is observed, we add one to both A_{ij} and A_{ji}. In the example, there is a change between a K and a T on the branch between nodes 2 and 3. If the root of the tree is as shown in Fig. 4.5, then this change is from K to T; however, if the root had been drawn such that sequence G were the outgroup then the change would be from T to K. We thus count one change in both directions. In the example there are a total of seven substitutions and the following A_{ij} matrix results:

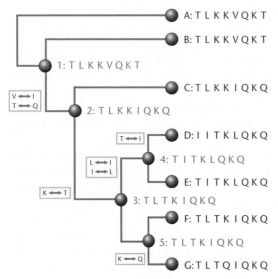

Fig. 4.5 A short protein sequence alignment and the phylogenetic tree obtained for these sequences using the parsimony method. Internal nodes 1–5 are labeled with the deduced amino acid sequence at each point. Amino acid substitutions are labeled on the branch where they occur. Trees like this are the first stage of derivation of the PAM model.

$$
\begin{array}{c|cccccc}
 & I & K & L & Q & T & V \\
\hline
I & - & - & 2 & - & 1 & 1 \\
K & - & - & - & 1 & 1 & - \\
L & 2 & - & - & - & - & - \\
Q & - & 1 & - & - & 1 & - \\
T & 1 & 1 & - & 1 & - & - \\
V & 1 & - & - & - & - & - \\
\end{array}
\qquad (4.21)
$$

The A_{ij} matrix obtained by Dayhoff *et al.* (1978) involved all 20 amino acids and had 1572 substitutions. In some cases it is equally parsimonious for more than one different amino acid to occur on an internal node. Where the internal node sequences were ambiguous, fractional substitutions were

counted for each of the possible changes. Jones, Taylor, and Thornton (1992) carried out a similar method using all the sequences then available in Swiss-Prot. They automatically clustered sequences into groups of >85% similarity and used a distance matrix method to obtain phylogenetic trees (see Section 8.3). Their A_{ij} matrix contains 59,190 substitutions and is reproduced in Fig. 4.6 (above the diagonal). All possible interchanges between amino acids are observed at least once in this large data set.

Another reason why each substitution is counted in both directions when calculating the A_{ij} matrix is that we are aiming to create a time-reversible model of evolution, as discussed for DNA models in Section 4.1.2. The frequencies of the amino acids π_i are assumed to be constant through time on average. Therefore we would expect to see equal numbers of substitutions in both directions between any two amino acids, i.e., we would expect $A_{ij} = A_{ji}$ on average. The number of observed substitutions from i to j is proportional to the probability π_i of being in state i multiplied by the rate r_{ij} of change from i to j given that we are in state i. If we begin with a symmetric matrix where A_{ij} is set equal to A_{ji} then the evolutionary model derived from it will satisfy Eq. (4.15) and will be time reversible.

4.2.2 Defining an evolutionary model

As with the DNA models, $P_{ij}(t)$ is the probability of being in state j at time t given that we were in state i at time 0. We will define the PAM1 matrix as $M_{ij} = P_{ij}(\delta t)$, where δt is a small time value known as 1 PAM unit. For small times, we can assume that the substitution probabilities are proportional to the substitution rates (i.e., the probabilites vary linearly with time for small times). Our estimate of the substitution rate is proportional to the number of observed substitutions, A_{ij}, divided by the total number of times N_i that amino acid i is seen in the data. Hence,

$$
M_{ij} = \lambda \frac{A_{ij}}{N_i} \quad \text{(for } i \neq j) \qquad (4.22)
$$

where λ is a constant of proportionality to be determined. We have adopted the convention that M_{ij} and

	A	R	N	D	C	Q	E	G	H	I	L	K	M	F	P	S	T	W	Y	V
A	2	247	216	386	106	208	600	1183	46	173	257	200	100	51	901	2413	2440	11	41	1766
R	-1	5	116	48	125	750	119	614	446	76	205	2348	61	16	217	413	230	109	46	69
N	0	0	3	1433	32	159	180	291	466	130	63	758	39	15	31	1738	693	2	114	55
D	0	-1	2	5	13	130	2914	577	144	37	34	102	27	8	39	244	151	5	89	127
C	-1	-1	-1	-3	11	9	8	98	40	19	36	7	23	66	15	353	66	38	164	99
Q	-1	2	0	1	-3	5	1027	84	635	20	314	858	52	9	395	182	149	12	40	58
E	-1	0	1	4	-4	2	5	610	41	43	65	754	30	13	71	156	142	12	15	226
G	1	0	0	1	-1	-1	0	5	41	25	56	142	27	18	93	1131	164	69	15	276
H	-2	2	1	0	0	2	0	-2	6	26	134	85	21	50	157	138	76	5	514	22
I	0	-3	-2	-3	-2	-3	-3	-3	-3	4	1324	75	704	196	31	172	930	12	61	3938
L	-1	-3	-3	-4	-3	-2	-4	-4	-2	2	5	94	974	1093	578	436	172	82	84	1261
K	-1	4	1	0	-3	2	1	-1	1	-3	-3	5	103	7	77	228	398	9	20	58
M	-1	-2	-2	-3	-2	-2	-3	-3	-2	3	3	-2	6	49	23	54	343	8	17	559
F	-3	-4	-3	-5	0	-4	-5	-5	0	0	2	-5	0	8	36	309	39	37	850	189
P	1	-1	-1	-2	-2	0	-2	-1	0	-2	0	-2	-2	-3	6	1138	412	6	22	84
S	1	-1	1	0	1	-1	-1	1	-1	-1	-2	-1	-1	-2	1	2	2258	36	164	219
T	2	-1	1	-1	-1	-1	-1	-1	-1	1	-1	-1	0	-2	1	1	2	8	45	526
W	-4	0	-5	-5	1	-3	-5	-2	-3	-2	-3	-3	-1	-4	-3	-4	15	41	27	
Y	-3	-2	-1	-2	2	-2	-4	-4	4	-2	-1	-3	-2	5	-3	-1	-3	0	9	42
V	1	-3	-2	-2	-2	-3	-2	-2	-3	4	2	-3	2	0	-1	-1	0	-3	-3	4

Fig. 4.6 Above the diagonal – numbers of observed substitutions, A_{ij}, between each pair of amino acids in the data of Jones, Taylor, and Thornton (1992). On and below the diagonal – log-odds scoring matrix corresponding to PAM250 calculated from these data. This is calculated as $S_{ij} = 10 \log_{10} R_{ij}$ and rounded to the nearest integer. Cells shaded gray have positive scores, meaning that these amino acids are more likely to interchange than would be expected by chance. Values written in white on a black box correspond to amino acid substitutions that are possible via a single nucleotide substitution at one position in the codon.

r_{ij} represent substitutions from i to j in order to be consistent with the definition of DNA models used in the previous section, and also with the way the data of Jones *et al.* (1992) are presented in Fig. 4.6, and with recent authors such as Adachi and Hasegawa (1996) and Müller and Vingron (2000). Unfortunately the opposite convention was used by Dayhoff *et al.* (1978); therefore some of the formulae used here have their indices reversed with respect to the original paper.

The frequency of amino acid i in the data is $\pi_i = N_i/N_{tot}$, where N_{tot} is the total number of amino acids in the data set. To determine λ we adopt the convention that 1 PAM unit is the time such that an average of 1% of amino acids have changed. The fraction of sites that have changed is

$$\sum_i \pi_i \sum_{j\neq i} M_{ij} = \sum_i \sum_{j\neq i} \pi_i \lambda \frac{A_{ij}}{N_{tot}\pi_i} = \frac{\lambda A_{tot}}{N_{tot}} = 0.01$$

$$(4.23)$$

where A_{tot} is the total of all the elements in the A_{ij} matrix. Hence

$$\lambda = 0.01 N_{tot}/A_{tot} \tag{4.24}$$

To complete the specification of the PAM1 matrix we need to define the probability M_{ii} that amino acid i remains unchanged in a time of 1 PAM unit. This is simply

$$M_{ii} = 1 - \sum_{j\neq i} M_{ij} \tag{4.25}$$

The PAM1 matrix obtained by Jones *et al.* (1992) is shown in Fig. 4.7. Values have been multiplied by 100,000 for convenience. This means that the sum of the elements on each row is 100,000. Diagonal elements are all slightly less than one, and off-diagonal elements are all very small. The two highest non-diagonal elements in each row have been

	A	R	N	D	C	Q	E	G	H	I	L	K	M	F	P	S	T	W	Y	V
A	98759	27	24	42	12	23	66	129	5	19	28	22	11	6	99	**264**	**267**	1	4	193
R	41	98962	19	8	21	**125**	20	102	74	13	34	**390**	10	3	36	69	38	18	8	11
N	43	23	98707	**284**	6	31	36	58	92	26	12	150	8	3	6	**344**	137	0	23	11
D	63	8	**235**	98932	2	21	**478**	95	24	6	6	17	4	1	6	40	25	1	15	21
C	44	52	13	5	99450	4	3	41	17	8	15	3	10	28	6	**147**	28	16	**68**	41
Q	43	154	33	27	2	98955	**211**	17	130	4	64	**176**	11	2	81	37	31	2	8	12
E	82	16	25	**398**	1	**140**	99042	83	6	6	9	102	4	2	10	21	19	2	2	31
G	**135**	70	33	66	11	10	70	99369	5	3	6	16	3	2	11	**129**	19	8	2	32
H	17	164	171	53	15	**223**	15	15	98867	10	49	31	8	18	58	51	28	2	**189**	8
I	28	12	21	6	3	3	7	4	4	98722	**212**	12	113	31	5	28	149	2	10	**630**
L	24	19	6	3	3	29	6	5	12	**122**	99328	9	90	101	53	40	16	8	8	**117**
K	28	**334**	108	14	1	**122**	107	20	12	11	13	99101	15	1	11	32	57	1	3	8
M	36	22	14	10	8	19	11	10	8	**253**	**350**	37	98845	18	8	19	123	3	6	201
F	11	3	3	2	14	2	3	4	11	41	**230**	1	10	99357	8	65	8	8	**179**	40
P	**150**	36	5	7	3	66	12	16	26	5	97	13	4	6	99278	**190**	69	1	4	14
S	**297**	51	214	30	44	22	19	139	17	21	54	28	7	38	140	98548	**278**	4	20	27
T	**351**	33	100	22	9	21	20	24	11	134	25	57	49	6	59	**325**	98670	1	6	76
W	7	**65**	1	3	23	7	7	41	3	7	**49**	5	5	22	4	21	5	99684	24	16
Y	11	12	30	23	43	10	4	4	**134**	16	22	5	4	**222**	6	43	12	11	99377	11
V	**226**	9	7	16	13	7	29	35	3	**504**	161	7	71	24	11	28	67	3	5	98772

	A	R	N	D	C	Q	E	G	H	I	L	K	M	F	P	S	T	W	Y	V
π_i	0.077	0.051	0.043	0.052	0.020	0.041	0.062	0.074	0.023	0.053	0.091	0.059	0.024	0.040	0.051	0.069	0.059	0.014	0.032	0.066
m_i	1.241	1.038	1.293	1.068	0.550	1.045	0.958	0.631	1.133	1.273	0.672	0.899	1.155	0.643	0.722	1.452	1.330	0.316	0.623	1.228

Fig. 4.7 PAM1 matrix calculated by Jones, Taylor, and Thornton (1992). Values are multiplied by 10^5 for convenience. M_{ij} is the probability that the amino acid in row i changes to the amino acid in column j in a small time corresponding to 1 PAM unit. The two highest non-diagonal elements in each row are highlighted in black. These are the two most rapid substitution rates for each amino acid. Frequencies π_i and relative mutabilities m_i of each amino acid are shown at the bottom of the figure.

highlighted in black. These are the two highest substitution rates for each amino acid, e.g., A is more likely to change to S and T than to the other amino acids. We will consider more fully why some rates are much larger than others in the section on log odds matrices. For now we note that, unlike the A_{ij} matrix, the M_{ij} matrix is not symmetric. This is because the frequencies of the amino acids are not equal. The values of the amino acid frequencies in the data of Jones *et al.* are given in Fig. 4.7. As this model is time reversible we have $\pi_i M_{ij} = \pi_j M_{ji}$. So for amino acids A and V, for example, we have $0.077 \times 0.00193 = 0.066 \times 0.00226 = 0.000149$.

Since PAM1 corresponds to an average probability of 0.01 of amino acids changing, we would

expect all the diagonal elements to be 0.99 (i.e., 99,000 on the figure) if each amino acid changed at the same rate. However, some amino acids are more likely to undergo substitutions than others. Amino acids that change more rapidly than average will have a probability lower than 0.99 of remaining unchanged after a time of 1 PAM, whereas those that change more slowly than average will have a probability higher than 0.99 of remaining unchanged. Using Eq. (4.22), the total probability of substitution from amino acid i to any other amino acid in a time of 1 PAM unit is $1 - M_{ii}$. If we divide this by the mean substitution probability 0.01, we obtain the relative mutability m_i of amino acid i:

$$m_i = (1 - M_{ii})/0.01 \qquad (4.26)$$

The relative mutabilities are given at the foot of Fig. 4.7. A relative mutability of 1 corresponds to an amino acid that changes at the average rate. The least mutable amino acid is tryptophan (W) with mutability $m_W = 0.316$, which is less than a third of the average rate. The most mutable amino acid is serine (S) with $m_S = 1.452$. The mutability values reflect important properties of the amino acids (see Section 4.3.2 below). Note that Jones et $al.$ (1992) and Dayhoff et $al.$ (1978) used a different scale for relative mutabilities where the mutability of alanine is set to 100. The scale used here seems more logical because there is nothing special about alanine.

4.2.3 Extrapolating the model to higher distances

The PAM1 matrix in Fig. 4.7 is the equivalent of one of the evolutionary models for DNA discussed in Section 4.1, except that the M_{ij} matrix represents a transition probability in a small discrete unit of time, whereas the rate matrices r_{ij} represent rates of change in a continuous time model. For two amino acids that are not equal, the probability of substitution for a small time δt is

$$P_{ij}(\delta t) = \delta t \times r_{ij} = M_{ij} \qquad (4.27)$$

while the probability of remaining unchanged is

$$P_{ii}(\delta t) = 1 - \sum_{j \neq i} \delta t \times r_{ij} = M_{ii} \qquad (4.28)$$

Choosing the value of λ above is equivalent to choosing a value of δt so that there is a probability of 1% of an amino acid changing in this time.

Now suppose we want to obtain the transition probabilities for time $2\delta t$. It is not true that $P_{ij}(2\delta t) = 2M_{ij}$. In order to calculate this probability we need to consider all the intermediate states that could have existed at time δt, and sum over them:

$$P_{ij}(2\delta t) = \sum_k M_{ik} M_{kj} \qquad (4.29)$$

Thus if we want to get from A to V in two steps, we can go A→A→V, or A→V→V, or A→R→V, or any

of the other 17 possibilities. To calculate $P_{ij}(n\delta t)$ we have to sum over the possible amino acids that could occur at the $n-1$ intermediate time points. To do this we have to multiply the M matrix by itself n times to obtain the matrix M^n and take the ij element of this matrix. This is written:

$$P(n\delta t) = M_{ij}^n = \sum_k \sum_l \cdots \sum_m M_{ik} M_{kl} \cdots M_{mj} \qquad (4.30)$$

where there are $n-1$ indices to be summed over.

This completes the derivation of the PAM model. The substitution probability matrices known as PAM100, PAM250, etc. are simply the PAM1 matrix M multiplied by itself 100 or 250 times. It is straightforward to calculate these matrices by computer. The number of times we multiply the matrix is called the PAM distance. Box 4.2 gives more information about PAM distances.

The method for deriving the PAM model began with groups of closely related sequences. This is important for several reasons. When sequences are similar, it is easy to align them because there are almost no gaps. In this case the alignment is very insensitive to the scoring matrix we use – if we have not yet derived a good scoring matrix we do not want the alignments used to be sensitive to the details of the rather poor scoring matrix we use to align them. In addition, constructing phylogenetic trees is generally more reliable for more closely related sequences. The parsimony method weights each type of amino acid substitution equally when choosing the tree. Again, this avoids using an evolutionary model that has yet to be calculated. When counting the substitutions, it is assumed that no more than one substitution has occurred per site on any branch. For example, the K↔T substitution between nodes 2 and 3 in Fig. 4.5 is assumed to be a direct change and not via an intermediate route such as K↔Q↔T. If sequences are similar to one another, this is a reasonable assumption, but for more distant sequences, there is no reason this should be true and there would be little justification for using parsimony. Likelihood-based methods of estimating substitution matrices that avoid this problem are discussed in Section 11.2.2.

BOX 4.2
PAM distances

For two proteins at a PAM distance of 1, there have been an average of 0.01 substitutions per site, and the probability that two sites in the aligned pair of proteins will differ is $D = 0.01$. For two proteins at a PAM distance n, there have been an average of $0.01 \times n$ substitutions per site. To keep parallel with DNA models, where evolutionary distances are usually defined as $d =$ number of substitutions per site, we will define $d = 0.01 \times n$ for the PAM model. So PAM100 corresponds to an average of $d = 1$ substitution per site. However, it does not follow that all sites have changed exactly once, because there will have been more than one change in some places and no changes in other places. In fact, the fraction of sites that differ at a PAM distance n is

$$D = \sum_i \pi_i (1 - M_{ii}^n) \qquad (4.31)$$

This is just the probability that the amino acid in the first sequence is an i multiplied by the probability that the amino acid in the second sequence has not remained an i after n PAM units of time. This formula gives us a relationship between the observed fraction of sites that differ, D, and the evolutionary distance, d. It is thus the equivalent of Eq. (4.4) for the Jukes–Cantor distance in the model of DNA evolution. In Fig. 4.8 we have plotted this relationship using the data in Table 23 of Dayhoff *et al.* (1978). The curve has a similar shape to the curve for the DNA models in Fig. 4.4: it is linear for small D, and diverges for larger D. The value of D for unrelated proteins tends to about 0.89. This figure tells us that at $d = 1$ (or PAM distance 100) $D = 0.57$ or, in other words, there are still 43% of residues that will be the same even though there has been an average of one substitution per site.

As it stands, the distance formula (4.31) is much less useful than the one for the JC model. If we have two aligned proteins, we can easily calculate D, but it is not immediately obvious how to calculate the PAM distance d using this equation. One solution is to use the graph in Fig. 4.8 as a "calibration curve". For example, if we observe a fraction $D = 0.5$, then we can deduce that $d = 0.8$ substitutions per site (i.e., PAM distance = 80). To avoid having to use the graph, Kimura (1983) showed that the relationship
$$d = -\ln(1 - D - 0.2D^2) \qquad (4.32)$$

is extremely close to the curve calculated from the Dayhoff data over most of the range of PAM distances. Thus, the dashed line in Fig. 4.8 is almost indistinguishable from the solid line for D less than about 0.75. This equation can be used directly to calculate distances between proteins because it is "the right way round".

However, both Eqs. (4.31) and (4.32) apply only to very long sequences where the frequencies of all the amino acids are assumed to be equal to the average frequencies in the Dayhoff data. For two real proteins of finite length, the amino acid frequencies will not be exactly equal to the average frequencies. A more precise way to estimate distance uses maximum likelihood. For two sequences of length L, let a_k and b_k be the amino acids at the kth site in sequences 1 and 2, respectively. The likelihood of sequence 1 evolving into sequence 2 in n PAM units is

$$L(n) = \prod_{k=1}^{L} \pi_{a_k} M_{a_k b_k}^n \qquad (4.33)$$

The maximum likelihood value of the PAM distance is the value of n that maximizes $L(n)$, and the maximum likelihood value of the number of substitutions per site is $d = n/100$. As an example, we calculated evolutionary distances between a set of hexokinase sequences using the "protdist" program in the Phylip package (Felsenstein 2001). These sequences will be discussed in more detail in Section 6.5 on multiple alignment. For distance estimates, an alignment of 393 sites was used with all gap regions eliminated. In Fig. 4.8, the data points show the maximum likelihood evolutionary distance between one sequence (HXK1_RAT) and each of the other sequences in the alignment. These data points lie approximately on the average curve but not exactly.

There are many ways of estimating distances between proteins, and it is necessary to read the details of the method employed in any program that one uses. Evolutionary models calculated from different sequence sets give slightly different distances from those using the Dayhoff *et al.* (1978) PAM model. It is also worth noting that the distances given by the Tree-Puzzle program (Schmidt *et al.* 2000) are not exactly the same as those given by the Phylip protdist program, even when the same Dayhoff PAM model is used. This is because Phylip assumes the equilibrium amino acid frequencies are fixed as in the original data, whereas Tree-Puzzle re-estimates these frequencies from the data.

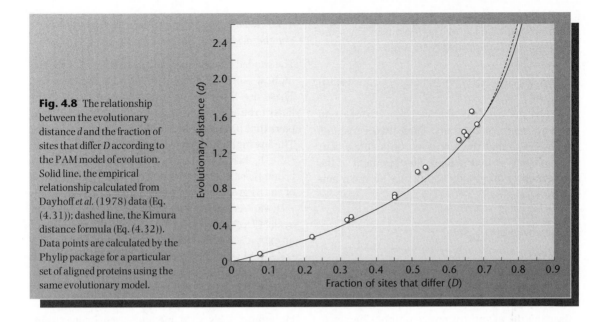

Fig. 4.8 The relationship between the evolutionary distance d and the fraction of sites that differ D according to the PAM model of evolution. Solid line, the empirical relationship calculated from Dayhoff *et al.* (1978) data (Eq. (4.31)); dashed line, the Kimura distance formula (Eq. (4.32)). Data points are calculated by the Phylip package for a particular set of aligned proteins using the same evolutionary model.

4.3 LOG-ODDS SCORING MATRICES FOR AMINO ACIDS

4.3.1 PAM scoring matrices

We will now describe the way in which the PAM model of evolution can be used to derive scoring matrices for use in protein sequence alignment. Consider two very long proteins that are n PAM units apart. The fraction of sites at which the first sequence has amino acid i and the second has amino acid j is $\pi_i M_{ij}^n$. Now suppose we randomly reshuffle the amino acids in each sequence. The frequency of the amino acids remains the same but there is no longer any evolutionary relationship between the sequences. The fraction of sites at which the first sequence is i and the second sequence is j is now just $\pi_i \pi_j$. We define R_{ij} as the ratio of these two values:

$$R_{ij} = \frac{\pi_i M_{ij}^n}{\pi_i \pi_j} = \frac{M_{ij}^n}{\pi_j} \qquad (4.34)$$

Thus R_{ij} is the ratio of the number of times that an i is aligned with a j in proteins evolving according to the PAM model to the number of times that an i would be aligned with a j in random protein sequences with

the same amino acid frequencies. If $R_{ij} > 1$, then amino acids i and j are more likely to be aligned with each other according to the PAM model than they would be by chance. Because the PAM model is time reversible, the matrix of R values is symmetric ($R_{ij} = R_{ji}$). This is common sense because the probability that an i and a j are aligned with each other cannot depend on which of the two sequences we call sequence 1 and which we call sequence 2. Let the amino acids at the kth site in the two proteins be a_k and b_k, as in Box 4.2. The relative likelihood of this pair of aligned sequences arising according to the PAM model to the likelihood of the pair arising if the amino acids were chosen at random is

$$\text{relative likelihood} = \prod_{k=1}^{L} R_{a_k b_k} \qquad (4.35)$$

When deciding how to align two proteins, we should choose the alignment that maximizes this relative likelihood, i.e., we should choose the alignment that is most likely to arise according to our evolutionary model. Algorithms for calculating the optimum alignment will be discussed in Chapter 6. For these algorithms, it is convenient to have an alignment score that is a sum over sites, rather than a product over sites like Eq. (4.35). To derive these scores, we just take logs of the R_{ij} values. The score for alignment of amino acid i with amino acid j is

$$S(i,j) = C \log_B R_{ij} \qquad (4.36)$$

B is the base of logarithm that we choose to take, and C is any arbitrary constant. These two constants are chosen so that the scores are on a convenient scale. We can write the score for alignment of the two sequences (forgetting about gaps for the moment) as

$$\text{alignment score} = \sum_{k=1}^{L} S(a_k, b_k) \qquad (4.37)$$

The alignment that maximizes the sum in Eq. (4.37) is the same as the alignment that maximizes the relative likelihood in Eq. (4.35).

Leaving the alignment problem until Chapter 6, let us now think more carefully about what these scores mean. A matrix of scores calculated according to Eq. (4.36) is called a log-odds matrix. The log-odds matrix calculated for PAM250 using the data of Jones *et al.* (1992) is shown in Fig. 4.6 (below the diagonal). In this case both B and C are 10, and the values shown are rounded to the nearest integer. Positive scores indicate that the two amino acids are more likely to be aligned with one another than they would be by chance, and negative scores mean that they are less likely to be aligned with one another than they would be by chance. Positive scores are shaded gray. Note that all the diagonal elements are shaded, meaning that each amino acid is more likely to be aligned with itself than it would be by chance. The highest score is $S(W,W) = 15$. To see what this means, we can invert Eq. (4.36) to give $R_{ij} = B^{S(i,j)/C}$. In this case we have $R_{WW} = 10^{15/10} = 31.6$, meaning that a W is 31.6 times more likely to be aligned with another W in proteins at PAM distance 250 than it would be by chance. This reflects the fact that W has a very low relative mutability – see Eq. (4.26) above. Similarly, amino acids with high mutability like S have low scores on the diagonal, meaning that an S is only slightly more likely to be aligned with itself than it would be by chance.

4.3.2 Relationship to the physico-chemical properties of amino acids

The shaded off-diagonal elements in Fig. 4.6 are the highest scores between non-identical amino acids. These are the pairs of amino acids that are most likely to be aligned with one another. They tend to be pairs that have similar physico-chemical properties. The few highest-scoring pairs are listed here.
• Y↔F. Tyrosine and phenylalanine both have an aromatic side group and differ only by the presence of an OH group on the benzene ring in tyrosine.
• D↔E. Aspartate and glutamate both have an acidic COO⁻ group and differ only by the presence of an extra carbon in the side chain of glutamate.
• K↔R. Lysine and arginine are both basic and have relatively long side chains ending in positively charged NH_3^+ or NH_2^+ groups.
• V↔I. Valine and isoleucine are both hydrophobic amino acids with medium-sized hydrocarbon side chains. They differ by a single extra carbon in isoleucine.

In contrast, amino acids that have different physico-chemical properties tend to have negative scores. For example, most scores between hydrophilic and hydrophobic amino acids are negative. It is also noticeable that the two most slowly evolving amino acids, tryptophan and cysteine, have rather unusual physico-chemical properties. Tryptophan is the largest amino acid with a very bulky side chain. If it is substituted by another side chain it is likely that this will cause a substantial change in the 3D structure of the protein. Hence it will most likely be a deleterious mutation that will be eliminated by selection. Cysteine residues have the ability to form disulfide bonds, C–S–S–C, between different parts of protein chains. The disruption of these bonds would also be likely to cause a major effect on the protein structure. Thus, the observed rate of substitutions involving both W and C is small. These amino acids have high scores on the diagonal but negative scores for alignment with almost everything else.

We spent a long time discussing the properties of amino acids in Chapter 2. The plot from principal component analysis (Fig. 2.10) is a good way of summarizing the differences in the amino acid properties. Almost all the pairs of amino acids with high PAM scores are close to one another on the PCA plot. To see this, try making a copy of Fig. 2.10 and joining all the pairs of points corresponding to the shaded cells in Fig. 4.6. This confirms the fact that evolution does take notice of amino acid properties. It also confirms what we said in Chapter 3 regarding the stabilizing action of natural selection, i.e., selection is acting against mutations that change the amino acid properties too drastically, and preventing these changes reaching fixation. The only significantly positive scores in the PAM matrix that appear quite far from one another on the PCA plot are CW and CY. We already said that cysteine is an unusual amino acid that cannot easily be replaced by anything else. The explanation for the high PAM scores for CW and CY is probably due to the proximity of these amino acids in the genetic code, as we now describe, rather than because of similarities in physico-chemical properties.

Examination of Fig. 4.6 reveals that the structure of the genetic code also has a role in the rates of amino acid substitution. As can be seen from the genetic code diagram in Table 2.1, it is possible for some substitutions of amino acids to occur by the substitution of a single nucleotide at one of the three positions in the codon, whereas other amino acid substitutions require more than one nucleotide substitution. We would expect the amino acid substitutions that require only one nucleotide substitution to occur more rapidly. In Fig. 4.6, the scores for the amino acid pairs that can be substituted with a single nucleotide change are highlighted in a black box. It can be seen that all the pairs with positive scores (gray cells) are also highlighted in black. Therefore, there is no substitution that is frequently observed that cannot be achieved by a single nucleotide change. However, there are many zero or negative scores that are also shaded black. Thus, not all the substitutions that are possible via a single nucleotide change are frequently observed. This shows that the effect of the genetic code is moderated by the action of natural selection on the amino acid properties. For example, the substitution from valine (V) to glutamate (E) is possible by a single nucleotide

substitution from T to A at the second position. This presumably occurs frequently as a mutation in the population. However, selection is likely to eliminate most of these changes because the physico-chemical properties of the amino acids are too different. The observed rate of V↔E substitutions is therefore low because most of these mutations do not go to fixation, and the score $S(V,E)$ is negative.

As an interesting aside, it appears that the structure of the genetic code is far from random. It is often the case that amino acids with similar properties are next to one another in the diagram (Table 2.1) and therefore that they are accessible by a single nucleotide change. As a result the average effect of random point mutations on the physico-chemical properties of a protein is less than one might expect. This has been shown by comparing the real code with many possible randomly reshuffled genetic codes (Freeland et al. 2000). It is therefore argued that the code has evolved by reshuffling the assignment of codons to amino acids in order to minimize errors. Although the canonical code is thought to have arisen very early in evolution (before the split of archaea, bacteria, and eukaryotes), and although the process of codon reassignment is potentially very disruptive to an organism, there are nevertheless many small changes to the code that have occurred in isolated groups of organisms or organelles since the canonical code was established (Knight, Freeland, and Landweber 2001). Thus reshuffling of the code due to selection for error minimization in the early stages of life on earth is not impossible.

Returning to scoring matrices, we note that a log-odds matrix can be derived for any PAM distance. The PAM250 matrix shown in Fig. 4.6 represents a relatively large distance. This means that there is a substantial chance that an amino acid has changed and therefore there are a fairly large number of positive scores in the off-diagonal elements. For a smaller PAM distance like 100 or less, there would be a higher probability that each amino acid had remained unchanged. Therefore the scores on the diagonal of the matrix would be larger, and most of the off-diagonal scores would be negative. The PAM scoring matrices are used for pairwise sequence alignments and for database searches. The matrices are most effective at spotting similarities between sequences that are at evolutionary distances comparable with the PAM distance of the matrix. Low PAM number matrices are used to spot matches with a high percentage identity between the sequences. High PAM number matrices are used to spot matches between distantly related sequences, where the residues may not be identical but the physicochemical properties of the amino acids are conserved. Examples of alignments and database search results using different PAM matrices are discussed in Chapters 6 and 7.

4.3.3 BLOSUM scoring matrices

The BLOSUM matrices (Henikoff and Henikoff 1992) are another set of log-odds scoring matrices that have much in common with PAM matrices, although they are derived in a different way. In particular, the scoring matrix is derived directly from sequence alignments without the use of a phylogenetic tree, and without deriving an evolutionary model.

The basics of the method can be explained using the sequence alignment in Fig. 4.5. First we count the number of times each amino acid appears and hence obtain the frequencies π_i of each amino acid in the data set. Then we count the number of times A_{ij} that each amino acid is aligned with each other amino acid. In Fig. 4.5 there are seven sequences. The total number of ways of picking a pair of amino acids from one column of the alignment is $7 \times 6 = 42$. The seventh column of the alignment is all K. This column contributes 42 to A_{KK}. In the first column of the alignment we have six Ts and one I. This column contributes $6 \times 5 = 30$ to A_{TT}, six to A_{TI} and six to A_{IT}. When the totals from all the columns are added, the following matrix A_{ij} results:

$$
\begin{array}{c|cccccc}
 & I & K & L & Q & T & V \\
\hline
I & 8 & - & 16 & - & 6 & 6 \\
K & - & 78 & - & 6 & 12 & - \\
L & 16 & - & 22 & - & - & 4 \\
Q & - & 6 & - & 62 & 10 & - \\
T & 6 & 12 & - & 10 & 44 & - \\
V & 6 & - & 4 & - & - & 2 \\
\end{array}
\tag{4.38}
$$

This differs from the A_{ij} matrix calculated by the PAM method (Eq. (4.21)) in that we are counting pairs of aligned amino acids instead of numbers of substitutions. Thus we have non-zero entries on the diagonal. There are also some entries like A_{VL} that are non-zero in Eq. (4.38) but are zero in Eq. (4.21). This is because V and L amino acids are aligned with each other at site 5, whereas the parsimony method used in Fig. 4.5 found that there were substitutions V↔I and I↔L at site 5 but no direct substitution V↔L. The two methods treat the same alignment data in different ways and thus result in different scoring matrices.

We can now obtain the fraction of aligned pairs that are of type ij:

$$
q_{ij} = \frac{A_{ij}}{A_{tot}}
\tag{4.39}
$$

where A_{tot} is the total of all the elements in the A_{ij} matrix. From this we can directly obtain the relative frequency of ij pairs in the aligned proteins compared with what we would expect for randomly reshuffled proteins with the same base frequencies:

$$
R_{ij} = \frac{q_{ij}}{\pi_i \pi_j}
\tag{4.40}
$$

This quantity R_{ij} is directly comparable to the R_{ij} that we calculated for the PAM model in Eq. (4.34). We can use it to calculate a log-odds scoring system using Eq. (4.36) in the same way.

The mathematics of this method is much simpler than for the PAM method because we can go directly from the data to the log-odds matrix without needing to derive an evolutionary model (M_{ij}) first. There are nevertheless some disadvantages. The result that we obtain is sensitive to the presence of groups of very closely related sequences in the alignment. These sequences will be almost identical and will contribute a large amount to the numbers on the diagonal of the A_{ij} matrix. In contrast, groups of almost identical sequences do not affect the PAM method because these appear very close to one another on the phylogenetic tree and very few substitutions are required on these branches of the tree. Real sequence data are usually obtained in a very "patchy" way, with a lot of sequences available for a few well-studied groups and only a few sequences available from other groups. It is therefore necessary to develop a systematic way of accounting for the distribution of distances between the sequences in the alignments used for the BLOSUM method.

Henikoff and Henikoff (1992) used alignments of protein domains from the Blocks database (see Section 5.5.5). These are reliably aligned regions of proteins without gaps. Within each alignment sequences were grouped into clusters that have a percentage identity greater than a certain cut-off value (e.g., 80% or 62%). When counting the number of paired amino acids for the A_{ij} matrix, sequences in the same cluster are not counted. When counting pairs between clusters, sequences in the same cluster are weighted as a single sequence (i.e., if there were two clustered sequences, each would count half the weight of a single distinct sequence). The log-odds matrix that results from clustering sequences at a given percentage is known as the corresponding number BLOSUM matrix. Hence BLOSUM80 only includes substitutions between sequences that are less than 80% similar, and BLOSUM62 only includes substitutions between sequences that are less than 62% similar. Lower BLOSUM number therefore represents lower similarity, whereas lower PAM number represents greater similarity.

The scaling of the scores used in BLOSUM62 is $S(i,j) = 2 \log_2 R_{ij}$. This means that the scores are in "half-bit" units. A bit is a factor of two, and a score of 1 in the matrix corresponds to a factor of $\sqrt{2}$ in relative likelihood. The highest score is $S(W,W) = 11$, which corresponds to $R_{WW} = 2^{11/2} = 45.2$. This is not too different from what we got with PAM250. Table 4.1 compares three log-odds scoring systems derived by different methods and using different sets of sequence alignments. The numerical details of the scoring systems are different, hence if we use them for sequence alignments, we may obtain slightly different results. However, the most important features of these scoring matrices are the same. The sets of the most conserved and least conserved amino acids and the most significant positive scores between non-identical amino acids are very similar in the three cases. This gives us confidence that the scoring systems are really calculating something fundamental about the process of protein sequence evolution and the way in which natural selection acts on the physico-chemical properties of the amino acids.

When we are searching for similarities between sequences, it is easy to spot sequences that are very similar. The difficulty arises in spotting relationships in the "twilight zone" of sequence similarity where the resemblance between the sequences is only slight. It has been argued that the BLOSUM series of matrices is more effective for use in database search algorithms than the PAM series because it is based directly on comparison of distant sequences (Henikoff and Henikoff 1993). The PAM method, on the other hand, counts substitutions in closely related sequences only, and then predicts what will happen for more distant sequences by extrapolation of the evolutionary model to longer times. It may be that this extrapolation is not very reliable (Benner, Cohen, and Gonnet 1994). The disadvantage of BLOSUM is that since it is not based on an evolutionary model it cannot be used for calculating evolutionary distances and phylogenetic trees. The reason that the PAM matrix only uses similar sequences is because it relies on parsimony to count the substitutions. Maximum likelihood methods of estimating evolutionary models do not have this restriction (see Section 11.2.2 for further details).

Table 4.1 Important features of amino acid substitution matrices.

Matrix	JTT250*	BLOSUM62**	VT160***
Most conserved amino acids	1. W 2. C 3. Y 4. F	1. W 2. C 3. H 4. Y, P	1. W 2. C 3. Y 4. F, H, P
Least conserved amino acids	1. S, A, T 2. N 3. I, V	1. S, A, I, V, L	1. S, A 2. T, I, V, L
Most significant non-identical matches	1. Y↔F 2. V↔I, K↔R, D↔E, Y↔H 3. M↔I, M↔L	1. Y↔F, V↔I 2. K↔R, D↔E, Y↔H, L↔I, M↔L, E↔Q, W↔Y	1. Y↔F 2. V↔I, K↔R 3. D↔E, Y↔H, M↔I, M↔L, H↔Q

The most/least conserved amino acids are those with the highest/lowest scores in the diagonals of the log-odds matrices. The most significant non-identical matches are the highest off-diagonal scores in the log-odds matrices. Entries are ranked equally when the rounding scheme leads to identical values in the published matrices. * Jones, Taylor, and Thornton (1992) version of PAM250 matrix; ** Henikoff and Henikoff (1992) BLOSUM62; *** Müller and Vingron (2000) Variable Time VT160 matrix.

SUMMARY

The models discussed in this chapter are used to describe the evolution of a sequence along a branch of a phylogenetic tree. In phylogenetic studies we are not usually interested in sequence variation within a single species and we are using a single sequence to represent the whole species. Therefore the rates of substitution in the evolutionary models describe the changes in the representative sequence that arise due to fixation of new alleles. Substitution rates are hence dependent on mutation, selection, and drift, and are not simply the rates of mutation in individual sequences.

DNA sequence evolution can be described by models based on rate matrices having four states: A, C, G, and T. The simplest of these models is the Jukes–Cantor model, where all four bases are assumed to have equal frequency and where all possible substitutions occur at equal rate. More complicated models relax these assumptions. The general reversible model has four parameters for the base frequencies and six parameters for substitution rates. It is the most general four-state model that satisfies the principle of time reversibility (i.e., that average properties of sequences do not change with time along an evolutionary lineage). The numerical values of these parameters can be determined by fitting the model to

sequence data, and may not be the same for different types of sequence.

When comparing two sequences it is straightforward to count the fraction of sites that differ between them, D. However, it is useful to have a measure of evolutionary distance between sequences that increases linearly with time at all times. The usual distance measure is d, the average number of substitutions per site. Although d is not directly observable by comparing sequences, it is possible to calculate d from quantities that are observable. This calculation depends on the model of sequence evolution that is being used. The lengths of branches on phylogenetic trees are often measured in units of substitutions per site.

Protein sequence evolution can be described by rate matrices involving 20 states for the 20 amino acids. It requires large sets of many sequence alignments to determine all the parameters in these large matrices. The PAM model was the first protein evolution model of this type. The PAM1 substitution matrix M_{ij} is the probability of substitution from amino acid i to j in a time of 1 PAM unit. This is the time in which there is an average of 1% sequence divergence. The probabilities of substitution in a time of n PAM units can be obtained by multiplying this matrix by itself n times. The PAM distance between two proteins is an estimate of the number of substitutions per site, as with DNA sequence distances, although it is conventionally mutiplied by 100. Thus a PAM distance of n corresponds to $n/100$ substitutions per site.

The PAM substitution matrices can be used to calculate log-odds scoring matrices for use with protein sequence alignment algorithms. The scores $S(i,j)$ are defined as the logarithm of the ratio of the probability that amino acids i and j are aligned in sequences evolving according to the model to the probability that they would be aligned in random sequences. Positive scores indicate pairs of amino acids that substitute for one another more frequently than would be expected by chance. The scores are influenced by the physico-chemical properties of the amino acids and also by the structure of the genetic code. The BLOSUM scoring matrices are another set of log-odds matrices that are calculated directly from sets of sequence alignments without the use of an evolutionary model. These matrices can be used for sequence alignments but not for phylogenetics.

REFERENCES

Adachi, J. and Hasegawa, M. 1996. A model of amino acid substitution in proteins encoded by mitochondrial DNA. *Journal of Molecular Evolution*, **42**: 459–68.

Benner, S.A., Cohen, M.A., and Gonnet, G.H. 1994. Amino acid substitution during functionally constrained divergent evolution of protein sequences. *Protein Engineering*, **7**: 1323–32.

Dayhoff, M.O., Schwartz, R.M., and Orcutt, B.C. 1978. A model of evolutionary change in proteins. In *Atlas of Protein Sequence and Structure*, **5**(3): 345–52. Washington DC: National Biomedical Research Foundation.

Felsenstein, J. 2001. PHYLIP Phylogeny Inference Package version 3.6. Available from http://evolution.genetics.washington.edu/phylip.html.

Freeland, S.J., Knight, R.D., Landweber, L.F., and Hurst, L.D. 2000. Early fixation of an optimal genetic code. *Molecular Biology and Evolution*, **17**: 511–18.

Hasegawa, M., Kishino, H., and Yano, T.A. 1985. Dating of the human–ape splitting by a molecular clock of mitochondrial DNA. *Journal of Molecular Evolution*, **22**: 160–74.

Henikoff, S. and Henikoff, J.G. 1992. Amino acid substitution matrices from protein blocks. *Proceedings of the National Academy of Sciences USA*, **89**: 10915–19.

Henikoff, S. and Henikoff, J.G. 1993. Performance evaluation of amino acid substitution matrices. *PROTEINS: Structure, Function and Genetics*, **17**: 49–61.

Jones, D.T., Taylor, W.R., and Thornton, J.M. 1992. The rapid generation of mutation data matrices from protein sequences. *CABIOS*, **8**: 275–82.

Jukes, T.H. and Cantor, C.R. 1969. Evolution of protein molecules. In H.N. Munro (ed.), *Mammalian Protein Metabolism*, pp. 21–123. New York: Academic Press.

Kimura, M. 1983. A simple method of estimating evolutionary rates of base substitutions through comparative studies of nucleotide sequences. *Journal of Molecular Evolution*, **16**: 111–20.

Knight, R.D., Freeland, S.J., and Landweber, L.F. 2001. Rewiring the keyboard: Evolvability of the genetic code. *Nature Reviews Genetics*, **2**: 49–58.

Müller, T. and Vingron, M. 2000. Modeling amino acid replacement. *Journal of Computational Biology*, **7**: 761–76.

Schmidt, H.A., Strimmer, K., Vingron, M., and von Haeseler, A. 2000. Tree-Puzzle version 5.0 available from http://www.tree-puzzle.de/.

Yang, Z. 1996. Among-site rate variation and its impact on phylogenetic analyses. *Trends in Ecology and Evolution*, **11**: 367–72.

PROBLEMS

4.1 In Section 4.3.1 we itroduced the Kimura two-parameter model of DNA sequence evolution. The aim of this problem is to guide you through the calculation of the evolutionary distance in this model in a step-by-step way. The result was already given in the text – Eq. (4.13)

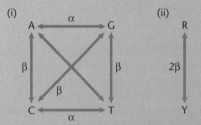

(i)   (ii)

The rate model is defined as in (i). The rate of transitions = α, and the rate of transversions = β.

This means that the net rate of transition from purines (R) to pyrimidies (Y) is 2β, as in (ii). Throughout this problem $P_{ij}(t)$ is the probability that a site is in state j at time t given that it was in state i at time 0.

(a) From diagram (ii) we have

$$\frac{dP_{RR}}{dt} = -2\beta P_{RR} + 2\beta P_{RY}$$

We also know $P_{RY} = 1 - P_{RR}$. Write down the differential equation for $P_{RR}(t)$ and find the solution that satisfies the initial condition $P_{RR} = 1$, at $t = 0$. You need to assume a general solution of the form $P_{RR}(t) = A + Be^{-4\beta t}$ and calculate the constants A and B.

(b) Now we know both $P_{RR}(t)$ and $P_{RY}(t)$. We also know that $P_{RR} = P_{AA} + P_{AG}$, and that $P_{RY} = P_{AC} + P_{AT}$. By symmetry we know $P_{AC} = P_{AT}$. What is $P_{AC}(t)$?

(c) From diagram (i) we have

$$\frac{dP_{AA}}{dt} = -(\alpha + 2\beta)P_{AA} + \alpha P_{AG} + \beta(P_{AC} + P_{AT})$$

Use what you know to eliminate P_{AG}, P_{AC}, and P_{AT} from the equation and get a differential equation involving only P_{AA}.

(d) Solve this equation. You need to assume a general solution of the form

$$P_{AA}(t) = A + Be^{-4\beta t} + Ce^{-2(\alpha+\beta)t}$$

and calculate the constants.

(e) Now consider a pair of sequences that diverged at time t in the past. Let S be the fraction of sites that differ by transitions and V be the fraction of sites that differ by transversions. We know that $S = P_{AG}(2t)$ and $V = P_{RY}(2t)$. Why is this?

(f) Rearrange the equation for V to show that

$$8\beta t = -\ln(1 - 2V)$$

(g) Obtain an equation for $V + 2S$ as a function of time. Rearrange this equation to show that

$$4(\alpha + \beta)t = -\ln(1 - V - 2S)$$

(h) The mean number of substitutions per site occurring on the branches between the two sequences is $d = 2t(\alpha + \beta)$. (Why?) Now find an equation for the evolutionary distance d as a function of S and V. (This is what we set out to do! The answer should be Eq. (4.13)). Show that this equation is the same as the result from the Jukes–Cantor model (Eq. (4.4)) if $\alpha = \beta$.

4.2 Consider a long section of DNA that is evolving neutrally (i.e., in absence of selection). The rates of mutation from A, C, and G to each of the other types of base are all equal to u, while the rates of mutation from T to each of the other bases are equal to $2u$. What would you expect to be the frequencies of the four bases in the sequence?

SELF-TEST
Molecular evolution

This test covers material in Chapters 3 and 4.

1 The rate of substitutions in a certain region of DNA of length 1000 bases is estimated as 10^{-9} per base per year. If two species diverged approximately 10 million years ago, the fraction of sites that differ between them should be approximately
A: 1%
B: 2%
C: 20%
D: 75%

2 For the same sequences as above, which of these statements is true?
A: The probability that the two sequences will not differ at all is greater than 50%.
B: The probability that the two sequences will not differ at all is less than one in a million.
C: If there were considerable variation in the substitution rate between sites, the fraction of sites that differ would be greater than if all sites changed at the average rate.
D: The two sequences should be almost completely randomized with respect to each other after 100 million years.

3 It is expected that synonymous sites will change more rapidly than non-synonymous sites because
A: The fraction of transversions is larger at synonymous sites.
B: Stabilizing selection reduces the rate of substitutions at non-synonymous sites.
C: The mutation rate at synonymous sites is higher.

D: Natural selection favors new variants arising at synonymous sites.

4 A population of fixed size, N, is evolving according to the coalescent theory, with no selection acting. Let T be the time since the last common ancestor of two individuals that are chosen randomly from the population. Which of the following is true?
A: The mean value of T is dependent on the product of the mutation rate, u, and the population size, N.
B: T has a normal distribution, with a mean value equal to N generations.
C: T will always be very close to N generations when N is very large.
D: T depends on the particular positions of the branches in the tree, and therefore fluctuates greatly from one population to another.

5 It is believed that all present-day copies of human mitochondrial DNA descended from a single person, "Eve," living in Africa around 200,000 years ago. Which of the following is true?
A: Differences in mitochondrial DNA sequences between different African populations are larger than differences between non-African populations.
B: There are no fossil human remains prior to 200,000 years ago.
C: Mitochondrial sequences in present-day African populations have changed **less** since the time of Eve than have the sequences in non-African populations.
D: Mitochondrial sequences in present-day African populations have changed **more** since the time of Eve than have the sequences in non-African populations.

6 A new mutant allele has just arisen in a population. Which statement is true?
A: If the mutant is neutral with respect to the original allele, there is a 50% probability that the mutant allele will replace the original allele.
B: It is very likely to disappear in a few generations due to random drift.
C: It will only become fixed in the population if there is a strong selective advantage.
D: If the mutant allele reaches a frequency of 50%, it will almost always go on to fixation.

7 If two sequences evolve according to the Jukes–Cantor model, and they are observed to differ at 20% of sites, which of the following is true?
A: The Jukes–Cantor distance is 0.18.
B: The Jukes–Cantor distance is 0.20.
C: The Jukes–Cantor distance is 0.23.
D: The Jukes–Cantor distance cannot be calculated without further information.

8 Which of the following statements is correct?
A: A reversible rate matrix is a symmetrical matrix.
B: A reversible rate matrix is only used for calculations with rooted trees.
C: A reversible rate matrix can be used to describe DNA but not protein evolution.
D: A reversible rate matrix assumes frequencies of different nucleotides are constant in time.

	A	C	D	E	F	G	H	I	K	L	M	N	P	Q	R	S	T	V	W	Y
A	4	0	-2	-1	-2	0	-2	-1	-1	-1	-1	-2	-1	-1	-1	1	0	0	-3	-2
C		9	-3	-4	-2	-3	-3	-1	-3	-1	-1	-3	-3	-3	-3	-1	-1	-1	-2	-2
D			6	2	-3	-1	-1	-3	-1	-4	-3	1	-1	0	-2	0	-1	-3	-4	-3
E				5	-3	-2	0	-3	1	-3	-2	0	-1	2	0	0	-1	-2	-3	-2
F					6	-3	-1	0	-3	0	0	-3	-4	-3	-3	-2	-2	-1	1	3
G						6	-2	-4	-2	-4	-3	0	-2	-2	-2	0	-2	-3	-2	-3
H							3	-3	-1	-3	-2	1	-2	0	0	-1	-2	-3	-2	2
I								4	-3	2	1	-3	-3	-3	-3	-2	-1	3	-3	-1
K									5	-2	-1	0	-1	1	2	0	-1	-2	-3	-2
L										4	2	-3	-3	-2	-2	-2	-1	1	-2	-1
M											5	-2	-2	0	-1	-1	-1	1	-1	-1
N												6	-2	0	0	1	0	-3	-4	-2
P													7	-1	-2	-1	-1	-2	-4	-3
Q														5	1	0	-1	-2	-2	-1
R															5	-1	-1	-3	-3	-2
S																4	1	-2	-3	-2
T																	5	0	-2	-2
V																		4	-3	-1
W																			11	2
Y																				7

9 Which of the following statements is correct?

A: The PAM250 log-odds matrix applies to sequences that are more distant from one another than the PAM100 matrix.

B: The BLOSUM85 log-odds matrix applies to sequences that are more distant from one another than the BLOSUM62 matrix.

C: The BLOSUM85 log-odds matrix applies to sequences that are more distant from one another than the PAM250 matrix.

D: All three of the above.

10 Which of the following statements concerning the BLOSUM62 matrix above is correct?

A: Alanine is aligned with arginine more often than expected by chance.

B: Alanine never changes to cysteine.

C: Tryptophan evolves the slowest.

D: The off-diagonal elements are proportional to the rates of substitution from one amino acid to another.

11 The scores in BLOSUM62 are measured in half-bits. The score of D against E is 2. Therefore:

A: D is aligned with E twice as often as expected by chance.

B: D is aligned with E four times as often as expected by chance.

C: D is aligned with E ln 2 times as often as expected by chance.

D: D is aligned with E e^2 times as often as expected by chance.

Information resources for genes and proteins

CHAPTER

5

CHAPTER PREVIEW

In this chapter, we describe some of the major sequence databases that are available for nucleic acids and proteins. We begin by discussing how biological databases first arose from the need to store sequence information, and how this led to the need for consistent file formats and data organization. We describe the primary DNA and protein databases, and then the secondary databases that focus on protein family data. In each case, we focus on the nature of the information included and the structure of the database files.

family and functional relationships among proteins, and to provide diagnostic tools for sequence classification. Such resources now play key roles in genome annotation.

With the advent of automated, high-throughput techniques, whole-genome sequencing projects became commonplace. Because of the need to document information relating to specific organisms, and to better manage the data, the fruits of these activities were soon harvested in species-specific databases. At the same time, scientists saw the importance of placing the genomic data in wider biological and medical contexts: e.g., to facilitate comparative analysis of genomes and to help assign gene functions more precisely, knowledge of molecular and cellular biology was captured in metabolic pathway and molecular interaction databases; and, to examine aspects of disease systematically, the structure of human genes was cataloged in compendia of genes and genetic disorders.

The deluge of available genomic information is set to have a major impact on biological and medical research. The accumulated data are already allowing scientists to search for novel genes and proteins, to evaluate their roles as drug targets, and to investigate why individuals respond in different ways to the same drug regimes. The scale of information gathering world-wide is daunting; the challenge will be to integrate it all into a coherent picture, to use the

5.1 WHY BUILD A DATABASE?

Historically, databases have arisen to satisfy diverse needs, whether to address a biological question of interest to an individual scientist, to better serve a particular section of the biological community, to coordinate data from sequencing projects, or to facilitate drug discovery in pharmaceutical companies. The workhorses of modern biology, databanks now number in the hundreds and house information of all kinds. Indeed, a special issue of the journal *Nucleic Acids Research*, with an online molecular biology database catalog (Baxevanis 2003), is devoted every year to the documentation of ongoing and new database initiatives, and there is even a database of databases, DBCat (Discala *et al.* 2000).

So how did this avalanche of data arise? The first databases that emerged concentrated on collecting and annotating nucleotide and protein sequences generated by the early sequencing techniques. Once sequence repositories were established, more analytical resources were developed, both to catalog genome sequences of humans and other species to help us understand the complexities of biology, and ultimately to provide insights into how individual genetic variations result in disease.

Clearly, sequencing entire genomes of diverse organisms and acquiring a rough draft of the human genome represent major technical achievements. But merely increasing the amount of information we collect does not in itself bestow an increase in knowledge, or endow us with a miraculous understanding of genomes. A lot of work is required to turn raw data into biological insight: e.g., genes must be located and the coding regions translated to yield their protein products; functions must be assigned; and, if the structure is known, the function must be rationalized in structural terms, otherwise prediction or modeling techniques must be used to derive feasible models. All of these tasks depend on having computational access to biological databases in which the data are both reliable and comprehensive.

The databases in common use today each have different scopes and different objectives. Primary DNA sequence repositories, such as EMBL, GenBank, and DDBJ (described in Section 5.3), are those that attempt to keep a comprehensive record of all sequenced DNA as it becomes available. As primary databases are all-inclusive, they are inevitably fairly shallow in terms of the information they contain. By contrast, secondary (or protein family) databases aim to combine information from several different primary database entries and to provide added information not present in their primary sources. Protein family databases (some of which are described in Section 5.5) are based on the recognition of conserved regions (or motifs) in multiple sequence alignments – Chapter 9 describes the methods used to recognize these patterns of conservation. There is a list of Web addresses for the resources referred to in this chapter at the end of this book.

5.2 DATABASE FILE FORMATS

Database entries have well-defined file formats: this is important so that data can be read by computer and extracted automatically. Computers need to know what type of information to expect in an entry and where to find it. Some examples of file formats are discussed in the rest of this chapter: all have a few things in common. Some sort of "accession number" is always included; this is a unique identifier, assigned when the entry is originally added to the database, and should not change. Accession numbers are easily searchable and indexable by computers, but are meaningless to most people. Some databases therefore have a second identifying (ID) code that is more comprehensible (e.g., the Swiss-Prot codes discussed in Section 5.4). ID numbers don't stop there, though. For example, biological sequences may have entries in several different databases, each with their own numbering system – cross-referencing the accession numbers between them is therefore important; some sequences have multiple parts and may need ID numbers for the separate sections, such as parts of a sequenced genome containing more than one gene, or eukaryotic genes containing multiple introns and exons. Furthermore, some sequences change: e.g., if they are resequenced by a second group, or mistakes are found and later corrected. We need a way of keeping track of all this, and the result is usually a stack of IDs at the top of the file.

Another thing we need in a database is a description of what the entry is. In a sequence database, at the very least, we need the name of the gene, but we would probably also like to know the organism it came from, who sequenced it and when, perhaps some information about its structure and function, and whether there are papers describing research related to it. Different databases handle this type of information in different ways. There is no single right way to do it, but it is important to be consistent. The better organized the database file format, the easier it will be to use the information at a later stage.

In many sequence databases, it is necessary to scroll way down the file before reaching the sequence itself! Usually, there is a specified number of characters per line, and sometimes lines are broken into groups of 10 characters. Numbers at the beginning and end of lines help us to know where we are in a sequence. Many bioinformatics software packages also have their own sequence formats. One of the simplest is FASTA format, which simply includes a line with ">" preceding the sequence accession number/ID/description, and another line with the actual sequence. The human prion protein in FASTA format looks like this:

```
>gi|190469|gb|M13667.1|HUMPRP0A
Human prion protein 27-30 mRNA
CGAGCAGCCAAGGTTCGCCATAATGACTGCTCTCGGTCGTGAGGAGA
GGAGAAGCTCGCGGCGCCGCGGCTGCTGGATGCTGGTTCTCTTTGT
GGCCACATGGAGTGACCTGGGCCTCTGCAAGAAGCGCCGAAGCCTG
```
+ lots more lines of sequence . . .

In this example, the first line contains various sequence ID numbers, but this is not obligatory. In fact, you can put more or less anything you like after the > and most bioinformatics software will still read it as a FASTA sequence. In contrast, GenBank and Swiss-Prot formats are much more complicated (see Figs. 5.1 and 5.3). As programmers soon find out, it is much easier to write a program to read a FASTA sequence than it is to extract a sequence from a GenBank file. However, simple formats have disadvantages for long sequences – having a sequence thousands of bases long with no gaps, carriage returns, or numbering system is not necessarily a good idea. Different formats have different objectives. For databases, we want readability and a lot of information associated with the sequence. For sequence analysis software, we need ease of input and output.

There are usually certain key features that are obligatory in sequence files in order for software to read them and recognize them for what they are: e.g., the initial > is obligatory in FASTA format; and EMBL and GenBank records must have // to terminate each entry. Small errors in file formats are a prime reason why people have problems getting bioinformatics software to work. There are surprisingly many different formats; so many, in fact, that specialist software has been written just to convert sequences from one format to another – Don Gilbert's Readseq program is available on the Web from several places (e.g., http://www.ebi. ac.uk/readseq/index.html). If you are writing your own programs, then it is a good idea not to invent any more sequence formats, unless you absolutely have to!

The preceding discussion about file formats refers to what are called "flatfiles" by computer scientists. This just means that the information is stored in simple text files. A program that wants to use information in a file has to read through it, find the right line and the right piece of text, and put this information into an internal variable that can be used by the program. This process is termed "parsing", and is where the file format is essential. If you know that the sequence will begin on a line following the label SQ (as it does in Swiss-Prot and EMBL), then it is easier to find it. If you know that the organism name is on the line beginning ORGANISM (as it is in GenBank), this helps when trying to write a program to locate sequences from a particular species. In fact, retrieval of information from databases is a non-trivial process, and specialized systems are available for doing it, such as the SRS Sequence Retrieval System (Etzold, Ulyanov, and Argos 1996), and the Entrez system at the NCBI (Schuler 1996). We turn now to details of some of the most commonly used databases.

5.3 NUCLEIC ACID SEQUENCE DATABASES

Nucleic acid sequences offer a starting point for understanding the structure, function, and development of genetically diverse organisms. A testament to their central importance in modern biology has been the simultaneous effort in different parts of the world to collect, process, and disseminate them. As mentioned in Chapter 1, biological databanks began to emerge in the early 1980s, to store information generated by applications of the then new sequencing techniques. The first of these collected and annotated nucleotide sequences, namely EMBL and GenBank in 1982. These were followed in 1986 by the DNA Data Bank of Japan (DDBJ) and their collaborative repository, the International Nucleotide Sequence Database (INSD). In the following sections, we will review these resources and, to give a flavor of the type of information they house, we will examine the format of a typical GenBank entry. We begin with the world's first nucleotide sequence repository, EMBL.

5.3.1 EMBL

The EMBL database, also known as EMBL-Bank (Kulikova *et al.* 2004), is Europe's primary collection of nucleotide sequences. Its first public release was in June 1982, with 568 entries. The database is maintained at the European Bioinformatics Institute (EBI) in Hinxton, the UK outpost of the European Molecular Biology Laboratories (EMBL), whose headquarters are in Heidelberg, Germany (note that future references to "EMBL" will signify the database, while "the EMBL" will denote its parent organization). To populate the database, data are collected from genome sequencing centers, individual scientists, the European Patent Office (EPO), and via exchange from partners of the INSD (see Section 5.3.7).

Initially, EMBL used sequences published in the scientific literature as its principal source of information. Today, however, electronic submissions via the Web are the more usual practice. Web submission is important, as the protocols for data entry help standardization and error minimization. The vast majority of data are transferred directly from major sequencing centers, and the database is consequently growing at a staggering rate – in February 2004, it contained 30,351,263 entries (comprising 36,042,464,651 nucleotides), representing >150,000 organisms, but with model organisms dominating.

Owing to its enormous size and to ease its management, EMBL is split into divisions, most of which are taxonomic (e.g., prokaryotes, fungi, plants, mammals); others are based on the types of data being held, such as expressed sequence tags (ESTs), sequence tagged sites (STSs), genome survey sequences (GSSs), high-throughput genomic (HTG) data, and unfinished high-throughput cDNA (HTC) sequences – see Table 5.1.

Table 5.1 The divisions of GenBank.

Division	Sequence subset
PRI	Primate
ROD	Rodent
MAM	Other mammalian
VRT	Other vertebrate
INV	Invertebrate
PLN	Plant, fungal, algal
BCT	Bacterial
VRL	Viral
PHG	Bacteriophage
SYN	Synthetic
UNA	Unannotated
EST	EST (expressed sequence tag)
PAT	Patent
STS	STS (sequence tagged site)
GSS	GSS (genome survey sequence)
HTG	HTG (high-throughput genomic sequence)
HTC	HTC (high-throughput cDNA)

5.3.2 The structure of EMBL entries

Entries in the database are structured so as to be both human and computer readable. Free-text descriptions are held within information fields that are structured systematically for ease of access by computer programs – query software need not then search the full file, but can be directed to those fields that are specific to the nature of the query. Data included in an entry are stored on different lines, each beginning with a two-character code indicating the type of information contained in the line. There are more than 20 different line types, storing for example, the entry name, taxonomic division, and sequence length (the ID line); a unique accession number (AC line), the primary and only stable means of identifying sequences from release to release; a description (DE) of the sequence that includes designations of genes for which it codes, the region of the genome from which it is derived, or other information that helps to identify the sequence; literature references (RN, RP, etc.); database cross-references (DR) that link to related information in other resources; free-text comments (CC); keywords (KW) that highlight functional, structural, or other defining characteristics that can be used to generate cross-references; and the Feature Table (FT), which houses sequence annotations, including signals or other characteristics reported in the literature, locations of protein coding sequences (CDS), ambiguities or features noted during data preparation, and so on.

Translations of protein-coding regions included as CDS features are automatically added to TrEMBL, from which curators then create annotated Swiss-Prot entries. EMBL is thus cross-referenced to both TrEMBL and Swiss-Prot, both of which share EMBL's general format (more detailed examples of these databases are described in Section 5.4).

Information can be retrieved from EMBL using SRS (Etzold, Ulyanov, and Argos 1996); this links the principal DNA and protein sequence databases with a variety of specialist resources (which house motif, structure, mapping, and other information), and also provides links to the biomedical literature. The system allows searches of a number of different fields, including sequence annotations, keywords, and author names. In addition to text-based interrogation, EMBL may also be searched with query sequences via the EBI's Web interfaces to BLAST, FASTA, and other rapid search programs.

5.3.3 GenBank

GenBank (Benson *et al.* 2004) is the genetic sequence database maintained at the National Center for Biotechnology Information (NCBI), Bethesda, USA. Its first public release was in December 1982, with 606 entries. Sequence information is incorporated from: direct author submissions; large-scale sequencing projects; the Genome Sequence Data Base, Santa Fe, USA; the United States Patent and Trademark Office (USTPO) and other international patent offices; and via exchange from partners of the INSD.

The database is growing at a prodigious rate, largely through inclusion of EST and other high-throughput data from sequencing centers: for example, ESTs constituted ~63% of Release 139 in February 2004, which contained 30,968,418 sequence records (36,553,368,485 bases). In this release, *Homo sapiens* was the most highly represented species, the next most represented species, in terms of the number of bases, being *Mus musculus*, *Rattus norvegicus*, *Danio rerio*, and *Zea mays*. Owing to its size, and the diversity of data sources available, GenBank files are split into sections that roughly correspond to taxonomic groups, but divisions for ESTs, GSSs, HTGs, and so on are also included, as summarized in Table 5.1. Although perhaps somewhat artificial groupings, separation of the information into discrete divisions can be useful for various reasons: e.g., it facilitates fast, specific searches by restricting queries to particular database subsets; and it allows searches to be directed to the higher quality annotated sequence sections, avoiding contamination of results with lower quality high-throughput data.

Information can be retrieved from GenBank using NCBI's Entrez retrieval system, which combines data from DNA and protein sequence databases with, for example, genome mapping, phylogenetic, gene expression, and protein structure information. It also gives access to MEDLINE, whose citations and abstracts are the primary component of the National Library of Medicine's PubMed database. This integrated system, with its direct links to the literature and additional sequence sources, adds enormously to GenBank's richness as a biological data repository. In addition to text-based interrogation, GenBank may also be searched with query sequences using NCBI's Web interface to the BLAST suite of programs.

A GenBank release includes the sequence files, indices created on different database fields (e.g., author, reference), and information derived from the database (e.g., GenPept, a repository of translated CDSs in FASTA format). For convenience, the data were once released on CD-ROM, but as the release grew, the number of CDs required to contain it became unwieldy. Thus, today, GenBank is available solely via FTP.

5.3.4 The structure of GenBank entries

Of the files distributed in each release, the most commonly used is probably the sequence file, containing

```
LOCUS       HUMPRPOA                 2420 bp    mRNA    linear   PRI 13-JUL-1994
DEFINITION  Human prion protein 27-30 mRNA, complete cds.
ACCESSION   M13667
VERSION     M13667.1  GI:190469
KEYWORDS    amyloid; prion protein; sialoglycoprotein.
SOURCE      Human, cDNA to mRNA, clones lambda [3,6,7].
  ORGANISM  Homo sapiens Eukaryota; Metazoa; Chordata; Craniata; Vertebrata; Euteleostomi;
            Mammalia; Eutheria; Primates; Catarrhini; Hominidae; Homo.
REFERENCE   1  (bases 1 to 2420)
  AUTHORS   Liao,Y.C., Lebo,R.V., Clawson,G.A. and Smuckler,E.A.
  TITLE     Human prion protein cDNA: molecular cloning, chromosomal mapping, and biological
implications
  JOURNAL   Science 233 (4761), 364-367 (1986)
  MEDLINE   86261778
  PUBMED    3014653
COMMENT     A single prion protein gene is found on chromosome 20 per haploid genome.
FEATURES             Location/Qualifiers
     source          1..2420
                     /organism="Homo sapiens"
                     /db_xref="taxon:9606"
     gene            1..2420
                     /gene="PRNP"
     mRNA            <1..2420
                     /gene="PRNP"
                     /product="PrP mRNA"
     CDS             77..814
                     /gene="PRNP"
                     /note="prion protein"
                     /codon_start=1
                     /protein_id="AAA19664.1"
                     /db_xref="GI:190470"
                     /translation="MLVLFVATWSDLGLCKKRPKPGGWNTGGSRYPGQGSPGGNRYPP
                     QGGGWGQPHGGGWGQPHGGGWGQPHGGGWGQPHGGGWGQGGGTHSQWNKPSKPKTNM
                     KHMAGAAAGAVVGGLGGYMLGSAMSRPIIHFGSDYEDRYYRENMHRYPNQVYYRPMDE
                     YSNQNNFVHDCVNITIKQHTVTTTTKGENFTETDVKMMERVVEQMCITQYERESQAYY
                     QRGSSMVLFSSPPVILLISFLIFLIVG"
BASE COUNT       669 a      500 c      583 g      668 t
ORIGIN          171 bp upstream of SmaI site; chromosome 20.
        1 cgagcagcca aggttcgcca taatgactgc tctcggtcgt gaggagagga gaagctcgcg
       61 gcgccgccggc tgctggatgc tggttctctt tgtggccaca tggagtgacc tgggcctctg
      121 caagaagcgc ccgaagcctg gaggatggaa cactgggggc agccgatacc cggggcaggg
      . . . . .
     2341 tgcatgttct tgttttgtta tataaaaaaa ttgtaaatgt ttaatatctg actgaaatta
     2401 aacgagccaa gatgagcacc
//
```

Fig. 5.1 Example GenBank entry for the human prion protein, illustrating the use of keywords, sub-keywords, and the Feature Table. For convenience, the nucleotide sequence has been abbreviated (. . .).

the sequence itself and associated annotation. Many Web systems link to this file, so it is instructive to examine its structure in some detail.

Each entry consists of a number of keywords, associated sub-keywords, and an optional Feature Table. In Fig. 5.1, the keywords are LOCUS, DEFINITION, ACCESSION, VERSION, KEYWORDS, SOURCE, REFERENCE, COMMENT, FEATURES, BASE COUNT, and ORIGIN. The LOCUS keyword introduces a short label for the entry (here, HUMPRPOA); this line summarizes other relevant facts, including the number of bases, source of sequence data (mRNA), section of database (PRI), and date of submission. The DEFINITION line contains a concise description of the sequence (in this example, human prion protein).

Following this, the ACCESSION line gives the accession number, a unique, constant code assigned to each entry (here M13667), and the associated VERSION line indicates the version of any sequence revisions. This line also includes a nucleotide identifier (GI:190469), intended to provide a unique reference to the current version of the sequence information; this allows the sequence to be revised while still being associated with the same locus name and accession number.

The KEYWORDS line introduces a list of short phrases, assigned by the author, describing gene products and other relevant information about the entry (in this example, amyloid; prion protein; sialoglycoprotein). The SOURCE record provides information on the source from which the data have been derived; the sub-keyword ORGANISM illustrates the biological classification of the source organism (*Homo sapiens*, Eukaryota, etc., as shown in the figure). REFERENCE records indicate the portion of sequence data to which the cited literature refers; sub-keywords, AUTHORS, TITLE, and JOURNAL, provide a structure for the citation; the MEDLINE sub-keyword is a pointer to an online medical literature information resource, which allows the abstract of a given article to be viewed.

The COMMENT field is a free-text section that allows any additional annotations to be added to the entry (here, the copy number of the gene and its chromosomal location are described). The FEATURES keyword marks the Feature Table, whose purpose is to describe the properties of the sequence in detail, such as the gene name (PRNP), coordinates of its coding sequence (77..814), and so on. Within the Table, database cross-references are made through the "/db_xref" qualifier (here, we see links to a taxonomic database (taxon:9606) and to its sequence translation in GenBank's peptide database (GI:190470)). This example is not exhaustive, but indicates the type of information that can be represented in the Feature Table.

The entry continues with the BASE COUNT record, which details the frequency of the different base types in the sequence (here, 669A, 500C, 583G, 668T). The ORIGIN line notes, where possible, the location of the first base of the sequence in the genome. The nucleotide sequence itself follows, and the entry terminates with //.

5.3.5 dbEST

Expressed sequence tags (ESTs) are the product of automated partial DNA sequencing on randomly selected complementary DNA (cDNA) clones. ESTs are an important research tool, as they can be used to discover new genes, to map genes to particular chromosomes, and to identify coding regions in genomic sequences. This fast approach to cDNA characterization has been particularly valuable for tagging human genes (at a fraction of the cost of complete genomic sequencing) and for providing new genetic markers.

Given their importance as a biological resource and research tool, a significant portion of GenBank is devoted to the storage of ESTs – its EST division is termed dbEST (Boguski, Lowe, and Tolstoshev 1993). In February 2004, dbEST contained more than 20 million sequences from more than 580 different organisms; the top five organisms represented in the database were *H. sapiens* (~5.5 million records), *M. musculus* (~4.0 million records), *R. norvegicus* (580,000 records), *Triticum aestivum* (500,000 records), and *Ciona intestinalis* (490,000 records). The 20 most abundant organisms, in terms

Table 5.2 Top 20 organisms in dbEST in January 2004.

Organism	No. of ESTs
Homo sapiens (human)	5,469,433
Mus musculus + domesticus (mouse)	4,030,839
Rattus sp. (rat)	558,402
Triticum aestivum (wheat)	549,915
Ciona intestinalis	492,511
Gallus gallus (chicken)	451,655
Danio rerio (zebrafish)	405,962
Zea mays (maize)	391,145
Xenopus laevis (African clawed frog)	357,038
Hordeum vulgare + subsp. *vulgare* (barley)	348,282
Glycine max (soybean)	344,524
Bos taurus (cattle)	331,139
Silurana tropicalis	297,086
Drosophila melanogaster (fruit fly)	267,332
Oryza sativa (rice)	266,949
Saccharum officinarum	246,301
Sus scrofa (pig)	240,001
Caenorhabditis elegans (nematode)	215,200
Arabidopsis thaliana (thale cress)	196,904
Medicago truncatula (barrel medic)	187,763

of their contributions to dbEST in January 2004, are listed in Table 5.2.

One of the drawbacks of the increasing levels of automation demanded by high-throughput sequencing approaches is the tendency to generate errors. In fact, errors are a serious problem in **all** databases, not least because once they find their way into a resource, they tend to propagate – and the more popular the database, the more widespread the propagation!

A notable example of the incorporation of errors into dbEST occurred between August 1996 and February 1997, when it was discovered that up to 1.5% of 892,075 ESTs submitted by Washington University Genome Sequencing Center were mislabeled as to species origin. More than 100 96-well plates were found in which cDNAs from both mouse and human were present or that contained cDNAs incorrectly labeled as human or mouse. As a result, ESTs generated from these plates were tagged with a warning, and a further 657 apparently incorrectly labeled reads from eight plates were removed completely. This was an important reminder of the pitfalls of automation, and sequence validation software was subsequently implemented to minimize the possibility of future errors of this type.

5.3.6 DDBJ

The DNA Data Bank of Japan (Miyazaki *et al.* 2004) began in 1986 at the National Institute of Genetics (NIG) in Mishima. DDBJ functions as part of the INSD and receives its principal input from a variety of sequencing centers, but especially from the international human genome sequencing consortium. The database is thus growing rapidly – the number of entries processed in 1999 alone exceeded the total number processed in the preceding 10 years, and the database doubled in size between July 2000 and July 2001! To cope with the deluge, the curators use in-house software for mass sequence submission and data processing, which helps both to improve consistency and to reduce the proliferation of errors. The Web is also used to provide standard search tools, such as FASTA and BLAST.

In February 2004, DDBJ held 30,405,173 entries (36,079,046,032 bases). To better manage its increasing size, the database has been organized into species-oriented divisions, which facilitate more efficient, species-specific information retrieval. In addition, the database includes an independent division for the patent data collected and processed by the Japanese Patent Office, the USPTO, and the EPO.

Finally, just as translations of CDSs in EMBL are automatically added to TrEMBL, from which curators then create Swiss-Prot entries, so translations of DDBJ are automatically added to the Japan International Protein Information Database (JIPID), which in turn feeds the International PIR-PSD (see Section 5.4).

5.3.7 The INSD

Sequence data are being generated on such a massive scale that it is impossible for individual groups to collate them. As a consequence, in February 1986, EMBL and GenBank, together with DDBJ in 1987, joined forces to streamline and standardize the pro-

Fig. 5.2 The tripartite International Nucleotide Sequence Database (INSD), comprising EMBL (Europe), GenBank (USA), and DDBJ (Japan).

cesses of data collection and annotation. Each now collects a portion of the total sequence data reported world-wide and, to achieve optimal synchronization, new and updated entries are exchanged between them daily via the Internet.

The result of this tripartite collaboration is the International Nucleotide Sequence Database (INSD) (Fig. 5.2). This ensures that the participating databanks share virtually the same quantity and quality of data, and means that users need only submit to one of the resources to have their sequence reflected in all of the others. The databases, which primarily collect data via direct submission from individual laboratories and large-scale sequencing projects, now incorporate DNA sequences from over 150,000 different organisms, and new species are being added at a rate of more than 1400 per month.

As we have seen, within each of the databases, to add value to the raw data, characteristics of the sequences are stored in Feature Tables. To improve data consistency and reliability, and to facilitate interoperation and data exchange, a major goal of the INSD was therefore to devise a common Feature Table format and common standards for annotation practice. The types of feature documented in the common format include regions that: perform particular functions; affect the expression of function or the replication of a sequence; interact with other molecules; have secondary or tertiary structure; and so on. Regulating the content, vocabulary, and syntax of such feature descriptions ensures that the data are released in a form that can be exchanged efficiently and that is readily accessible to analysis software.

5.4 PROTEIN SEQUENCE DATABASES

5.4.1 History

As we have seen from the above discussions, as soon as sequences began to emerge from applications of the earliest sequencing methods, scientists began to collect and analyze them. In fact, the first published collection of sequences was Margaret Dayhoff's 1965 *Atlas of Protein Sequence and Structure* (Dayhoff 1965). By 1981, the *Atlas* listed 1660 proteins, but this was a "database" in paper form only, and scientists wishing to use the information had to type the data into computers by hand. Burdensome though this task was, two important databanks evolved from this approach.

In 1984, Dayhoff's *Atlas* was released electronically, changing its name to the Protein Sequence Database (PSD) of the Protein Identification Resource (PIR). The first release contained 859 entries; by June 2002, the collection had grown more than 300-fold to almost 300,000 entries. To cope with this vast expansion and ease its maintenance, a collaboration was formed between PIR, the Munich Information Center for Protein Sequences (MIPS), and JIPID. The resulting PIR-International PSD is now one of the most comprehensive compendia of sequences publicly available, drawing its data from many sources, including sequencing centers, the literature, and directly from authors.

During this period, Amos Bairoch had begun to establish an archive of protein sequences, initially based on Dayhoff's *Atlas* and later on its electronic version. EMBL had become available in 1982, and Bairoch obtained a version in 1983 containing 811 sequences. His innovation was to couple the structured format of EMBL entries (Section 5.3.2) with sequences from the PIR-PSD. This painstaking endeavor eventually gave rise to the first public release of Swiss-Prot in 1986, which contained over 4,000 sequences; by February 2004, a 36-fold increase in size saw that number rise to over 143,790.

In the following sections, we will briefly review some of these resources and, because it is based closely on the structure of EMBL, we will take a closer look at the format of a typical Swiss-Prot entry.

5.4.2 PIR

The Protein Sequence Database was developed at the National Biomedical Research Foundation (NBRF) in the 1960s, as mentioned above, essentially as a byproduct of Dayhoff's studies on protein evolution. Since 1988, the PSD has been maintained collaboratively by PIR-International, an association of macromolecular sequence data collection centers that includes PIR, JIPID, and MIPS.

More recently, PIR-PSD (Wu *et al.* 2003) has been incorporated into an integrated knowledge base of databases and analytical tools: this includes PIR-NREF, a comprehensive database for sequence searching and protein identification – by February 2004, PIR-NREF contained 1,485,025 non-redundant sequences from PIR-PSD, Swiss-Prot, TrEMBL, RefSeq, GenPept, and PDB. An interesting feature of PIR-PSD is its emphasis on protein family classification, a reflection of its origins in Dayhoff's evolutionary work. The database is thus organized hierarchically around this family concept, its constituent proteins being automatically clustered based on sequence similarity. Within superfamilies, sequences have similar overall length and share a common architecture (i.e., contain the same number, order, and types of domain). The automated classification system, which places new members into existing superfamilies, and defines new clusters using parameters such as sequence identity, overlap length, and domain arrangement, is augmented by manual annotation – this provides superfamily names, brief descriptions, bibliographies, lists of representative and seed members, and domain and motif architectures characteristic of the superfamily.

Linking sequences to experimentally verified data in the literature helps both to improve the quality of annotation provided by the database and to avoid propagation of errors that may have resulted from large-scale genome-sequencing projects.

Accordingly, a bibliography submission system has been developed to help the scientific community to submit, categorize, and retrieve literature for PSD entries. This, and many other tools for database interrogation, data mining and sequence analysis, is provided via the PIR Web site.

5.4.3 MIPS

The Munich Information Center for Protein Sequences (MIPS) supports national and European sequencing and functional analysis projects, develops and maintains genome-specific databases (e.g., for *Arabidopsis thaliana*, *Neurospora crassa*, yeast), develops classification schemes for protein sequence annotation, and provides tools for sequence analysis (Mewes *et al.* 2002). One of its central activities is to collect and process sequence data for the tripartite PIR-International PSD.

As with PIR, a feature of the database is its automatic clustering of proteins into families and superfamilies based on sequence similarity. For family classification, an arbitrary cut-off of 50% identity is used – this means that resulting "families" need not necessarily have strict biological significance. Regions of local similarity within otherwise unrelated proteins are annotated as domains. Clusters (whether at family, superfamily, or domain level) are annotated (e.g., with protein names, EC numbers, or keywords), and are accompanied by their alignments, which are also annotated (e.g., with domains, sequence motifs, active sites, post-translational modifications (PTMs)). All of these resources, and a variety of analysis tools, are accessible via the MIPS Web server.

5.4.4 Swiss-Prot

From its inception in 1986, Swiss-Prot was produced collaboratively by the Department of Medical Biochemistry at the University of Geneva and the EMBL. After 1994, the database transferred to the EMBL's UK outstation, the EBI, and April 1998 saw further change with a move to the Swiss Institute of Bioinformatics (SIB). The database is thus now maintained collaboratively by the SIB and the EBI (Boeckmann *et al.* 2003).

Sequence information is housed in Swiss-Prot in a structured manner, adhering closely to the EMBL format. The most important features of the resource lie in the minimal level of redundancy it contains, the level of cross-referencing to other databases, and the quality of its annotations. In addition to the sequence itself, the annotations include bibliographic references and taxonomic data, and, wherever possible, the function of the protein, PTMs, domain structure, functional sites, associated diseases, similarities to other proteins, and so on. Much of this information is gleaned from the primary literature and review articles, but use is also made of a panel of experts who have particular knowledge of specific families.

A great deal of useful information about the characteristics of the sequence is held in the Feature Table. This includes details of any known or predicted transmembrane domains, glycosylation or phosphorylation sites, metal or ligand-binding sites, enzyme active sites, internal repeats, DNA-binding or protein–protein interaction domains, and so on. Specially written browsers allow this information to be viewed graphically, yielding a bird's-eye view of the most important sequence features.

Swiss-Prot offers added value by providing links to over 30 different databases, including those that store nucleic acid or protein sequences, protein families or structures, and specialized data collections. As illustrated in Fig. 5.3, it is therefore possible, say, to retrieve the nucleic acid sequence (EMBL) that codes for a protein, information on associated genetic diseases (OMIM), information specific to the protein family to which it belongs (PROSITE, PRINTS, InterPro), and its 3D structure (PDB). Swiss-Prot thus provides a hub for database interconnectivity. The extent of its annotations sets Swiss-Prot apart from other protein sequence resources and has made it the database of choice for most research purposes. By February 2004, the database contained 143,790 entries.

5.4.5 The structure of Swiss-Prot entries

An example Swiss-Prot entry is shown in Fig. 5.3. As with EMBL, each line is flagged with a two-letter

```
ID    PRIO_HUMAN     STANDARD;      PRT;    253 AA.
AC    P04156;
DT    01-NOV-1986 (Rel. 03, Created)
DT    01-NOV-1986 (Rel. 03, Last sequence update)
DT    15-JUN-2002 (Rel. 41, Last annotation update)
DE    Major prion protein precursor (PrP) (PrP27-30) (PrP33-35C) (ASCR) (CD230 antigen).
GN    PRNP.
OS    Homo sapiens (Human).
OC    Eukaryota; Metazoa; Chordata; Craniata; Vertebrata; Euteleostomi;
OC    Mammalia; Eutheria; Primates; Catarrhini; Hominidae; Homo.
OX    NCBI_TaxID=9606;
RN    [1]
RP    SEQUENCE FROM N.A.
RX    MEDLINE=86300093; PubMed=3755672; [NCBI, ExPASy, EBI, Israel, Japan]
RA    Kretzschmar H.A., Stowring L.E., Westaway D., Stubblebine W.H.,
RA    Prusiner S.B., Dearmond S.J.;
RT    "Molecular cloning of a human prion protein cDNA.";
RL    DNA 5:315-324(1986).
...
CC    -!- FUNCTION: THE FUNCTION OF PRP IS NOT KNOWN. PRP IS ENCODED IN THE HOST GENOME AND IS
CC        EXPRESSED BOTH IN NORMAL AND INFECTED CELLS.
CC    -!- SUBUNIT: PRP HAS A TENDENCY TO AGGREGATE YIELDING POLYMERS CALLED "RODS".
CC    -!- SUBCELLULAR LOCATION: Attached to the membrane by a GPI-anchor.
CC    -!- POLYMORPHISM: THE FIVE TANDEM OCTAPEPTIDE REPEATS REGION IS HIGHLY UNSTABLE. INSERTIONS CC
CC        OR DELETIONS OF OCTAPEPTIDE REPEAT UNITS ARE ASSOCIATED WITH PRION DISEASE.
CC    -!- DISEASE: PRP IS FOUND IN HIGH QUANTITY IN THE BRAIN OF HUMANS AND ANIMALS INFECTED WITH CC
CC        NEURODEGENERATIVE DISEASES KNOWN AS TRANSMISSIBLE SPONGIFORM ENCEPHALOPATHIES OR PRION
CC        DISEASES, LIKE: CREUTZFELDT-JAKOB DISEASE (CJD), GERSTMANN-STRAUSSLER SYNDROME (GSS),
CC        FATAL FAMILIAL INSOMNIA (FFI) AND KURU IN HUMANS; SCRAPIE IN SHEEP AND GOAT; BOVINE
CC        SPONGIFORM ENCEPHALOPATHY (BSE) IN CATTLE; TRANSMISSIBLE MINK ENCEPHALOPATHY (TME);
CC        CHRONIC WASTING DISEASE (CWD) OF MULE DEER AND ELK; FELINE SPONGIFORM ENCEPHALOPATHY
CC        (FSE) IN CATS AND EXOTIC UNGULATE ENCEPHALOPATHY (EUE) IN NYALA AND GREATER KUDU. THE
CC        PRION DISEASES ILLUSTRATE THREE MANIFESTATIONS OF CNS DEGENERATION: (1) INFECTIOUS (2)
CC        SPORADIC AND (3) DOMINANTLY INHERITED FORMS. TME, CWD, BSE, FSE, EUE ARE ALL THOUGHT TO CC
CC        OCCUR AFTER CONSUMPTION OF PRION-INFECTED FOODSTUFFS.
...
DR    EMBL; M13667; AAA19664.1; -. [EMBL / GenBank / DDBJ] [CoDingSequence]
DR    PIR; S14078; S14078.
DR    PDB; 1QM3; 16-DEC-99. [ExPASy / RCSB]
DR    Ensembl; P04156.
DR    MIM; 606688; -. [NCBI / EBI]
DR    InterPro; IPR000817; Prion.
DR    Pfam; PF00377; prion; 1.
DR    PRINTS; PR00341; PRION.
DR    PROSITE; PS00291; PRION_1; 1.
DR    PROSITE; PS00706; PRION_2; 1.
DR    ProDom [Domain structure / List of seq. sharing at least 1 domain]
DR    BLOCKS; P04156.
...
KW    Prion; Brain; Glycoprotein; GPI-anchor; Repeat; 3D-structure; Polymorphism; Disease mutation.
FT    SIGNAL         1   22
FT    CHAIN         23  230 MAJOR PRION PROTEIN.
FT    PROPEP       231  253 REMOVED IN MATURE FORM (BY SIMILARITY).
FT    LIPID        230  230 GPI-ANCHOR (BY SIMILARITY).
FT    CARBOHYD     181  181 N-LINKED (GLCNAC...) (PROBABLE).
FT    CARBOHYD     197  197 N-LINKED (GLCNAC...) (PROBABLE).
FT    DISULFID     179  214
FT    DOMAIN        51   91 5 X 8 AA TANDEM REPEATS OF P-H-G-G-G-W-G-Q.
FT    REPEAT        51   59 1.
FT    REPEAT        60   67 2.
FT    REPEAT        68   75 3.
FT    REPEAT        76   83 4.
FT    REPEAT        84   91 5.
FT    VARIANT      102  102 P -> L (IN GSS AND EOAD).
FT                         /FTId=VAR_006464.
FT    VARIANT      238  238 P -> S.
FT                         /FTId=VAR_008754.
FT    CONFLICT     118  118 MISSING (IN REF. 3).
SQ    SEQUENCE     253 AA;  27661 MW;  43DB596BAAA66484 CRC64;
      MANLGCWMLV LFVATWSDLG LCKKRPKPGG WNTGGSRYPG QGSPGGNRYP PQGGGGWGQP
      HGGGWGQPHG GGWGQPHGGG WGQPHGGGWG QGGGTHSQWN KPSKPKTNMK HMAGAAAAGA
      VVGGLGGYML GSAMSRPIIH FGSDYEDRYY RENMHRYPNQ VYYRPMDEYS NQNNFVHDCV
      NITIKQHTVT TTTKGENFTE TDVKMMERVV EQMCITQYER ESQAYYQRGS SMVLFSSPPV
      ILLISFLIFL IVG
//
```

Fig. 5.3 Excerpt from the Swiss-Prot entry for the human prion protein, illustrating the EMBL-like structured format, with extensive annotations and database cross-references (note: dotted lines denote points at which, for convenience, material has been excised). Compare the GenBank entry for human prion protein illustrated in Fig. 5.1. The characteristic tandem octapeptide-repeat region thought to be associated with various prion diseases has been highlighted in bold.

code. The first is an identification (ID) line, and the last a // terminator. The ID codes, which attempt to be informative and people-friendly, take the form PROTEIN_ SOURCE, where PROTEIN is an acronym that denotes the type of protein, and SOURCE encodes the organism name. The protein in this example, PRIO_HUMAN, is clearly of human origin and, with the eye of experience, we can deduce that it is a prion protein.

Unfortunately, ID codes are not fixed, so an additional identifier, an accession number, is also provided, which should remain static between database releases. The accession number is provided on the AC line, here P04156, which, although meaningless to most normal humans, is nevertheless computer readable (some database curators also have a disturbing capacity to be able to read accession numbers!). If several numbers appear on the same AC line, the first, so-called primary accession number, is the most current.

Next, the DT lines provide information about the date of entry of the sequence to the database, and details of when it was last modified. The description (DE) line, or lines, then informs us of the name, or names, by which the protein is known – here Major prion protein precursor (PrP), PrP27-30, PrP33-35C, ASCR and CD230 antigen. The following lines give the gene name (GN), the organism species (OS), and organism classification (OC) within the biological kingdoms.

The next part of the file provides a list of references: these can be from the literature, unpublished information submitted directly from sequencing projects, data from structural or mutagenesis studies, and so on. The database thus houses information that is difficult, or impossible, to find elsewhere. For human prion protein, 30 references are provided, but for convenience, only one of these is shown in Fig. 5.3.

Following the references are found comment (CC) lines. These are divided into themes describing the FUNCTION of the protein, its SUBUNIT structure, SUBCELLULAR LOCATION, DISEASE associations, and so on. Where such information is available, the CC lines also indicate any known SIMILARITY or relationship with particular protein families. In this example, we see that the function of the prion protein is unknown, it tends to polymerize into rod-shaped fibrils, and its sequence is characterized by tandem octapeptide repeats that are believed to be associated with various degenerative diseases.

Database cross-reference (DR) lines follow the comment field. These provide links to other resources, including for example, the primary DNA repositories, protein family databases, and specialist databases. For human prion protein, we find links to nucleotide and protein sequence databases (EMBL, PIR), to the protein structure databank (PDB), to the online Mendelian Inheritance in Man (MIM) mutation database, and to the InterPro, PRINTS, Pfam, PROSITE, etc., family resources.

Directly after the DR lines, keywords (KW) are listed, followed by the Feature Table (FT). The latter highlights regions of interest in the sequence, including signal sequences, lipid attachment sites, disulfide bridges, repeats, known sequence variants, and so on. Each line includes a key (e.g., REPEAT), the location in the sequence of the feature (e.g., 51–59), and a comment, which might, for example, indicate the level of confidence of a particular annotation. For the prion example, the lipid attachment site is flagged BY SIMILARITY, indicating that the assignment has been made *in silico* rather than *in vitro* and must therefore be viewed as an unverified prediction.

The final section of the file includes the sequence itself (SQ), encoded for the sake of efficiency using the single-letter amino acid code. The stored sequence data correspond to the precursor form of the protein, before post-translational processing – thus information concerning the size or molecular weight will not necessarily correspond to values for the mature protein. The extent of mature proteins or peptides may be deduced from the Feature Table, which indicates regions of a sequence that correspond to the signal (SIGNAL), transit (TRANSIT), or pro-peptide (PROPEP). The keys CHAIN and PEPTIDE are used to denote the location of the mature form.

The above is not an exhaustive example, but should provide a flavor of the richness of Swiss-Prot annotations. For a more comprehensive guide to the type of information typically included in the database, and for details of the differences between the Swiss-Prot and EMBL formats, readers are referred to Appendix C of the Swiss-Prot user manual available on the ExPASy Web server.

5.4.6 TrEMBL

Owing to the extent of its annotations, Swiss-Prot has become one of the most popular and widely used databases. However, the price paid for its manual approach is a relatively slow rate of growth (it is less than half the size of PIR-PSD). Therefore, to increase its sequence coverage, a computer-generated supplement was introduced in 1996, TrEMBL, based on translations of CDSs in EMBL (Boeckmann *et al.* 2003). In February 2004, TrEMBL contained 1,075,779 entries.

TrEMBL has two main sections: (i) SP-TrEMBL (Swiss-Prot-TrEMBL), which contains entries that will eventually be incorporated into Swiss-Prot, but have not yet been manually annotated (these are given Swiss-Prot accession numbers); and (ii) REM-TrEMBL (REMaining-TrEMBL), which contains sequences that are not destined to be included in Swiss-Prot – these include immunoglobulins and T-cell receptors, fragments of fewer than eight amino acids, synthetic sequences, patented sequences, and codon translations that do not encode real proteins.

SP-TrEMBL is partially redundant, as many of its entries are just additional reports of proteins already in Swiss-Prot. To reduce redundancy, and automatically add annotation, a rule-based system has been designed that uses existing Swiss-Prot entries as a template. This approach can only be applied to sequences with obvious relatives in Swiss-Prot (more sophisticated expert systems are required to increase the coverage by automatic annotation); nevertheless, it provides a basic level of information that offers a stepping-stone for further information retrieval.

TrEMBL shares the Swiss-Prot format described above, and was designed to address the need for a well-structured Swiss-Prot-like resource that would allow rapid access to sequence data from genome projects, without having to compromise the quality of Swiss-Prot itself by incorporating sequences with insufficient analysis and annotation. A typical TrEMBL entry is illustrated in Fig. 5.4 – comparison

```
ID   O46593      PRELIMINARY;   PRT;    257 AA.
AC   O46593;
DT   01-JUN-1998 (TrEMBLrel. 06, Created)
DT   01-JUN-1998 (TrEMBLrel. 06, Last sequence update)
DT   01-DEC-2001 (TrEMBLrel. 19, Last annotation update)
DE   Prion protein.
GN   PRP.
OS   Canis familiaris (Dog).
OC   Eukaryota; Metazoa; Chordata; Craniata; Vertebrata; Euteleostomi;
OC   Mammalia; Eutheria; Carnivora; Fissipedia; Canidae; Canis.
OX   NCBI_TaxID=9615;
RN   [1]
RP   SEQUENCE FROM N.A.
RA   Doyle D., Rogers M.S.;
RT   "Dog prion protein gene.";
RL   Submitted (JAN-1998) to the EMBL/GenBank/DDBJ databases.
DR   EMBL; AF042843; AAB99743.1; -. [EMBL / GenBank / DDBJ] [CoDingSequence]
DR   HSSP; P04925; 1AG2. [HSSP ENTRY / SWISS-3DIMAGE / PDB]
DR   InterPro; IPR000817; Prion.
DR   InterPro; Graphical view of domain structure.
DR   Pfam; PF00377; prion; 1.
DR   PRINTS; PR00341; PRION.
DR   PROSITE; PS00291; PRION_1; 1.
DR   PROSITE; PS00706; PRION_2; 1.
DR   ProDom [Domain structure / List of seq. sharing at least 1 domain]
...
SQ   SEQUENCE   257 AA;   27793 MW;   D4EA77764519676C CRC64;
     MVKSHIGGWI LLLFVATWSD VGLCKKRPKP GGWNTGGGSR YPGQGSPGGN RYPPQGGGGW
     GQPHGGGWGQ PHGGGWGQPH GGGWGQPHGG GGWGQGGGSH SQWGKPNKPK TNMKHVAGAA
     AAGAVVGGLG GYMLGSAMSR PLIHFGNDYE DRYYRENMYR YPEQVYYRPV DQYSNQNNFV
     RDCVNITVKQ HTVTTTTKGE NFTETDMKIM ERVVEQMCVT QYQKESEAYY QRGASAILFS
     PPPVILLISL LILLIVG
//
```

Fig. 5.4 Excerpt from the TrEMBL entry for the canine prion protein, illustrating the EMBL-like structured format, with automatically generated annotations and database cross-references – there are no annotation-rich CC or FT fields (note: dotted lines denote points at which several DR lines have been excised). Compare the Swiss-Prot entry for human prion protein illustrated in Fig. 5.3.

with the (truncated) Swiss-Prot entry shown in Fig. 5.3 highlights the difference in the extent of annotation provided by each of the resources. Note that there is no comment (CC) field and no Feature Table, the hallmarks of richly annotated Swiss-Prot entries. As shown, the bulk of the automatically added annotation is largely in the form of bibliographic and database cross-references.

5.4.7 PIR-NRL3D

NRL3D (Garavelli *et al.* 2001) is produced by PIR-International from sequences and annotations extracted from 3D structures deposited in the Protein Databank (PDB). The titles, biological sources, and keywords of its entries have been modified from the PDB format to conform to the nomenclature standards used in the other PIR databases. Bibliographic references and MEDLINE cross-references are included, together with secondary structure, active site, binding site, and modified site annotations and, whenever possible, details of experimental methods, resolution, R-factor, etc.

The format of a typical PIR-NRL3D entry is illustrated in Fig. 5.5. The example shows the sequence, together with relevant literature references and

```
ENTRY 2PRP
            #type complete
TITLE     prion protein 90 231 - golden hamster
ALTERNATE_NAMES sha rprp90-231
PDB_TITLE solution NMR structure of recombinant syrian hamster prion protein rprp(90-231),
            15 structures
ORGANISM
            #formal_name Mesocricetus auratus
            #common_name golden hamster
            #cross-references taxon:10036
            #note expressed in E.coli, strain 27c7, ATCC: 55244, genentech derived vector system
REFERENCE A68653
            #authors James, T.L.; Liu, H.; Ulyanov, N.B.; Farr-jones, S.
            #submission submitted to the Brookhaven Protein Data Bank, October 1997
            #cross-references PDB:2PRP
REFERENCE TN050926
            #authors James, T.L.; Liu, H.; Ulyanov, N.B.; Farr-jones, S.; Zhang, H.; Donne, D.G.; Kaneko, K.;
            Groth, D.; Mehlhorn, I.; Prusiner, S.B.; Cohen, F.E.
            #journal Proc. Natl. Acad. Sci. U.S.A. (1997) 94:10086
            #title Solution  structure  of  a  142-residue  recombinant  prion  protein  corresponding  to  the
            infectious fragment of the scrapie isoform.
DETERMINATION NMR
            #resolution not applicable
KEYWORDS brain; glycoprotein; prion; scrapie
FEATURES
55-67
            #region helix (right hand alpha)\
85-104
            #region helix (right hand alpha)\
111-137
            #region helix (right hand alpha)\
40-42, 72-74
            #region beta sheet\
90-125
            #disulfide_bonds
SUMMARY
            #length 142
            #molecular_weight 16243
SEQUENCE

                5         10        15        20        25        30
        1 G Q G G G T H N Q W N K P S K P K T N M K H M A G A A A A G
       31 A V V G G L G G Y M L G S A M S R P M M H F G N D W E D R Y
       61 Y R E N M N R Y P N Q V Y Y R P V D Q Y N N Q N N F V H D C
       91 V N I T I K Q H T V T T T T K G E N F T E T D I K I M E R V
      121 V E Q M C T T Q Y Q K E S Q A Y Y D G R R S
```

Fig. 5.5 PIR-NRL3D entry for the hamster prion protein, showing the sequence, together with relevant bibliographic references and structural annotations. Note that the sequence is truncated by comparison with those in the GenBank, Swiss-Prot, and TrEMBL entries illustrated in Figs. 5.1, 5.3, and 5.4.

structural annotations from PDB entry 2PRP. By comparison with some of the other sequence databases (e.g., as shown in Figs. 5.1, 5.3, and 5.4), the extent of annotation is limited; nevertheless, it provides a direct link between sequence, structure, experimental data, and the literature.

PIR-NRL3D is a valuable resource, as it makes data in the PDB available both for keyword interrogation and for similarity searches. Sequence information is extracted only for residues with resolved 3D coordinates (as represented in PDB "ATOM" records) and not for those that are structurally undefined. Accordingly, the sequence shown in Fig. 5.5 represents a C-terminal fragment of the prion protein, without the disordered N-terminal 90 residues, which contain the characteristic octapeptide repeat (cf. Figs. 5.1, 5.3, and 5.4).

5.4.8 UniProt

In an effort both to rationalize the currently available protein sequence data and to create the ultimate, comprehensive catalog of protein information, the curators of Swiss-Prot, TrEMBL, and PIR have recently pooled their efforts to build UniProt, the Universal Protein Resource (Apweiler *et al.* 2004). UniProt comprises three components, each of which has been tailored for specific uses: the **UniProt Knowledgebase (UniProt)** is the central point of access for annotated protein information (including function, classification, cross-references, and so on); the **UniProt Non-redundant Reference (UniRef)** databases combine closely related sequences into a single record to speed searches; and the **UniProt Archive (UniParc)** is a comprehensive repository that accurately reflects the history of all protein sequences. This unified endeavor, which brings together the major protein sequence data providers, is a significant achievement and should make sequence analysis, for all levels of user, far easier in the future. UniProt is accessible via the EBI's Web interface for text and similarity searches.

5.5 PROTEIN FAMILY DATABASES

5.5.1 The role of protein family databases

In addition to the numerous sequence databases, there are also many protein family databases (sometimes termed pattern or secondary databases) derived from the fruits of analyses of the sequences in the primary sources. Because there are several different sources, and a variety of ways of analyzing sequences, the information housed in each of the family databases is different – and their formats reflect these disparities. Although this appears to present a confusing picture, Swiss-Prot and TrEMBL have become the most popular data sources, and most family databases now use them as their basis. Some of the main resources are listed in Table 5.3.

From inspection of Table 5.3, it is evident that the type of information stored in each of the family databases is different. Nevertheless, these resources have arisen from a common principle: namely, that homologous sequences may be gathered together in multiple alignments, within which are conserved

Table 5.3 Some of the major protein family databases: in each case, the data source is noted, together with the type of data stored.

Family database	Data source	Stored information
PROSITE	Swiss-Prot	Regular expressions (patterns)
Profiles	Swiss-Prot	Weighted matrices (profiles)
PRINTS	Swiss-Prot and TrEMBL	Aligned motifs (fingerprints)
Pfam	Swiss-Prot and TrEMBL	Hidden Markov models (HMMs)
Blocks	InterPro/PRINTS	Aligned motifs (blocks)
eMOTIF	Blocks/PRINTS	Permissive regular expressions (patterns)

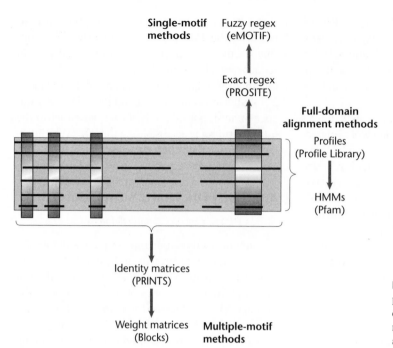

Single-motif methods

Fuzzy regex
(eMOTIF)

Exact regex
(PROSITE)

Full-domain alignment methods

Profiles
(Profile Library)

HMMs
(Pfam)

Identity matrices
(PRINTS)

Weight matrices
(Blocks)

Multiple-motif methods

Fig. 5.6 Illustration of the three principal methods for building family databases, based on the use of single motifs, multiple motifs, and full-domain alignments.

regions that show little or no variation (Fig. 5.6). Such regions, or motifs (Attwood 1997), usually reflect some vital biological role (i.e., are somehow crucial to the structure or function of the protein).

Alignments have been exploited in various ways to build diagnostic signatures, or discriminators, for different protein families, as illustrated in Fig. 5.6 (Attwood 2000a,b, Attwood and Parry-Smith 1999). An unknown query sequence may then be searched against a database of such signatures to determine whether it contains any of the predefined characteristics, and hence whether it can be assigned to a particular family. As these resources are derived from multiple sequence information, searches of them are often better able to identify distant relationships than are searches of sequence databases; and, if the structure and function of the family are known, they theoretically offer a fast track to the inference of biological function. However, none of these databases is complete; they should therefore be used to augment rather than to replace sequence database searches. Some of the major family databases in common use are outlined in the following sections.

5.5.2 PROSITE

The first protein family database to have been developed was PROSITE, in 1989, with 202 entries. It is maintained collaboratively at the University of Geneva and the Swiss Institute for Experimental Cancer Research (ISREC) (Hulo *et al.* 2004). In February 2004, it documented 1245 families, domains, and functional sites. The rationale behind its development was that a protein family could be characterized by the single most conserved motif within a multiple alignment of its member sequences, as this would likely encode a key biological feature (e.g., an enzyme active site, ligand- or metal-binding site). Searching a database of such features should therefore help to determine to which family a new sequence might belong, or which domain/s or functional site/s it might contain.

Within PROSITE, motifs are encoded as regular expressions (or regexs), often referred to as patterns. Sometimes, a complete protein family cannot be characterized by a single motif, usually because its members are too divergent. In these cases, additional regexs are designed to encode other well-

```
ID    PRION_1; PATTERN.
AC    PS00291;
DT    APR-1990 (CREATED); DEC-1992 (DATA UPDATE); JUL-1998 (INFO UPDATE).
DE    Prion protein signature 1.
PA    A-G-A-A-A-A-G-A-V-V-G-G-L-G-G-Y.
NR    /RELEASE=40.7,103373;
NR    /TOTAL=44(44); /POSITIVE=44(44); /UNKNOWN=0(0); /FALSE_POS=0(0);
NR    /FALSE_NEG=0; /PARTIAL=0;
CC    /TAXO-RANGE=??E??; /MAX-REPEAT=1;
DR    P40245, PRIO_AOTTR, T; P40246, PRIO_ATEGE, T; P51446, PRIO_ATEPA, T;
DR    P10279, PRIO_BOVIN, T; P40247, PRIO_CALJA, T; P40248, PRIO_CALMO, T;
DR    P79141, PRIO_CAMDR, T; O46501, PRIO_CANFA, T; P52113, PRIO_CAPHI, T;
DR    P40249, PRIO_CEBAP, T; P40250, PRIO_CERAE, T; Q95145, PRIO_CERAT, T;
DR    P79142, PRIO_CEREL, T; Q95172, PRIO_CERMO, T; Q95174, PRIO_CERPA, T;
DR    Q95176, PRIO_CERTO, T; P27177, PRIO_CHICK, T; P40251, PRIO_COLGU, T;
DR    Q60506, PRIO_CRIGR, T; Q60468, PRIO_CRIMI, T; O18754, PRIO_FELCA, T;
DR    P40252, PRIO_GORGO, T; P04156, PRIO_HUMAN, T; P40254, PRIO_MACFA, T;
DR    P40255, PRIO_MANSP, T; P04273, PRIO_MESAU, T; P04925, PRIO_MOUSE, T;
DR    P52114, PRIO_MUSPF, T; P40244, PRIO_MUSVI, T; P47852, PRIO_ODOHE, T;
DR    P40253, PRIO_PANTR, T; P49927, PRIO_PIG  , T; P40256, PRIO_PONPY, T;
DR    P40257, PRIO_PREFR, T; Q95211, PRIO_RABIT, T; P13852, PRIO_RAT  , T;
DR    P40258, PRIO_SAISC, T; P23907, PRIO_SHEEP, T; Q9Z0T3, PRIO_SIGHI, T;
DR    Q95270, PRIO_THEGE, T; P51780, PRIO_TRIVU, T; P40242, PRP1_TRAST, T;
DR    Q01880, PRP2_BOVIN, T; P40243, PRP2_TRAST, T;
3D    1DX0; 1DX1; 1QLX; 1QLZ; 1QM0; 1QM1; 1B10;
DO    PDOC00263;
//
```

Fig. 5.7 Example PROSITE entry, showing one of two data files for the prion protein family.

conserved parts of the alignment. When a set of regexs is achieved that is capable of capturing all, or most, of the characterized family from a given version of Swiss-Prot, without matching too many, or any, false positives, the results are manually annotated prior to deposition in PROSITE.

Entries are deposited in PROSITE in two distinct files. The first of these houses the regex and lists all matches in the parent version of Swiss-Prot; as shown in Fig. 5.7, the data are structured in a manner reminiscent of Swiss-Prot entries, where each field relates to a specific type of information.

Each entry contains both an identifier (ID), which is usually some sort of acronym for the family (here, PRION_1, indicating that more than one regex has been derived), and a unique accession number (AC), which takes the form PS00000. The ID line also indicates the type of discriminator to expect in the file – the word PATTERN here tells us to expect a regular expression. A title, or description of the family, is contained in the DE line, and the regex itself resides on PA lines. The following NR lines provide technical details about the derivation and diagnostic performance of the regex (for this reason, they are probably the most important lines to inspect when first viewing a PROSITE entry – large numbers of false-positive and false-negative results are indicative of a poorly performing regex). In the

example shown in Fig. 5.7, we learn that the regex was derived from release 40.7 of Swiss-Prot, which contained 103,373 sequences; it matched a total of 44 sequences, all of which are true-positives – in other words, this is a good regex, with no false matches.

The comment (CC) lines provide information on the taxonomic range of the family (defined here as eukaryotes), and the maximum number of observed repeats of the pattern (here just one). Following the comments are lists of the accession numbers and Swiss-Prot ID codes of all the true matches to the regex (denoted by T), and any "possible" matches (denoted by P), which are often fragments, of which there are none in this example. There are no false-positive or false-negative matches here, but when these do occur, they are listed after the T and P matches and are denoted by the letters F and N, respectively (the number of false and missed matches is also documented in the NR lines).

Where structural data are available, 3D lines are used to list all relevant PDB identifiers (e.g., 1DX0, 1DX1, etc.). Finally, the DO line points to the associated family documentation file (here PDOC00263), and the file ends with a // terminator.

The second type of PROSITE file is the documentation file, which provides details of the family (its structure, function, disease associations, etc.) and,

```
{PDOC00263}
{PS00291; PRION_1}
{PS00706; PRION_2}
{BEGIN}
*****************************
* Prion protein signatures *
*****************************
```

Prion protein (PrP) [1,2,3] is a small glycoprotein found in high quantity in the brains of humans or animals infected with a number of degenerative neurological diseases such as Kuru, Creutzfeldt-Jacob disease (CJD), scrapie or bovine spongiform encephalopathy (BSE). PrP is encoded in the host genome and expressed both in normal and infected cells. It has a tendency to aggregate yielding polymers called rods.

Structurally, PrP consists of a signal peptide, followed by an N-terminal domain containing tandem repeats of a short motif (PHGGGWGQ in mammals, PHNPGY in chicken), itself followed by a highly conserved domain of ~140 residues that contains a disulfide bond. Finally comes a C-terminal hydrophobic domain post-translationally removed when PrP is attached to the extracellular side of the cell membrane by a GPI-anchor. The structure of PrP is shown schematically below:

```
+---+----------------+-******---------------------****-----+-----+
|Sig| Tandem repeats |                    C         C   S|    |
+---+----------------+----------------------|--------|----|+-----+
                                            +--------+    |
                                                         GPI
```

'C': conserved cysteine involved in a disulfide bond.
'*': position of the patterns.

As signature pattern for PrP, we selected a perfectly conserved alanine- and glycine-rich region of 16 residues as well as a region centered on the second cysteine involved in the disulfide bond.

```
-Consensus pattern: A-G-A-A-A-A-G-A-V-V-G-G-L-G-G-Y
-Sequences known to belong to this class detected by the pattern: ALL.
-Other sequence(s) detected in SWISS-PROT: NONE.

-Consensus pattern: E-x-[ED]-x-K-[LIVM](2)-x-[KR]-[LIVM](2)-x-[QE]-M-C-x(2)-Q-Y
                    [C is involved in a disulfide bond]
-Sequences known to belong to this class detected by the pattern: ALL.
-Other sequence(s) detected in SWISS-PROT: NONE.

-Last update: November 1997 / Text revised.
```

[1] Stahl N., Prusiner S.B.
     FASEB J. 5:2799-2807(1991).
[2] Brunori M., Chiara Silvestrini M., Pocchiari M.
     Trends Biochem. Sci. 13:309-313(1988).
[3] Prusiner S.B.
     Annu. Rev. Microbiol. 43:345-374(1989). [E1]
     http://bioinformatics.weizmann.ac.il/hotmolecbase/entries/prp.htm
{END}

Fig. 5.8 Example PROSITE entry, showing the documentation file for the prion protein family.

where known, a description of the biological role of the chosen motif/s and a supporting bibliography; as shown in Fig. 5.8, this is a free-format text file. As evident from the figure, the structure of the documentation file is much simpler than that of the data file. Each entry is identified by its own accession number (which takes the form PDOC00000), and provides cross-references to the accession number and identifier of its associated data file or files – in this example, two regexs have been derived for the family, and hence there are links to two data files

(PS00291 and PS00706). A free-format description of the family and its regex(s) then follows, and the file concludes with relevant bibliographic references. The database is accessible for keyword and sequence searching via the ExPASy Web server.

5.5.3 PRINTS

From inspection of sequence alignments, it is clear that most protein families are characterized not by one, but by several conserved motifs. It therefore makes sense to use many, or all, of these to build "fingerprints" of family membership. The technique of fingerprinting was first developed in 1990 – the rationale behind its development was that single-motif approaches in general, and regexs in particular, are often not sufficiently powerful to characterize certain families unambiguously. Fingerprints inherently offer improved diagnostic reliability over methods of this type by virtue of the mutual context provided by motif neighbors: in other words, if a query sequence fails to match all the motifs in a given fingerprint, the pattern of matches formed by the remaining motifs still allows the user to make a reasonably confident diagnosis.

Fingerprinting was readily applicable to a diverse range of proteins, and effectively complemented the results stored in PROSITE, especially in situations where particular regexs failed to provide adequate discrimination. As a result, a prototype database was developed in 1990; this matured during the years that followed, such that, in 1993, the first public release was made of a compendium of protein fingerprints, containing 107 entries – this was PRINTS (Attwood and Beck 1994). PRINTS is now maintained at the University of Manchester, UK, and in February 2004 contained 1850 entries (Attwood *et al.* 2003). The database is released in major and minor versions: minor releases reflect updates, bringing the contents in line with the current version of the source database (a Swiss-Prot/TrEMBL composite); major releases denote the addition of new material to the resource. At each major release, 50 new families are added to the database and, where necessary, old entries are updated.

Within PRINTS, motifs are encoded as ungapped, unweighted local alignments derived from an iterative database-scanning procedure (the methods used to derive fingerprints and regexs differ markedly, and are described in detail in Chapter 9). When a set of motifs is achieved that is capable of capturing all, or most, of the family from a given version of Swiss-Prot and TrEMBL, without matching too many, or any, false positives, the results are manually annotated prior to deposition in PRINTS.

Today, PRINTS is the most comprehensive fully manually annotated protein family database available. Nevertheless, overall the database is still small relative to the number of protein families that exist, largely because the detailed documentation of entries is extremely time consuming. However, the extent of manually crafted annotations sets it apart from the growing number of automatically derived resources, for which there is little or no biological documentation and/or result validation, and in which family groupings may change between database releases.

5.5.4 The structure of PRINTS entries

PRINTS is built as a single text file – see Fig. 5.9. The contents are separated into specific fields, relating to general information, bibliographic references, text, lists of matches, and the motifs themselves – each line is assigned a distinct two-letter code, allowing the database to be indexed for rapid querying (Attwood *et al.* 1997). Each entry begins with a 12-character identification code (an acronym that describes the family – here, simply PRION), and an accession number, which takes the form PR00000. Following this is a description of the type of entry – the term "COMPOUND" indicates that the fingerprint contains several motifs, the number being indicated in parentheses (8 in this example). Details of the creation and latest update information are then given, followed by a descriptive title, and cross-references to other databases (e.g., InterPro, PROSITE, etc.), allowing users to access further information about the family in related resources. References are then provided, which relate to an abstract of the family describing its function and structure (where known), its disease associations, evolutionary relationships, and so on. Each abstract also contains a technical description of how the fingerprint was derived, including, where possible, details of the structural and/or functional relevance of the motifs – here, the motifs encode a number of hydrophobic regions and the characteristic octapeptide repeats.

Fingerprint diagnostic performance is indicated via a summary that lists how many sequences made complete matches and how many failed to match one or more motifs – the fewer the partial matches, the better the fingerprint (in this example, 37

```
gc; PRION
gx; PR00341
gn; COMPOUND(8)
ga; 19-OCT-1992; UPDATE 07-JUN-1999
gt; Prion protein signature
gp; INTERPRO; IPR000817
gp; PROSITE; PS00291 PRION_1; PS00706 PRION_2
gp; PFAM; PF00377 prion
bb;
gr; 1. STAHL, N. AND PRUSINER, S.B.
gr; Prions and prion proteins.
gr; FASEB J. 5 2799-2807 (1991).
gr;
gr; 2. BRUNORI, M., CHIARA SILVESTRINI, M. AND POCCHIARI, M.
gr; The scrapie agent and the prion hypothesis.
gr; TRENDS BIOCHEM.SCI. 13 309-313 (1988).
gr;
gr; 3. PRUSINER, S.B.
gr; Scrapie prions.
gr; ANNU.REV.MICROBIOL. 43 345-374 (1989).
bb;
bb;
gd; Prion protein (PrP) is a small glycoprotein found in high quantity in the brain of animals infected with
gd; certain degenerative neurological diseases, such as sheep scrapie and bovine spongiform encephalopathy (BSE),
gd; and the human dementias Creutzfeldt-Jacob disease (CJD) and Gerstmann-Straussler syndrome (GSS). PrP is
gd; encoded in the host genome and is expressed both in normal and infected cells. During infection, however, the
gd; PrP molecules become altered and polymerise, yielding fibrils of modified PrP protein.
gd;
gd; PrP molecules have been found on the outer surface of plasma membranes of nerve cells, to which they are
gd; anchored through a covalent-linked glycolipid, suggesting a role as a membrane receptor. PrP is also expressed
gd; in other tissues, indicating that it may have different functions depending on its location. The primary
gd; sequences of PrP's from different sources are highly similar: all bear an N-terminal domain containing multiple
gd; tandem repeats of a Pro/Gly rich octapeptide; sites of Asn-linked glycosylation; an essential disulphide bond;
gd; and 3 hydrophobic segments. These sequences show some similarity to a chicken glycoprotein, thought to be an
gd; acetylcholine receptor-inducing activity (ARIA) molecule. It has been suggested that changes in the octa-
gd; peptide repeat region may indicate a predisposition to disease, but it is not known for certain whether the
gd; repeat can be used as a fingerprint to indicate susceptibility.
gd;
gd; PRION is an 8-element fingerprint that provides a signature for the prion proteins. The fingerprint was derived
gd; from an initial alignment of 5 sequences: the motifs were drawn from conserved regions spanning virtually the
gd; full alignment length, including the 3 hydrophobic domains and the octapeptide repeats (WGQPHGGG). Two
gd; iterations on OWL18.0 were required to reach convergence, at which point a true set comprising 9 sequences was
gd; identified. Several partial matches were also found: these include a fragment (PRIO_RAT) lacking part of the
gd; sequence bearing the first motif, and the PrP homologue found in chicken - this matches well with only 2 of the
gd; 3 hydrophobic motifs (1 and 5) and one of the other conserved regions (6), but has an N-terminal signature
gd; based on a sextapeptide repeat (YPHNPG) rather than the characteristic PrP octapeptide.
gd;
gd; An update on SPTR37_9f identified a true set of 37 sequences, and 1 partial match.
bb;
bb;
si; SUMMARY INFORMATION
si; -------------------
sd;   37 codes involving  8 elements
sd;    0 codes involving  7 elements
sd;    0 codes involving  6 elements
sd;    0 codes involving  5 elements
sd;    0 codes involving  4 elements
sd;    1 codes involving  3 elements
sd;    0 codes involving  2 elements
bb;
bb;
ci; COMPOSITE FINGERPRINT INDEX
ci; ---------------------------
cr;
cd;  8|  37   37   37   37   37   37   37   37
cd;  7|   0    0    0    0    0    0    0    0
cd;  6|   0    0    0    0    0    0    0    0
cd;  5|   0    0    0    0    0    0    0    0
cd;  4|   0    0    0    0    0    0    0    0
cd;  3|   1    0    0    0    1    1    0    0
cd;  2|   0    0    0    0    0    0    0    0
cd; --+----------------------------------------
cd;   |   1    2    3    4    5    6    7    8
bb;
bb;
tp; PRIO_COLGU      PRIO_MACFA      PRIO_CEREL      PRIO_ODOHE
tp; PRIO_GORGO      PRIO_PANTR      PRIO_HUMAN      O46648
tp; PRIO_SHEEP      PRIO_CALJA      PRIO_BOVIN      PRP2_BOVIN
tp; PRIO_ATEPA      PRIO_SAISC      PRIO_PREFR      PRIO_PONPY
tp; O75942          PRIO_CAPHI      PRIO_CEBAP      PRIO_CAMDR
tp; PRIO_FELCA      PRP1_TRAST      PRIO_RABIT      PRP2_TRAST
tp; PRIO_PIG        PRIO_CANFA      PRIO_CRIGR      PRIO_CRIMI
tp; Q15216          PRIO_RAT        PRIO_CERAE      PRIO_MUSPF
tp; PRIO_MUSVI      PRIO_MESAU      PRIO_MOUSE      O46593
tp; PRIO_TRIVU
bb;
sn; Codes involving 3 elements
st; PRIO_CHICK
bb;
bb;
```

```
tt; PRIO_COLGU      MAJOR PRION PROTEIN PRECURSOR (PRP) (PRP27-30) (PRP33-35C) - COLOBUS GUEREZA
tt; PRIO_MACFA      MAJOR PRION PROTEIN PRECURSOR (PRP) (PRP27-30) (PRP33-35C) - MACACA FASCICULA
tt; PRIO_CEREL      MAJOR PRION PROTEIN PRECURSOR (PRP) - CERVUS ELAPHUS (RED DEER)
tt; PRIO_ODOHE      MAJOR PRION PROTEIN PRECURSOR (PRP) - ODOCOILEUS HEMIONUS (MULE DEER) (BLACK-
tt; PRIO_GORGO      MAJOR PRION PROTEIN PRECURSOR (PRP) (PRP27-30) (PRP33-35C) - GORILLA GORILLA
tt; PRIO_PANTR      MAJOR PRION PROTEIN PRECURSOR (PRP) (PRP27-30) (PRP33-35C) - PAN TROGLODYTES
tt; PRIO_HUMAN      MAJOR PRION PROTEIN PRECURSOR (PRP) (PRP27-30) (PRP33-35C) (ASCR) - HOMO SAPI
.
.
tt;
tt; PRIO_CHICK      MAJOR PRION PROTEIN HOMOLOG PRECURSOR (PR-LP) (ACETYLCHOLINE RECEPTOR-INDUCIN
bb;
bb;
sh; SCAN HISTORY
sh; -----------
dn; OWL18_0      2      30 NSINGLE
dn; OWL19_1      1      30 NSINGLE
dn; OWL26_0      1     160 NSINGLE
dn; OWL29_1      1     150 NSINGLE
dn; SPTR37_9f    2     134 NSINGLE
bb;
bb;
im; INITIAL MOTIF-SETS
im; ----------------
ic; PRION1
il; 16
it; Prion protein motif I - 1
id; WMLVLFVATWSDLGLC              PRIO_HUMAN     7     7
id; WILVLFVAMWSDVGLC              PRIO_BOVIN     9     9
id; WILVLFVAMWSDVGLC              PRIO_SHEEP     9     9
id; WILVLFVAMWSDVGLC              PRIP_BOVIN     9     9
id; WLLALFVAMWTDVGLC              PRIO_MESAU     7     7
id; WLLALFVTMWTDVGLC              PRIO_MOUSE     7     7
bb;
```

Fig. 5.9 Example PRINTS entry, showing the fingerprint for the prion protein family. For convenience, only the first motif is depicted. The two-letter code in the left-hand margin separates the information into specific fields (relating to text, references, motifs, etc.), which allows indexing of the data for rapid querying.

sequences matched all eight elements of the fingerprint, and one sequence matched only three motifs, indicating that this is a good fingerprint). The table that follows the summary breaks down this result to indicate how well individual motifs have performed, from which it is possible to deduce which motifs are missing from any partial matches – here, we discover that the reported partial hit failed to match motifs 2, 3, 4, 7, and 8.

After the summary are listed the ID codes of all full and partial true- and false-positive matches, followed by their database titles. In this case, the data show that the partial match is a chicken prion protein – referring to the technical description reveals that the reason it fails to match the fingerprint completely is that its sequence is characterized by a different type of tandem repeat. The scan history then indicates which version of the source database was used to derive the fingerprint, and on which versions it has been updated, how many iterations were required, what hit-list length was used, and the scanning method employed (Parry-Smith and Attwood 1992).

The final field relates to the motifs themselves, listing both the seed motifs used to generate the fingerprint, followed by the final motifs (not shown) that result from the iterative database search procedure. Each motif is identified by its parent ID code with the number of that particular motif appended – for convenience, only initial motif 1 (PRION1) is shown in Fig. 5.9. After the code, the motif length is given, followed by a short description, which indicates the relevant iteration number (for the initial motifs, of course, this will always be 1). The aligned motifs themselves are then provided, together with the corresponding source database ID code of each of the constituent sequence fragments (here only sequences from Swiss-Prot were included in the initial alignment, which contained a total of six sequences). The location in the parent sequence of each fragment is then given, together with the interval (i.e., the number of residues) between the fragment and its preceding neighbor – for motif 1, this value is the distance from the N terminus.

An important consequence of storing the motifs in this "raw" form is that, unlike with regexs or other such abstractions, no sequence information is lost. This means that a variety of scoring methods may be superposed onto the motifs, providing different scoring potentials for, and hence different perspectives on, the same underlying data. PRINTS thus provides the raw material for a number of automatically derived databases that make use of such approaches. As a consequence of being generated using more automated procedures, however, these resources provide little or no family annotation. They therefore further exploit PRINTS, by linking to its manually generated family documentations, which help to place conserved sequence information in structural or functional contexts. This is vital for the end user, who not only wants to discover whether a sequence matches a set of predefined motifs, but also needs to understand their biological significance. A number of these, and other, automatically derived databases are described in the following sections. In its original form, PRINTS is accessible for keyword and sequence searching via the University of Manchester's Bioinformatics Web server.

5.5.5 Blocks

In 1991, the diagnostic limitations of regexs led to the creation of another multiple-motif resource, at the Fred Hutchinson Cancer Research Center (FHCRC) in Seattle – this was the Blocks database (Henikoff et al. 2000). Originally, Blocks was based on protein families stored in PROSITE; today, however, its motifs, or blocks, are created automatically by finding the most highly conserved regions of families held in InterPro (or by using PRINTS' motifs directly – see Section 5.5.6). Different motif-detection algorithms are used and, when a consensus set of blocks is achieved, they are calibrated against Swiss-Prot to ascertain the likelihood of a chance match. Various technical details are noted, and the blocks, encoded as ungapped local alignments, are then deposited in Blocks.

Figure 5.10 illustrates a typical block. The structure of the database entry is compatible with that used in PROSITE. An ID line provides a general code by which the family may be identified (here Prion); this line also indicates the type of discriminator to expect in the file – in this case, the word BLOCK tells us to expect a block. Each block has an accession number, which takes the form IPB000000X (where IPB indicates that the block is derived from a family in InterPro, and X is a letter that specifies which the block is within the family's set of blocks, e.g., IPB000817E is the fifth prion block). The AC line also provides an indication of the minimum and maximum distances of the block from its preceding neighbor, or from the N terminus if it is the first in a set of blocks.

A title, or description of the family, is contained in the DE line. This is followed by the BL line, which provides an indication of the diagnostic power and some physical details of the block: these include an amino acid triplet (here CQY), which are the conserved residues used as anchors by one of the motif-detection algorithms; the width of the block and the number of sequences it contains; and two scores that quantify the diagnostic power of the block – strong blocks are more effective than weak blocks (strength less than 1100) at separating true-positives from true-negatives. Following this information comes the block itself, which indicates: the Swiss-Prot and TrEMBL IDs and accession numbers of the constituent sequences; the start position of the fragment; the sequence fragment itself; and a score, or weight, that provides a measure of the closeness of the relationship of that sequence to others in the block (100 being the most distant). The file ends with a // terminator.

In February 2004, Blocks contained 4944 entries from InterPro 6.0, indexed to Swiss-Prot 41.0 and TrEMBL 23.0; Blocks-format PRINTS, of course, contains the same number of entries as the version of PRINTS from which it was derived – see Section 5.5.6. Both databases are accessible for keyword and sequence searching via the Blocks Web server.

5.5.6 Blocks-format PRINTS

In addition to the Blocks database, the FHCRC Web server provides a version of the PRINTS database in Blocks format. In this resource, the scoring methods that underlie the derivation of blocks have been applied to each of the aligned motifs in PRINTS.

```
ID Prion; BLOCK
AC IPB000817E; distance from previous block=(1,2)
DE Prion protein
BL CQY; width=30; seqs=63; 99.5%=1496; strength=1395
PRIO_CAMDR|P79141 ( 218)    ITQYQREYQASYGRGASVIFSSPPVILLIS 41
PRIO_MOUSE|P04925 ( 214)    VTQYQKESQAYYDGRRSSSTVLFSSPPVIL 29
PRIO_TRIVU|P51780 ( 220)    ITQYQAEYEAAAQRAYNMAFFSAPPVTLLF 44
PRIO_CHICK|P27177 ( 238)    VQQYREYRLASGIQLHPADTWLAVLLLLT 38
PRIO_BOVIN|P10279 ( 226)    ITQYQRESQAYYQRGASVILFSSPPVILLI 6
PRP2_BOVIN|Q01880 ( 218)    ITQYQRESQAYYQRGASVILFSSPPVILLI 6
PRIO_CAPHI|P52113 ( 218)    ITQYQRESQAYYQRGASVILFSPPPVILLI 7
PRIO_CANFA|O46501 ( 217)    ITQYQRESEAYYQRGASVILFSSPPVILLV 11
PRIO_CEREL|P79142 ( 218)    ITQYQRESEAYYQRGASVILFSSPPVILLI 6
PRIO_FELCA|O18754 ( 218)    VTQYQKESEAYYQRRASAILFSSPPVILLI 8
PRIO_HUMAN|P04156 ( 215)    ITQYERESQAYYQRGSSMVLFSSPPVILLI 6
PRIO_AOTTR|P40245 ( 207)    ITQYERESQAYYQRGSSMVLFSSPPVILLI 6
PRIO_ATEGE|P40246 ( 199)    ITQYERESQAYYQRGSSMVLFSSPPVILLI 6
PRIO_ATEPA|P51446 ( 214)    ITQYERESQAYYQRGSSMVLFSSPPVILLI 6
PRIO_CALJA|P40247 ( 214)    ITQYEKESQAYYQRGSSMVLFSSPPVILLI 6
PRIO_CALMO|P40248 ( 208)    ITQYEKESQAYYQRGSSMVLFSSPPVILLI 6
PRIO_CEBAP|P40249 ( 214)    ITQYERESQAYYQRGSSMVLFSSPPVILLI 6
PRIO_CERAE|P40250 ( 207)    ITQYEKESQAYYQRGSSMVLFSSPPVILLI 6
PRIO_CERAT|Q95145 ( 200)    ITQYEKESQAYYQRGSSMVLFSSPPVILLI 6
PRIO_CERMO|Q95172 ( 208)    ITQYEKESQAYYQRGSSMVLFSSPPVILLI 6
PRIO_CERPA|Q95174 ( 208)    ITQYEKESQAYYQRGSSMVLFSSPPVILLI 6
PRIO_CERTO|Q95176 ( 208)    ITQYEKESQAYYQRGSSMVLFSSPPVILLI 6
PRIO_COLGU|P40251 ( 215)    ITQYEKESQAYYQRGSSMVLFSSPPVILLI 6
PRIO_GORGO|P40252 ( 215)    ITQYERESQAYYQRGSSMVLFSSPPVILLI 6
PRIO_PANTR|P40253 ( 215)    ITQYERESQAYYQRGSSMVLFSSPPVILLI 6
PRIO_MACFA|P40254 ( 215)    ITQYERESQAYYQRGSSMVLFSSPPVILLI 6
PRIO_MANSP|P40255 ( 208)    ITQYERESQAYYQRGSSMVLFSSPPVILLI 6
PRIO_PONPY|P40256 ( 215)    ITQYERESQAYYQRGSSMVLFSSPPVILLI 6
PRIO_PREFR|P40257 ( 215)    ITQYEKESQAYYQRGSSMVFFSSPPVILLI 7
PRIO_SAISC|P40258 ( 222)    ITQYEKESQAYYQRGSSMVLFSSPPVILLI 6
PRIO_MUSPF|P52114 ( 219)    VTQYQQESEAYYQRGASAILFSPPPVILLI 9
PRIO_MUSVI|P40244 ( 219)    VTQYQRESEAYYQRGASAILFSPPPVILLI 8
PRIO_ODOHE|P47852 ( 218)    ITQYQRESQAYYQRGASVILFSSPPVILLI 6
PRIO_PIG|P49927 ( 219)      ITQYQKEYEAYAQRGASVILFSSPPVILLI 13
PRIO_RABIT|Q95211 ( 214)    ITQYQQESQAAYQRAAGVLLFSSPPVILLI 17
PRIO_SHEEP|P23907 ( 218)    ITQYQRESQAYYQRGASVILFSSPPVILLI 6
PRIO_THEGE|Q95270 ( 200)    ITQYQKESQAYYQRGSSIVLFSSPPVILLI 10
PRP1_TRAST|P40242 ( 226)    ITQYQRESEAYYQRGASVILFSSPPVILLI 6
PRP2_TRAST|P40243 ( 218)    ITQYQRESEAYYQRGASVILFSSPPVILLI 6
PRIO_CRIGR|Q60506 ( 215)    VTQYQKESQAYYDGRRSSAVLFSSPPVILL 15
PRIO_CRIMI|Q60468 ( 215)    VTQYQKESQAYYDGRRSSAVLFSSPPVILL 15
PRIO_MESAU|P04273 ( 215)    TTQYQKESQAYYDGRRSSAVLFSSPPVILL 18
PRIO_RAT|P13852 ( 215)      VTQYQKESQAYYDGRRSSAVLFSSPPVILL 15
Q9UP19 ( 215)               ITQYERESQAYYQRGSSMVLFSSPPVILLI 6
O75942 ( 247)               ITQYERESQAYYQRGSSMVLFSSPPVILLI 6
Q15216 ( 207)               ITQYERESQAYYQRGSSMVLFSSPPVILLI 6
O46593 ( 219)               VTQYQKESEAYYQRGASAILFSPPPVILLI 8
Q9TV01 ( 218)               ITQYQRESQAYYQRGASVILFSSPPVILLI 6
Q9TU07 ( 218)               ITQYQRESQAYYQRGASVILFSSPPVILLI 6
Q9TTU5 ( 218)               ITQYQRESQAYYQRGASVILFSSPPVILLI 6
Q9TSF8 ( 216)               ITQYQQESQAAYQRAAGVLLFSSPPVILLI 17
O46648 ( 218)               ITQYQRESQAYYQRGASVILFSSPPVILLI 6
O02841 ( 218)               ITQYQRESQAYYQRGASVILFSSPPVILLI 6
O62670 ( 218)               ITQYQRESEAYYQRGASVILFSSPPVILLI 6
Q9MZU8 ( 218)               ITQYQRESQAYYQRGASVILFSSPPVILLI 6
Q9MZU6 ( 226)               ITQYQRESQAYYQRGASVILFSSPPVILLI 6
P97895 ( 204)               TTQYQKESQAYYDGRRSSAVLFSSPPVILL 18
Q9Z0T3 ( 215)               VTQYQKESQAYYDGRRSSAVLFSSPPMILL 19
Q9QYT9 ( 214)               VTQYQKESQAYYDGRRSSSTVLFSSPPVIL 29
Q9I9F3 ( 232)               VQQYREYRLASGIQLHPADTWLAVLLLLTT 41
Q9I9F2 ( 232)               VQQYREYRLASGIQLHPADTWLAVLLLLTT 41
Q9I9C0 ( 237)               MQQYQQYQLASGVKLLSDPSLMLIIMLVIF 100
Q9I8G1 ( 238)               VQQYREYRLASGIQLHPADTWLAVLLLLTT 41
//
```

Fig. 5.10 Example Blocks entry, showing the fifth block used to characterize the prion protein family.

```
ID PRION; BLOCK
AC PR00341C; distance from previous block=(1,4)
DE Prion protein signature
BL adapted; width=16; seqs=37; 99.5%=717; strength=1708
PRIO_COLGU|P40251 ( 57)      WGQPHGGGWGQPHGGG 9
PRIO_MACFA|P40254 ( 57)      WGQPHGGGWGQPHGGG 9
PRIO_CEREL|P79142 ( 60)      WGQPHGGGWGQPHGGG 9
PRIO_ODOHE|P47852 ( 60)      WGQPHGGGWGQPHGGG 9
PRIO_GORGO|P40252 ( 57)      WGQPHGGGWGQPHGGG 9
PRIO_PANTR|P40253 ( 57)      WGQPHGGGWGQPHGGG 9
PRIO_HUMAN|P04156 ( 57)      WGQPHGGGWGQPHGGG 9
O46648 ( 60)                 WGQPHGGGWGQPHGGG 9
PRIO_SHEEP|P23907 ( 60)      WGQPHGGGWGQPHGGG 9
PRIO_CALJA|P40247 ( 56)      WGQPHGGGWGQPHGGG 9
PRIO_BOVIN|P10279 ( 60)      WGQPHGGGWGQPHGGG 9
PRP2_BOVIN|Q01880 ( 60)      WGQPHGGGWGQPHGGG 9
PRIO_ATEPA|P51446 ( 56)      WGQPHGGGWGQPHGGG 9
PRIO_SAISC|P40258 ( 56)      WGQPHGGGWGQPHGGG 9
PRIO_PREFR|P40257 ( 57)      WGQPHGGGWGQPHGGG 9
PRIO_PONPY|P40256 ( 57)      WGQPHGGGWGQPHGGG 9
O75942 ( 57)                 WGQPHGGGWGQPHGGG 9
PRIO_CAPHI|P52113 ( 60)      WGQPHGGGWGQPHGGG 9
PRIO_CEBAP|P40249 ( 56)      WGQPHGGGWGQPHGGG 9
PRIO_CAMDR|P79141 ( 60)      WGQPHGGGWGQPHGGG 9
PRIO_FELCA|O18754 ( 60)      WGQPHGGGWGQPHGGG 9
PRP1_TRAST|P40242 ( 60)      WGQPHGGGWGQPHGGG 9
PRIO_RABIT|Q95211 ( 57)      WGQPHGGGWGQPHGGG 9
PRP2_TRAST|P40243 ( 60)      WGQPHGGGWGQPHVGG 18
PRIO_PIG|P49927 ( 60)        WGQPHGGGWGQPHGGG 9
PRIO_CANFA ( 60)             WGQPHGGGWGQPHGGG 9
PRIO_CRIGR|Q60506 ( 57)      WGQPHGGGWGQPHGGG 9
PRIO_CRIMI|Q60468 ( 57)      WGQPHGGGWGQPHGGG 9
Q15216 ( 57)                 WGQPHGGGWGQPHGGG 9
PRIO_RAT|P13852 ( 57)        WGQPHGGGWGQPHGGG 9
PRIO_CERAE|P40250 ( 57)      WGQPHGGGWGQPHGGG 9
PRIO_MUSPF|P52114 ( 60)      WGQPHGGGWGQPHGGG 9
PRIO_MUSVI|P40244 ( 60)      WGQPHGGGWGQPHGGG 9
PRIO_MESAU|P04273 ( 57)      WGQPHGGGWGQPHGGG 9
PRIO_MOUSE|P04925 ( 56)      WGQPHGGGWGQPHGGS 23
O46593 ( 60)                 WGQPHGGGWGQPHGGG 9
PRIO_TRIVU|P51780 ( 60)      HPQGGGTNWGQPHPGG 100
//
```

Fig. 5.11 Example Blocks-format PRINTS entry, showing the third block used to characterize the prion protein family.

Figure 5.11 illustrates a typical motif in Blocks format. The structure of the entry is identical to that used in Blocks, with only minor differences occurring on the AC and BL lines. On the AC line, the PRINTS accession number is given, with an appended letter to indicate which component of the fingerprint it is (in the example, PR00341C indicates that this is the third motif). On the BL line, the triplet information is replaced by the word "adapted", indicating that the motifs have been taken from another resource.

Because the Blocks databases are derived automatically, the blocks are not annotated. Nevertheless, relevant documentation may be accessed via links to the corresponding InterPro and PRINTS entries. A further important consequence of the direct derivation of the Blocks databases from InterPro and PRINTS is that they offer no further family coverage. However, as different methods are used to construct the blocks in each of the databases, it is worthwhile searching both these and the parent resources.

5.5.7 Profiles

An alternative philosophy to the motif-based approach of protein family characterization adopts the principle that the variable regions between conserved motifs also contain valuable information. Here, the complete sequence alignment effectively becomes the discriminator. The discriminator, termed a profile, is weighted to indicate where insertions and deletions are allowed, what types of residue are allowed at what positions, and where the most conserved regions are. Profiles (alternatively known as weight matrices) provide a sensitive means of detecting distant relationships, where only few residues are conserved – in these circumstances, regexs cannot provide good discrimination, and will either miss too many true-positives or will catch too many false ones.

The limitations of regexs in identifying distant homologs led to the creation of a compendium of profiles at the Swiss Institute for Experimental Cancer Research (ISREC) in Lausanne. Each profile has separate data and family-annotation files whose formats are compatible with PROSITE data and documentation files. This allows results that have been annotated to an appropriate standard to be made available for keyword and sequence searching as an integral part of PROSITE via the ExPASy Web server (Hulo *et al.* 2004). Those that have not yet achieved the standard of annotation necessary for inclusion in PROSITE are made available for searching directly via the ISREC Web server.

Figure 5.12 shows an excerpt from a profile data file. The file structure is based on that of PROSITE, but with obvious differences. The first change is seen on the ID line, where the word MATRIX indicates that the type of discriminator to expect is a profile. Pattern (PA) lines are replaced by matrix (MA) lines, which list the various parameter specifications used to derive the profile: they include details of the alphabet used (i.e., whether nucleic acid {ACGT} or amino acid {ABCDEFGHIKLMNPQRSTVWYZ}); the length of the profile; cut-off scores (which are designed, as far as possible, to exclude random matches); insertion and deletion penalties, insert–match–delete transition penalties; and so on. The I

and M fields contain position-specific profile scores for insert and match positions, respectively.

As in PROSITE, the pattern itself is followed by NR lines that depict the diagnostic performance of the pattern in terms of the number of true- and false-positive and false-negative results. In the example in Fig. 5.12, which illustrates part of the profile entry for WD-40 repeats, 1367 matches were made to the profile, from a total of 352 proteins – note that the number of matches differs from the number of proteins (indicated in parentheses) because sequences may contain multiple copies of the WD-40 domain. Of the 1367 matches, 1362 (from 348 proteins) are known to be true-positives; one is known to be false; 25 true WD-40-containing proteins were missed; one sequence made a partial match; and for four matches (from three proteins) it is unknown whether or not they are true family members.

Following the NR lines, a comment field (CC) provides further details about the profile, such as the author (here K. Hofmann), the taxonomic range of the family, the maximum number of observed repeats (here 15), keywords that describe the type of matrix (protein_domain), and so on. The familiar database cross-reference (DR) field then tabulates the database IDs and accession numbers of all true-positive (T), true-negative (N), unknown (?), and false-positive (F) matches to the profile in the version of Swiss-Prot indicated in the NR lines (in this case, 40.7).

Where structural data are available, 3D lines list all relevant PDB identifiers (e.g., 1SCG, 1AOR, etc.). Finally, the DO line points to the associated documentation file (here PDOC00574), which is a text file with identical format to that already described for PROSITE (Fig. 5.8). The file ends with a // terminator.

In addition to protein families and domains, profiles can be used to represent a number of other sequence features. These "special" profiles are distinguished by additional information on the MA and/or CC lines. For example, MA TOPOLOGY= CIRCULAR and CC /MATRIX_TYPE=repeat_region: these lines identify a circular profile that will produce one match for a repeat region consisting of several tandem repeated units. Profiles of this type

```
ID   WD_REPEATS_2; MATRIX.
AC   PS50082;
DT   DEC-2001 (CREATED); DEC-2001 (DATA UPDATE); DEC-2001 (INFO UPDATE).
DE   Trp-Asp (WD) repeats profile.
MA   /GENERAL_SPEC: ALPHABET='ABCDEFGHIKLMNPQRSTVWYZ'; LENGTH=42;
MA   /DISJOINT: DEFINITION=PROTECT; N1=5; N2=38;
MA   /NORMALIZATION: MODE=1; FUNCTION=LINEAR; R1=1.0511; R2=0.03341820; TEXT='-LogE';
MA   /CUT_OFF: LEVEL=0; SCORE=222; N_SCORE=8.5; MODE=1; TEXT='!';
MA   /CUT_OFF: LEVEL=-1; SCORE=163; N_SCORE=6.5; MODE=1; TEXT='?';
MA   /DEFAULT: D=-20; I=-20; B1=-50; E1=-50; MI=-105; MD=-105; IM=-105; DM=-105; M0=-10;
MA   /I: B1=0; BI=-105; BD=-105;
MA   /M: SY='L'; M=-9,-18,-20,-21,-14,6,-23,-12,4,-18,10,4,-15,-20,-14,-14,-14,-6,1,-16,3,-15;
MA   /M: M=-5,-2,-21,-2,0,-17,-13,-6,-16,-1,-15,-9,0,-11,0,-1,0,-1,-13,-25,-10,-1;
MA   /M: SY='G'; M=2,-5,-22,-6,-7,-22,18,-12,-24,-10,-20,-13,-1,-14,-9,-12,2,-7,-17,-24,-19,-9;
MA   /M: SY='H'; M=-14,0,-27,0,0,-19,-16,57,-25,-7,-19,-5,6,-15,5,-2,-6,-13,-24,-27,8,-1;
MA   /M: SY='S'; M=-4,1,-21,0,3,-20,-11,-8,-18,0,-17,-11,2,-9,-1,-1,4,3,-14,-28,-13,1;
MA   /M: SY='B'; M=-2,4,-21,4,1,-22,-1,-6,-22,-3,-20,-13,4,-13,-1,-6,3,-4,-17,-27,-15,-1;
MA   /M: M=-2,-7,-23,-7,-3,-16,-9,-11,-16,-6,-17,-12,-5,-5,-5,-7,-1,-3,-14,-17,-11,-5;
MA   /M: SY='V'; M=-3,-26,-12,-30,-26,-1,-29,-26,25,-21,11,9,-23,-25,-23,-21,-12,-3,30,-26,-7,-26;
MA   /M: SY='T'; M=-7,-5,-17,-10,-9,-7,-16,-8,-11,-7,-10,-7,0,-19,-7,-4,-1,4,-9,-16,-4,-8;
MA   /M: SY='S'; M=2,-2,0,-4,-6,-19,-7,-12,-20,-10,-19,-14,-1,-17,-7,-11,8,2,-11,-31,-17,-7;
MA   /M: SY='V'; M=-3,-25,-10,-29,-23,2,-27,-23,18,-23,17,11,-23,-26,-21,-20,-14,-5,20,-24,-6,-23;
MA   /M: SY='A'; M=5,-1,-15,-3,-2,-20,-11,-10,-17,-2,-16,-11,-2,-14,-2,-4,4,-1,-10,-25,-14,-2;
MA   /M: SY='F'; M=-12,-26,-24,-31,-24,25,-24,-19,0,-22,1,-2,-22,-25,-23,-18,-19,-11,-3,25,16,-22;
MA   /M: SY='S'; M=-2,3,-16,1,-3,-17,-9,3,-17,-8,-18,-12,8,-15,-3,-8,9,2,-14,-31,-11,-3;
MA   /M: SY='P'; M=-6,-6,-28,-3,1,-23,-12,-8,-20,-4,-22,-15,-5,24,-3,-8,-2,-6,-21,-27,-18,-3;
MA   /M: SY='D'; M=-8,14,-22,16,4,-21,-10,-2,-22,-3,-20,-16,10,-12,-2,-6,3,-1,-18,-30,-13,1;
MA   /I: I=-12; MI=-15; MD=-25; IM=-15; DM=-25;
MA   /M: SY='G'; M=-2,-1,-19,-1,-3,-19,13,-8,-20,-7,-18,-12,2,-8,-5,-8,2,-6,-16,-21,-15,-4; D=-5;
MA   /I: DM=-25;
MA   /M: SY='N'; M=-6,1,-20,-1,0,-17,-9,-4,-17,0,-16,-10,4,-8,1,1,-2,-15,-22,-10,0; D=-5;
MA   /I: DM=-25;
MA   /M: SY='Y'; M=-9,-15,-21,-18,-13,1,-23,-7,0,-10,3,2,-12,-22,-9,-6,-12,-5,-1,-11,6,-12; D=-5;
MA   /I: DM=-25;
MA   /M: SY='L'; M=-8,-27,-18,-31,-23,13,-29,-21,19,-25,23,13,-23,-26,-22,-20,-19,-7,15,-16,2,-23;
MA   /M: SY='V'; M=8,-21,-13,-27,-21,0,-19,-21,10,-20,8,4,-19,-22,-19,-20,-8,-3,14,-19,-6,-20;
MA   /M: SY='T'; M=8,-6,-8,-10,-9,-13,-9,-15,-11,-12,-15,-11,-1,-13,-9,-13,20,21,-1,-32,-15,-9;
MA   /M: SY='G'; M=10,-10,-9,-14,-16,-21,22,-20,-22,-16,-19,-14,-5,-18,-16,-18,6,-5,-11,-26,-23,-16;
MA   /M: SY='S'; M=5,-3,-13,-5,-8,-19,11,-12,-23,-13,-23,-16,2,-15,-8,-14,16,3,-14,-29,-17,-8;
MA   /M: M=-4,-2,-21,-2,-1,-16,-10,-7,-17,-4,-14,-9,-2,-14,-3,-4,-1,-4,-14,-18,-8,-2;
MA   /M: SY='D'; M=-16,39,-28,52,16,-35,-8,0,-35,0,-28,-26,19,-11,1,-8,2,-7,-27,-37,-19,9;
MA   /M: SY='G'; M=-5,-1,-22,-4,-6,-22,11,-5,-27,-1,-23,-13,6,-16,-4,1,1,-8,-22,-24,-16,-6;
MA   /M: SY='T'; M=-3,-2,-17,-6,-4,-14,-16,-8,-11,-4,-12,-7,0,-15,-2,-4,6,11,-7,-27,-9,-3;
MA   /M: SY='I'; M=-2,-27,-14,-32,-25,0,-30,-26,27,-24,17,12,-23,-24,-22,-23,-15,-5,26,-23,-6,-25;
MA   /M: SY='R'; M=-10,-7,-23,-10,-3,-15,-19,-5,-16,13,-14,-5,-3,-17,2,18,-7,-6,-12,-20,-5,-2;
MA   /M: SY='I'; M=-7,-27,-18,-31,-24,6,-30,-22,23,-24,19,13,-23,-26,-21,-20,-17,-6,22,-20,-1,-24;
...
NR   /RELEASE=40.7,103373;
NR   /TOTAL=1367(352); /POSITIVE=1362(348); /UNKNOWN=4(3); /FALSE_POS=1(1);
NR   /FALSE_NEG=25; /PARTIAL=1;
CC   /MATRIX_TYPE=protein_domain;
CC   /SCALING_DB=reversed;
CC   /AUTHOR=K_Hofmann;
CC   /TAXO-RANGE=??EP?; /MAX-REPEAT=15;
CC   /FT_KEY=REPEAT; /FT_DESC=WD;
DR   O15491, AN1H_HUMAN, T; Q92747, AR1A_HUMAN, T; O15143, AR1B_HUMAN, T;
DR   Q9WV32, AR1B_MOUSE, T; O88656, AR1B_RAT , T; P78774, AR41_SCHPO, T;
DR   P26449, BUB3_YEAST, T; Q9FMU5, CG48_ARATH, T; Q9Y5J1, CG48_HUMAN, T;
DR   P78750, CG48_SCHPO, T; O14011, CWF8_SCHPO, T; Q92466, DDB2_HUMAN, T;
DR   Q09019, DMWD_HUMAN, T; Q08274, DMWD_MOUSE, T; Q16959, DYI2_ANTCR, T;
DR   Q62871, DYI2_RAT , T; Q9UI46, DYI3_HUMAN, T; Q24246, DYIN_DROME, T;
DR   O45487, EMAL_CAEEL, T; Q9VUI3, EMAL_DROME, T; P57775, FBW4_HUMAN, T;
DR   Q9JMJ2, FBW4_MOUSE, T; Q99698, LYST_HUMAN, T; P33215, NED1_MOUSE, T;
...
DR   P25382, YCW2_YEAST, T; Q09990, YSS1_CAEEL, T; Q12220, DIP2_YEAST, T;
DR   O14727, APAF_HUMAN, T; Q9UNX4, WDR3_HUMAN, T; P74442, Y143_SYNY3, T;
DR   O55029, COPP_MOUSE, P;
DR   P38328, AR41_YEAST, N; P50851, CC4H_HUMAN, N; P42000, CG48_CAEEL, N;
DR   Q9V7P1, CG48_DROME, N; O14576, DYI1_HUMAN, N; O88485, DYI1_MOUSE, N;
DR   Q63100, DYI1_RAT , N; Q39578, DYI2_CHLRE, N; Q13409, DYI2_HUMAN, N;
DR   O88487, DYI2_MOUSE, N; Q16960, DYI3_ANTCR, N; P27766, DYI3_CHLRE, N;
DR   P54703, DYIN_DICDI, N; Q9UKB7, FBW3_HUMAN, N; P25365, SED4_YEAST, N;
DR   Q03010, UME1_YEAST, N; P93231, VP41_LYCES, N; Q9P2S5, WDR8_HUMAN, N;
DR   Q9JM98, WDR8_MOUSE, N; Q9BZH6, WDRB_HUMAN, N; Q12363, WTM1_YEAST, N;
DR   Q12206, WTM2_YEAST, N; Q10437, YDE3_SCHPO, N; P43601, YFJ1_YEAST, N;
DR   Q9RRK5, YO84_DEIRA, N;
DR   P40960, PA11_YEAST, ?; P22219, VP15_YEAST, ?; P38163, SNI2_YEAST, ?;
DR   P13699, NCAP_LASSJ, F;
3D   1SCG; 1A0R; 1B9X; 1B9Y; 1GG2; 1GP2; 1TBG;
DO   PDOC00574;
//
```

```
HMMER2.0   [2.2g]
NAME   prion
ACC    PF00377
DESC   Prion protein
LENG   244
ALPH   Amino
RF     no
CS     no
MAP    yes
COM    hmmbuild -F HMM_ls.ann SEED.ann
COM    hmmcalibrate --seed 0 HMM_ls.ann
NSEQ   3
DATE   Wed May  1 19:45:02 2002
CKSUM  4026
GA     -61.0 -61.0
TC     -60.7 -60.7
NC     -69.3 -69.3
XT     -8455 -4 -1000 -1000 -8455 -4 -8455 -4
NULT   -4   -8455
NULE    595 -1558 85 338 -294 453 -1158 197 249 902 -1085 -142 -21 -313 45 531 201 384 -1998 -644
EVD    -152.902084    0.151793
HMM
...
//
```

Fig. 5.13 Excerpt from a Pfam entry, illustrating some of the technical parameters stored for the prion protein family.

include: ANK_REP_REGION, for ankyrin repeats; COLLAGEN_REP, for the G–X–X collagen repeat; and PUM_REPEATS, for the pumilio RNA-binding domain. In addition, CC/MATRIX_TYPE=composition denotes a profile for compositionally biased regions, e.g., PRO_RICH, proline-rich region (such profiles exist for each amino acid).

5.5.8 Pfam

Just as there are different ways of using motifs to characterize protein families (e.g., depending on the scoring scheme used), so there are different methods of using full sequence alignments to build family discriminators. An alternative to the use of profiles is to encode alignments using hidden Markov models (HMMs). These are statistically based mathematical treatments, consisting of linear chains of match, delete, or insert states that attempt to encode the sequence conservation within aligned families (more details in Chapter 10).

In 1994, a collection of HMMs for a range of protein domains was created at the Sanger Institute – this was the Pfam database. Today, Pfam (Bateman

et al. 2004) comprises two distinct sections: (i) Pfam-A, which contains automatically generated (but often hand-edited) seed alignments of representative family members, and full alignments containing all members of the family detected with an HMM constructed from the seed alignment using the HMMER2 software; and (ii) Pfam-B, which is a fully automatically generated supplement derived from the ProDom database (Corpet et al. 2000). Full alignments in Pfam-A can be large (many contain thousands of sequences) and, as with Pfam-B, because they are generated fully automatically, are not always reliable. In February 2004, Pfam contained 7316 entries.

The methods that generate the best full alignments vary for different families; the parameters are therefore saved so that the result can be reproduced. This information, coupled with minimal annotations (sometimes just a descriptive family title), database and literature cross-references, etc., seed and full alignments, and the HMMs themselves, form the backbone of Pfam-A. Some of the technical documentation for the prion entry is illustrated in Fig. 5.13.

Fig. 5.12 (opposite) Excerpt from a PROSITE profile entry, illustrating part of the profile used to characterize WD-40 repeats.

The first line of the file indicates the version of the software used to create the HMM, and the NAME tag provides a code by which the entry can be identified (here prion). Each entry also has an accession number, which takes the form PF00000, and a descriptive title (DESC) indicating the name of the protein domain or family. The length of the domain is also given (LENG) and the alphabet specified (here, Amino).

Details of how the HMM was constructed are provided on the COM lines, the number of sequences used in the seed alignment on the NSEQ line (here 3), and the date of its creation on the DATE line. A variety of cut-off values are stored on the GA (gathering), TC (trusted), and NC (noise) lines, respectively. Various statistical parameters then follow (including data relating to the extreme value distribution (EVD)) prior to the HMM itself (not shown). The file ends with a // terminator.

Although entries in Pfam-A carry some annotation (e.g., including brief details of the domain, its structure and function (where known), links to other databases), extensive family annotations are not provided. However, where available, additional information is incorporated directly from InterPro (much of which is derived from PROSITE and PRINTS). Pfam is available for keyword and sequence searching via the Sanger Institute Web server.

5.5.9 eMOTIF

Another automatically generated database is eMOTIF, produced at Stanford University (Huang and Brutlag 2001). The method used to create this resource is based on the generation of regexs from ungapped motifs stored in the Blocks and PRINTS databases. Rather than encoding the exact information observed at each position in these motifs, eMOTIF adopts a "permissive" approach in which alternative residues are tolerated according to a set of prescribed groupings, as illustrated in Table 5.4 (Neville-Manning, Wu, and Brutlag 1998). These groups correspond to various biochemical properties, such as charge and size, theoretically ensuring that the resulting motifs have sensible biochemical interpretations.

Table 5.4 Sets of amino acids and their properties used in eMOTIF.

Residue property	Residue groups
Small	Ala, Gly
Small hydroxyl	Ser, Thr
Basic	Lys, Arg
Aromatic	Phe, Tyr, Trp
Basic	His, Lys, Arg
Small hydrophobic	Val, Leu, Ile
Medium hydrophobic	Val, Leu, Ile, Met
Acidic/Amide	Asp, Glu, Asn, Gln
Small/Polar	Ala, Gly, Ser, Thr, Pro

eMOTIF is created as a single text file, in which each line corresponds to a unique record. Within these records, the fields denote: the expected false-positive frequency of the regex (i.e., the probability that the regex will match a sequence of the same length by chance); the accession number of the parent motif and its descriptive name; the regex itself; and its sensitivity (i.e., the percentage of sequences matched from the parent motif). eMOTIF entries are sorted by their expected false-positive frequencies, such that the regex with the smallest value resides at the beginning of the file.

Although eMOTIF's use of residue groups is more flexible than the exact-residue matching characteristic of PROSITE, its inherent permissiveness brings with it an inevitable signal-to-noise trade-off: i.e., the regexs not only have the potential to make more true-positive matches, but will also match more false-positives. Note should therefore be made of the various stringency values when interpreting the results of eMOTIF searches. The database is accessible from Stanford University's Biochemistry Department Web server.

5.6 COMPOSITE PROTEIN PATTERN DATABASES

5.6.1 InterPro

While there is some overlap between them, the contents of PROSITE, PRINTS, Profiles, and Pfam are different. Furthermore, diagnostically, the resources have different areas of optimum application, owing to the different strengths and weaknesses of their underlying analysis methods: e.g., regular expressions are unreliable in the identification of members of highly divergent superfamilies (where HMMs and profiles excel); fingerprints perform relatively poorly in the diagnosis of very short motifs (where regexs do well); and profiles and HMMs are less likely to give specific subfamily diagnoses (where fingerprints excel). Thus, while all of the resources share a common interest in protein sequence classification, some focus on divergent domains, some focus on functional sites, and others focus on families, specializing in hierarchical definitions from superfamily down to subfamily levels in order to pinpoint specific functions. Therefore, in building a search strategy, it is sensible to include all of the available databases, to ensure both that the analysis is as comprehensive as possible and that it takes advantage of a variety of search methods.

Accessing these disparate resources, gathering their different outputs, and arriving at a consensus view of the results can be quite a challenge. Therefore, in 1998, in an effort to provide a one-stop shop for protein family diagnosis and hence make sequence analysis more straightforward, the curators of PROSITE, Profiles, PRINTS, Pfam, and ProDom began working together to create a unified database of protein families. The aim was to create a single resource, based primarily on the comprehensive annotation in PROSITE and PRINTS, wherein each entry would point to relevant discriminators in the parent databases – this was InterPro (Mulder *et al.* 2003). From its first release in November 1999, InterPro proved tremendously popular, and further databases have since joined the consortium – recent additional partners include SMART (Letunic *et al.* 2002) and TIGRFAMs (Haft *et al.* 2001).

Integrating data from so many different databases, each of which uses different terminology and concepts to describe families and domains, is not straightforward. Files submitted by each of the groups must therefore be systematically merged and dismantled before being incorporated into InterPro. Where relevant, family annotations are amalgamated, and all method-specific annotation separated out.

The amalgamation process is complicated by the relationships that can exist between entries in the same database, and between entries in different databases. When investigating these relationships more closely, different types of parent–child relationships became evident; these were subsequently designated "sub-types" and "sub-strings". A substring means that a motif (or motifs) is contained within a region of sequence encoded by a wider pattern (e.g., a PROSITE regex is typically contained within a PRINTS fingerprint, or a fingerprint might be contained within a Pfam domain). A sub-type means that one or more motifs are specific for a subset of sequences captured by another more general pattern (e.g., a superfamily fingerprint may contain several family- and subfamily-specific fingerprints; or a generic Pfam domain may include several family fingerprints). A further relationship also exists between InterPro entries – the "contains/found in" relationship. This arises in the case of domains that are found in several structurally and functionally discrete families.

Having classified the different relationships of contributed entries, all recognizably distinct entities are assigned unique accession numbers. In doing this, the guiding principle is that parents and children with sub-string relationships should have the same accession numbers, while sub-type parent–child and contains/found in relationships warrant their own accessions. In February 2004, InterPro contained 10,403 entries, including 7901 families, 2239 domains, 197 repeats, 26 active sites, 20 binding sites, and 20 PTMs, built from PRINTS 37.0, PROSITE 18.1, Pfam 11.0, ProDom 2002.1, SMART 3.4, TIGRFAMs 3.0, PIR SuperFamily 2.3, and SUPERFAMILY 1.63, indexed to Swiss-Prot 42.5 and TrEMBL 25.5.

5.6.2 *The structure of InterPro entries*

InterPro entries contain one or more signatures from the individual member databases, together with merged annotation and lists of matched proteins from specified versions of Swiss-Prot and

```
Accession       IPR000817; Prion (matches 144 proteins)
Name            Prion protein
Type            Family
Dates           08-OCT-1999 (created)
                26-JAN-2001 (last modified)
Signatures      PR00341; PRION (130 proteins)
                PS00291; PRION_1 (110 proteins)
                PS00706; PRION_2 (99 proteins)
                PF00377; prion (113 proteins)
                SM00157; PRP (109 proteins)
Abstract        Prion protein (PrP) [1, 2, 3] is a small glycoprotein found in high quantity in the brain
                of animals infected with certain degenerative neurological diseases, such as sheep scrapie
                and bovine spongiform encephalopathy (BSE), and the human dementias Creutzfeldt-Jacob
                disease (CJD) and Gerstmann-Straussler syndrome (GSS). PrP is encoded in the host genome
                and is expressed both in normal and infected cells. During infection, however, the PrP
                molecules become altered and polymerise, yielding fibrils of modified PrP protein. PrP
                molecules have been found on the outer surface of plasma membranes of nerve cells, to
                which they are anchored through a covalent-linked glycolipid, suggesting a role as a
                membrane receptor. PrP is also expressed in other tissues, indicating that it may have
                different functions depending on its location. Structurally, PrP is a protein consisting
                of a signal peptide, followed by an N-terminal domain that contains tandem repeats of a
                Pro/Gly rich octapeptide. This is followed by a highly conserved domain of about 140
                residues that contains a disulfide bond. Finally comes a C-terminal hydrophobic domain
                post-translationally removed when PrP is attached to the extracellular side of the cell
                membrane by a GPI-anchor. The sequences also contain sites of Asn-linked glycosylation.
                These PrP sequences show some similarity to a chicken glycoprotein, thought to be an
                acetylcholine receptor-inducing activity (ARIA) molecule. It has been suggested that
                changes in the octapeptide repeat region may indicate a predisposition to disease, but it
                is not known for certain whether the repeat can meaningfully be used as a fingerprint to
                indicate susceptibility.
Examples
                O46501 PRIO_CANFA: Dog
                P04925 PRIO_MOUSE: Mouse
                P10279 PRIO_BOVIN: Bovine
                P04156 PRIO_HUMAN: Human

References
                Prusiner S.B.
                Scrapie prions.
                Annu. Rev. Microbiol. 43: 345-374(1989). [MEDLINE:90024956]
                Stahl N., Prusiner S.B.
                Prions and prion proteins.
                FASEB J. 5: 2799-2807(1991). [MEDLINE:92008960]
                Brunori M., Chiara Silvestrini M., Pocchiari M.
                The scrapie agent and the prion hypothesis.
                Trends Biochem. Sci. 13: 309-313(1988). [MEDLINE:91134832]
Database links  PROSITE doc; PDOC00263
                Blocks; IPB000817
```

Fig. 5.14 Excerpt from an InterPro entry, illustrating some of the annotation stored for the prion protein family.

TrEMBL. An example is shown in Fig. 5.14. Each entry has a unique accession number (which takes the form IPR000000), a short name, or code, by which the entry may be identified, and a descriptive title. An abstract derived from merged annotation from the member databases describes the family, domain, repeat or PTM, and literature references used to create the abstract are stored in a reference field. Examples of representative family members are also provided. Where relationships exist between entries, these are displayed in a "parent", "child", "contains", or "found in" field.

Additional annotation is available for some entries in the form of mappings to Gene Ontology (GO) terms (The Gene Ontology Consortium 2001). The GO project is an effort to provide a universal system of terminology for describing gene products across all species. We discuss GO in Section 13.6, and consider more carefully what is meant by the word "ontology". For the moment, we note that GO is divided into three sections describing "molecular function", "biological process", and "cellular component". As many of these concepts are central to InterPro annotation, wherever possible its entries

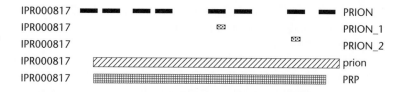

Fig. 5.15 Example graphical output from InterPro, showing matches to the PRINTS (black), PROSITE (spotted), Pfam (striped), and SMART (checked) prion protein signatures.

have been mapped to GO terms to provide an automatic means of assigning GO terms to their constituent proteins.

In addition to cross-referencing the member database signatures and GO terms, a separate field provides cross-references to other databases. Included here are links to corresponding Blocks entries; PROSITE documentation; and the Enzyme Commission (EC) Database, where the EC number(s) for proteins matching the entry are shared. Where applicable, there may also be links to specialized resources, e.g., the CArbohydrate-Active EnZymes (CAZy) site, which describes families of related catalytic and carbohydrate-binding modules of enzymes that act on glycosidic bonds.

A particular advantage of InterPro is that each sequence within a family is linked to a graphical output that conveniently summarizes where matches occur to each of the signatures in the source databases. An example of such graphical output is illustrated in Fig. 5.15, which shows matches from PRINTS, PROSITE, Pfam, and SMART on the human prion sequence. This type of output also neatly illustrates the difference between motif-based methods, which match small regions of the sequence, and profile-based methods, which more or less span the sequence completely. InterPro is available for

keyword and sequence searching via the EBI Web server.

5.7 PROTEIN STRUCTURE DATABASES

A discussion of the repertoire of biological databases available for sequence analysis would not be complete without some consideration of protein structure-based resources. These are limited to the relatively few 3D structures generated by crystallographic and spectroscopic studies, but their impact will increase as more structures become available, notably as a result of ongoing high-throughput structural genomics initiatives. In general, these resources can be divided into those that house the actual 3D coordinates of solved structures, and those that classify and summarize the structures of those proteins.

5.7.1 The PDB

The PDB is the single world-wide archive of biological macromolecular structures (Bourne *et al.* 2004). In February 2004, it contained a total of 24,248 entries, of which 21,948 were protein, peptide, and virus structures deduced by means of X-ray diffraction,

Table 5.5 PDB protein structure holdings in February 2004.

Experimental method	Proteins, peptides, and viruses	Protein/Nucleic acid complexes	Nucleic acids	Carbohydrates
X-ray diffraction, etc.	19,014	898	719	14
NMR	2,934	96	569	4
Total	21,948	994	1,288	18

NMR, or modeling (see Table 5.5); the remainder were nucleic acid, nucleic acid/protein complex, and carbohydrate structures.

What is not evident from these figures is that the PDB is a massively redundant resource. Owing to the nature of research, where particular proteins become the focus of repeated structure determinations, the contents of PDB are skewed by relatively small numbers of such commonly studied proteins (e.g., hemoglobin, myoglobin, lysozyme, immunoglobulins). Consequently, the number of unique structures within the PDB is probably little more than 3000.

Nevertheless, the PDB is immensely valuable. Studies of protein structure archives have revealed that many proteins share structural similarities, often reflecting common evolutionary origins, via processes that involve substitutions, insertions, and deletions in the underlying sequences. For distantly related proteins, such changes can be extensive, yielding folds in which the numbers and orientations of secondary structures vary considerably. However, where the functions of proteins are conserved, the structural environments of critical active-site residues are also generally conserved. In an attempt to better understand sequence/structure relationships, and the evolutionary processes that give rise to different fold families, a variety of structure classification schemes have been established.

The nature of the information presented by structure classification schemes is largely dependent on the methods used to identify and evaluate similarity. Families derived, for example, using algorithms that search and cluster on the basis of common motifs will be different from those generated by procedures based on global structure comparison; and the results of such automatic procedures will differ again from those based on visual inspection, where software tools are used essentially to render the task of classification more manageable. A number of structure-classification resources are outlined in the sections that follow.

5.7.2 SCOP

The SCOP (Structural Classification of Proteins) database classifies proteins of known structure according to their evolutionary and structural relationships (Andreeva *et al.* 2004). Domains in SCOP are grouped by species and hierarchically classified into families, superfamilies, folds, and classes, a task complicated by the fact that structures exhibit great variety, ranging from small, single domains to vast multi-domain assemblies, so that sometimes it makes sense to classify by individual domains, and other times at the multi-domain level. Because automatic structure comparison tools cannot identify all such relationships reliably, SCOP has been constructed using a combination of manual inspection and automated methods.

Proteins are classified in a hierarchical fashion to reflect their structural and evolutionary relatedness. Within the hierarchy there are many levels, but principally these describe the family, superfamily, and fold. The boundaries between these levels may be subjective, but the higher levels generally reflect the clearest structural similarities:

• **Families** generally represent clear evolutionary relationships, with sequence identities greater than or equal to 30%. This is not an absolute measure, as it is sometimes possible to infer common descent in the absence of significant sequence identity (e.g., some members of the globin family share only 15% identity).

• **Superfamilies** represent groups of proteins with low sequence identity that probably share common evolutionary origins on the basis of similar structural and functional characteristics.

• **Folds** denote major structural similarities (i.e., the same major secondary structures in the same arrangement and with the same topology), irrespective of evolutionary origin (e.g., similarities could arise as a result of physical principles that favor particular packing arrangements and fold topologies).

The seven main classes in SCOP 1.65 contain 40,452 domains organized into 2327 families, 1294 superfamilies, and 800 folds. These domains correspond to 20,619 entries in the PDB. Analyzing the contents more closely (Table 5.6), it is clear that the total number of families, superfamilies, and folds (~4,400) is significantly less than the total number of protein structures deposited in the PDB. SCOP is

Table 5.6 SCOP statistics for PDB release 1 August 2003 (excluding nucleic acids and theoretical models).

Class	No. of folds	No. of superfamilies	No. of families
All-α proteins	179	299	480
All-β proteins	126	248	462
α and β proteins (α/β)	121	199	542
α and β proteins (α + β)	234	349	567
Multi-domain proteins	38	38	53
Membrane and cell surface proteins	36	66	73
Small proteins	66	95	150
Total	800	1,294	2,327

available for interrogation from the MRC Laboratory of Molecular Biology Web server.

5.7.3 CATH

The CATH (Class, Architecture, Topology, Homology) database is a hierarchical domain classification of protein structures (Pearl *et al.* 2003). The resource is largely derived using automatic methods, but manual inspection is necessary where automatic methods fail. Different categories within the classification are identified by means of both unique numbers (by analogy with the EC system for enzymes) and descriptive names. Such a numbering scheme allows efficient computational manipulation of the data. There are four levels within the hierarchy:

• **Class** denotes gross secondary structure content and packing, including domains that are: (i) mainly α, (ii) mainly β, (iii) α – β (alternating α/β or α + β structures), and (iv) those with low secondary structure content.

• **Architecture** describes the gross arrangement of secondary structures, ignoring connectivity, and is assigned manually using simple descriptions, such as barrel, roll, sandwich, etc.

• **Topology** assigns the overall shape **and** secondary structure connectivity by means of structure comparison algorithms – structures in which at least 60% of the larger protein matches the smaller are assigned to the same level.

• **Homology** clusters domains that share greater than or equal to 35% sequence identity and are thought to share a common ancestor (i.e., are

homologous) – similarities are identified both by sequence- and structure-comparison algorithms. CATH is accessible for keyword interrogation via the UCL Web server.

5.7.4 PDBsum

A useful resource for accessing structural information is PDBsum (Laskowski 2001), a Web-based compendium of largely pictorial summaries and analyses of all structures in the PDB. Each summary gives an at-a-glance overview of the contents of a PDB entry in terms of resolution and *R* factor, numbers of protein chains, ligands, metal ions, secondary structure, fold cartoons, ligand interactions, and so on. This is helpful, not only for visualizing the structures concealed in PDB files, but also for drawing together information at both sequence and structure levels. Resources of this type will become more and more important as visualization techniques improve, and new-generation software allows more direct interaction with their contents. In February 2004, PDBsum contained 25,632 entries. The resource is accessible for keyword interrogation via the UCL Web server.

5.8 OTHER TYPES OF BIOLOGICAL DATABASE

This chapter has focused on databases of nucleic acid and protein sequences, and of protein families and protein structures (their Web addresses are listed in the Web Address Appendix at the end of this book). Beyond these is a huge variety of more specialized databases, with their own aims and subject coverage, and it is impossible to do justice to them all. However, we will highlight two types of resource in this necessarily brief final section: first, species-specific databases, and second, databases that include information about complex biological processes and interactions.

Many species for which genome sequencing is complete, or well advanced, have their own specific databases. Sequence information *per se* may not be their primary thrust – often, the goal is to present a more integrated view of a particular biological system, in which sequence data represent just one level of abstraction, and higher levels lead to an overall understanding of the genome organization. Important genome databases of model organisms include *Saccharomyces cerevisiae* (SGD; Dwight *et al.* 2002), *Caenorhabditis elegans* (WormBase; Stein *et al.* 2001), *Drosophila melanogaster* (FlyBase; The FlyBase Consortium 2002), *Mus musculus* (MGD; Blake *et al.* 2003), and *Arabidopsis thaliana* (TAIR; Rhee *et al.* 2003). The curators of these and other databases have, in recent years, worked together as the Gene Ontology Consortium (see Section 13.6), which aims to develop a consistent form of annotation for gene functions in different organisms.

There are several databases that aim to capture some of the complexity of the processes within cells, such as metabolic pathways, and gene signaling and protein–protein interaction networks. These types of information are fundamentally different from sequence data and, arguably, much more complex. They thus pose challenges in terms of how to organize and store the data, and require us to think carefully about which information is important to us and what we want to do with it. Important examples in this category are KEGG (Kanehisa *et al.* 2002), WIT (Overbeek *et al.* 2000), and EcoCyc (Karp *et al.* 2002), UM-BBD (Ellis *et al.* 2001), and BIND (Bader, Betel, and Hogue 2003). Another type

of biological information results from high-throughput experimental techniques, like microarrays and proteomics. There is also an increasing awareness that storage and sharing of this type of experimental data is important, if comparisons are to be made between results from different laboratories. We discuss databases of this type specifically in Section 13.6.

Other resources of which readers should be aware include the Ensembl genome database (Hubbard *et al.* 2002), which provides a comprehensive source of stable automatic annotation of the human genome sequence; UniGene (Wheeler *et al.* 2002), which attempts to provide a transcript map by utilizing sets of non-redundant gene-oriented clusters derived from GenBank sequences (the collection represents genes from many organisms, including human genes); TDB, which provides a substantial suite of databases containing DNA and protein sequence, gene expression, cellular role, and protein family information, and taxonomic data for microbes, plants, and humans (Quackenbush *et al.* 2001); OMIM (Hamosh *et al.* 2002), a compendium of genes and genetic disorders that allows systematic study of aspects of human disease; CDD, a compilation of multiple sequence alignments representing protein domains conserved in molecular evolution (Marchler-Bauer *et al.* 2002); and hundreds more!

We apologize to the curators of the many databases we have had to leave out. As mentioned earlier, a more comprehensive overview of molecular biology databases is given by Baxevanis in the annual *Nucleic Acids Research* (NAR) Database Issue (Baxevanis 2003) – short, searchable summaries and updates of each of the databases are available through the NAR Web site. A still more comprehensive list is available from DBcat, which enumerated more than 500 biological databases in 2000 (Discala *et al.* 2000). We hope that you now feel brave enough to go searching in today's forest of bio-databases: happy hunting!

SUMMARY

Primary DNA sequence databases maintain complete records of nucleotide sequences of all kinds, and database deposition of new sequences is usually a prerequisite of their publication in a scientific paper. The principal nucleotide databases are EMBL, GenBank, and DDBJ. These databases now collaborate as part of the International Nucleotide Sequence Database (INSD), and have an agreed system of sharing information, so that sequences submitted to one will automatically appear in the others.

The most widely used primary databases of protein sequences are PIR-PSD and Swiss-Prot. Swiss-Prot's notoriety arises from the extent of manual annotation of its entries. It is therefore smaller than other protein sequence repositories and, for this reason, its curators developed TrEMBL, an annotation-poor resource compiled automatically from translations of coding sequences in EMBL. UniProt is an integrated resource created by the Swiss-Prot and PIR-PSD curators, providing the most comprehensive repository of protein sequences worldwide.

Protein family databases aim to group together proteins with similar sequences, structures, and functions. Proteins may be diagnosed as belonging to particular families by virtue of sharing conserved sequence motifs or domains. Family databases store these diagnostic features, with annotations that relate the patterns of conservation to known structural and functional characteristics. Databases of this type include PROSITE, PRINTS, and Pfam, which store regular expressions and profiles, fingerprints, and HMMs, respectively. InterPro is a collaboration between the protein family databases that pools the discriminators and their annotation in an integrated way, creating one of the most sophisticated current systems for identification and classification of proteins.

There are also databases of protein structure and structure-based information. PDB is the principal archive of molecular structures derived from X-ray crystallography and NMR. SCOP and CATH use these structures to classify proteins hierarchically according to similarities between their 3D folds. Sequences and structures are probably the most commonly used type of information in biological databases at present, but they are only the tip of an enormous iceberg of possibilities. A vast range of databases now exists, capturing information on protein–protein interactions, metabolic reaction pathways, gene signaling networks, and information related to the genome projects of individual species. Integration of these different types of biological information is now a key issue in bioinformatics.

In our discussions of databases, we emphasized the format of their entries. A format has to be readable by both humans and computers. Although there are many different formats, most databases have some things in common. For example, they require a unique identifier for each entry, termed an accession number, which is assigned at the time the entry is added to the database and should remain static between database releases. Accession numbers are a quick and accurate way for computer programs to retrieve specified entries, and also provide an easy way to link entries between different databases. Most formats also contain information on the submitting authors, together with relevant scientific publications – depending on the type of information they house, some databases also include relevant annotation of sequences, their families, functions, structures, and so on. This can vary from full manual annotation to brief automatically derived annotation.

REFERENCES

Andreeva, A., Howorth, D., Brenner, S.E., Hubbard, T.J.P., Chothia, C., and Murzin, A.G. 2004. SCOP database in 2004: Refinements integrate structure and sequence family data. *Nucleic Acids Research*, **32**: D226–9.

Apweiler, R., Bairoch, A., Wu, C.H., Barker, W.C., Boeckmann, B., Ferro, S., Gasteiger, E., Huang, H., Lopez, R., Magrane, M., Martin, M.J., Natale, D.A., O'Donovan, C., Redaschi, N., and Yeh, L.S. 2004. UniProt: the Universal Protein Knowledgebase. *Nucleic Acids Research*, **32**(1): D115–19.

Attwood, T.K. 1997. Exploring the language of bioinformatics. In H. Stanbury (ed.), *Oxford Dictionary of Biochemistry and Molecular Biology*, pp. 715–23. Oxford, UK: Oxford University Press.

Attwood, T.K. 2000a. The quest to deduce protein function from sequence: The role of pattern databases. *International Journal of Biochemistry Cell Biology*, **32**(2): 139–55.

Attwood, T.K. 2000b. The role of pattern databases in sequence analysis. *Briefings in Bioinformatics*, **1**: 45–59.

Attwood, T.K., Avison, H., Beck, M.E., Bewley, M., Bleasby, A.J., Brewster, F., Cooper, P., Degtyarenko, K., Geddes, A.J., Flower, D.R., Kelly, M.P., Lott, S., Measures, K.M., Parry-Smith, D.J., Perkins, D.N., Scordis, P., and Scott, D. 1997. The PRINTS database of protein fingerprints: A novel information resource for computational molecular biology. *Journal of Chemical Information and Computer Sciences*, **37**: 417–24.

Attwood, T.K. and Beck, M.E. 1994. PRINTS – A protein motif fingerprint database. *Protein Engineering*, **7**(7): 841–8.

Attwood, T.K., Blythe, M.J., Flower, D.R., Gaulton, A., Mabey, J.E., Maudling, N., McGregor, L., Mitchell, A.L., Moulton, G., Paine, K., and Scordis, P. 2003. PRINTS and PRINTS-S shed light on protein ancestry. *Nucleic Acids Research*, **30**(1): 239–41.

Attwood, T.K. and Parry-Smith, D.J. 1999. *Introduction to Bioinformatics*. Harlow, UK: Addison Wesley Longman.

Bader, G.D., Betel, D., and Hogue, C.W.V. 2003. BIND: The Biomolecular Interaction Database. *Nucleic Acids Research*, **31**: 248–50.

Bateman, A., Coin, L., Durbin, R., Finn, R.D., Hollich, V., Griffiths-Jones, S., Khanna, A., Marshall, M., Moxon, S., Sonnhammer, E.L.L., Studholme, D.J., Yeats, C., and Eddy, S.R. 2004. The Pfam protein families database. *Nucleic Acids Research*, **32**: D138–44.

Baxevanis, A.D. 2003. The Molecular Biology Database Collection: 2003 update. *Nucleic Acids Research*, **31**: 1–12.

Benson, D.A., Karsch-Mizrachi, I., Lipman, D.J., Ostell, J., and Wheeler, D.L. 2004. GenBank: Update. *Nucleic Acids Research*, **32**: D23–6.

Boeckmann, B., Bairoch, A., Apweiler, R., Blatter, M.C., Estreicher, A., Gasteiger, E., Martin, M.J., Michoud, K., O'Donovan, C., Phan, I., Pilbout, S., and Schneider, M. 2003. The Swiss-Prot protein knowledge base and its supplement TrEMBL in 2003. *Nucleic Acids Research*, **31**(1): 365–70.

Boguski, M.S., Lowe, T.M., and Tolstoshev, C.M. 1993. dbEST – database for "expressed sequence tags". *Nature Genetics*, **4**(4): 332–3.

Blake, J.A., Richardson, J.E., Bult, C.J., Kadin, J.A., Eppig, J.T., and the Mouse Genome Database Group. 2003. MGD: The Mouse Genome Database. *Nucleic Acids Research*, **31**: 193–5.

Bourne, P.E., Addess, K.J., Bluhm, W.F., Chen, L., Deshpande, N., Feng, Z., Fleri, W., Green, R., Merino-Ott, J.C., Townsend-Merino, W., Weissig, H., Westbrook, J., and Berman, H.M. 2004. The distribution and query

systems of the RCSB Protein Data Bank. *Nucleic Acids Research*, **32**: D223–5.

Corpet, F., Servant, F., Gouzy, J., and Kahn, D. 2000. ProDom and ProDom-CG: Tools for protein domain analysis and whole genome comparisons. *Nucleic Acids Research*, **28**(1): 267–9.

Dayhoff, M.O. 1965. *Atlas of Protein Sequence and Structure*. Silver Spring, MD: National Biomedical Research Foundation.

Discala, C., Benigni, X., Barillot, E., and Vaysseix, G. 2000. DBCat: A catalog of 500 biological databases. *Nucleic Acids Research*, **28**(1): 8–9.

Dwight, S.S., Harris, M.A., Dolinski, K., Ball, C.A., Binkley, G., Christie, K.R., Fisk, D.G., Issel-Tarver, L., Schroeder, M., Sherlock, G., Sethuraman, A., Weng, S., Botstein, D., and Cherry, M.J. 2002. *Saccharomyces* Genome Database (SGD) provides secondary gene annotation using the Gene Ontology (GO). *Nucleic Acids Research*, **30**(1): 69–72.

Ellis, L.B.M., Hershberger, C.D., Bryan, E.M., and Wackett, L.P. 2001. The University of Minnesota Biocatalysis/ Biodegradation Database: Emphasizing enzymes. *Nucleic Acids Research*, **29**(1): 340–3.

Etzold, T., Ulyanov, A., and Argos, P. 1996. SRS – Information-retrieval system for molecular-biology databanks. *Methods in Enzymology*, **266**: 114–28.

Garavelli, J.S., Hou, Z., Pattabiraman, N., and Stephens, R.M. 2001. The RESID Database of protein structure modifications and the NRL-3D Sequence–Structure Database. *Nucleic Acids Research*, **29**(1): 199–201.

Haft, D.H., Loftus, B.J., Richardson, D.L., Yang, F., Eisen, J.A., Paulsen, I.T., and White, O. 2001. TIGRFAMs: A protein family resource for the functional identification of proteins. *Nucleic Acids Research*, **29**(1): 41–3.

Hamosh, A., Scott, A.F., Amberger, J., Bocchini, C., Valle, D., and McKusick, V.A. 2002. Online Mendelian Inheritance in Man (OMIM), a knowledge base of human genes and genetic disorders. *Nucleic Acids Research*, **30**(1): 52–5.

Henikoff, J.G., Greene, E.A., Pietrokovski, S., and Henikoff, S. 2000. Increased coverage of protein families with the Blocks database servers. *Nucleic Acids Research*, **28**(1): 228–30.

Huang, J.Y. and Brutlag, D.L. 2001. The eMOTIF database. *Nucleic Acids Research*, **29**(1): 202–4.

Hubbard, T., Barker, D., Birney, E., Cameron, G., Chen, Y., Clark, L., Cox, T., Cuff, J., Curwen, V., Down, T., Durbin, R., Eyras, E., Gilbert, J., Hammond, M., Huminiecki, L., Kasprzyk, A., Lehvaslaiho, H., Lijnzaad, P., Melsopp, C.,

Mongin, E., Pettett, R., Pocock, M., Potter, S., Rust, A., Schmidt, E., Searle, S., Slater, G., Smith, J., Spooner, W., Stabenau, A., Stalker, J., Stupka, E., Ureta-Vidal, A., Vastrik, I., and Clamp, M. 2002. The Ensembl genome database project. *Nucleic Acids Research*, **30**(1): 38–41.

Hulo, N., Sigrist, C.J., Le Saux, V., Langendijk-Genevaux, P.S., Bordoli, L., Gattiker, A., De Castro, E., Bucher, P., and Bairoch, A. 2004. Recent improvements to the PROSITE database. *Nucleic Acids Research*, **32**(1): D134–7.

Kanehisa, M., Goto, S., Kawashima, S., and Nakaya, A. 2002. The KEGG databases at GenomeNet. *Nucleic Acids Research*, **30**(1): 42–6.

Karp, P.D., Riley, M., Saier, M., Paulsen, I.T., Collado-Vides, J., Paley, S.M., Pellegrini-Toole, A., Bonavides, C., and Gama-Castro, S. 2002. The EcoCyc Database. *Nucleic Acids Research*, **30**(1): 56–8.

Kulikova, T., Aldebert, P., Althorpe, N., Baker, W., Bates, K., Browne, P., van den Broek, A., Cochrane, G., Duggan, K., Eberhardt, R., Faruque, N., Garcia-Pastor, M., Harte, N., Kanz, C., Leinonen, R., Lin, Q., Lombard, V., Lopez, R., Mancuso, R., McHale, M., Nardone, F., Silventoinen, V., Stoehr, P., Stoesser, G., Tuli, M.A., Tzouvara, K., Vaughan, R., Wu, D., Zhu, W., and Apweiler, R. 2004. The EMBL Nucleotide Sequence Database. *Nucleic Acids Research*, **32**: D27–30.

Laskowski, R.A. 2001. PDBsum: Summaries and analyses of PDB structures. *Nucleic Acids Research*, **29**(1): 221–2.

Letunic, I., Goodstadt, L., Dickens, N.J., Doerks, T., Schultz, J., Mott, R., Ciccarelli, F., Copley, R.J., Ponting, C.P., and Bork, P. 2002. Recent improvements to the SMART domain-based sequence annotation resource. *Nucleic Acids Research*, **30**(1): 242–4.

Marchler-Bauer, A., Panchenko, A.R., Shoemaker, B.A., Thiessen, P.A., Geer, L.Y., and Bryant, S.H. 2002. CDD: A database of conserved domain alignments with links to domain three-dimensional structure. *Nucleic Acids Research*, **30**(1): 281–3.

Mewes, H.W., Frishman, D., Güldener, U., Mannhaupt, G., Mayer, K., Mokrejs, M., Morgenstern, B., Münsterkötter, M., Rudd, S., and Weil, B. 2002. MIPS: A database for genomes and protein sequences. *Nucleic Acids Research*, **30**(1): 31–4.

Miyazaki, S., Sugawara, H., Gojobori, T., and Tateno, Y. 2004. DDBJ in the stream of various biological data. *Nucleic Acids Research*, **32**: D31–4.

Mulder, N.J., Apweiler, R., Attwood, T.K., Bairoch, A., Bateman, A., Binns, D., Biswas, M., Bradley, P., Bork, P., Bucher, P., Copley, R.R., Courcelle, E., Das, U., Durbin, R., Falquet, L., Fleischmann, W., Griffith-Jones, S., Haft, D., Harte, N., Hermjakob, H., Hulo, N., Kahn, D., Kanapin, A., Krestyaninova, M., Lopez, R., Letunic, I., Lonsdale, D., Silventoinen, V., Orchard, S., Pagni, M., Peyruc, D., Ponting, C.P., Servant, F., Sigrist, C.J.A., Vaughan, R., and Zdobnov, E. 2003. The InterPro database – 2003 brings increased coverage and new features. *Nucleic Acids Research*, **31**(1): 315–18.

Neville-Manning, C.G., Wu, T.D., and Brutlag, D.L. 1998. Highly specific protein sequence motifs for genome analysis. *Proceedings of the National Academy of Sciences USA*, **95**: 5865–71.

Overbeek, R., Larsen, N., Pusch, G.D., D'Souza, M., Selkov, E., Jr., Kyrpides, N., Fonstein, M., Maltsev, N., and Selkov, E. 2000. WIT: Integrated system for high-throughput genome sequence analysis and metabolic reconstruction. *Nucleic Acids Research*, **28**(1): 123–5.

Parry-Smith, D.J. and Attwood, T.K. 1992. ADSP – A new package for computational sequence analysis. *CABIOS*, **8**(5): 451–9.

Pearl, F.M.G., Bennett, C.F., Bray, J.E., Harrison, A.P., Martin, N., Shepherd, A., Sillitoe, I., Thornton, J., and Orengo, C.A. 2003. The CATH database: An extended protein family resource for structural and functional genomics. *Nucleic Acids Research*, **31**: 452–5.

Quackenbush, J., Cho, J., Lee, D., Liang, F., Holt, I., Karamycheva, S., Parvizi, B., Pertea, G., Sultana, R., and White, J. 2001. The TIGR Gene Indices: Analysis of gene transcript sequences in highly sampled eukaryotic species. *Nucleic Acids Research*, **29**(1): 159–64.

Rhee, S.Y. *et al.* 2003. The *Arabidopsis* Information Resource (TAIR): A model organism database providing a centralized, curated gateway to *Arabidopsis* biology, research materials and community. *Nucleic Acids Research*, **31**: 224–8.

Stein, L., Sternberg, P., Durbin, R., Thierry-Mieg, J., and Spieth, J. 2001. WormBase: Network access to the genome and biology of *Caenorhabditis elegans*. *Nucleic Acids Research*, **29**(1): 82–6.

Schuler, G.D. 1996. Entrez: Molecular biology database and retrieval system. *Methods in Enzymology*, **266**: 141–62.

The FlyBase Consortium. 2002. The FlyBase database of the *Drosophila* genome projects and community literature. *Nucleic Acids Research*, **30**(1): 106–8.

The Gene Ontology Consortium. 2001. Creating the gene ontology resource: Design and implementation. *Genome Research*, **11**: 1425–33.

Wheeler, D.L., Church, D.M., Lash, A.E., Leipe, D.D., Madden, T.L., Pontius, J.U., Schuler, G.D., Schriml, L.M., Tatusova, T.A., Wagner, L., and Rapp, B.A. 2002. Database resources of the National Center for Biotechnology Information: 2002 update. *Nucleic Acids Research*, **30**(1): 13–16.

Wu, C.H., Yeh, L-S.L., Huang, H., Arminski, L., Castro-Alvear, J., Chen, Y., Hu, Z.Z., Ledley, R.S., Kourtesis, P., Suzek, B.E., Vinayaka, C.R., Zhang, J., and Barker, W.C. 2003. The Protein Information Resource. *Nucleic Acids Research*, **31**(1): 345–7.

Sequence alignment algorithms

CHAPTER

6

CHAPTER PREVIEW

We present the dynamic programming algorithms used for pairwise sequence alignment. We consider variations in the algorithm that account for different types of gap cost functions and for global and local alignments. Examples are given that show the sensitivity of the results to the scoring parameters used. We then discuss heuristic algorithms for multiple sequence alignment, focusing on the commonly used progressive alignment technique.

6.1 WHAT IS AN ALGORITHM?

The sequence alignment problem is fundamental to bioinformatics. Alignments show us which bits of a sequence are variable and which bits are conserved. They indicate the positions of insertions and deletions. They allow us to identify important functional motifs. They are the starting point for evolutionary studies using phylogenetic methods. They form the basis of database search methods. You can't get far in gene sequence analysis without an alignment. Before describing the computational methods used to produce alignments, we need to explain what is meant by the term algorithm.

An algorithm is a series of instructions that explain how to solve a particular problem. It is a recipe that describes what steps to take to obtain the desired answer. It consists of a series of logical or mathematical statements that explain what to do in what circumstances. An algorithm is not itself a computer program, but a good programmer should easily be able to translate any properly defined algorithm into a program in any programming language.

As a simple example, suppose we have a list of numbers in a random order and we wish to sort them into ascending order (see Fig. 6.1(a)). One easy algorithm for doing this is called Insertion Sort, which can be described in the following way:

• If the second element is less than the first element, move the second element one place to the left (in this case, swap 22 with 71).

• If the third element is less than the second, move the third element to the left. If this element is also less than the first element, move it another place to the left (in this case, 9 moves to the beginning of the list).

• Consider each subsequent element and move it to the left until the element preceding it is lower than it or until it reaches the beginning of the list (e.g., 18 is moved to second position).

• Continue until all elements are in ascending order. It would be easy to write a computer program to carry out this algorithm. You could also do it yourself by hand (e.g., if you were sorting a pack of playing cards into ascending numerical value).

An important property of algorithms is how long they take to carry out or, more specifically, how the time taken scales with the size of the problem, N. In this example, N is the number of elements in the list to be sorted. Since there are N elements to move, the time taken must at least be proportional to N. However, each element must be compared with several elements to its left in order to slide it into its correct place. The number of comparisons to be made is typically proportional to N. Thus, the total time taken is proportional to N^2. We say that the time taken for this algorithm is $O(N^2)$ – "of order N squared". The actual time taken will depend on how good the programmer is, what language the program is written in, and what computer runs the program. Nevertheless, the time will always scale with N^2. Thus, it will always take four times as long to sort a list of 200 numbers as it does to sort a list of 100 numbers using the Insertion Sort algorithm.

Sorting the elements of an array is something that quite often crops up in real applications. For example, suppose you write a search program to measure similarity between sequences in a database and a query sequence. It would be natural to sort the output into descending order of sequence similarity, so that the highest scoring match is at the top of the list. Although Insertion Sort is simple to understand, it is actually a rather inefficient algorithm for sorting large arrays. There are many other methods that run faster, although their algorithms are more complex to describe. The book by Press *et al.* (1992) has a chapter describing several practical methods that is worth consulting if you ever need a sorting program. We are not interested in the details of these methods here. However, the sorting problem does

illustrate an important point about algorithms in general. Often, there are several algorithms that will produce the same result. In this case, there is a trade-off between time and memory. Fast algorithms often require additional working memory in the computer to store quantities determined at intermediate stages of the calculation. The Counting Sort algorithm described by Cormen, Leiserson, and Rivest (1990) runs in a time $O(N)$ and is therefore much faster than Insertion Sort, but it requires two further storage arrays of size $O(N)$, in addition to the $O(N)$ memory used to store the data. Insertion Sort only requires a single spare variable to use as swap space when exchanging pairs of numbers. Counting Sort therefore uses roughly three times as much memory as Insertion Sort. In cases where algorithms for complex problems require memory storage that is large compared to the input data for the problem, the scaling of the required memory space with N can be an issue. When memory is scarce, a slow algorithm with smaller memory requirements may be preferable.

Figure 6.1(b) illustrates another famous problem in computer science: the traveling salesman problem (TSP). The problem is defined in the following way: a salesman must make a trip to visit each one of a set of cities exactly once and then return to his

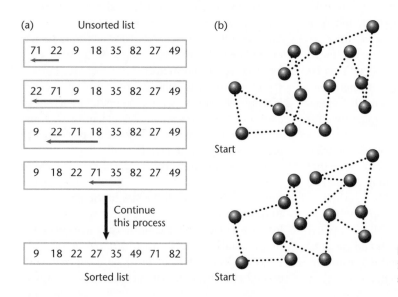

(a)   Unsorted list

71 22 9 18 35 82 27 49

22 71 9 18 35 82 27 49

9 22 71 18 35 82 27 49

9 18 22 71 35 82 27 49

Continue this process

9 18 22 27 35 49 71 82

Sorted list

(b)

Start

Start

Fig. 6.1 (a) Sorting a list of numbers into ascending order. (b) The traveling salesman problem.

starting point. The distance between each pair of cities is specified in advance. The problem is to find the route that minimizes the total distance traveled. The upper figure shows one rather long route that wastes time by crossing over itself in several places. The lower figure shows a more plausible route that probably does not go too much farther than necessary; however, it may not be the best. If there are N cities, then there are $N!$ possible routes. One way of solving the TSP that is guaranteed to give the best answer is by exact enumeration, i.e., a computer could be asked to consider every one of the possible orders in turn and save the best one. The time required would also increase as $N!$, which is huge (see Section M.2). So exact enumeration would only be possible for very small N.

There are many computational problems where the number of cases to be considered increases either factorially (as $N!$) or exponentially (as α^N for some constant α). In both cases, exact enumeration of all the cases becomes impossible for large values of N. Nevertheless, there are often algorithms that can solve the problem in "polynomial time", i.e., in a time that is $O(N^\beta)$, for some constant β. The best algorithms have β as small as possible. For example, in the sorting problem there are also $N!$ possible orderings of the N elements, but we do not need to check all these orderings one by one. The Insertion Sort method is a polynomial time algorithm with $\beta = 2$. On the other hand, the TSP is an example of a problem for which there is no known algorithm that can solve the problem exactly in polynomial time. Computationally difficult problems like this are known as NP-complete (where NP stands for nondeterministic polynomial – see Cormen *et al.* 1992). Where we cannot find exact algorithms that are guaranteed to give the correct answer, heuristic algorithms are used. A heuristic algorithm is one that we believe is likely to give a fairly good solution most of the time, even though we cannot prove that the solution given is really the best one.

One possible heuristic algorithm for the TSP would simply be always to travel to the nearest city that you have not been to before. This is likely to produce a route that is not too bad – certainly it will be much better than a random route. However, it may

lead to a situation where the last city visited is very far from the starting point, and hence there will be a very large distance to travel home. This type of heuristic is known as a greedy algorithm. Greedy algorithms are shortsighted, i.e., they always do what is best in the short term. Often, greedy algorithms can be proposed as first solutions to a problem, and can then be improved by better heuristic algorithms. There is a large literature on the development of heuristic algorithms for the TSP. Both exact and heuristic algorithms will arise in our discussion of alignment methods in this chapter. In molecular phylogenetics, we also require heuristic algorithms to search through very large numbers of possible evolutionary trees (see Section 8.5).

6.2 PAIRWISE SEQUENCE ALIGNMENT – THE PROBLEM

We will begin with the basic problem of aligning two sequences. A pairwise alignment of two DNA sequences looks something like this:

```
CAGT-AGATATTTACGGCAGTATC----
CAATCAGGATTTT--GGCAGACTGGTTG
```

Gaps have been inserted in both the top and the bottom sequence in order to slide the sequences along so that the regions of similarity between them are apparent. When we align two sequences, we usually do so because they are homologous, i.e., we believe they have evolved from a common ancestor. The alignment above means that the first two sites CA have remained unchanged, while there has been a mutation at the third site from a G to an A or from an A to a G (we don't know which). The gap at the fifth site means that either there has been a deletion in the top sequence or there has been an insertion of a C in the bottom sequence (again we don't know which). When we look for the "best" alignment of two sequences, we are really looking for the alignment that we think is most likely to have occurred during evolution.

In order to obtain the best alignment, we need a scoring system. We know that the sequences to be aligned are composed of letters from an alphabet of allowed characters. For DNA, the alphabet is just A, C, G, and T. For proteins, it is the 20-letter alphabet corresponding to the amino acids. The first part of the scoring system is to define a score, $S(\alpha,\beta)$, for every pair of letters in the allowed alphabet. For DNA, the scoring system is usually a very simple one: we assign $S(\alpha,\beta) = 1$ whenever α and β are the same nucleotide, and $S(\alpha,\beta) = 0$ whenever they are different nucleotides. For proteins, we could also assign 1 for identical amino acids and 0 for different amino acids if we wished; however, this is not a very useful scoring system for proteins in practice. A good scoring system needs to reflect the fact that some amino acids are similar to one another and some are different. We want to assign a high positive score to two identical amino acids, a slightly positive score to two similar amino acids (say, D and E, or I and L), and a slightly negative score for two amino acids that are very different from one another (say, D and I). In Chapter 4, we discussed the way scoring matrices, such as PAM and BLOSUM, are calculated. Any matrix of scores like this offers a possible scoring system for aligning protein sequences. The alignment algorithm works in the same way whatever scores we use, but the optimal alignment produced by the algorithm will depend on the scoring system.

The second part of the scoring system is the penalty for gaps. In general, let us define $W(l)$ as the penalty for a gap of length l characters. The simplest function we can choose is a linear gap penalty: $W(l) = gl$. This mean that each individual gap character has a cost g, and the cost of the full gap is just g times the length. However, in reality, it may not be that a gap of three residues is three times as unlikely as a gap of a single residue. It could be that a single event created a gap of three residues. The likelihood of this occurring therefore depends on the underlying mutational process, i.e., the error in DNA replication that created the gap in this gene. It also depends on the selection process. We will usually be aligning sequences that are real working molecules. Many insertions or deletions may lead to molecules that are nonfunctional (e.g., frame-shift mutations in the coding regions of DNA) and those sequences will be eliminated by selection. There are certain regions of

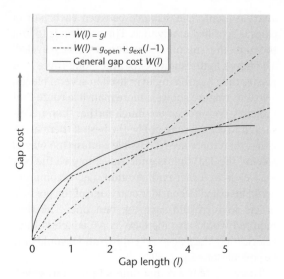

Fig. 6.2 Gap cost functions $W(l)$ – linear, affine, and general gap functions are shown.

sequences where we tend to find gaps and certain regions where we don't. For example, unstructured "loop" regions in proteins tend to be much more variable than regions with well-defined structures, such as helices in membrane proteins. Taking all these effects into account, we might expect that the cost function $W(l)$ should look something like the curve in Fig. 6.2. It should increase steeply at first, because making a gap is an unlikely evolutionary event, but then it should increase less steeply as l gets larger, because if we already know that a gap is possible at this position, then it may not be much more unlikely to have a bigger gap than a smaller one.

There is no real theory to tell us what function to use for gap penalties, although some progress has been made in analyzing the empirical distribution of gap sizes (Benner, Cohen, and Gonnet 1993). For practical reasons, the gap-scoring system is usually kept fairly simple. The most widely used gap-scoring system in real applications is the Affine Gap Penalty function. Here, we suppose that the penalty for opening a new gap is g_{open}, and the penalty for extending the gap for each subsequent step is g_{ext}. Hence the penalty for a gap of length l is $W(l) = g_{open} + g_{ext}(l - 1)$. This is shown as the dashed line in Fig. 6.2. Usually $g_{ext} < g_{open}$ for the reasons given above. In fact, the Affine Gap Penalty function is not that far from the general curved shape that we might expect. An extreme case would be to set $g_{ext} = 0$, in which case the penalty of a gap would be just g_{open}, independent of its length.

Having defined our scoring system, it is possible to calculate a score for any pairwise alignment, such as the one at the beginning of this section:

$$Score = \sum_{\substack{aligned \\ pairs}} S(\alpha, \beta) - \sum_{gaps} W(l) \qquad (6.1)$$

The principle of pairwise alignment algorithms is that we consider every possible way of aligning two sequences by sliding one with respect to the other and inserting gaps in both sequences, and we determine the alignment that has the highest score. The number of alignments is huge (it increases factorially with the lengths of the sequences), hence we cannot simply enumerate all possible alignments. However, polynomial time algorithms are known for solving this problem using a method known as dynamic programming.

6.3 PAIRWISE SEQUENCE ALIGNMENT – DYNAMIC PROGRAMMING METHODS

6.3.1 Algorithm 1 – Global alignment with linear gap penalty

Suppose we have two sequences with lengths N_1 and N_2. The letters in the first sequence will be denoted a_i (where $1 \le i \le N_1$) and the letters in the second sequence will be denoted b_j (where $1 \le j \le N_2$). We will define $H(i,j)$ as the score of the best alignment that can be made using the first i letters of the first sequence and the first j letters of the second sequence. We want to calculate the score $H(N_1, N_2)$ for aligning the full sequences. To do this, we will have to calculate $H(i,j)$ for each value of i and j and gradually build up to the full length.

The first algorithm we consider is known as the Needleman–Wunsch algorithm after its originators (Needleman and Wunsch 1970). We wish to calculate $H(i,j)$, the score for the partial alignment ending

in a_i and b_j. There are only three things that can happen at the end of this partial alignment: (i) a_i and b_j are aligned with each other; (ii) a_i is aligned with a gap; or (iii) b_j is aligned with a gap. Since we want the highest alignment score, $H(i,j)$ is the maximum of the scores corresponding to these three possibilities. Hence we may write

$$H(i,j) = \max \begin{cases} H(i-1, j-1) + S(a_i, b_j) & \text{diagonal} \\ H(i-1, j) - g & \text{vertical} \\ H(i, j-1) - g & \text{horizontal} \end{cases}$$

$$(6.2)$$

In option (i), we add the score $S(a_i, b_j)$ for aligning a_i and b_j to the score $H(i-1, j-1)$ for aligning the sequences up to a_{i-1} and b_{j-1}. In option (ii), we have created a gap; therefore, we pay a cost, g, which we subtract from the score $H(i-1, j)$ for aligning the sequences up to a_{i-1} and b_j. Similarly, in option (iii), we subtract g from the score for aligning the sequences up to a_i and b_{j-1}.

The scores $H(i,j)$ are the elements of a matrix where i labels the rows and j labels the columns. Option (i) corresponds to making a diagonal step from the top left of the matrix towards the bottom right. Option (ii) corresponds to making a step vertically downwards, and option (iii) corresponds to making a step horizontally from left to right. Equation (6.2) says that if we want to know the score for the cell (i, j) in the matrix, then we need to know the score for the diagonal, vertical, and horizontal neighbors of this cell. The key point about Eq. (6.2) is that it works for every value of i and j. We can use the same formula to work out the scores for the three neighbors in terms of **their** neighbors. This formula is called a recursion relation. We can use it recursively for each cell in the matrix, starting with small values of i and j and building up to large ones.

In order to get started with this recursion, we need to know some initial conditions. In this case, we know that $H(i,0) = -gi$. This means that the first i letters of sequence 1 are aligned with nothing at all from sequence 2. Hence there is a gap of size i at the beginning, which costs $-gi$. Similarly, $H(0,j) = -gj$, which corresponds to the first j letters of sequence 2

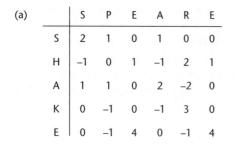

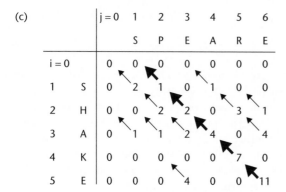

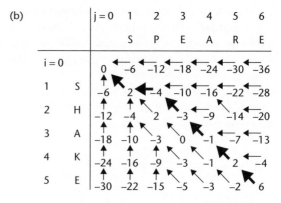

Fig. 6.3 Pairwise alignment of SHAKE and SPEARE. (a) Pairwise amino acid scores taken from the PAM250 matrix. (b) Alignment scores $H(i,j)$ using algorithm 1 and $g = 6$. (c) Alignment scores $H(i,j)$ using algorithm 2 and $g = 6$. The pathways through the matrix corresponding to the optimal alignments in (b) and (c) are indicated by the thick arrows.

being aligned with nothing in sequence 1. It is also necessary to define $H(0,0) = 0$. This means that the score is 0 before we begin the alignment.

As an example, we will align the two protein sequences SHAKE and SPEARE using the PAM250 scoring matrix that was already shown in Fig. 4.6. From this, we may calculate a score $S(a_i,b_j)$ for each pair of amino acids that occur in this example. For convenience, these scores are shown in Fig. 6.3(a). We now draw out the matrix $H(i,j)$, as in Fig. 6.3(b), including a row and column corresponding to $i = 0$ and $j = 0$. Initially, all the numbers in the cells are unknown; however, we can start right away by filling in the cells corresponding to the initial conditions. Let the gap cost be $g = 6$. All the cells around the edge of the matrix will have scores that are multiples of −6. For example, $H(3,0) = -18$, and $H(0,2) = -12$. These correspond to the two partial alignments

```
SHA     and     --
---             SP
```

We can now proceed with the row $i = 1$. From Eq. (6.2), the options for cell $(1,1)$ are

```
S(0+2=2) or -S(-6-6=-12) or S-(-6-6=-12)
S            S-               -S
```

Clearly, the first option is the best of these three. Therefore, we enter 2 in cell $(1,1)$. A diagonal arrow has been drawn from this cell in Fig. 6.3(b) to indicate that the best option was to move diagonally to this cell. We are now in a position to calculate cell $(1,2)$. The three options are:

```
-S(-6+1=-5) or --S(-12-6=-18) or S-(2-6=-4)
SP             SP-                SP
```

The third option is the best. Therefore, we enter −4 in this cell and draw a horizontal arrow from this cell to indicate this. It is possible to continue using the recursion relation to fill up all the cells in the matrix, as shown in Fig. 6.3(b) (try doing this yourself to make sure you understand it).

Having filled in the matrix, we know that the optimal alignment of SHAKE and SPEARE has a score of $H(5,6) = 6$. More importantly, however, we need to know **what the alignment is** that has the score of 6. To construct the alignment, we start in the bottom right-hand corner of the matrix and follow the arrows backwards to the top left-hand corner. Whenever the arrow is diagonal, then the corresponding two letters are aligned with each other. Whenever there is a vertical move, we align a letter from sequence 1 with a gap, and whenever there is a horizontal move, we align a letter from sequence 2 with a gap. This procedure is known as backtracking. The result in our case is

```
S-HAKE
SPEARE
```

Sometimes, Shakespeare spelled his name without the final E, so it is interesting to know what is the optimal alignment of SHAKE and SPEAR. We might guess that it would be

```
S-HAKE
SPEAR-
```

but the matrix in Fig. 6.3(b) tells us this is not true. If we ignore the final column and trace back the arrows from cell (5,5), we obtain

```
SHAKE
SPEAR
```

which has a score of -2. The original guess contains positive scores for matches, $+2$ for (A,A) and $+3$ for (K,R), that are lost in the second alignment. However, the original guess has two costly gaps, and its net score is -4, which is worse than the second alignment. When the final E was present, we gained an extra $+4$ from the (E,E) pair, which helped to compensate for the gap opposite the P in the original optimal alignment. Also, the two sequences had unequal length when the E was present, so we had to put a gap somewhere, even if it was costly.

Let us consider how much memory is required by this algorithm. Suppose both sequences are of equal length, N. The number of cells in the $H(i,j)$ matrix is $(N+1)^2$. For scaling arguments, we are only interested in the leading order of N, so this is $O(N^2)$. Thus, we need $O(N^2)$ memory to store the scores. We also need a similar amount of memory to store the backtracking information. On a computer, you cannot draw arrows, so one way of storing this information is to define a backtracking matrix, $B(i,j)$, whose elements are set to 0 for a diagonal move, $+1$ for a vertical move, and -1 for a horizontal move.

The scaling of the time with N is also important. There are $O(N^2)$ cells to calculate in the matrix $H(i,j)$. Each of these requires calculating three numbers and taking the best. There is no N dependence in these calculations; therefore, the time required is $O(N^2)$. Hence, this algorithm has fairly modest requirements in terms of both memory and time, and is a practical way of aligning long sequences.

6.3.2 Algorithm 2 – Local alignment with linear gap penalty

A global alignment algorithm (such as algorithm 1) aligns the whole of one sequence with the whole of another. This is appropriate if we believe that the two sequences in question are similar along their whole length. However, in some cases, it may be that two sequences share a common domain but are not similar outside this domain. Therefore, it would be inappropriate to try to align the whole sequences. It might also happen that we have a fragment of a gene from one species and we want to align this to the matching part of a complete gene from another species. In both these cases, we would use a local alignment algorithm that looks for high-scoring matches between any part of one sequence and any part of the other, and neglects parts of the sequences for which there is not a good match.

The simplest local alignment algorithm just adds a fourth option to the recursion relation of algorithm 1. If the best of the three scores for the diagonal, vertical, and horizontal moves is negative, then we simply assign 0 to this cell. This means that we are not to bother to align these parts of the sequences and that we are to start another alignment after this point. The recursion is therefore:

$$H(i,j) = \max \begin{cases} H(i-1,j-1) + S(a_i, b_j) & \text{diagonal} \\ H(i-1,j) - g & \text{vertical} \\ H(i,j-1) - g & \text{horizontal} \\ 0 & \text{start again} \end{cases}$$

$$(6.3)$$

It is also necessary to change the initial conditions to reflect the fact that gaps at the beginning and end of an alignment are to be ignored (e.g., if we align a fragment of a gene with a whole gene). This is done by setting the cells around the edges of the matrix, $H(i,0)$ and $H(0,j)$, to zero. The resulting matrix for the alignment of SHAKE and SPEARE is shown in Fig. 6.3(c). The backtracking procedure for local alignments begins at the highest-scoring point in the matrix, and follows the arrows back until a 0 is reached. In our case, the highest-scoring cell is the bottom right-hand corner, but in examples with longer sequences, it could be anywhere in the matrix. The optimal local alignment is

```
SHAKE
PEARE
```

which scores 11 with this scoring system. Another possible local alignment with a score of 4 is

```
SHA
ARE
```

6.3.3 Algorithm 3 – General gap penalty

So far, we considered only the case of linear gap penalty $W(l) = gl$. We will now give an algorithm that works for any general gap penalty function, $W(l)$, of the shape given in Fig. 6.2.

$$H(i,j) = \max \begin{cases} H(i-1,j-1) + S(a_i, b_j) & \text{diagonal} \\ \max_{1 \le l \le i}(H(i-1,j) - W(l)) & \text{vertical} \\ \max_{1 \le l \le i}(H(i,j-l) - W(l)) & \text{horizontal} \end{cases}$$

$$(6.4)$$

The diagonal option corresponds to aligning a_i with b_j, as before. The second option corresponds to aligning a_i with a gap in the second sequence, but this

time we need to know the length l of the gap. For any given l, the cost of the gap is $W(l)$ and the score for aligning the previous part of the chain up to a_i and b_{j-l} is $H(i,j-l)$. In terms of moving through the matrix, this corresponds to moving l spaces vertically. The size of the gap can be anywhere between 1 and i. Thus, there are i different cases to evaluate, and we need the highest score among these. Similarly, the third option in Eq. (6.4) corresponds to making a horizontal jump of size l in the matrix, and l can be anywhere in the range 1 to j. The final score to be entered for $H(i,j)$ is the maximum of the single dia-gonal move, the i vertical moves, and the j horizontal moves. There are thus $O(N)$ calculations to be made in order to fill in one cell in the matrix (whereas there were only three calculations to be made in algorithm 1). The number of cells in the matrix is still $O(N^2)$. Hence the total time required for this algorithm is $O(N^3)$.

Both global and local versions of the general gap penalty algorithm can be written. In the global version, the initial conditions are $H(i,0) = -W(i)$, $H(0,j) = -W(j)$. In the local version, the initial conditions are $H(i,0) = H(0,j) = 0$, and we need to add a zero-score option into the recursion (6.4), as in Eq. (6.3). There is one further subtlety: Eq. (6.4) only makes sense if

$$W(l_1 + l_2) \le W(l_1) + W(l_2) \qquad (6.5)$$

for all values of l_1 and l_2. In other words, the penalty for one long gap has to be less than the penalty for two short gaps that add up to the same length. If this is not true, a long gap will be scored as though it were a series of smaller gaps.

The local version of this algorithm is the one proposed by Smith and Waterman (1981). However, it was pointed out by Gotoh (1982) that when the affine gap penalty function is used, the algorithm can be rewritten more efficiently, so that the time is only $O(N^2)$. The more efficient version is described below as algorithm 4. In practice, it is this that is usually referred to as "Smith–Waterman" alignment.

6.3.4 Algorithm 4 – Affine gap penalty

The affine gap penalty function, $W(l) = g_{open} +$

$g_{ext}(l-1)$, was introduced in Section 6.2. In practice, this is the most commonly used gap function because it is more flexible and realistic than the linear gap penalty and because the time-efficient algorithm given here can be used with this form of gap function but not with more general functions.

Let us define $M(i,j)$ as the score of the best alignment up to point i on the first sequence and j on the second, with the constraint that a_i and b_j are aligned with each other. Let $I(i,j)$ be the score of the best alignment up to this point, with the constraint that a_i is aligned with a gap, and let $J(i,j)$ be the score of the best alignment up to this point, with the constraint that b_j is aligned with a gap. The score of the best alignment to this point, with no restrictions, is $H(i,j)$, as before, which can be written

$$H(i,j) = \max(M(i,j), I(i,j), J(i,j)) \quad (6.6)$$

In this algorithm there are three separate matrices to be calculated, so the memory requirement is three times as large, but the time taken is speeded up by an order of magnitude. We can now write a recursion for $M(i,j)$:

$$M(i,j) = \max \begin{cases} M(i-1,j-1) + S(a_i, b_j) \\ I(i-1,j-1) + S(a_i, b_j) \\ J(i-1,j-1) + S(a_i, b_j) \end{cases} \quad (6.7)$$

Each of these three options corresponds to matching a_i with b_j, and, therefore, all include the score $S(a_i, b_j)$; however, they cover the three different possible configurations of the residues a_{i-1} and b_{j-1}. For $I(i,j)$, there are just two options:

$$I(i,j) = \max \begin{cases} M(i-1,j) - g_{open} \\ I(i-1,j) - g_{ext} \end{cases} \quad (6.8)$$

The first case is where a_{i-1} is aligned with b_j, and a_i is aligned with the beginning of a new gap; therefore, there is a cost of g_{open}. The second case is where a_{i-1} is already aligned with a gap and a_i is aligned with an extension of the same gap; therefore, there is a cost g_{ext} (the cost of the first gap is already included in $I(i,j-1)$). Similarly, there are two options for $J(i,j)$:

$$J(i,j) = \max \begin{cases} M(i,j-1) - g_{open} \\ J(i,j-1) - g_{ext} \end{cases} \quad (6.9)$$

This version of the algorithm disallows configurations where a gap in one sequence is immediately followed by a gap in the other sequence. This simplifies the number of cases to consider, and usually it makes no difference to the optimal alignment. In the following example,

(i) `S--HAKE`   (ii) `S-HAKE`
     `SPE-ARE`   `SPEARE`

alignment (ii) has a higher score than alignment (i) if $S(H,E) > -g_{open} - g_{ext}$. It follows that if $S(\alpha, \beta) > -g_{open} - g_{ext}$ for any two letters, α and β then configurations like (i) will never occur, so we do not need to include them in the recursion relations. For configurations where the gaps in both sequences are of length 1, the inequality would be $S(\alpha, \beta) > -2g_{open}$, but this inequality is automatically satisfied if the previous one is satisfied, and therefore only the previous one is relevant.

As with algorithm 3, it is possible to produce both local and global versions of algorithm 4 by adding in the zero-score option and changing the initial conditions. We will not give details here. This treatment has followed that given in Durbin *et al.* (1998) quite closely, which is a useful source of further information.

All the alignment algorithms described in this section fall into the category of dynamic programming algorithms. The essential feature of dynamic programming is that it uses recursion relations to break down the problem into a number of smaller problems. Many slightly different alignment algorithms will be found in the literature. Sometimes, the alignment problem is expressed as the minimization of an editing distance between two sequences rather than the maximization of a score (e.g., Clote and Backofen 2000, Gotoh 1982). The recursions are then similar, but they all contain the minimum of a number of alternatives rather than the maximum.

6.4 THE EFFECT OF SCORING PARAMETERS ON THE ALIGNMENT

All the algorithms above are exact in the sense that

they are guaranteed to give the highest-scoring alignment for a given scoring system. However, this does not mean that they are guaranteed to give the "correct" alignment. In fact, we can never know what the correct alignment is, since we cannot go back in time to follow the process by which the two sequences evolved from their common ancestor. By choosing scoring systems that reflect what we know about molecular evolution, we can hope to obtain alignments that are evolutionarily meaningful. However, there are several different ways of deriving substitution matrices for amino acids, and different sets of matrices, such as PAM and BLOSUM, are commonly used in alignments (as we described in Section 4.3). The alignment we get with any two sequences will depend on the details of the substitution matrix used, as well as on the details of the algorithm. The only way to tell which of these matrices is best is to look at the alignments produced for real sequences, and to use some biological intuition to decide which one seems to make the most sense.

The values of the gap penalties affect the properties of the alignments produced in important ways. To illustrate this, we have chosen hexokinase, which is the first enzyme in the glycolytic pathway, and is responsible for converting glucose into glucose-6-phosphate. Table 6.1 shows the accession numbers of the hexokinase sequences referred to in this chapter. The hexokinase sequence HXK_SCHMA from the blood fluke parasite *Schistosoma mansoni* was aligned with the sequence HXKP_HUMAN, which is one of several hexokinase genes in the human genome. The program ClustalX was designed to align many sequences, and is discussed more fully below. In this example, we used ClustalX to carry out pairwise alignment of these two sequences. The parameters were initially left at their defaults: the substitution matrix was Gonnet 250, the gap-opening cost was 10, and the gap extension cost was 0.1. The resulting alignment is shown in Fig. 6.4(a) – to save space, only part of the alignment is shown. There is a fairly high degree of similarity across the whole of the sequence, as seen by the fairly large number of sites with identical residues (denoted with an asterisk), or residues with similar properties (denoted with a colon or a dot). Gaps arise in several

Table 6.1 Accession numbers of hexokinase sequences.

Organism	Swiss-Prot ID	GenBank accession number
Plasmodium falciparum	HXK_PLAFA	Q02155
Saccharomyces cerevisiae	HXKA_YEAST	P04806
	HXKB_YEAST	P04807
	HXKG_YEAST	P17709
Schistosoma mansoni	HXK_SCHMA	Q26609
Drosophila melanogaster	HXK1_DROME	Q9NFT9
	HXK2_DROME	Q9NFT7
Human	HXK1_HUMAN	P19367
	HXK2_HUMAN	P52789
	HXK3_HUMAN	P52790
	HXKP_HUMAN*	P35557
Rat	HXK1_RAT	P05708
	HXK2_RAT	P27881
	HXK3_RAT	P27926
	HXKP_RAT*	P17712

* These two sequences were renamed HXK4_HUMAN and HXK4_RAT in September 2003 after preparation of the figures for this book. This is a reminder to us that ID codes and sequence annotation in databases can be updated, whereas accession numbers are intended to remain constant.

```
HXKP_HUMAN    ELVRLVLLRLVDENLLFHGEASEQLRTRGAFETRFVSQVESDTGDRKQIYNILSTLGLRP
HXK_SCHMA     ELVRHIIVYLVEQKILFRGDLPERLKVRNSLLTRYLTDVERDPAHLLYNTHYMLTDDLHV
              ****  :::  **:::**:*:  .*:*:..*.:: **:::** *...     : : * .*:

HXKP_HUMAN    ---STTDCDIVRRACESVSTRAAHMCSAGLAGVINRMRESRSEDVMRITVGVDGSVYKLH
HXK_SCHMA     PVVEPIDNRIVRYACEMVVKRAAYLAGAGIACILRRIN--RSE----VTVGVDGSLYKFH
               .. *  *** *** *  .***::..**:* ::.*:.  ***    :*******:**:*

HXKP_HUMAN    PSFKERFHASVRRLTP-SCEITFIESEEGSRGAALVSAVACKKACMLGQ
HXK_SCHMA     PKFCERMTDMVDKLKPKNTRFCLRLSEDGSGKGAAAIAASCTRQN-----
              *.* **:    * :*.* . .: :  **:***:*** ::* . .::
```

```
HXKP_HUMAN    ELVRLVLLRLVDENLLFHGEASEQLRTRGAFETRFVSQVESDTGDRKQIYNILSTLGLRP
HXK_SCHMA     ELVRHIIVYLVEQKILFRGDLPERLKVRNSLLTRYLTDVERDPAHLLYNTHYMLTDDLHV
              ****  :::  **:::**:*:  .*:*:..*.:: **:::** *...     : : * .*:

HXKP_HUMAN    ---STTDCDIVRRACESVSTRAAHMCSAGLAGVINRMRESRSEDVMRITVGVDGSVYKLH
HXK_SCHMA     PVVEPIDNRIVRYACEMVVKRAAYLAGAGIACILRRINRSE------VTVGVDGSLYKFH
               .. *  *** *** *  .***::..**:* ::.*:...*.      :*******:**:*

HXKP_HUMAN    PSFKERFHASVRRLTPSCEITFIESEEGSRGAALVSAVACKKACMLGQ
HXK_SCHMA     PKFCERMTDMVDKLKPKNTRFCLRLSEDGSGKGAAAIAASCTRQN----
              *.* **:    * :*.*       :: ..*... .* . *.:*.:
```

```
HXKP_HUMAN    ELVRLVLLRLVDENLLFHGEASEQLRTRGAFETRFVSQVESDTGDRKQIYN---ILSTLG
HXK_SCHMA     ELVRHIIVYLVEQKILFRGDLPERLKVRNSLLTRYLTDVERDPA--HLLYNTHYMLTD-D
              ****  :::  **:::**:*:  .*:*:..*.:: **:::** *..   : :**   :*:   .

HXKP_HUMAN    LR-PSTTDCD--IVRRACESVSTRAAHMCSAGLAGVINRMRESRSEDVMRITVGVDGSVY
HXK_SCHMA     LHVPVVEPIDNRIVRYACEMVVKRAAYLAGAGIACILRRIN--RSE----VTVGVDGSLY
              *: *  .    *  *** *** *  .***::..**:* ::.*:  .***    :*******:*

HXKP_HUMAN    KLHPSFKERFHASVRRLTPSCEITF-IE-SEEGSRGAALVSAVACKKACMLGQ-
HXK_SCHMA     KFHPKFCERMTDMVDKLKPK-NTRFCLRLSEDGSGKGAAAIAA-SCTR-----QN
              *:**.* **:    * :*.*. :  * :. **:***:*** ::* :*.:      *
```

Fig. 6.4 Global pairwise alignments of hexokinase proteins from human and *Schistosoma mansoni* using an affine gap penalty function. The three parameters used for the three alignments differ only in the value of the gap opening parameter. Regions of alignments (b) and (c) that differ from alignment (a) are written in bold.

places, indicating that insertions and deletions have occurred during evolution.

For the sake of comparison, we performed two further alignments of the same sequences, keeping all the parameters fixed except for the gap-opening cost. In (b), the cost has been increased to 50. Regions of alignment (b) that differ from (a) are highlighted. The short sequence RSE in the *Schistosoma* sequence has been shifted two places to the left in (b) with respect to (a). This changes two short gaps into one longer one, and thus increases the alignment score, even though the score of the exact match on RSE itself is lost. A long section at the end of the human sequence has also been shifted one place to the left in (b), which eliminates a single gap. However, by doing this, a large number of identical pairs in alignment (a) have been lost. In (c), the gap-opening penalty has been decreased to 2. There are four highlighted regions in (c), all showing small differences from (a). Reducing the gap-opening cost in (c) leads to the introduction of a number of additional short gaps. These tend to occur in regions where there are few exactly matching residues – i.e., rapidly evolving regions of the sequence. In such regions, the alignment will depend on the details of the scores used for non-identical residues (i.e., on which PAM or BLOSUM matrix is used), as well as on the size of the gap cost. In strongly conserved regions, the alignment tends to come out the same for almost any scoring system used, because all reasonable scoring systems assign a high positive score to identical residues.

Which of these three alignments would you say is the best? The answer seems rather subjective, but there are clues in the results (e.g., the hallmark of lax gap penalties is the stretching out of a sequence by inclusion of large numbers of gaps between small numbers of aligned residues). Let us take a closer look. The alignment of the final part of the sequences appears better in (a) than (b). This suggests that the gap cost of 50 used in (b) is too high, and that the algorithm has been forced to remove gaps at the expense of reducing the quality of the alignment elsewhere. However, without additional information, it is hard to say that (b) is obviously wrong – had we only done the alignment with the parameters used in (b), we might have been satisfied with the result. For the differences between (a) and (c), it is difficult to decide which is better, but the "gappiness" in (c) gives the impression that the algorithm has been forced to introduce too many gaps in order to maximize identities and similarities. There is not much information in the sequences to decide conclusively; nevertheless, on balance, (a) feels better than (c). In spite of the "exact" nature of the pairwise align-

ment algorithms, there will always be an element of subjectivity involved in choosing parameters and choosing between resulting alignments. Thus, it is always worth trying different scoring systems when aligning real sequences to see what alternative alignments arise. Another useful way to check alignments is to view them using a color alignment editor, such as CINEMA (Parry-Smith *et al.* 1997). The coloring scheme can make errors appear more obvious. On a positive note, we should not be too pessimistic about sequence alignment programs: there are large sections of unhighlighted sequence that are the same in all three alignments. Often, it is the conserved regions that contain the important structural and functional motifs that we are interested in. It is these conserved regions that are most likely to be correctly aligned.

6.5 MULTIPLE SEQUENCE ALIGNMENT

6.5.1 The progressive alignment method

In the previous sections of this chapter, we discussed pairwise sequence alignment. However, we often wish to align sets of many sequences. Exact dynamic programming algorithms have been proposed for small numbers of sequences that find optimal alignments using scoring matrices and gap penalties, as for pairwise alignments. The recursion relations are considerably more complicated and the algorithms take correspondingly longer to run. The time would be at least $O(N^S)$ for S sequences with the simplest linear gap penalty. Carillo and Lipman (1988) considered efficient ways of implementing an exact multiple alignment algorithm. Nevertheless, this method is still very slow and is not practical for more than a few sequences of realistic length. In fact, aligning three or four sequences exactly is not really very useful. For most applications, we are interested either in just two sequences or in many sequences at the same time.

The most common approach to multiple sequence alignment is known as progressive alignment. This uses the fact that we are usually interested in aligning families of sequences that are evolutionarily related. The principle is to construct an approximate phylogenetic tree for the sequences to be aligned and then to build up the alignment by progressively adding sequences in the order specified by the tree.

As an example, we use the hexokinase sequences HXKP_HUMAN and HXK_SCHMA used in Fig. 6.4, together with 13 related sequences from human, rat, yeast, *Drosophila melanogaster*, and *Plasmodium falciparum* (see Table 6.1 for the accession numbers of these sequences). A phylogenetic tree calculated from these sequences is shown in Fig. 6.5. We note in passing that the tree tells us several interesting things about the evolution of this gene family. Hexokinases are found in a wide range of eukaryotes. More than one hexokinase is found in the mammals, in *Drosophila*, and in yeast. The sequences from each of these three groups are more closely related to each other than they are to the sequences from the other groups. This suggests that there have been independent gene-duplication events occurring in mammals, insects, and fungi that occurred since the divergence of these species. In the case of the mammals, the gene duplications must have occurred prior to the divergence of human and rat, since each of the human genes has a recognizable homolog in the rat.

Progressive alignment routines begin by alignment of closely related pairs of sequences, e.g., the yeast HXKA and HXKB sequences, or any of the four human/rat pairs. A global version of a pairwise alignment algorithm would be used for this (as described in Section 6.3). Additional single sequences can then be added to form clusters of three sequences, e.g., the yeast HXKG with the other two yeast sequences. Clusters can also be aligned with each other to form larger clusters, e.g., the human/rat HXK1 pair can be aligned with the human/rat HXK2 pair. Progressively larger clusters are built up in the order specified by the tree until all the sequences are combined into a single alignment.

The algorithm for aligning two clusters (or for one sequence and one cluster) is basically the same as the usual pairwise alignment algorithm. The simplest way of scoring the alignment of sites between clusters is to average the individual scores for the amino acid pairs that can be formed between the

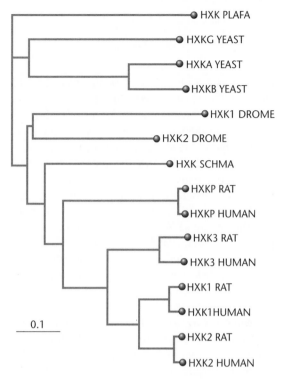

Fig. 6.5 Phylogenetic tree of hexokinase sequences from human, rat, *Schistosoma mansoni*, *Drosophila melanogaster*, *Saccharomyces cerevisiae*, and *Plasmodium falciparum*. This tree is produced by Clustal and used as a guide tree during progressive multiple alignment.

clusters. For example, the score for aligning $_A^P$ with $_R^I$ would be $(S(P,I) + S(P,R) + S(A,I) + S(A,R)/4)$.

Gap penalties are used in the same way as for pairwise alignments. If a gap is inserted into a cluster, then it must be inserted into the same position in every sequence in that cluster. The relative alignment of sequences already in a cluster is not altered by this process. The progressive alignment algorithm is not exact because it does not consider all possible alignments. For example, suppose we have already aligned the pair

```
S-HAKE
SPEARE
```

and the next sequence to be added is THEATRE. We might get the alignment

```
S-HA-KE
SPEA-RE
THEATRE
```

where a gap has been added at the same point in SHAKE and SPEARE. Looking in retrospect at these three sequences, we might have preferred to shift the H in SHAKE to align with the H in THEATRE. However, this possibility is never considered by the progressive alignment method because the first two sequences are fixed prior to the alignment of the third sequence. It is because possibilities like this are omitted that this is a heuristic method, not an exact one. (We note that the mutant sequence THEATER, found in some parts of the world, would not align so well!)

In order to begin a progressive alignment, we need to construct a guide tree. A pairwise alignment is first done for each pair of sequences in the set, and these alignments are used to calculate distances. The simplest way to calculate a distance is to ignore the sites with gaps and to calculate the fraction D of the non-gap sites where the two residues are different. A correction could be made for multiple substitutions per site if desired (e.g., using the Kimura distance formula, Eq. (4.29)). Feng and Doolittle (1987) suggested a formula for the distance between two sequences calculated directly from the alignment score, S, for those sequences:

$$d = -100 \ln \left(\frac{S - S_{rand}}{S_{ident} - S_{rand}} \right) \qquad (6.9)$$

where S_{rand} is the average score for alignment of two random sequences, and S_{ident} is the average score for alignment of two identical sequences. Thus, by one means or another, we can obtain a matrix of distances between all sequences and this can be used as input to one of the distance matrix phylogenetic methods that we will describe in Chapter 8. The widely used Clustal multiple alignment software (Thompson, Higgins, and Gibson 1994, Thompson *et al.* 1997) uses the neighbor-joining method to construct the tree and the midpoint method to determine the root. The guide tree for the hexokinases shown in Fig. 6.5 was constructed in this way.

Figure 6.6 shows part of the multiple alignment for the hexokinases constructed by Clustal, using the guide tree in Fig. 6.5. The progressive nature of the alignment is evident from comparing the gap positions with the guide tree. For example, all the human and rat sequences have a gap inserted between Q and V in the upper line and two gaps inserted between the two Ds in the middle line. These were introduced at the point when the eight human and rat sequences were aligned with the *Schistosoma* sequence. The sequences HXKP_HUMAN and HXK_SCHMA used in the pairwise alignment example appear together in the middle of the multiple alignment. The default alignment parameters have been used in Fig. 6.6, so that this is directly comparable with Fig. 6.4(a). The highlighted regions of these two sequences in Fig. 6.6 differ from the pairwise alignment in Fig. 6.4(a) even though the same scoring system is being used. This is because, in the multiple alignment, the *Schistosoma* sequence is being aligned with all eight mammal sequences at the same time, and not just with the HXKP_HUMAN sequence. In principle, this means that the alignment of the *Schistosoma* sequence to the cluster of mammal sequences should be more reliable than the pairwise alignment to any one of these eight sequences would be. Evolutionarily, the *Schistosoma* sequence is equally distantly related to all eight sequences and it makes sense to use the sequence information from all eight when calculating the alignment.

The highlighted region of the two *Drosophila* and two yeast sequences in the second line of Fig. 6.6 is also a visible relic of the guide tree. When the two *Drosophila* sequences are aligned with the mammals plus *Schistosoma* cluster, the algorithm decides to place the gap in the same place as the gap that is already present in the *Schistosoma* sequence. When the three yeast genes are aligned, the gap in the HXKA and HXKB pair is put in a slightly different place. When these two clusters are combined, a staggered arrangement of gaps arises, as shown. In contrast, if a multiple alignment is done with just the *Drosophila* and yeast sequences, then the final stage of the clustering is to combine the two *Drosophila* sequences with the three yeast sequences. In this case, Clustal puts the gap in the *Drosophila* sequences

```
CLUSTAL X (1.81) multiple sequence alignment


HXK1_RAT        EIVRNILIDFTKKGFLFR-----GQISEPLKTRGIFETKFLSQIESDRLALLQ-VRAILQ
HXK1_HUMAN      EIVRNILIDFTKKGFLFR-----GQISETMKTRGIFETKFLSQIESDRLALLQ-VRAILQ
HXK2_RAT        EIVRNILIDFTKRGLLFR-----GRISERLKTRGIFETKFLSQIESDCLALLQ-VRAILR
HXK2_HUMAN      EIVRNILIDFTKRGLLFR-----GRISERLKTRGIFETKFLSQIESDCLALLQ-VRATLQ
HXK3_RAT        EIVRHILLHLTSLGVLFR-----GQKTQCLQTRDIFKTKFLSEIESDSLALRQ-VRAILE
HXK3_HUMAN      EIVRHILLHLTSLGVLFR-----GQQIQRLQTRDIFKTKFLSEIESDSLALRQ-VRAILE
HXKP_RAT        ELVRLVLLKLVDENLLFH-----GEASEQLRTRGAFETRFVSQVESDSGDRKQ-IHNILS
HXKP_HUMAN      ELVRLVLLRLVDENLLFH-----GEASEQLRTRGAFETRFVSQVESDTGDRKQ-**IYNILS**
HXK_SCHMA       ELVRHIIVYLVEQKILFR-----GDLPERLKVRNSLLTRYLTDVERDPAHLLY**NTHYMLT**
HXK1_DROME      ELVRIIVLRLMKSGAIFA-----EDRRDYIGIQWKLDMVSLIEIVSDPPGVYTKAQEVMD
HXK2_DROME      ELVRLVITDMIAKGFMFH-----GIISEKIQERWSFKTAYISDVESDAPGEYRNCNKVLS
HXKA_YEAST      ELLRLVLLELNEKGLMLK-----DQDLSKLKQPYIMDTSYPARIEDDPFENLEDTDDIFQ
HXKB_YEAST      EILRLALMDMYKQGFIFK-----NQDLSKFDKPFVMDTSYPARIEEDPFENLEDTDDLFQ
HXKG_YEAST      EVLRNILVDLHSQGLLLQQYRSKEQLPRHLTTPFQLSSEVLSHIEIDDSTGLRETELSLL
HXK_PLAFA       EIVRRFMVNVLQS-----------ACSKKMWISDSFNSESGSVVLNDTSKNFEDSRKVAK
                *::*     :     .                     :         :       :  *


HXK1_RAT        QLGLNSTCD--DSILVKTVCGVVSKRAAQLCGAGMAAVVEKIRENRGLDHLNVTVGVDGT
HXK1_HUMAN      QLGLNSTCD--DSILVKTVCGVVSRRAAQLCGAGMAAVVDKIRENRGLDRLNVTVGVDGT
HXK2_RAT        HLGLESTCD--DSIIVKEVCTVVARRAAQLCGAGMAAVVDKIRENRGLDNLKVTVGVDGT
HXK2_HUMAN      HLGLESTCD--DSIIVKEVCTVVARRAAQLCGAGMAAVVDRIRENRGLDALKVTVGVDGT
HXK3_RAT        DLGLTLTSD--DALMVLEVCQAVSRRAAQLCGAGVAAVVEKIRENRGLQELTVSVGVDGT
HXK3_HUMAN      DLGLPLTSD--DALMVLEVCQAVSQRAAQLCGAGVAAVVEKIRGNRGLEELAVSVGVDGT
HXKP_RAT        TLGLRPSVT--DCDIVRRACESVSTRAAHMCSAGLAGVINRMRESRSEDVMRITVGVDGS
HXKP_HUMAN      **TLGLRPSTT--**DCDIVRRACESVSTRAAHMCSAGLAGVINRMR**ESRSEDVMR**ITVGVDGS
HXK_SCHMA       **DDLHVPVVEPI**DNRIVRYACEMVVKRAAYLAGAGIACILRRINRS------EVTVGVDGS
HXK1_DROME      KFRIRHCKER-DLAALKYICDTVTNRAAMLVASGVSCLIDR**MRLP**------QISIAVDGG
HXK2_DROME      ELGILGCQEP-DKEALRYICEAVSSRSAKLCACGLVTIINK**MNIN**------EVAIGIDGS
HXKA_YEAST      KDFGVKTTLP-ERKLIRRLCELIGTRAARLAVCGIAAICQK------**RGYK**TGHIAADGS
HXKB_YEAST      NEFGINTTVQ-ERKLIRRLSELIGARAARLSVCGIAAICQK------**RGYK**TGHIAADGS
HXKG_YEAST      QSLRLPTTPT-ERVQIQKLVRAISRRSAYLAAVPLAAILIK**TNALNKRYHG**EVEIGCDGS
HXK_PLAFA       AAWDMDFTDE-QIYVLRKICEAVYNRSAALARGTIAAIAKRIKIIEHS---KFTCGVDGS
                      :      :         :   *:*  :    : :   :          .  **


HXK1_RAT        LYKLHPHFSRIMHQTVKELS------PK-CTVSFLLSEDGSGKGAALITAVGVRLRGDPS
HXK1_HUMAN      LYKLHPHFSRIMHQTVKELS------PK-CNVSFLLSEDGSGKGAALITAVGVRLRTEAS
HXK2_RAT        LYKLHPHFAKVMHETVRDLA------PK-CDVSFLESEDGSGKGAALITAVACRIREAGQ
HXK2_HUMAN      LYKLHPHFAKVMHETVKDLA------PK-CDVSFLQSEDGSGKGAALITAVACRIREAGQ
HXK3_RAT        LYKLHPHFSRLVSVTVRKLA------PQ-CTVTFLQSEDGSGKGAALVTRVACRLTQMAC
HXK3_HUMAN      LYKLHPRFSSLVAATVRELA------PR-CVVTFLQSEDGSGKGAALVTAVACRLAQLTR
HXKP_RAT        VYKLHPSFKERFHASVRRLT------PN-CEITFIESEEGSGRGAALVSAVACKKACMLA
HXKP_HUMAN      VYKLHPSFKERFHASVRRLT------PS-CEITFIESEEGSGRGAALVSAVACKKACMLG
HXK_SCHMA       LYKFHPKFCERMTDMVDKLK------PKNTRFCLRLSEDGSGKGAAAIAASCTRQN----
HXK1_DROME      IYRLHPTFSTVLNKYTRLLA------DPNYNFEFVITQDSCGVGAAIMAGMAHANKYKTD
HXK2_DROME      VYRFHPKYHDMLQYHMKKLL------KPGVKFELVVSEDGSGRGAALVAATAVQAKSKL-
HXKA_YEAST      VYNKYPGFKEAAAKGLRDIYGWTGDASKD-PITIVPAEDGSGAGAAVIAALSEKRIAEGK
HXKB_YEAST      VYNRYPGFKEKAANALKDIYGWTQTSLDDYPIKIVPAEDGSGAGAAVIAALAQKRIAEGK
HXKG_YEAST      VVEYYPGFRSMLRHALALSP---LGAEGERKVHLKIAKDGSGVGAALCALVA--------
HXK_PLAFA       LFVKNAWYCKRLQEHLKVILA-----DKAENLIIIPADDGSGKGAAITAAVIALNADIPQ
                :      .  :                  . :  :.::.* *** :
```

Fig. 6.6 Multiple alignment of hexokinase sequences constructed by Clustal using the guide tree in Fig. 6.5. Bold sections illustrate points discussed in the text.

in the same place as the gap in the two shorter yeast sequences (i.e., the MNIN is directly above the RGYK). The lesson to be drawn from this is that the details of a multiple alignment depend on exactly which sequences are included in the set and on the order in which the sequences were added to the alignment. In cases where one is not prepared to trust an automatically generated guide tree to determine the order, it is possible to carry out a progressive alignment by gradually combining sequences in an order that reflects the user's prior beliefs about the way the sequences are related. This example shows that it is the positioning of gaps within the most variable regions of sequences that tends to vary with the details of the alignment procedure, while more conserved regions tend to be quite reliably aligned. It is sometimes possible to improve the alignment on either side of gap regions by making small changes by eye. However, there can often be regions within sequences that are simply too variable to align reliably, and it is necessary to make a fairly arbitrary choice.

6.5.2 Improving progressive alignments

The Clustal software has developed through a number of incarnations. Recent versions include several modifications intended to improve the accuracy of the progressive alignment procedure (Thompson, Higgins, and Gibson 1994, Thompson *et al.* 1997). Alignments often contain groups of closely related sequences as well as more distant sequences. If the score for aligning clusters is simply the average score for all sequence pairs, as in the $_A^P$ and $_R^I$ example above, this tends to overemphasize the influence of closely related sequences. Closely related sequences are bound to be almost identical because there has been little time for substitutions to occur between them. Consider the point where the two *Drosophila* sequences are aligned with the mammals plus *Schistosoma* cluster following the guide tree in Fig. 6.5. The eight mammal sequences are likely to be relatively similar to one another because they share a lot of their evolutionary history, whereas the *Schistosoma* sequence contains independent information. This information would be swamped by the eight mammal

sequences if all the sequences were weighted equally when calculating the score. Thompson, Higgins, and Gibson (1994) introduced a weighting scheme that reduces the weight of closely related sequences using a rule that depends on the amount of shared evolutionary history of the sequences.

Another heuristic improvement is the introduction of position-specific gap penalties. In proteins, gaps occur more frequently in loop regions than in elements of secondary structure. This is because insertions and deletions in these regions are more readily tolerated without destroying the function of the molecule. The positioning of gaps in alignments tends to be more accurate for closely related sequences than for more distant sequences. Therefore, information from gap positions in clusters of closely related sequences that is determined in the early steps of a progressive alignment can be used to help position the gaps when more distant clusters are aligned in the later stages of the process. This is done by reducing the gap penalty in sites where there is already at least one gap, so that future gaps tend to occur in the same place. It is also known that different scoring matrices are appropriate for aligning sequences of different distances. Progressive alignments can therefore be improved by using a series of different PAM or BLOSUM matrices and choosing the most appropriate matrix at each step of the alignment according to the distance between the clusters to be aligned.

With sophisticated modifications such as these, progressive alignment is a practical tool that often produces quite reliable alignments. The method is also fairly rapid. The time scales as $O(SN^2)$ for S sequences of length N, which means that large numbers of long sequences can be easily dealt with. However, two words of warning are appropriate when using Clustal. First, the method uses **global** alignment and is therefore not appropriate when sequences have very different lengths. In this case, it is better to cut out the region from the longer sequences that is alignable with the shorter sequences and to align only these regions. Second, the guide tree should not be taken too seriously. The guide tree is calculated using a rather approximate distance measure and a relatively quick-and-dirty method of tree construction. It cannot therefore be relied on if we are really interested in the phylogeny of the sequences. The matrix of pairwise distances can be recalculated from the sequences using the relative sequence positions specified by the complete multiple alignment. If these distances are put back into a phylogenetic program, the resulting tree can sometimes be better than the guide tree. This point was emphasized by Feng and Doolittle (1987). Clustal has an option to recalculate the neighbor-joining tree from the alignment. However, there are many more sophisticated ways of calculating molecular phylogenies (see Chapter 8), and we would recommend using specialized programs in cases where the real interest is in the phylogeny.

Although, as just stated, the tree produced from a progressive multiple alignment is not necessarily the same as the guide tree used to produce the alignment, it is usually fairly similar. It has been shown that if alignments generated by following a guide tree are used as input to phylogenetic methods, which may well be much more sophisticated than the clustering algorithm used to generate the guide tree, the trees obtained tend to be strongly correlated with the initial guide tree (Lake 1991, Thorne and Kishino 1992). To some extent, this defeats the object of using more sophisticated phylogeny programs. A pragmatic way out of this problem is simply to recognize that automated multiple sequence alignments are never completely reliable and that effort must be put into adjusting alignments manually in order to get meaningful results. Where functional and structural information is available, it is possible to check that active sites and secondary structure elements are correctly aligned, and make appropriate adjustments if not. On the other hand, the problems of multiple alignment and tree construction are inextricably linked, and algorithms are under development for generating the alignment and the tree simultaneously (Vingron and von Haeseler 1997, Holmes and Bruno 2001). A statistician may argue that we should not rely on any single alignment as being the "true" one, but should consider all the possible pathways by which one sequence could evolve into another when we measure the evolutionary distance between them.

Algorithms of this type are currently under development and promise to give interesting results in the future (Hein *et al.* 2000), but they are beyond the scope of this book. The main reason for their complexity is the difficulty of constructing a probabilistic model of insertions and deletions. For the moment, phylogenetic studies almost always do the alignment first and the tree construction second.

6.5.3 Recent developments in multiple sequence alignment

The divide and conquer method of multiple alignment (Stoye 1998) is an alternative heuristic method that takes a different approach from progressive alignment. We can define a score for a multiple alignment as the sum of the pairwise alignment scores for all the sequences. If we like, we can weight the sequences in some way when calculating the multiple alignment score so as to reduce the influence of clusters of very similar sequences. We would like to find a multiple alignment that maximizes this score. Dynamic programming algorithms to exactly solve this problem are theoretically feasible, but impractical due to large time and memory requirements. However, for short sections of sequences, exact alignment is possible. The divide and conquer method divides long sequences into short sections, does exact alignments of the short sections, and then recombines the short alignments into one long one. The method implements a heuristic rule to decide at which positions to cut the sequences. The object is to choose places that are likely to affect the overall alignment as little as possible. Tests have shown that the method can give good results in some tricky cases with sets of around 10 sequences (Stoye 1998), although it is still a fairly slow algorithm.

Another recently developed heuristic for multiple alignment is known as T-Coffee (Notredame, Higgins, and Heringa 2000). This begins with a library of pairwise alignments for every pair of sequences in the set. Two different pairwise programs are used, one local and one global, so that every sequence pair has two alignments in the library. A weight is attached to each of these alignments that is intended to reflect the reliability of each alignment.

The initial value for the weight, W_{ij}, for two sequences i and j is set to the percentage identity of i and j, because alignments of highly similar sequences are likely to be more reliable. The second stage of the algorithm is called library extension. The object is to calculate a weight associated with the alignment of each pair of residues in two different sequences. Let X be a residue in sequence i, and Y be a residue in sequence j. If X and Y are aligned in the pairwise alignment of i and j, then the weight for the residue pair is set to $W(X,Y) = W_{ij}$, otherwise, it is set to zero. The algorithm then looks at all the other sequences in the set. If there is another sequence k with a residue Z such that X and Z are aligned in the pairwise alignment of i and k, and Y and Z are aligned in the pairwise alignment of j and k, then this indicates that the three pairwise alignments are consistent with each other and lends support for the alignment of X with Y. In this case, the weight $W(X,Y)$ is increased by the minimum of W_{ik} and W_{jk}. This is done for every sequence k that has a consistent alignment.

At the end of this procedure, weights are available for every pair of residues in every pair of sequences. These weights are then used as scores in the progressive multiple alignment method. Since all the weights are positive, gaps can be treated as zero score, and it is not necessary to introduce any extra gap penalty parameters. The problem with the standard progressive alignment method is that it is unable to use information from sequences added at a later stage to improve the alignments of sequences that were already aligned at an earlier stage. The idea behind the T-Coffee method is to use the library extension procedure to calculate scores for aligning pairs of sequences that reflect the information in all the sequences in the set, not just the information in the two sequences themselves. Hopefully, this should minimize the problems that may arise with progressive alignment. Notredame, Higgins, and Heringa (2000) give several examples where the T-Coffee method does indeed appear to perform better than alternatives, and our own experiences with the program have also been very promising.

SUMMARY

An algorithm is a logical description of a way of solving a problem. It is possible to translate an algorithm into a computer program written in any programming language. An exact algorithm is one that is guaranteed to find the correct solution to a problem. A heuristic algorithm is one that will most likely find a fairly good solution but cannot be guaranteed to find the best solution.

An important property of an algorithm is the way in which the time taken scales with the size of the problem, N (where N could mean the length of sequences to be aligned, or the number of objects to be sorted, or whatever quantity is appropriate to the problem being solved). Often this is a power law. We say that an algorithm is $O(N^\alpha)$ if the time scales as N^α. The lower the value of α, the more rapid the algorithm, and the more practical the algorithm will be with large data sets. Some problems are computationally hard and cannot be solved with any exact algorithm that scales as a power law of N. These are known as NP-complete problems.

Efficient, exact algorithms are possible that find the optimal alignment of two sequences according to a specified scoring system. Scoring systems give positive scores for aligning pairs of amino acids that are identical or similar in properties and negative scores for aligning pairs of amino acids that are very different from one another. Penalties are also assigned for insertion of gaps. The most commonly used are affine penalties, where the cost for a gap of length l is $W(l) = g_{open} + g_{ext}(l - 1)$. Pairwise alignment of sequences of length N scales as $O(N^2)$ using affine gap penalties.

Alignment algorithms use a technique called dynamic programming, which relies on writing a recursion relation that breaks the problem down into smaller sections. The solution to the full problem is built up progressively from the solution of smaller subproblems. Pairwise alignments are said to be global, if the whole of one sequence is aligned with the whole of another sequence, or local, if only the high scoring parts of sequences are aligned.

Although it is possible to write down exact algorithms for aligning more than two sequences, these are rarely used in practice because they are extremely slow. The most common algorithms for multiple alignment use a progressive method where a rough phylogenetic tree, or guide tree, is estimated for the sequences and then the sequences are aligned in progressively larger clusters in the order specified by the guide tree. A pairwise alignment algorithm is used to align the pairs of sequence clusters at each step of the multiple alignment process. Rearrangement of sequences within clusters is not permitted once the clusters have been formed. Thus, the algorithm is heuristic and is not guaranteed to give the optimal-scoring multiple alignment. Current multiple alignment programs have introduced many refinements to improve the functioning of progressive methods.

Pairwise alignment is sensitive to the scoring system used. Different scores will yield different alignments and there are no hard-and-fast rules to determine which alignment is the most meaningful biologically. In addition, multiple alignments are also sensitive to the details of the heuristic algorithm used. It is necessary to use common sense and biological experience when judging the quality of alternative alignments. Overzealous gap insertion is a common hallmark of alignment algorithms and, in many cases, improvements are necessary using manual alignment editors to correct the errors introduced by alignment programs.

REFERENCES

Benner, S.A., Cohen, M.A., and Gonnet, G.H. 1993. Empirical and structural models for insertions and deletions in the divergent evolution of proteins. *Journal of Molecular Biology*, **229**: 1065–82.

Carillo, H. and Lipman, D. 1988. The multiple sequence alignment problem in biology. *SIAM Journal of Applied Mathematics*, **48**: 1073–82.

Clote, P. and Backofen, R. 2000. *Computational Molecular Biology – An Introduction*. Chichester, UK: Wiley.

Cormen, T.H., Leiserson, C.E., and Rivest, R.L. 1990. *Introduction to Algorithms*. Cambridge, MA: MIT Press.

Durbin, R., Eddy, S.E., Krogh, A., and Mitchison, G. 1998. *Biological Sequence Analysis – Probabilistic Models of Proteins and Nucleic Acids*. Cambridge, UK: Cambridge University Press.

Feng, D.F. and Doolittle, R.F. 1987. Progressive sequence alignment as a prerequisite to correct phylogenetic trees. *Journal of Molecular Evolution*, **25**: 351–60.

Gotoh, O. 1982. An improved algorithm for matching biological sequences. *Journal of Molecular Biology*, **162**: 705–8.

Hein, J., Wiuf, C., Knudsen, B., Moller, M.B., and Wibling, G. 2000. Statistical alignment: Computational properties, homology testing and goodness-of-fit. *Journal of Molecular Biology*, **302**: 265–79.

Holmes, I. and Bruno, W.J. 2001. Evolutionary HMMs: A Bayesian approach to multiple alignment. *Bioinformatics*, **17**: 803–20.

Lake, J.A. 1991. The order of sequence alignment can bias the selection of the tree topology. *Molecular Biology and Evolution*, **8**: 378–85.

Needleman, S.B. and Wunsch, C.D. 1970. A general method applicable to the search for similarities in the amino acid sequence of two proteins. *Journal of Molecular Biology*, **48**: 443–53.

Nicholas, K.B. and Nicholas, H.B. Jr. 1997. Genedoc: A tool for editing and annotating multiple sequence alignments. http://www.psc.edu/biomed/genedoc.

Notredame, C., Higgins, D.G., and Heringa, J. 2000. T-Coffee: A novel method for fast and accurate multiple sequence alignment. *Journal of Molecular Biology*, **302**: 205–17.

Parry-Smith, D.J., Payne, A.W.R., Michie, A.D., and Attwood, T.K. 1997. CINEMA – A novel Colour Interactive Editor for Multiple Alignments. *Gene*, **211**: GC45–6 (Version 5 of CINEMA can be downloaded from http://aig.cs.man.ac.uk/utopia/download).

Press, W.H., Teukolsky, S.A., Vetterling, W.T., and Flannery, B.P. 1992. *Numerical Recipes in C*, 2nd edition. Cambridge, UK: Cambridge University Press.

Smith, T.F. and Waterman, M.S. 1981. Identification of common molecular subsequences. *Journal of Molecular Biology*, **147**: 195–7.

Stoye, J. 1998. Multiple sequence alignment with the divide and conquer method. *Gene*, **211**: GC45–56.

Thompson, J.D., Higgins, D.G., and Gibson, T.J. 1994. CLUSTAL W: Improving the sensitivity of progressive multiple sequence alignment through sequence weighting, position-specific gap penalties and weight matrix choice. *Nuclear Acids Research*, **22**: 4673–80.

Thompson, J.D., Gibson, T.J., Plewniak, F., Jeanmougin, F., and Higgins, D.G. 1997. The CLUSTAL X windows interface: Flexible strategies for multiple sequence alignment aided by quality analysis tools. *Nuclear Acids Research*, **25**: 4876–82.

Thorne, J.L. and Kishino, H. 1992. Freeing phylogenies from artifacts of alignment. *Molecular Biology and Evolution*, **9**: 1148–62.

Vingron, M. and von Haeseler, A. 1997. Towards integration of multiple alignment and phylogenetic tree construction. *Journal of Computational Biology*, **4**: 23–34.

6.1

(a) Allowing a score of 1 for a match, 0 for a mismatch, and a linear gap cost of 0.5 per gap, use the global alignment algorithm (Section 6.3.1) to complete the following scoring matrix:

		C	G	C	A	T	G
A							
C							
G							
A							
G							

(b) Using the filled matrix, determine the optimal alignment for the two sequences.

(c) Consider the following three alignments:

(i) -CGCATG (ii) CGCATG (iii) -CGCATG
   ACG-A-G   ACGA-G   ACG-AG

For a linear gap cost of g per gap, for what values of g is alignment (i) better than alignment (ii)? Show that alignment (iii) is never the optimal alignment.

(d) For an affine gap cost $W(l) = g_{open} + g_{ext}(l-1)$, are there any values of g_{open} and g_{ext} for which (iii) is the optimum alignment?

6.2 The best way to understand alignment algorithms is to try using them. If you have sequences that you are working on yourself, then try out some of the suggestions below on them. Otherwise you can download the following files from www.blackwellpublishing.com/higgs.

Hexokinase.txt – This file contains the amino acid sequences of the hexokinase proteins discussed in this Chapter (see Figs. 6.4, 6.5, and 6.6).

CytBProt.txt – This file contains the amino acid sequences of the cytochrome b proteins from the mitochondrial genome of 20 vertebrate species. The sequences are labeled by a short English name for the species and

by the full Latin name. For more details on the species, see Problem 8.3.

CytBDNA.txt – This file contains the DNA sequences of the same cytochrome b genes.

(a) Take any two of the hexokinase sequences and try aligning them with a pairwise alignment program. If you are going to be dealing with many alignments, you will probably want to install Clustal on your own machine (software available from http://www-igbmc.u-strasbg.fr/BioInfo/ClustalX/Top.html). Otherwise, a convenient one is the EMBOSS pairwise alignment program available on the EBI web site http://www.ebi.ac.uk/emboss/align/. You can paste any two of the sequences from the Hexokinase.txt file into the Web form. Try running the alignment with several different scoring matrices and/or gap penalties. How much difference is there?

(b) Try a comparison of global and local alignments of the same pair of sequences using the same scoring parameters. How much difference is there? Some of the sequences are much longer than others. Can you determine which pieces of the longer sequences are similar to the shorter ones? What happens if you do a global alignment of one of the shorter sequences with a longer sequence (problems may arise if the gap cost is low)?

(c) Try doing a multiple alignment of all the hexokinase sequences using Clustal. If you do not already have it on your own machine, a web version is available via the EBI web site http://www.ebi.ac.uk/clustalw/. Try several alignments using different scoring matrices and gap penalties. How much difference is there? You may want to remove parts of the longer sequences so that the remaining parts can be aligned across their whole length.

(d) Do a multiple alignment of the cytochrome b sequences from the CytBProt.txt file. How sensitive is this alignment to the scoring parameters? These sequences are all relatively similar to one another. They can probably be aligned with almost no gaps.

(e) Try aligning the DNA sequences from the CytBDNA.txt file. Is the alignment you get consistent with the alignment of the protein sequences for the same genes? Is the DNA alignment consistent with the triplet (codon) structure that you expect in coding sequences? If not, can you adjust the alignment slightly by sliding some of the sequences so that the protein and DNA alignments are equivalent? Two useful programs for manual editing of sequence alignments are GeneDoc (Nicholas and Nicholas 1997) and CINEMA (Parry-Smith et al. 1997). These will read in the output alignment from Clustal, and will also read in sequences in many other formats.

Searching sequence databases

CHAPTER 7

CHAPTER PREVIEW

We discuss the way in which pairwise alignment methods can be used to search databases for sequences that are similar to a query sequence. We describe the algorithms used in the heuristic search programs FASTA, BLAST, and PSI-BLAST. We discuss the distribution of scores expected in alignment methods and show that this is often of a form called an extreme value distribution. We discuss the meaning of the E values produced by database search tools as measures of the statistical significance of matches.

7.1 SIMILARITY SEARCH TOOLS

7.1.1 Smith–Waterman searching

It is natural to think of the score of a pairwise alignment as a measure of the similarity of the two sequences. Often, we have a sequence of interest and we want to know if there are any other sequences that are similar to it. Pairwise alignment algorithms can be straightforwardly extended to answer this question. We simply align the original sequence (we call this the query sequence) with each sequence in a database in turn, and calculate the alignment score for each one. We then rank the scores in descending order and return the details of the top few "hits" against the database. Several important practical search tools that work in this way will be described in this section.

The Smith–Waterman algorithm (local alignment with affine gap penalty) is implemented in the MPsrch facility on the EBI web site (http://www.ebi.ac.uk/MPsrch). It is possible to select from a number of different PAM scoring matrices, and

also to select gap costs, using either a linear or an affine gap penalty function. Although there are no absolutely correct values for the gap costs, it is important to use costs that are reasonable in comparison to the scores of pairing residues of different types. The most appropriate gap costs therefore depend on which PAM matrix is used. The MPsrch Web page allows the user to select values that are thought to be appropriate from previous trials after selecting the PAM matrix.

As an example, we chose the Swiss-Prot sequence PTP1_YEAST (accession number P25044), which is a protein tyrosine phosphatase (PTP) from *Saccharomyces cerevisiae*. These are enzymes that catalyze the removal of a phosphate group attached to a tyrosine residue. The sequence was queried against the Swiss-Prot database using the PAM300 matrix and affine gap penalties, with $g_{open} = 12$ and $g_{ext} = 2$. The results, shown in Fig. 7.1(a), are ranked by decreasing alignment score. The figures in the column marked "Pred. No." are the E values, which are the expected numbers of sequences in the database that would match the query with a score greater than or equal to the observed score. When $E \ll 1$, the match is highly significant, while if E is of order 1 or greater, then the match represents a weak level of similarity that could have occurred by chance. More details on alignment statistics are given later in this

```
Scoring table:    PAM 300          Gap open 12;   Gap extend 2

Result          %
   No.   Score  Match  Length  DB   ID          Description           Pred. No.
------------------------------------------------------------------------------------
      1   1390   98.5     335   1   PTP1_YEAST   Protein-tyrosine phosp   4.73e-207
      2    343   24.3     711   1   PYP2_SCHPO   Protein-tyrosine phosp   1.97e-33
      3    339   24.0     550   1   PYP1_SCHPO   Protein-tyrosine phosp   8.11e-33
      4    324   23.0    1442   1   PTPG_MOUSE   Protein-tyrosine phosp   1.61e-30
      5    321   22.7    2314   1   PTPZ_HUMAN   Receptor-type protein-   4.60e-30
      6    321   22.7    2316   1   PTPZ_RAT     Receptor-type protein-   4.60e-30
      7    316   22.4    1445   1   PTPG_HUMAN   Protein-tyrosine phosp   2.65e-29
      8    316   22.4    1422   1   PTPG_CHICK   Protein-tyrosine phosp   2.65e-29
      9    303   21.5    1457   1   PTPK_MOUSE   Receptor-type protein-   2.47e-27
     10    299   21.2     434   1   PTN1_CHICK   Protein-tyrosine phosp   9.91e-27

     90    174   12.3     750   1   PTP2_YEAST   Protein-tyrosine phosp   7.99e-09
     91    138    9.8     928   1   PTP3_YEAST   Protein-tyrosine phosp   3.15e-04

     95    110    7.8     478   1   YDIU_SHIFL   Hypothetical UPF0061 p   5.47e-01
     96    109    7.7     478   1   YDIU_ECO57   Hypothetical UPF0061 p   7.02e-01

    108    101    7.2     296   1   RN15_YEAST   mRNA 3'-end processing   4.91e+00

    121     98    6.9     341   1   YH10_YEAST   Hypothetical 37.9 kDa    9.92e+00
```

```
RESULT     90
ID    PTP2_YEAST        STANDARD;        PRT;    750 AA.
DE    Protein-tyrosine phosphatase 2 (EC 3.1.3.48) (PTPase 2).

  DB  1;   Score    174;  Match 26.9%;  QryMatch 12.3%;  Pred. No. 7.99e-09;
  Matches    76;  Conservative   69;  Mismatches   89;  Indels   49;  Gaps   11;


          ******.*.*.        *.*.. *****.*    * *.**..      *... **. . .
Db   465  NDYINANYLKLT----QINPDFKYIATQAPLPSTMDDFWKVI----TLNKVKVIISLNSD 516
Qy    77  ndyinasyvkvnvpgqsiepgy-yiatqgptrktwdqfwqmcyhncpldni-vivmvtpl 134


          * *   *    **    .  ..*... . **.  . *.      ..*    . . ...*
Db   517  DELNLRKWDIYWNNLSYSNHTIKLQNTWENICNINGCVLRVFQVKKTAPQNDNISQDCDL 576
Qy   135  veynrekcyqywprgg-vddtvriaskwespggandmtqfpsdlkiefvnvhkvkdyytv 193


          .    **.      *.     *  .. * *              . .*.  .  .*
Db   577  PHNGDLTSITMAVSEPFIVYQLQYKNWLDSCGVDMNDIIKLHKVKNSLLFNPQSFITSLE 636
Qy   194  tdi-kltptdplvgpvktvhhfyfdlwkd------------------mnkpeevvpime 233


            .*       .. *    . .*..****** **** *..** *.     ...  .
Db   637  KDVCKPDLIDDNNSELHLDTANSSPLLVHCSAGCGRTGVFVTLDFLL------SILSPTT 690
Qy   234  --lc---------ahshslnsrgnpiivhcsagvgrtgtfialdhlmhdtldfkniters 282


          **..    . *.*** ** .** **. ***    *..  *.*
Db   691  NHSNKIDVWNMTQDLIFIIVNELRKQRISMVQNLTQYIACYEA 733
Qy   283  rhsdrate-eytrdliieqivlqlrsqrmkmvqtkdqflfiyha 324
```

Fig. 7.1 (a) Selected parts of the output from MPsrch using the Swiss-Prot database and the sequence PTP1_YEAST as the query with scoring matrix PAM 300, g_{open} = 12, g_{ext} = 2.

```
Scoring table:    PAM 50          Gap open 40;   Gap extend 7
                    %
Result           Query
   No.   Score   Match  Length  DB   ID           Description              Pred. No.
-----------------------------------------------------------------------------------
     1    3660   100.0     335   1   PTP1_YEAST   Protein-tyrosine phosp   0.00e+00
     2     254     6.9    1445   1   PTPG_HUMAN   Protein-tyrosine phosp   6.39e-48
     3     251     6.9    1422   1   PTPG_CHICK   Protein-tyrosine phosp   5.98e-47
     4     248     6.8    1442   1   PTPG_MOUSE   Protein-tyrosine phosp   5.55e-46
     5     226     6.2     711   1   PYP2_SCHPO   Protein-tyrosine phosp   5.73e-39
     6     217     5.9    2316   1   PTPZ_RAT     Receptor-type protein-   3.81e-36
     7     208     5.7     595   1   PTN6_MOUSE   Protein-tyrosine phosp   2.36e-33
     8     208     5.7     613   1   PTN6_RAT     Protein-tyrosine phosp   2.36e-33
     9     208     5.7    1216   1   PTPO_HUMAN   Receptor-type protein-   2.36e-33
    10     207     5.7    1452   1   PTPM_MOUSE   Receptor-type protein-   4.79e-33

    31     181     4.9     550   1   PYP1_SCHPO   Protein-tyrosine phosp   3.43e-25

    69     151     4.1     750   1   PTP2_YEAST   Protein-tyrosine phosp   1.35e-16

    90     120     3.3     928   1   PTP3_YEAST   Protein-tyrosine phosp   1.83e-08
    91     117     3.2     171   1   VH01_RACVI   Dual specificity prote   9.89e-08
    92     117     3.2     171   1   DUSP_VACCV   Dual specificity prote   9.89e-08
    93     117     3.2     171   1   DUSP_VACCC   Dual specificity prote   9.89e-08

   105     108     3.0     551   1   CC14_YEAST   Probable protein-tyros   1.33e-05

   108      97     2.7     489   1   MSG5_YEAST   Protein-tyrosine phosp   3.62e-03

   110      93     2.5     468   1   YOPH_YERPS   Protein-tyrosine phosp   2.47e-02
   111      93     2.5     468   1   YOPH_YEREN   Protein-tyrosine phosp   2.47e-02

   140      84     2.3     312   1   DCTD_YEAST   Deoxycytidylate deamin   1.41e+00

RESULT    69
ID    PTP2_YEAST      STANDARD;       PRT;     750 AA.
DE    Protein-tyrosine phosphatase 2 (EC 3.1.3.48) (PTPase 2).

  DB  1;   Score     151;   Match 62.5%;   QryMatch 4.1%;   Pred. No. 1.35e-16;
  Matches    15;   Conservative    4;   Mismatches    5;   Indels   0;   Gaps   0;

          .*  ****** ****  *..** *.
Db    660 SPLLVHCSAGCGRTGVFVTLDFLL 683
Qy    246 npiivhcsagvgrtgtfialdhlm 269
```

Fig. 7.1 (b) Selected parts of the output from MPsrch using the Swiss-Prot database and the sequence PTP1_YEAST as the query with scoring matrix PAM 300, $g_{open} = 40$, $g_{ext} = 7$.

chapter. In this example, the top hit is PTP1_YEAST itself, as the query sequence is contained in the database. The top 10 hits are shown in full, and these include genes from the fission yeast *Schizosaccharomyces pombe*, and several vertebrates. All these have very small E values. In total, 91 hits were found with $E < 0.05$, and almost all of these are known to be PTPs. In this discussion, we will use the InterPro family IPR000387 (tyrosine-specific protein phosphatase and dual-specificity protein phosphatase) as a reference for which sequences are thought to be true PTPs. For more details on InterPro and other protein family databases, see Chapters 5 and 9.

The highest-scoring sequences of other *S. cerevisiae* genes, ranked 90 and 91, are PTP2_YEAST and PTP3_YEAST, which are also PTPs. The next two *S. cerevisiae* genes on the list are RN15_YEAST and YH10_YEAST, with E values greater than 0.1. Therefore, these probably do not represent a significant similarity to the query. The highest-scoring hits to bacterial sequences are also shown in Fig. 7.1(a). These are YDIU_SHIFL from *Shigella flexneri* and YDIU_EC057 from *Escherichia coli.* They have E values greater than 1, and most likely these are also chance matches between unrelated sequences. Both these sequences belong to a different family of proteins of unknown function (InterPro entry IPR003846) and not the PTP family.

The results of MPsrch also contain a printout of the highest scoring local alignments between the query and each of the related database sequences. Figure 7.1(a) shows the alignment between PTP1_YEAST and PTP2_YEAST. These two sequences are relatively divergent. The alignment shows several short regions of quite high similarity, separated by regions with rather little similarity. There are also some fairly long gaps inserted. The query sequence has length 335. As this is a local alignment, it does not necessarily cover the whole of the query sequence. The alignment extends from position 77 to position 324, which covers most, but not all, of this sequence. The PTP2 sequence is much longer (750), and the alignment goes from position 465 to 733. Thus, the algorithm has found that the majority of the PTP1 sequence has a significant level of similarity to a region of comparable length within PTP2.

Figure 7.1(b) shows the results of a search using the PAM50 matrix and associated gap penalties, $g_{open} = 40$, $g_{ext} = 7$. The same query sequence and the same Swiss-Prot database were used. The results are different from the previous case in some important respects. Most of the significant hits from the previous search are also significant hits this time; however, the ranking order is different. For example, the two *S. pombe* genes, previously ranked 2 and 3, are now ranked 5 and 31. This time there are 114 hits with $E < 0.05$, which is more than last time. The

two related *S. cerevisiae* sequences PTP2_YEAST and PTP3_YEAST are again found as significant hits, but this time two further *S. cerevisiae* sequences, CC14_YEAST and MSG5_YEAST, also turn up with E values considerably below 1. These are most likely true relatives of PTP1_YEAST that were missed in the previous search. Both are included in the list of protein matches to the InterPro entry for PTPs (IPR000387).

There are some other notable hits not found in the previous search. The three sequences ranked 91–93 are from Vaccinia and related viruses. At ranks 110 and 111, there are two significant matches to bacterial sequences from *Yersinia pestis* and *Y. enterocolitica.* These additional hits from viruses and bacteria are probably biologically meaningful, and they are included in the InterPro entry for PTPs. They share significant similarity to PTP1_YEAST over a very short domain of about 20 residues, but it is difficult to align the sequences outside this domain. Low PAM number matrices are appropriate for high similarity sequences. When such a matrix is combined with high gap costs, as in this example, it finds short matching regions of very high similarity and almost no gaps. This spots the short conserved tyrosine phosphatase domains in this example. High PAM number matrices are appropriate for sequences of lower similarity. Such matrices give less weight to exact amino acid matches and more weight to similar but non-identical amino acids. They are thus more effective at spotting relatively weak similarities that extend over longer sequence lengths. The local alignment of PTP1_YEAST and PTP2_YEAST with the PAM50 matrix is shown in Fig. 7.1(b). This is a conserved domain of 24 residues with no gaps. This same domain is contained within the alignment produced with the PAM300 matrix in Fig. 7.1(a).

It should be pointed out that all these results were obtained in December 2003, with a version of Swiss-Prot containing 138,922 sequences. You may like to repeat the same searches to see how much has changed since then. You should expect to see many of the same sequences, but the rankings and the significance values will have changed as more sequences have been added to the database.

7.1.2 Heuristic local alignment tools – FASTA and BLAST

Although dynamic programming routines for pairwise sequence alignment can be written very efficiently, the size of biological sequence databases continues to grow alarmingly rapidly. The more sequences in the database, the longer each run of a search program will take. Several heuristic search tools have been developed that sacrifice the guarantee of finding exactly optimal alignments in the interest of increasing the speed of the search. These methods provide a quick way of locating as many database sequences as possible that are in some way similar to the query. If an exact alignment is required, then we can always use a fast search tool initially to locate the sequences of interest, and then use an exact alignment algorithm on these sequences.

We are often only interested in sequences that are quite similar to the query, so there is no point in wasting time with exact alignment of sequences that are very different from the query. Also, the region of similarity may cover only a fraction of the sequences, so there is little point in aligning the more divergent regions. The FASTA program (Pearson and Lipman 1988; also see http://www.ebi.ac.uk/fasta33/fastadoc.html) begins by looking for subsequences of the database sequence that exactly match subsequences of the query and that are at least of length *ktup* (for protein sequences, the default is *ktup* = 2 amino acids, and for DNA sequences, the default is *ktup* = 6 nucleotides). It then looks for diagonal regions in the alignment matrix that contain as many of these *ktup* matches as possible with only small distances between them. The 10 highest-scoring regions are retained. These regions correspond to high-scoring local alignments without gaps. The algorithm then determines which of these initial regions can be joined by allowing gaps in the alignments of the sequences between them. The score obtained from these approximate alignments is used to rank the database sequences. The highest-scoring sequences are then aligned using a dynamic programming algorithm. Time is only spent on dynamic programming for sequences that are already known to be quite similar to the query according to the simplified scoring system of the first parts of the routine. As an additional time-saving feature, the program only considers pathways through the alignment matrix that remain within a band centered around the highest-scoring initial regions. This saves time in calculating the $H(i,j)$ scores for cells in the matrix that are very far from the presumed optimal alignment path.

The BLAST program, or Basic Local Alignment Search Tool, in its original version (Altschul *et al.* 1990), tries to find the highest-scoring ungapped local alignment between the query and a database sequence. It uses a word length parameter, *w*, similar to the *ktup* parameter in FASTA. Usually *w* is 3 for proteins and around 12 for DNA. BLAST scans the database for words of length *w* that score higher than a threshold *T* when aligned with words in the query. The local alignment is then extended outwards from these words. In each direction, the extension is stopped when the score falls more than a certain distance below the best score reached so far. The algorithm then returns the best-scoring local alignment within the region considered. As an example, we queried the protein

CAPTAINKIRKANDTHESTARSHIPENTERPRISECREW

against the non-redundant protein database using the BLAST algorithm available on the NCBI Website http://www.ncbi.nlm.nih.gov/BLAST/ and using the PAM30 scoring matrix. The top hit is to a protein from *Caulobacter crescentus*. The alignment looks like this:

```
Score = 30.3 bits (64), Expect = 2.4
Identities =  11/19 (57%), Positives
= 15/19 (78%)

Query: 17 ESTARSHIPENTERPRISE 35
          E+TAR H+PEN E  R++E
Sbjct: 43 ETTAREHLPENAEIARLTE 61
```

The line between the two sequences indicates identical and similar amino acids (the latter denoted by a +). The algorithm has found a high-scoring word (possibly PEN) and extended in both directions to give a high-scoring ungapped local alignment of length 19 residues.

More recent versions of BLAST (Altschul *et al.* 1997) have changed the initial step of the algorithm by requiring two high-scoring words close together on the same diagonal before initiating the extension procedure. In the above alignment, TAR and PEN are on the same diagonal because there are no gaps inserted between them. This avoids wasting time trying to extend alignments in a lot of cases that are just chance matches of a single high-scoring word.

The new algorithm also allows gaps in the extension procedure. The gapped extension is done using a version of the Smith–Waterman algorithm that stops if the score falls more than a certain distance below the highest score yet found. This is a heuristic rule that tries to avoid wasting time calculating low-scoring cells in the alignment matrix. If the CAPTAINKIRK protein is used as a query against the non-redundant NCBI database with the BLOSUM62 matrix (rather than the PAM30 matrix used above), the top hit is a gapped alignment to a capsid protein precursor from *Plautia stali* intestine virus.

```
Score = 32.7 bits (67), Expect = 0.54
Identities = 15/39 (38%), Positives = 20/39 (50%), Gaps = 4/39 (10%)

Query: 1   CAPTAINKIRKANDTHESTARSHIPENTERPRISECREW 39
           CAP  +N+ R A+D  E T    I +  ERPR+     W
Sbjct: 53  CAPQTMNESRPASDFREHT----IVDFLERPRVVATHIW 87
```

Database search tools are almost always able to find a sequence with at least some degree of similarity to the query sequence, even if the query is nonsense, like the CAPTAINKIRK protein. It is necessary to use common sense when looking at the output of a program like BLAST. One thing that can help to interpret the significance of results is the *E* value. In the two examples above, the *E* values are 2.4 and 0.54, which suggests that the matches are not significant (fortunately!).

BLAST is a frequently used program that has been implemented in several ways to deal with different types of sequence data. BLASTP compares a protein query to a protein database; BLASTN compares a DNA query to a DNA database; BLASTX takes a DNA query, translates it, and compares it to a protein database; and TBLASTN compares a protein query to a translated protein database. These programs all work in very similar ways.

7.1.3 PSI-BLAST

The basic version of BLAST looks for matches between a single query and sequences in a database. However, the more recent Position-Specific Iterated-BLAST (PSI-BLAST) uses information from sets of related sequences for database searches (Altschul *et al.* 1997). The method begins with a single query sequence and locates database sequences with significant matches (say *E* < 0.01) using the original BLAST algorithm. A multiple alignment is then constructed by placing all the locally aligned sections of the database sequences below the query sequence. In cases where the pairwise local alignment would put a gap in the query sequence, the corresponding residue from the database sequence is eliminated. This means that the multiple alignment has the same number of columns as the query sequence. This is a quick and approximate way of making a multiple alignment, as the algorithm is designed for speed and simplicity.

The alignment produced in the first BLAST run is then used as input to a second run. Scores are now assigned for matching a new sequence from the database with the set of sequences already aligned, rather than for just matching the sequence with the original query. The score depends on the frequencies of the residues in the columns of the alignment, i.e., a V residue is more likely to be aligned with a column that has several Vs in it already, or a column with lots of other hydrophobic residues, than with a column containing mostly, say, acidic amino acids. Scoring systems like this are known as position specific scoring matrices

```
                                                                        Score      E
Sequences producing significant alignments:                            (bits)  Value

gi|131557|sp|P25044|PTP1_YEAST    Protein-tyrosine phosphatase...        705    0.0
gi|417567|sp|P32586|PYP2_SCHPO    Protein-tyrosine phosphatase...        148    2e-35
gi|417568|sp|P32587|PYP3_SCHPO    Protein-tyrosine phosphatase...        142    1e-33
gi|33112421|sp|Q13332|PTNS_HUMAN   Receptor-type protein-tyro...         137    3e-32
gi|1709906|sp|P23468|PTPD_HUMAN   Protein-tyrosine phosphatas...         135    2e-31
gi|462551|sp|Q05909|PTPG_MOUSE    Protein-tyrosine phosphatase...        134    3e-31
gi|3183128|sp|Q62656|PTPZ_RAT   Receptor-type protein-tyrosin...         133    7e-31
gi|20455509|sp|P35992|PTP1_DROME   Protein-tyrosine phosphata...         132    8e-31
gi|462550|sp|P23470|PTPG_HUMAN    Protein-tyrosine phosphatase...        132    9e-31
gi|125978|sp|P10586|PTPF_HUMAN    LAR protein precursor (Leuko...        132    1e-30

gi|266860|sp|P29461|PTP2_YEAST    Protein-tyrosine phosphatase...         85    3e-16
gi|731478|sp|P40048|PTP3_YEAST    Protein-tyrosine phosphatase...         49    1e-05

gi|2499759|sp|P80994|VH01_RACVI   Dual specificity protein ph...          36    0.13

gi|138374|sp|P07239|DUSP_VACCV   Dual specificity protein pho...          35    0.18

gi|138373|sp|P20495|DUSP_VACCC   Dual specificity protein pho...          35    0.19
gi|418237|sp|P33064|DUSP_VARV   Dual specificity protein phos...          35    0.33

gi|1168807|sp|Q00684|CC14_YEAST    Probable protein-tyrosine p...        33    1.2
```

Fig. 7.2 Query of PTP1_YEAST against Swiss-Prot using BLASTP. The top 10 hits are shown, plus the hits to other yeast sequences and a few other sequences discussed in the text.

(PSSMs), and we will talk about them again in Section 9.4.

Searching the database using the PSSM may turn up further significant matches that were not found by the original query. If so, the new sequences can be added to the alignment, the scoring matrix can be updated, and the search can be repeated. The whole process is iterated until no additional sequences are found. PSI-BLAST can sometimes lead to the detection of distantly related sequences that would not be found by a straightforward BLAST search. This is because there is extra information in the aligned group of sequences that is not in any one sequence. However, it is necessary to be careful when adding sequences, because if the range of sequences becomes too broad, then the alignment can end up having little relationship to the original query, and future matches to the alignment will not represent biologically meaningful similarities. It is possible to manually edit the list of new hits to include in the PSSM at each iteration to avoid addition of spurious matches that do not appear to be relevant.

7.1.4 Comparison of search methods

FASTA and BLAST produce ranked lists of hits against a database in a similar way to MPsrch. The BLAST results for PTP1_YEAST are shown in Fig. 7.2. This was obtained using the NCBI BLAST facility with the BLOSUM62 scoring matrix and with Swiss-Prot as the specified database (per-formed in December 2003). This time, 77 significant hits were found with $E < 0.05$. The top few, shown in Fig. 7.2, include sequences from *Schizosaccharomyces pombe*, from mammals, and from *Drosophila*. The list is similar, but not identical to those produced by MPsrch (Fig. 7.1). The two highest-ranking *S. cerevisiae* sequences are PTP2_YEAST and PTP3_YEAST, as before. The only other *S. cerevisiae* sequence is CC14_YEAST, which is way down the list and is apparently not significant ($E = 1.2$). The MSG5_YEAST sequence that was found by MPsrch does not appear on this list, because it has an E value greater than the threshold, which was set to 10 in this example. BLAST can spot the similarity between these sequences, and would report it if the threshold were increased, or if a search were made against only *S. cerevisiae* sequences. We return to this point in Section 7.3, when we discuss the statistical significance of alignments more carefully. For now, we note that BLAST has also spotted the match to the sequences from viruses found using MPsrch with the PAM50 matrix.

Although the ranking of hits in BLAST output gives some information about the degree of relatedness of database sequences to a query, it should not be assumed that the top hit is necessarily the most meaningful evolutionarily. Several authors have found human genes whose top BLAST hits are to bacterial sequences. This has led to claims of horizontal transfer of genes from bacteria to vertebrates. However, more careful studies have shown that most of these cases are not supported by phylogenetic methods (Stanhope *et al.* 2001), and comparison of BLAST results with phylogenetic trees (Koski and Golding 2001) shows that frequently the top BLAST hit is not the closest relative. In Chapter 12, we discuss horizontal transfer in bacteria in more detail.

In order to compare the effectiveness of different search tools, we need to think more carefully about what we actually want from a database search. Usually, we have some idea of what we mean by a "real" relationship between sequences. For example, we might require that sequences should be descended from a common ancestor, or we might require that they should each contain at least one domain that is descended from a common ancestor, or we might require that they should share the same function. Unfortunately, alignment algorithms know nothing about evolutionary history or about protein function – they only know about similarity between characters in linear sequences. It is therefore necessary to find a measure of sequence similarity that reflects real biological similarities as closely as possible. Let us suppose that we have already identified all the sequences in a database that belong to a given protein family (by some combination of hard work and biological knowledge that we will not specify). We may now compare the ability of different algorithms to identify members of this family.

We take one member of the family and use this as a query against a database. The algorithm returns a ranked list of hits and stops at some predefined threshold score. Ideally, all family members will be at the top of the list with scores above the threshold, and all unrelated sequences will be at the bottom of the list with scores below the threshold. Unfortunately, this will not always be entirely true in reality, as often an algorithm will fail to match true family members and/or will incorrectly match unrelated sequences. We use the following definitions: True Positives are family members that are correctly identified by the algorithm with a score above the threshold; False Positives are unrelated sequences that are assigned a score above the threshold by the algorithm; False Negatives are family members that are missed by the algorithm because their score is below the threshold; and True Negatives are unrelated sequences that correctly fall below the threshold. The **sensitivity** of the algorithm is defined as the fraction of family members that are correctly identified:

$$\text{Sensitivity} = \frac{\text{True Positives}}{\text{True Positives} + \text{False Negatives}} \quad (7.1)$$

However, it is always possible to produce an algorithm with 100% sensitivity by simply reducing the threshold to a low enough value so that there are no False Negatives. This is counterproductive in practice, as reducing the threshold is likely to give rise to greater numbers of False Positives. Therefore, we need a second measure of effectiveness of the algorithm. We define the **selectivity** as the fraction of sequences with score above the threshold that are real family members:

$$\text{Selectivity} = \frac{\text{True Positives}}{\text{True Positives} + \text{False Positives}} \quad (7.2)$$

If the algorithm is any good at all, then at least the top few hits will be family members. Therefore, if we make the threshold high enough, we will achieve 100% selectivity, but in so doing we may obtain very few hits (i.e., low sensitivity). In an ideal case, where all the family members are ranked above any of the unrelated sequences, it is possible to find a threshold such that both sensitivity and selectivity are 100%. In reality, we need to choose a threshold that is a good compromise between the two quantities. A sensible rule of thumb is to choose the threshold where the two quantities are equal.

If we have a way of calculating the statistical significance of hits, then we could choose a threshold based on E values, e.g., all sequences with $E < 0.05$ could be counted as positives. However, estimates of significance are hard to calculate and are not always accurate. Using 0.05 as a cut-off with the BLAST example in Fig. 7.2, we obtained 100% selectivity and low sensitivity. Once again, we used the list of sequences in the InterPro PTP entry to define true family members. If the threshold were set at $E = 10$ (the end of the BLAST output in this case) we would still have virtually 100% selectivity and substantially higher sensitivity, so this is not a bad compromise in this example.

More importantly than simply moving the threshold up or down, we can change the scoring system so that the rank order of the hits changes and we can hope to increase both sensitivity **and** selectivity. One important issue is the choice of amino acid substitution matrix. Henikoff and Henikoff (1993) used BLAST and FASTA with many different matrices to compare the ability of different matrices to detect members of protein families. They found that BLOSUM62 was the most successful of the matrices under the test conditions. The BLOSUM series of matrices is based on counting substitutions between proteins that are greater than a given evolutionary distance apart, whereas the PAM matrices are based on counting substitutions between closely related proteins to give a PAM1 matrix, and then extrapolating this to higher PAM distances. Henikoff and Henikoff (1993) argue that this extrapolation does not work well in practice, and that the BLOSUM series is better at detecting distant sequence similarities because it is explicitly calculated from alignments of distant sequences. However, the original PAM matrices (Dayhoff 1978) were calculated at a time when there were few sequences available. More recent matrices, calculated using very similar methods (Jones, Taylor, and Thornton 1992, Gonnet, Cohen, and Benner 1992), performed better than the Dayhoff matrices because they were calculated from much larger sequence sets. It is known that

different matrices are appropriate for finding similarities at different evolutionary distances. When searching a database, we do not know in advance at which evolutionary distance matching sequences will be found. Several authors have therefore suggested doing searches with a set of different matrices in order to increase the likelihood of finding related sequences at all distances. The drawback of this is that an increased number of False Positives are likely to be found if searches are done with more than one matrix. For the criteria used by Henikoff and Henikoff (1993), using multiple matrices did not improve on the performance of the single BLOSUM62 matrix because the increase in sensitivity was outweighed by the loss of selectivity.

BLAST, FASTA, and Smith–Waterman searches can all be performed with any of the substitution matrices. It is therefore of interest to compare the performance of the different algorithms using the same substitution matrices. Pearson (1995) concluded that the full Smith–Waterman search performed better than either of the heuristic algorithms when optimized scoring systems were used. Nevertheless, BLAST is probably the most frequently used search tool because of its speed and its free availability on a number of Web sites. It also produces results that are very similar to those of a full Smith–Waterman search (as seen in the examples in Figs. 7.1 and 7.2). Closely related sequences will be identified in a search using any of these algorithms. However, it should be remembered that no search algorithm is guaranteed to find all the sequences of biological interest. The matches found by a search tool will depend on the details of the search algorithm and the scoring system used. When we are interested in finding as many distant matches as possible, it is worth using more that one algorithm and/or more than one set of parameters.

7.2 ALIGNMENT STATISTICS (IN THEORY)

7.2.1 Why bother with statistics?

If we take any two sequences and run them through a pairwise alignment program, there will always be some best alignment. Similarly, if we run a database search with any query sequence, there will always be some top-hit sequence in the database (remember the CAPTAINKIRK example discussed above). It is possible to find some degree of similarity between parts of two completely unrelated random sequences. So, how do we know that the result we get from an alignment or a database search is a biologically meaningful one, and that it does not just arise simply from chance similarities between unrelated sequences? If we have two biological sequences with a recent common ancestor, there will be a high degree of similarity and it will be easy to see that the matching regions in the alignment are real. If we are in the "twilight zone" of rather weak similarity, then it may not be easy to tell from just looking at the result whether it means anything. For this reason, it is important to estimate the statistical significance of any potential match that we find between sequences.

Sequence-matching algorithms assign scores to alignments between two sequences. To determine the significance of any particular score, it is necessary to calculate the distribution of scores that we would expect between pairs of random unrelated sequences. From this distribution, we can then obtain the p value, i.e., the probability that an alignment with a score greater than or equal to the observed score would occur between unrelated sequences. If p is small, this indicates that the alignment between the sequences is a significant one that is unlikely to have occurred by chance. In the rest of this section, we are therefore concerned with calculating the score distributions between pairs of random sequences. However, the section comes with a health warning – it is one of the most mathematically difficult in the book. Nevertheless, we include it because of its fundamental importance.

7.2.2 Simplest possible case – Pairwise scores

Calculating the distribution of scores between random sequences can be difficult for real alignment algorithms. For this reason, we begin this section with a very simplified version of the problem. Consider an ideal world, where all gene sequences have exactly the same length, N, and where sequence

evolution occurs with only point mutations and no insertions or deletions. In this case, it is obvious how to align two sequences – we simply put one below the other with no gaps, like this:

```
A C C G T T A G G G
A C T T T T A G G A
* *     * * * * *
```

A straightforward measure of similarity between two sequences is just the number of matching sites m. In this case, $m = 7$ and $N = 10$. Let the frequencies of the bases be π_A, π_C, π_G, and π_T. For any site in the alignment, the probability that two unrelated bases would both be A is π_A^2, and the probability that they are both C is π_C^2, etc. Thus, the probability that the two sequences match at this site, irrespective of what base is present, is

$$a = \pi_A^2 + \pi_C^2 + \pi_G^2 + \pi_T^2 \qquad (7.3)$$

The probability that they do not match is $1 - a$. If all bases have equal frequency, then $a = {}^1/_4$. The probability that there are m matching sites is given by a binomial distribution:

$$P(m) = C_m^N a^m (1 - a)^{N-m} \qquad (7.4)$$

(If you need a reminder, see Section M.9 in the Appendix.) The expected number of matches between two random sequences is the mean of this distribution $\bar{m} = Na$, and the variance of the distribution is given by $\sigma^2 = Na(1 - a)$. If N is large, then the binomial distribution can be well approximated by a normal distribution with the same mean and variance (see Section M.10). To calculate the significance of a given alignment with an observed number of matches, m_{obs}, we first calculate the z value, which is the number of standard deviations the observed value is above the expected value:

$$z = \frac{m_{obs} - \bar{m}}{\sigma} \qquad (7.5)$$

We know that z has a standard normal distribution, and hence that there is 95% probability that z

falls in the range $-1.96 \leq z \leq 1.96$ (again, see Section M.10). If the observed value of z is greater than 1.96, the probability of the alignment having a number of matches this high by chance is less than 2.5%. We can then reject the null hypothesis that the sequences are unrelated.

For example, if the base frequencies are all equal, then $\bar{m} = 0.25N$, and $\sigma^2 = 0.1875N$. Suppose we observe 120 base matches in a sequence of length 400. We calculate

$$z = (120 - 100)/\sqrt{75} = 2.309$$

This is slightly outside the confidence interval, and suggests that there are more matches than we would expect by chance. We can calculate the actual probability of observing greater than or equal to 120 matches between random sequences by evaluating the area under the tail of the normal distribution – this is the p value in Eq. (27) of the Appendix. For an observed z of 2.309, $p = 1.05\%$. Thus, somewhat surprisingly, this is a fairly significant match according to our model, even though the percentage similarity between the sequences is quite low (120/400 is only 30% similarity).

7.2.3 Simplest possible case – Database search scores

Now let us consider a database search algorithm in our ideal world where all sequences have the same length and there are no gaps. The database contains S sequences. The search algorithm calculates the number of matches m between a query sequence and each of the S sequences. It then returns the score m_{max} of the most closely matching sequence from the database. We want to know whether this value is significant. To determine this, we need to know the distribution of scores $F(m_{max})$ that we would obtain for matching a random query sequence against the database. We already calculated the distribution $P(m)$ of scores between pairs of random sequences. However, m_{max} is the maximum of a large number of different values of m, so it is likely to be considerably higher than the value of m obtained from just two typical sequences. Therefore $F(m_{max})$ is **not** the same distribution as $P(m)$. It can be shown that $F(m_{max})$ is an example of a type

of distribution known as an extreme value distribution (EVD) and can be written as:

$$F(m_{max}) = \lambda e^{-\lambda(m_{max}-u)} \exp(-e^{-\lambda(m_{max}-u)}) \quad (7.6)$$

This is a smooth function with a single peak skewed to one side. EVDs arise whenever we are dealing with the maximum value taken from a large number of independent alternatives. More details are given in Box 7.1.

To show that this distribution applies in our example, a database of 2000 random sequences of length 200 was generated. The sequences contained only C and G bases, each with frequency 50%. The number of matches, m, was counted for each pair of sequences in the database, and the distribution $P(m)$ is plotted in Fig. 7.3. This is a binomial distribution, with $N = 200$ and $a = 0.5$. We then considered each sequence in turn as a query sequence, and calculated m_{max}, the highest number of matches against any of the other sequences in the database. The distribution $F(m_{max})$ is shown in Fig. 7.3 as a solid line. A least-squares fitting routine was then written to fit the EVD function (Eq. (7.6)) to the simulated distribution. The dashed line is the best-fitting EVD, which has $\lambda = 0.497$ and $u = 123.2$. This fits the simulated data rather well. Note that typical values of m_{max} are in the range 120–130. This is much larger than the typical values of m, which are in the range 90–110.

We can now consider the problem of the significance of a match of a given query sequence against the database. If the observed value of the top-hit score for the query sequence is m_{obs}, then the probability of obtaining a value m_{max} greater than or equal to this score by chance is given by the area under the tail of the probability distribution (again see Box 7.1 for the details):

$$p(m_{obs}) = 1 - \exp(-e^{-\lambda(m_{obs}-u)}) \quad (7.7)$$

The smaller the value of p, the less likely the match is to arise by chance and therefore the greater the significance of the match. Suppose we run a query sequence against our database and the top hit has a score of $m_{obs} = 130$, then using the measured values of λ and u in Eq. (7.7) gives $p = 3.3\%$. (Note that

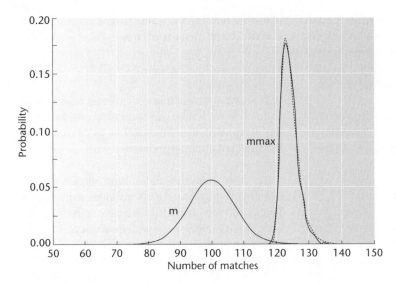

Fig. 7.3 Probability distributions for match scores using the Simplest Possible Case example of a database search tool. $P(m)$ is a binomial distribution (Eq. (7.4)). $F(m_{max})$ is an extreme value distribution. The solid curve is calculated from simulated data, and the dashed line is obtained by fitting Eq. (7.6) to the simulated data.

there is a small tail of the distribution to the right of 130 visible in Fig. 7.3 that contains 3.3% of the area under the curve.) Thus, a score of 130 is just big enough to be significant.

However, the significance of a given score depends on the size of the database. The more sequences there are in the database, the higher the score of the best match is likely to be. If the number of sequences in the database increases from S_1 to S_2, the new distribution of m_{max} is an EVD with the same value of λ, but a higher value of u. In other words, $F(m_{max})$ is the same shape but shifted to the right. Suppose we repeated our simulation with a database of 10,000 sequences instead of 2000. Equation (7.11) in Box 7.1 says that we would expect the new value of u to be $123.2 + \ln(5)/0.497 = 126.4$. As a result of this, there will now be a lower significance attached to the top hit (i.e., the value of p will be larger). Let us suppose the sequences in the smaller database are included in the larger one. The score of the query against the original top hit is still 130; however, with $u = 126.4$, we find $p = 15.4\%$, which is not statistically significant. This is slightly alarming the first time you see it. The score for the match between the sequences is still the same, but the significance has changed because the database size has increased. The result is nevertheless quite logical. The more sequences there are in the database, the higher the best score will be for a random

sequence, and the higher the score needs to be for a real sequence before we can conclude that the match is significantly better than would be expected by chance. We return to this point in an example with real sequence databases in Section 7.3.

7.2.4 Word-matching example

We will now consider a second example of an alignment algorithm, which is also very idealized, but has a lot in common with real algorithms like BLAST. Consider two sequences of length N and M. The simplest version of a local alignment algorithm is to look for the longest exactly matching word between the two sequences. We will say that the score for aligning the sequences is just the length l of this longest word. In the example below, $l = 6$, corresponding to the word GATATC.

```
G G A T A T C C A G C G C T C C T C T
                                *
A T C C G A T A T C T T G G
      *
```

Suppose we pick any random site in one sequence and any random site in the other sequence. Let n be the length of the matching word we get by starting at these two random sites. For example, if we start at

BOX 7.1
Extreme value distributions

Suppose we have some quantity, x, with a probability distribution $P(x)$ and we sample a large number, S, of independent values of x from this distribution. We want to know the probability $F(x_{max})$ that the maximum of these values is equal to x_{max}. First, the probability that any one value is less than x_{max} is

$$G(x_{max}) = \int_{-\infty}^{x_{max}} P(x)\, dx \qquad (7.8)$$

To calculate $F(x_{max})$, we note that $S - 1$ of the values must be less than x_{max} and one of them must equal x_{max}. There are S ways of choosing the one that is equal. Hence:

$$F(x_{max}) = SP(x_{max})G(x_{max})^{S-1} \qquad (7.9)$$

We can calculate a more explicit formula for the case where $P(x)$ is an exponential distribution:

$$P(x) = \lambda e^{-\lambda x}$$

From Eq. (7.9),

$$G(x) = 1 - e^{-\lambda x}$$

and therefore

$$F(x_{max}) = S\lambda e^{-\lambda x_{max}}(1 - e^{-\lambda x_{max}})^{S-1} \approx S\lambda e^{-\lambda x_{max}} \exp(-Se^{-\lambda x_{max}}).$$

Here, we have assumed that $S \gg 1$ and we have used the approximation in the Appendix (Eq. (M.35)). If we define the constant u by $u = \ln(S)/\lambda$, so that $S = e^{\lambda u}$, then we can rewrite the above equation as

$$F(x_{max}) = \lambda e^{-\lambda(x_{max}-u)} \exp(-e^{-\lambda(x_{max}-u)}) \qquad (7.10)$$

This function is known as an extreme value distribution (EVD) or sometimes as a Gumbel distribution. The function is unusual in that it contains an exponential of an exponential. The distribution is controlled by the two constants λ and u. The function has a single peak that occurs at $x_{max} = u$. The peak has a skew shape that decays more slowly on the right than on the left. The width of the peak is controlled by λ. If λ is increased, the distribution gets narrower, but the peak stays in the same place. Examples of EVD curves are shown in Figs. 7.3 and 7.4.

If u is changed, the peak is shifted along the axis without changing shape. This is what happens if we increase the number of samples used to generate x_{max} from S_1 to S_2. The peak of the distribution is moved from u_1 to u_2, where

$$u_2 = u_1 + \frac{\ln(S_2/S_1)}{\lambda} \qquad (7.11)$$

Although we derived the above formula for the case where $P(x)$ was an exponential, it turns out that Eq. (7.10) is also a good approximation for the distribution of x_{max} for many other distributions of x. The values of λ and u will depend on $P(x)$ but the functional form of the EVD will be the same. In many cases, we cannot calculate the values of λ and u. It is necessary to obtain them by fitting the function in Eq. (7.10) to empirical data, as shown in the examples in this chapter.

For statistical purposes, we are often interested in the probability that a quantity is greater than or equal to the observed value. If x_{max} is distributed according to Eq. (7.11), then the probability that an observed value x_{obs} is greater than or equal to this is

$$p(x_{obs}) = \int_{x_{obs}}^{\infty} F(x_{max})\, dx_{max} = 1 - \exp(-e^{-\lambda(x_{obs}-u)})$$

$$(7.12)$$

which follows from integration of Eq. (7.10).

---

the two points indicated by asterisks above, then we get a matching word TCC of length $n = 3$. If we pick two random sites that do not match at all, this corresponds to a word of length $n = 0$. Thus, there are many different places at which we can choose the start points, and l is the longest of all the different word lengths that we get from all the different start points. Once again, this is a situation where we are taking the maximum value of a large number of possible alternatives. For this reason, we expect that the distribution of $F(l)$ will be an EVD. In Box 7.2, we show in more detail why the EVD arises in this case.

To show that the derivation in Box 7.2 works, we used the same simulated database as in the previous example, with 2000 random GC sequences of length 200. The value of l was calculated for every pair of

BOX 7.2
Derivation of the extreme value distribution in the word-matching example

If we pick any random site in one sequence and any random site in the other sequence, the probability that the two bases match is a (given by Eq. (7.3)). Let n be the length of the matching word that starts at these two points. The probability that the word is at least of length l is $P(n \geq l) = a^l$, because l sites in a row must match. It will be convenient to define the constant

$$\lambda = -\ln(a) = \ln(1/a) \tag{7.13}$$

so that

$$P(n \geq l) = e^{-\lambda l} \tag{7.14}$$

We would like to calculate the expected number of matching words $E(l)$ that there will be between the two sequences that are of length at least l. Now there are N ways of choosing the first starting site and M ways of choosing the second starting site; therefore, there are NM ways of choosing the pair of sites together. Each of these choices has a given word length associated with it. If we consider each of these NM starting points to be independent, the expected number of matching words should be $E(l) = Nme^{-\lambda l}$. However, the words formed from different starting points overlap one another so they are not really independent. As a result of this, the expected number of matching words is slightly less:

$$E(l) = KMNe^{-\lambda l} \tag{7.15}$$

K is a factor less than one that accounts for the overlap between words starting at neighboring points.

From Eq. (7.14), the probability that a word is of length less than l is

$$G(l) = 1 - e^{-\lambda l} \tag{7.16}$$

If we cheat slightly by treating l as a continuous variable instead of an integer, it follows that the probability that the word is of length exactly l is

$$P(l) = \frac{dG}{dl} = \lambda e^{-\lambda l} \tag{7.17}$$

This is exactly the case we considered in Box 7.1, hence the probability $F(l)$ that the longest matching word is of length exactly l is

$$F(l) = \lambda e^{-\lambda(l-u)} \exp(-e^{-\lambda(l-u)}) \tag{7.18}$$

We suppose that the number of independent words (the equivalent of S in Box 7.1) is KMN. Therefore,

$$u = \frac{\ln KMN}{\lambda} \tag{7.19}$$

Note that although we know the value of λ from Eq. (7.13), we do not know the value of u exactly, because it depends on K, which we did not determine in this argument.

sequences in the database and the distribution $F(l)$ is plotted in Fig. 7.4. We know from Eq. (7.3) that $a = \frac{1}{2}$ and hence $\lambda = \ln 2$, from Eq. (7.13). In this example, $M = N = 200$. We do not know what u is, and therefore we estimate u by doing a least-squares fit to the simulated data. This gives $u = 13.6$. The fitted EVD function is shown as a dashed line in Fig. 7.4 – it is difficult to see because the fitted curve is almost exactly on top of the data!

So far, we only considered the distribution of scores for two sequences. If we use this scoring system in a database search algorithm, we will calculate l_{max}, the length of the top-hit word between the query sequence and any of the sequences in the database. The

easiest way to imagine this is to consider all the database sequences linked together into a single long sequence. The distribution can therefore be calculated in the same way as in Box 7.2, except that the length of the second sequence, M, is replaced by the total length of the sequences in the database ($M = 2000 \times 200 = 400,000$ in our example). To calculate the distribution of l_{max} in the simulated database, each sequence in turn was treated as the query and the top hit was found against all the remaining sequences. Using Eq. (7.11), we would expect that the peak of the distribution would be shifted from $u_1 = 13.6$ to $u_2 = 13.6 + \ln(400,000/200)/\ln(2) = 24.5$. From the least-squares fitting routine, we found $u = 24.4$, which is

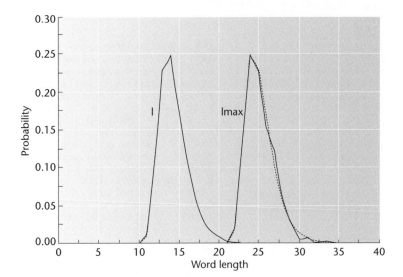

Fig. 7.4 Probability distributions for matching word lengths in the word-matching example. Solid curves show simulated data for the distributions of l and l_{max}. Both are extreme value distributions. Dashed curves are obtained by fitting Eq. (7.6) to the simulated data.

consistent with expectations. In Fig. 7.4, the fitted EVD is visible as a dashed line that closely approximates the simulated distribution of l_{max}.

There is an important difference between the simplest possible case example and the word-matching example in this section. For the word-matching example, the distribution of pairwise scores $F(l)$ is itself an EVD, because l is the longest matching word from many possible words that can be chosen from the two sequences. The distribution of the top-hit score, $F(l_{max})$, is an EVD because the longest word is calculated for each pair of sequences **and** because the best-scoring sequence in the database is returned as the top hit. This contrasts with the simplest possible case example, where the distribution $P(m)$ of pairwise scores is **not** an EVD (because m is not obtained by calculating the best of many alternatives), but the distribution $F(m_{max})$ of top-hit scores **is** an EVD.

7.3 ALIGNMENT STATISTICS (IN PRACTICE)

Real database searching algorithms have more complicated scoring systems than either of the above examples. BLAST works by looking for high-scoring local alignments, as described in Section 7.1.2, and allows for different paired scores between different

residues by using the BLOSUM matrices. In the earlier versions of BLAST, gaps were not allowed. In that case, it can be shown (Karlin and Altschul 1990) that the distribution of pairwise local alignment scores is an EVD. More recent versions of BLAST allow gaps, as do FASTA and Smith–Waterman algorithms. It is believed that for all these versions of local alignment algorithms, the distribution of scores between random sequences is also an EVD. The constants λ and u in the EVD formula cannot be calculated theoretically for these programs, but they can be obtained by running the algorithm against many random sequences in order to obtain an empirical probability distribution, and then fitting this distribution to an EVD (examples are given by Altschul *et al.* 1997). Accurate estimation of the parameters of EVDs from simulations with random sequences is a current area of research (Altschul *et al.* 2001).

Once the EVD parameters are known for a given search algorithm and a given scoring system, it is possible to estimate the significance of a given match against a database. Results of database search programs are usually presented in terms of E values, where E is the expected number of sequences in the database that will match the query sequence with a score greater than or equal to the observed score S. As in the word-matching example (Box 7.2, Eq. (7.15)), the E values decrease exponentially with the score:

(a)

```
                                                                 Score   E
gi|131557|sp|P25044|PTP1_YEAST   Protein-tyrosine phosphatase...   705   0.0
gi|266860|sp|P29461|PTP2_YEAST   Protein-tyrosine phosphatase...    85   2e-17
gi|731478|sp|P40048|PTP3_YEAST   Protein-tyrosine phosphatase...    49   6e-07
gi|1168807|sp|Q00684|CC14_YEAST  Probable protein-tyrosine p...     33   0.072
gi|1709121|sp|P38590|MSG5_YEAST  Protein-tyrosine phosphatas...     30   0.34
gi|417226|sp|P32895|KPR1_YEAST   RIBOSE-PHOSPHATE PYROPHOSPHO...     29   0.67
gi|1730719|sp|P53965|YND2_YEAST  HYPOTHETICAL 32.8 KD PROTEI...      28   1.6
gi|731615|sp|P38732|YHD9_YEAST   Hypothetical 67.5 kDa protei...     28   1.8
gi|416899|sp|P32872|DHAY_YEAST   Aldehyde dehydrogenase 2, mi...     28   2.4
gi|3914316|sp|Q07418|PEXJ_YEAST  Farnesylated protein PEX19 ...      27   3.0
gi|1730735|sp|P53949|YNF6_YEAST  HYPOTHETICAL 22.5 KD PROTEI...      27   5.0
gi|731579|sp|P38893|YH10_YEAST   Hypothetical 37.9 kDa protei...     27   5.5
gi|2497216|sp|Q03254|YM8K_YEAST  Hypothetical 83.4 kDa prote...      26   6.2
gi|731527|sp|P10356|YEY2_YEAST   Hypothetical 49.5 kDa protei...     26   7.3
gi|2494910|sp|Q04660|YMT9_YEAST  Hypothetical 91.7 kDa Trp-A...      26   7.5
```

(b)

```
   Saccharomyces cerevisiae (yeast, ...) [ascomycetes] taxid 4932
gi|6319971|ref|NP_010051.1| phosphotyrosine-specific prote...   705   0.0    =PTP1
gi|6324782|ref|NP_014851.1| protein tyrosine phosphatase; ...    84   2e-15  =PTP2
gi|172382|gb|AAB59323.1| tyrosine phosphatase                    80   3e-14  =PTP2
gi|6320919|ref|NP_010998.1| Protein tyrosine phosphatase; ...    49   9e-05  =PTP3
```

Fig. 7.5 BLAST results using PTP1_YEAST as query sequence. (a) Search made only against other yeast sequences in Swiss-Prot. The complete list of hits is shown. (b) Search made against all proteins in the non-redundant database. Only hits against yeast sequences are shown.

$E(S) = KMNe^{-\lambda S}$. Here, N is the length of the query sequence, M is the total length of all the sequences in the database, and K and λ have been determined from random sequences prior to running the search.

As an example of this, we chose the sequence PTP1_YEAST from *S. cerevisiae*, which was also used in Section 7.1. The BLASTP output for this sequence against the whole of Swiss-Prot was shown in Fig. 7.2. In Fig. 7.5(a), this query is repeated using the same search program, but with the option specified to search only the *S. cerevisiae* sequences in Swiss-Prot. The top hit is, of course, PTP1 itself, with a score of 705. The next is PTP2, with score 85, and $E = 2 \times 10^{-17}$, indicating a highly significant relationship to PTP1. When the search is made against the whole of Swiss-Prot (Fig. 7.2), the score of the hit against PTP2 is still 85, but the E value increases to 3×10^{-16}. This is an order of magnitude higher because the database is an order of magnitude larger. This does not matter much in this case, because the hit is obviously highly significant. However, with borderline cases, it might make a difference to our conclusions. The match to the CC14 sequence is of borderline significance with $E = 0.072$ in Fig. 7.5(a), which might suggest it is worth investigating further, whereas in Fig. 7.2 the same hit has $E = 1.2$, and we would probably ignore this if we had no further evidence.

The largest protein database available on the NCBI BLAST site is the "nr" database of non-redundant proteins. When PTP1_YEAST is run against this database, E values are increased with respect to the search against Swiss-Prot, because the nr database is larger. The output from this search was grouped into species using the taxon-specific summary facility. Figure 7.5(b) shows the hits against yeast sequences. With the larger database, the E value for the CC14 gene is increased from its value in Fig. 7.2, and this time it does not even appear in the output. The genes PTP2 and PTP3 appear as before, except that there are two hits against PTP2 which arise because the same gene has been sequenced and submitted to GenBank by different people. The two PTP2 sequences appear to be identical in length but differ at just a couple of amino acids.

Although the nr database is supposed to be non-redundant, it clearly contains examples of sequences that are almost identical (often, these are errors that have been corrected in Swiss-Prot, but reintroduced when nr's different source databases are merged). This means that the effective size of the database is smaller than it would be if all sequences were really independent, and the E values may be inaccurate for this reason. Which sequences we include in a database depends on our definition of non-redundant. This is a very gray area. What about alleles of the same gene in different organisms, or duplicate genes in the same genome, or related proteins in the same family, or pseudogenes, or copies of transposable elements? Significance values based on properties of random sequences cannot capture all these levels of biological complexity.

These examples show that alignment statistics need to be taken with a pinch of salt. We cannot draw a strict borderline at $E = 0.05$ (or anywhere else) and say that all hits below this value are significant, while all hits above this value are not. It is necessary to use common sense to decide whether we believe that a local alignment between sequences is biologically relevant or not. The E values we calculate depend on the assumptions underlying the statistical test and the null model that we use to calculate the distribution of scores between random sequences. For BLAST, the significance is calculated relative to the distribution of scores expected for random protein sequences composed of amino acids with the same frequency as real proteins but in random order. The null model for BLAST therefore assumes that the likelihood of any amino acid occurring is the same at any point in the sequence, and is independent of the amino acids next to it in the sequence. Real sequences are likely to have some degree of correlation between consecutive residues. For example, proteins tend to have hydrophilic and hydrophobic regions, so that the probability of a residue being hydrophobic is probably slightly larger if the previous residue was hydrophobic too. There are also particular short protein motifs that appear with high frequency (e.g., simple sequence repeats like polyglutamine regions). Although more complicated null models can be envisaged, it is difficult to account for all of these factors properly. The null model used for calculations is bound to be unrealistic to a certain extent, and the significance values we get from the model are therefore not as useful as we might like.

One approach is to avoid using random sequences for the null model. We can calculate the expected distribution of scores by aligning many pairs of unrelated proteins, and we can use this distribution to estimate the significance of a match in the case of a pair of putatively related proteins. It has the advantage that any unusual properties of real sequences like correlations and common motifs will automatically be accounted for in the distribution of scores. However, it has the disadvantage that we have to define in advance what we mean by related and unrelated proteins. This method is therefore rather circular.

SUMMARY

Tools are available for searching databases to find sequences that are similar to a user-supplied query sequence. MPsrch is an implementation of the Smith–Waterman alignment algorithm that does a pairwise local alignment of the query with every database sequence and returns a list of hits in descending order of alignment score. Although individual pairwise alignments are fairly rapid, doing full pairwise alignments with every database sequence can be slow. Therefore, heuristic search algorithms have been developed that try to find high-scoring local alignments as quickly as possible without using a full dynamic programming algorithm. The most com-monly used of these are FASTA, BLAST, and its extension PSI-BLAST.

Different search tools will usually produce similar but not identical lists of matching sequences to a given query. If it is important to locate as many low-scoring matches as possible, then it may be appropriate to use more than one search tool, or more than one set of scoring parameters within the same search tool. The ranking of matching sequences according to the scores from heuristic search tools is not always a reliable indicator of evolutionary relationships, i.e., it is dangerous to assume that the sequence with the top BLAST score is the most closely related one to the query. More accurate phylogenetic studies are necessary if this issue is important.

Similarities between sequences can arise by chance. When looking at the output of a database search program, it is useful to have a statistical measure of the significance of the match. If the score for a match is significantly higher than we would expect by chance, it can be concluded that the sequences are probably evolutionarily related, and the similarity is likely to mean something biologically. We are usually interested in the score of the top hit, i.e., the highest-scoring match in the database. When sequences are random (i.e., not evolutionarily related), it can be shown that the expected distribution of the top-hit score is an extreme value distribution. The shape of the EVD depends on two parameters, u and λ, that control the mean and the width of the distribution. The values of these parameters can be calculated theoretically in simple examples but, for realistic cases, they need to be determined by calculating the distribution of scores between random sequences using the real search algorithm.

When the parameters of the EVD are known, it is possible to return an E value, together with the match score against each of the top few hits in the output of a database search. This is the expected number of sequences in the database that would match the query by chance with a score at least as high as the real match score. When $E \ll 1$, the match is highly significant. The E value obtained for any given match score will depend on the size of the database. The larger the database, the higher E, and the less significant the result appears to be, even though the score remains the same.

REFERENCES

Altschul, S.F., Gish, W., Miller, W., Myers, E.W., and Lipman, D.J. 1990. Basic Local Alignment Search Tool. *Journal of Molecular Biology*, **215**: 403–10.

Altschul, S.F., Madden, T.L., Schäffer, A.A., Zhang, J., Zhang, Z., Miller, W., and Lipman, D.J. 1997. Gapped BLAST and PSI-BLAST: A new generation of protein database search programs. *Nucleic Acids Research*, **25**: 3389–402.

Altschul, S.F., Bundschuh, R., Olsen, R., and Hwa, T. 2001. The estimation of statistical parameters for local alignment score distributions. *Nucleic Acids Research*, **29**: 351–61.

Dayhoff, M. 1978. *Atlas of Protein Sequence and Structure*. Vol. 5, suppl. 3. Washington, DC: National Biomedical Research Foundation.

Gonnet, G.H., Cohen, M.A., and Benner, S.A. 1992. Exhaustive matching of the entire protein sequence database. *Science*, **256**: 1443–5.

Henikoff, S. and Henikoff, J.G. 1993. Performance evaluation of amino acid substitution matrices. *Proteins: Structure, Function and Genetics*, **17**: 49–61.

Jones, D.T., Taylor, W.R., and Thornton, J.M. 1992. The rapid generation of mutation data matrices from protein sequences. *CABIOS*, **8**: 275–82.

Karlin, S. and Altschul, S.F. 1990. Methods for assessing the statistical significance of molecular sequence features by using general scoring schemes. *Proceedings of the National Academy of Sciences USA*, **87**: 2264–8.

Koski, L.B. and Golding, G.B. 2001. The closest BLAST hit is often not the nearest neighbour. *Journal of Molecular Evolution*, **52**: 540–2.

Pearson, W.R. 1995. Comparison of methods for searching protein sequence databases. *Protein Science*, **4**: 1145–60.

Pearson, W.R. and Lipman, D.J. 1988. Improved tools for biological sequence comparisons. *Proceedings of the National Academy of Sciences USA*, **85**: 2444–8.

Stanhope, M.J., Lupas, A., Italia, M.J., Koretke, K.K., Volker, C., and Brown, J.R. 2001. Phylogenetic analyses do not support horizontal transfer from bacteria to vertebrates. *Nature*, **411**: 940–4.

PROBLEMS

7.1 The best way to understand database search tools is to use them. Try taking one of the hexokinase sequences from the file Hexokinase.txt on the website (www.blackwellpublishing.com/Higgs) and using it as a query sequence. Hopefully you will find all the other hexokinases in the file, but you should also find many others from different species. How does the set of sequences that you find depend on the algorithm used? Try BLAST, FASTA, MPsrch, and PSI-BLAST.

SELF-TEST
Alignments and database searching

This test covers material in Chapters 6 and 7.

1 The time taken for a sequence alignment algorithm is thought to scale as $O(N^3)$, where N is the sequence length. For sequences of length 500 it takes approximately 2.5 seconds. Which of the following statements is true?
A: The algorithm must be a local alignment algorithm rather than a global alignment algorithm.
B: The algorithm will be impractical to run because the memory storage required will be too large.
C: For sequences of length 2000 it is likely to run in approximately 10 seconds.
D: For sequences of length 2000 it is likely to take more than 2.5 minutes.

2 Consider the two DNA sequences CAGCAT and CGACAT. These sequences are aligned using a global alignment algorithm where the score is 1 for a match and 0 for a mismatch and there is a penalty of 0.2 for each gap. Which statement is true?
A: There are no gaps in the optimal alignment.
B: The score of the optimal alignment is 4.0.
C: There are two optimal alignments with equal score.
D: The score of the optimal alignment is 4.8.

3 The affine gap penalty function used in sequence alignments has a cost g_{open} for gap opening and a cost g_{ext} for gap extension. Which of these statements is true?
A: When $g_{open} \gg g_{ext}$, the gaps should be fewer but longer than when $g_{open} = g_{ext}$.
B: The distribution of gap lengths produced by the alignment algorithm is independent of g_{open}.
C: When the affine gap penalty is used instead of the linear gap penalty (where the total cost of a gap is proportional to the length of the gap) the scaling of the global alignment algorithm becomes $O(N^3)$ instead of $O(N^2)$.
D: An affine gap penalty function would be appropriate for aligning DNA sequences but not protein sequences.

4 Which of these statements about multiple alignments is correct?
A: It is not possible to define a dynamic programming algorithm to align more than two sequences.
B: The guide tree in CLUSTALW is produced using a distance matrix method.
C: The guide tree in CLUSTALW does not influence the final alignment.
D: All three of the above statements.

5 The expected distribution of scores from the BLAST algorithm is an extreme value distribution because
A: it uses ungapped alignments.
B: it is derived as an approximation to a dynamic programming algorithm.
C: it returns the highest-scoring match from a database.
D: it uses a probabilistic alignment model.

6 The distribution $F(S)$ of top-hit scores for a database search algorithm based on local sequence alignments follows an extreme value distribution $F(S) = \lambda e^{-\lambda(S-u)} \exp(-e^{-\lambda(S-u)})$ with $\lambda = 0.69$ and $u = 22$. A couple of years later, the database has doubled in size. The distribution of top-hit scores should now be given by an extreme value distribution with
A: $\lambda = 0.81$ and $u = 22$.
B: $\lambda = 1.38$ and $u = 24$.
C: $\lambda = 0.69$ and $u = 44$.
D: $\lambda = 0.69$ and $u = 23$.

7 Which of the following is correct?
A: If a database search algorithm is set to return a large number of results with high E values, it is likely to have high sensitivity and low selectivity.
B: The E value of a database search algorithm is sensitive to the statistical model used to calculate the probability of chance matches.
C: The z score is not useful as a measure of significance of search algorithms based on local sequence alignment because the z scores apply to normal distributions not extreme value distributions.
D: All of the above.

Phylogenetic methods

CHAPTER PREVIEW

We discuss the principal methods used in the construction of phylogenetic trees from sequence data, including distance matrix, maximum-parsimony, and maximum-likelihood methods. An example with small subunit ribosomal RNA sequences from primate mitochondrial genomes is used throughout the chapter to illustrate the differences that arise when using different methods.

8.1 UNDERSTANDING PHYLOGENETIC TREES

Molecular phylogenetics is the use of molecular sequences to construct evolutionary trees. Typically, we study a family of related sequences that we know evolved from a common ancestor and we want to know in which order these sequences diverged from one another. This information is usually presented in the form of a phylogenetic tree diagram. There are numerous methods for constructing trees but, before considering these methods in detail, we will discuss the different ways in which trees are presented, and how the information in a tree can be interpreted.

The simplest type of tree to understand is a rooted tree with a time axis – Fig. 8.1(a). The diagram shows that species A diverged from species B, C, and D 30 million years ago (Ma), that species B diverged from C and D 22 Ma, and that species C and D diverged from each other 7 Ma. The root of the tree is the most recent common ancestor of all the species shown. It is the earliest point in time shown on the tree. As the divergence times are known in Fig. 8.1, the lengths of the vertical branches can be drawn proportional to time. Thus, we get a clear impression from the diagram that species C and D are evolutionarily close to one another whereas species A and D are distant. The lengths of the horizontal lines in this form of tree do not signify anything. They are merely drawn to space the species out conveniently.

It is often said that trees are like hanging mobiles. You can imagine suspending them from the root and allowing the horizontal lines to swing around. Figure 8.1(b) shows one possible arrangement that

(a)

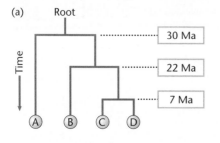

(b)

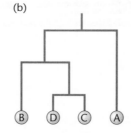

Fig. 8.1 Rooted trees with a time axis. Tree (a) can be converted to tree (b) by swinging around the horizontal branches like mobiles. Hence (a) and (b) are equivalent to one another.

could arise from allowing the tree in (a) to swing around. There are, of course, many others. It is important to realize that although these trees look different, in fact they are the same. We say that these trees have the same topology. In this context, the word topology means the branching order of the species on the tree. When we draw the tree, we are entitled to draw any version of the tree we like (by swinging the mobile around) as long as it does not change the topology. Although there are many rearrangements of the tree that **can** be made, there are also many that **cannot**. For example, in Fig. 8.1a, we can swap C and D around, but we cannot swap B and C.

It is often difficult to estimate divergence times from molecular data. Sometimes, the timing of the events is not known, but the order of events is. In this case, a rooted tree can be drawn with branch lengths that are not to scale. When looking at a tree, it is therefore important to recognize whether the authors have drawn branches to scale in order to avoid getting a false impression about the relative distance between different species.

In Chapter 4, we discussed evolutionary distance. If the sequences under study all evolve at the same rate, then the evolutionary distance between each pair of sequences is proportional to the time since divergence. In this case, the branch lengths can again be drawn to scale and the branch points can be labeled with the evolutionary distance, rather than with time in Ma. The UPGMA method described below assumes that the sequences evolved at the same rate and produces scaled rooted trees of this form. If the sequences do not all evolve at the same rate, it is not possible to have a well-defined time axis on the tree. It is nevertheless possible to draw a rooted tree with branch lengths scaled according to the amount of evolutionary change thought to have happened on each branch, as in Fig. 8.2. The main difference from Fig. 8.1(a) is that the present-day species on the tips of the tree do not all line up level with each other. For example, the branch leading to C in Fig. 8.2 is longer than the branch leading to D. This means that there has been more evolutionary change (i.e., a faster rate of molecular evolution) on the line leading to C than the line leading to D. When the time axis is strict, as

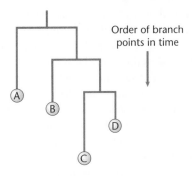

Fig. 8.2 A rooted tree with branches scaled according to the amount of evolutionary change.

in Fig. 8.1(a), the branch length leading to C must be equal to that leading to D, because both are drawn proportional to the length of time since the divergence of C and D. Note that even though the time axis in Fig. 8.2 is **not** strict, there is still some information about the **order** of events in time. The tree is still rooted: this means that we can still say that species A was the earliest species to diverge, and we can still say that C and D diverged from each other more recently than any other pair of species. Trees like this contain a lot of information. They tell us about the temporal order of events, and which species have high rates of molecular evolution (i.e., which species have long branches). For these reasons, this type of tree is probably the most useful way of representing the results of molecular phylogenetic studies.

Having said this, most phylogenetic methods produce **unrooted** trees. Figure 8.3(a) shows the same four species drawn on an unrooted tree. The four present-day species for which we have sequence data are shown on the tips of the tree. The internal nodes of the tree (labeled *i* and *j*) represent ancestors for which we do not have sequence data. The branch lengths are drawn scaled with evolutionary distance. There has been a lot of change along the branch from *i* to A, and rather little change along the branch from *j* to D, for example. Other than saying that the internal nodes must have occurred earlier in time than the tips of the tree, we cannot conclude anything related to time from this unrooted tree. We cannot say whether the branching at *i* occurred earlier than that at *j*. We cannot say which of the species was the earliest to diverge. The length of a branch tells us how much evolutionary change has happened along that branch but we cannot tell in which **direction** the change was happening.

Unrooted trees are thus much less informative than rooted ones, and it is desirable to convert unrooted trees to rooted trees, where possible. To do this, we need to know where the root is. Usually, we have to use prior knowledge to tell us. This could come from the fossil record, from previous phylogenetic studies using morphological data, or from plain biological "common sense". The earliest branching species (or group of species) in a tree is called the **outgroup**. It is common to include a species that we know is the outgroup in the set of species to be analyzed. For example, if we are interested in the phylogeny of Eutherian mammals, then we might add a marsupial as an outgroup. Our prior biological knowledge tells us that the Eutherian mammals are all more closely related to each other than to the marsupial. Hence the marsupial is the outgroup, and the root of the tree must lie on the branch connecting the marsupial to the rest of the tree.

There are many different ways to position the root on an unrooted tree. In Fig. 8.3(a), if we know that A is the outgroup, we can place the root on the branch from i to A. We can then bend the tree around to produce the rooted tree shown in Fig. 8.2. On the other hand, if B is the outgroup, then the resulting rooted tree is shown in Fig. 8.3(b); but we could equally well place the root on the internal branch from i to j, in which case we obtain Fig. 8.3(c). It is clear that Figs. 8.2, 8.3(b), and 8.3(c) are all different from each other. They imply different things about evolutionary history. For example, in Fig. 8.3(c), species A and B form a related group that are closer to each other than to either C or D, whereas in Fig. 8.2, species A is just as distant from B as from C and D. Nevertheless, all three rooted trees correspond to the same unrooted tree.

In cases where we want to turn an unrooted tree into a rooted tree and we do not know a definite outgroup, it is possible to use what is known as the "midpoint method". This consists of finding a point on the tree such that the mean distance measured along the tree to the sequences on one side of the

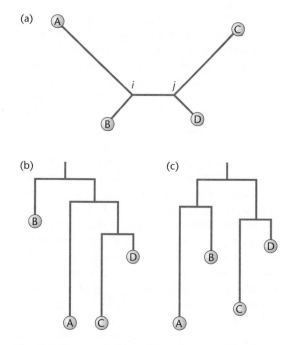

Fig. 8.3 The unrooted tree in (a) can be converted to the rooted trees in (b) and (c) and to the tree in Fig. 8.2 by placing the root in different positions.

point is equal to the mean distance to the sequences on the other side. This amounts to assuming that the average rate of evolution in the two halves of the tree is the same.

When a tree is drawn with branches spreading radially, as in Fig. 8.3(a), it is obvious that it is unrooted. However, occasionally trees are drawn with all branches vertical, as in Fig. 8.3(b), even though they are unrooted. This can be confusing. It is therefore important to read the small print in order to make sure whether the author of a tree intends it to be read as rooted or unrooted. When drawing trees, the default option on some tree-drawing programs is to draw a rooted tree with the first species in the list of sequences as the outgroup, unless otherwise specified. This can lead to misinterpretation of results if the tree is really an unrooted one. When the root is not known, our advice is to draw a radially branching tree, with the species labeled around the edge, as in Fig. 8.3(a). When the root is known, it is clearly better to draw a rooted tree, but it is important to ensure that the drawing program has put it in the right place.

A **clade** (or **monophyletic** group) is the term for the complete set of species that descends from an ancestral node. In Fig. 8.3(b), the group C + D forms a clade, and so does the group A + C + D. However, the group A + C is not a monophyletic group because there is no ancestral node that has only A and C as descendants and no other species. Another way of defining a clade is to say that it is a subset of species that are all more closely related to each other than to any other species in the tree that is not included in the subset.

All the trees discussed so far are bifurcating trees. This means that there is a two-way split at each branch point. Sometimes, trees are drawn with trifurcations (three-way splits), or multifurcations (many-way splits). This may indicate that the author believes there was a rapid radiation of these taxa at the same time, so that the multifurcation cannot be split into individual bifurcations. More usually, it is an indication that the author is unsure of the precise ordering of the branch points. A multifurcating tree contains less information than a strictly bifurcating tree, but it avoids giving a false sense of certainty to the reader. Nearly all phylogenetic programs produce bifurcating trees. It is straightforward to draw the bifurcating tree that is deemed "best" by the method used. Sometimes, there will be alternative trees that are almost as good. Information on alternative trees is sometimes hidden in the small print of papers: space usually prevents drawing more than one or two trees. There are ways of indicating how reliable the different branch points in a tree are deemed to be by using bootstrapping (Section 8.4) or posterior probabilities (Section 8.8). Often, some parts of a tree are extremely reliable while others are very uncertain, and it is important to be aware of this. When interpreting published trees, a healthy element of skepticism is required!

8.2 CHOOSING SEQUENCES

When choosing the sequences to study, we are usually motivated by a biological question. If we are interested in the phylogeny of a particular group of organisms, then we are entitled to choose any particular gene sequence we want from those organisms that we think will be informative about the pattern of divergence of the species. On the other hand, we may be interested in the evolution of a particular gene, or of a related family of genes. In this case, we will probably want to choose sequences of this type from as wide a range of species as possible. In general, we are likely to pick a gene where there are sequences available from many species, and where the degree of sequence variation within the species under consideration is neither too high nor too low. If the gene evolves too slowly, there will be very little variation among the sequences, and there will be too little information to construct a phylogeny. If the gene evolves too rapidly, it will be difficult to get a reliable alignment of the sequences. It is also difficult to get accurate estimates of evolutionary distances when sequences are highly divergent. Thus, phylogenetic methods are likely to become unreliable if the sequences are too different from one another, and this should be borne in mind when the choice of gene sequences is made initially.

Phylogenetic programs usually require a multiple sequence alignment as input. By specifying the alignment, we are implying that all the nucleotides (or amino acids) in a given column of the alignment are homologous to one another. If this is not true, the whole basis from which we construct the phylogeny is invalid. The results of phylogeny programs are sensitive to changes in the alignment; therefore, it is worth spending some time in getting an accurate alignment before beginning. It is also important to consider which bits of the sequence to include in the phylogenetic analysis. Some bits of the sequence may be very variable and prone to large insertions and deletions. It is usual to eliminate sections like this from the analysis because they do not contain a strong phylogenetic signal, and they can create noise that adds to the uncertainty of the results. As emphasized in Chapter 6, automated methods of multiple alignment cannot be relied on to get all the details correct, and time spent in manual checking of alignments before beginning phylogenetic work is likely to pay off.

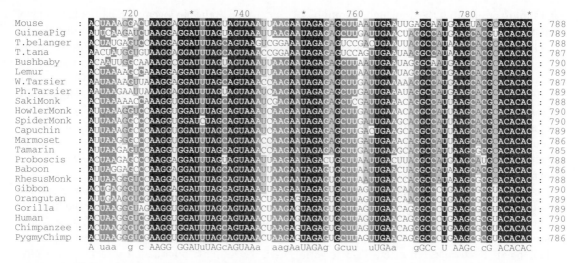

```
           720          *         740         *         760          *        780          *
Mouse       : ACUAAACGACUAAGGAGGAUUUAGUAGUAAAUUAAGAAUAGAGAGCUUAAUUGAAUUGAGCAUGAAGUACGCACACAC : 788
GuineaPig   : AUUCAAGAUCUAAGGAGGAUUUAGUAGUAAAUUAAGAAUAGAGAGCUUGAUUGAACUAGGCCAUGAAGCACGUACACAC : 789
T.belanger  : ACUAUGAGUCAAGGAGGAUUUAGCAGUAACUCGGAAAUAGAGAGUCCGACUGAAUUAGGCCAUAAAGCACGUACACAC  : 788
T.tana      : AACUAUGCGUGUAAGGAGGAUUUAGCAGUAAACCGGAAAUAGAGAGUCCAGUUGAAUAAGGCCAAUAAGCACGUACACAC : 787
Bushbaby    : ACAAUUGCAAAAGGCGGAUUUAGUAGUAAAUUAAGAAUAGAGAGCUUAAUUGAAUAGGGCAAUGAAGCACGCACACAC  : 790
Lemur       : ACUAAAAGCCAAAGGAGGAUUUAGCAGUAAAUUAAGAAUAGAGAGCUUAAUUGAAUAGGGCAUGAAGCACGCACACAC  : 789
W.Tarsier   : AUUAAAAGUUAAAGGAGGAUUUAGUAGUAAACCAAGAAUAGAGAGCUUAAUUGAAAAAGGCCAUGAAGCACGCACACAC : 787
Ph.Tarsier  : AAUAAGAAUUAAAGGAGGAUUUAGUAGUAAAUUAAGAAUAGAGAGCUUGAUUGAAUAUAGGCCAUGAAGCACGCACACAC : 789
SakiMonk    : ACUAAAAACCAAAGGUGGAUUUAGCAGUAAAUCGAGAAUAGAGAGCUCGAUUGAAACAGGCCAUUAAGCACGCACACAC : 788
HowlerMonk  : AUUAAACGUCCAAGGUGGAUUUAGCAGUAAAUCAAGAAUAGAGAGCUUGAUUGAAACAGGCCAUUAAGCACGCACACAC : 790
SpiderMonk  : AUUAAACGCCAAGGUGGAUCUAGCAGUAAAUCAAGAAUAGAGAGCUUGAUUGAAGCAGGCCAUUAAGCACGCACACAC  : 790
Capuchin    : AUUAAACGUCCAAGGUGGAUUUAGCAGUAAAUCAAGAAUAGAGAGCUUGACUGAAGCAGGCCAUUAAGCACGCACACAC : 789
Marmoset    : AUUAAAGGGCCAAGGUGGAUUUAGCAGUAAACCAAGAAUAGAGAGCUUGAUUGAAACGGGCCAUUAAGCACGCACACAC : 786
Tamarin     : AUUAAAGGUCAAAGGUGGAUUUAGCAGUAAACCAAGAAUAGAGAGCUUGACUGAAGCAGGCCAUUAAGCGCGCACACAC : 785
Proboscis   : ACUAAGAGCCCAAGGAGGAUUUAGUAGUAAAUUAAGAAUAGACUGCUUAAUUGAACUUAGGCCAUCAAGCAUGCACACAC : 788
Baboon      : AUUAGGAGCCCAAGGAGGAUUUAGCAGUAAAUUAAGAAUAGAGUGCUUAAUUGAACUAGGCCAUAAAGCACGCACACAC : 786
RhesusMonk  : AUUAAAGCGUCCAAGGAGGAUUUAGCAGUAAAUUAAGAAUAGAGUGCUUAAUUGAACCAGGCCAUAAAGCGCGCACACAC : 788
Gibbon      : ACUGAGGGUCCAAGGAGGAUUUAGCAGUAAAUUAAGAAUAGAGUGCUUAGUUGAAGUUGGCCAUGAAGCGCGUACACAC : 790
Orangutan   : AUUGAACGUCCAAGGUGGAUUUAGCAGUAAACUAAGAGUAGAGUGCUUAGUUGAACAAGGCCAUGAAGCGCGUACACAC : 789
Gorilla     : ACUAAGCGUACAAGGUGGAUUUAGCAGUAAAUUAAGCAUAGAGUGCUUAGUUGAACAGGGCCAUGAAGCGCGUACACAC : 789
Human       : ACUAAAGCGUCCAAGGUGGAUUUAGCAGUAAACUAAGCAUAGAGUGCUUAGUUGAACAGGGCCCUGAAGCGCGUACACAC : 790
Chimpanzee  : ACUAAAGCGUCCAAGGUGGAUUUAGCAGUAAACUAAGCAUAGAGUGCUUAGUUGAACAGGGCCCUGAAGCGCGUACACAC : 789
PygmyChimp  : ACUAAAGGGUCCAAGGUGGAUUUAGCAGUAAACUAAGCAUAGAGUGCUUAGUUGAACAGGGCCCUGAAGCGCGUACACAC : 786
              A uaa   g c AAGG GGAUuUAGcAGUAAa   aagAaUAGAg Gcuu    uUGAa    gGCc U AAGc cG ACACAC
```

Fig. 8.4 Part of the alignment of the mitochondrial small subunit rRNA gene from primates, tree shrews, and rodents.

As an example for this chapter, we will consider a set of sequences for the small subunit rRNA gene taken from the mitochondrial genome of various primate species. The gene has a total length of around 950 bases. A short section from the middle of the multiple sequence alignment is shown in Fig. 8.4. These sequences are relatively conserved within the group of species chosen. There are many sites that do not vary (shaded in black). This means we can be fairly sure about the alignment. There are, nevertheless, several variable positions that contain useful phylogenetic information. There are also a few very variable parts of the molecule that we could not reliably align, and these were removed from the alignment prior to beginning the phylogenetic analysis.

Table 8.1 lists the species used in our example. Representatives of the major primate groups have been chosen. The taxonomic groups follow the classification used in the NCBI taxonomy browser (http://www.ncbi.nlm.nih.gov/Taxonomy/taxono-myhome.html/). The main primate groups are the Strepsirhini (lemurs and related species), the Tarsii (tarsiers), the Platyrrhini (new-world monkeys), and the Catarrhini (old-world monkeys and apes). In addition, two tree shrew species have been included. These are members of the order Scandentia, which is

usually thought to be closely related to primates. As outgroups, we have added two rodent species. These are more distantly related, although recent studies of mammalian phylogenies place rodents, tree shrews, and primates together in one of the four major clades of mammals (see Chapter 11). As these species have been well studied, we can be reasonably sure of the true evolutionary relationship between most of the species. This illustrative example includes both straightforward relationships that are easy for the methods to get right and difficult relationships that do not always come out correctly.

8.3 DISTANCE MATRICES AND CLUSTERING METHODS

8.3.1 Calculating distances

In Chapter 4, we discussed different models for calculating evolutionary distances. The choice of evolutionary model is important in phylogenetic studies. For this example, we will use the Jukes–Cantor (JC) model, because it is the simplest, and this makes it straightforward to compare results from different methods. There is a fairly strong phylogenetic signal in this data set that can be seen with the simplest model. We should note, however, that

Table 8.1 Classification of species used in the examples in this chapter.

Taxon name	Species
Rodentia	*Mus musculus* (Mouse)
	Cavia porcellus (Guinea pig)
Scandentia	*Tupaia belangeri* (Northern tree shrew)
	Tupaia tana (Large tree shrew)
Primates	
Strepsirhini (Prosimians)	*Lemur catta* (Ringtailed lemur)
	Otolemur crassicaudatus (Thick-tailed bushbaby)
Tarsii (Tarsiers)	*Tarsius syrichta* (Philippine tarsier)
	Tarsius bancanus (Western tarsier)
Platyrrhini (New-world monkeys)	
Cebidae	*Pithecia pithecia* (White-faced Saki monkey)
	Cebus apella (Brown capuchin)
	Alouatta seniculus (Howler monkey)
	Ateles sp. (Spider monkey)
Callitrichidae	
	Callithrix pygmaea (Pygmy marmoset)
	Leontopithecus rosalia (Golden lion tamarin)
Catarrhini	
Cercopithecidae (Old-world monkeys)	*Nasalis larvatus* (Proboscis monkey)
	Macaca mulatta (Rhesus monkey)
	Papio hamadryas (Baboon)
Hylobatidae	*Hylobates lar* (Gibbon)
Hominidae	*Pongo pygmaeus* (Orangutan)
	Gorilla gorilla (Gorilla)
	Pan troglodytes (Chimpanzee)
	Pan paniscus (Pygmy chimpanzee)
	Homo sapiens (Human)

the JC model is not often used in modern studies. The implicit assumption of the JC model that the four bases have equal frequencies is definitely not true in the rRNA genes studied here, so some of the more complex models discussed in Chapter 4 would almost certainly give better fits to the data than the JC model.

From a multiple alignment, we can easily calculate the JC distance between each pair of sequences using Eq. (4.4). We can then construct a matrix whose elements, d_{ij}, are the distances between each pair of species, i and j. The matrix for the sequences from the Catarrhini is shown in Fig. 8.5. The matrix is symmetric (i.e., d_{ij} must be equal to d_{ji}) and the diagonal elements, d_{ii}, are zero, because the distance of a sequence from itself must be zero. To get an idea of

scale, the human to chimpanzee distance is 2.77%. The smallest distance is 1.49% between chimpanzee and pygmy chimpanzee, and the largest distance in the figure is the orangutan to proboscis monkey distance, which is close to 20%. The distances from the mouse to the primates (not shown) are around 26–30%. In fact, mitochondrial sequences are relatively rapidly evolving. If we had used a nuclear gene, we would have obtained smaller distances for the same set of species. It should also be remembered that we have removed some of the most rapidly changing regions of the molecule from the alignment prior to calculating these distances because these regions were deemed to be unreliably aligned. If these parts had not been removed, all the distances would have come out slightly higher.

	Probos	Baboon	Rhesus	Gibbon	Orang	Gorilla	Human	Chimp	Pyg.ch.
Proboscis M.	0.0000	0.1284	0.1458	0.1781	0.2062	0.1856	0.1853	0.1784	0.1790
Baboon	0.1284	0.0000	0.0994	0.1671	0.1817	0.1674	0.1559	0.1617	0.1581
Rhesus M.	0.1458	0.0994	0.0000	0.1738	0.1755	0.1655	0.1738	0.1627	0.1577
Gibbon	0.1781	0.1671	0.1738	0.0000	0.1280	0.1068	0.1106	0.1003	0.1072
Orangutan	0.2062	0.1817	0.1755	0.1280	0.0000	0.0875	0.0875	0.0900	0.0891
Gorilla	0.1856	0.1674	0.1655	0.1068	0.0875	0.0000	0.0371	0.0395	0.0372
Human	0.1853	0.1559	0.1738	0.1106	0.0875	0.0371	0.0000	0.0277	0.0277
Chimpanzee	0.1784	0.1617	0.1627	0.1003	0.0900	0.0395	0.0277	0.0000	0.0149
Pygmy chimp	0.1790	0.1581	0.1577	0.1072	0.0891	0.0372	0.0277	0.0149	0.0000

Fig. 8.5 The JC distance matrix for the some of the rRNA sequences in Fig. 8.4. The matrix has been divided into sections in order to indicate the most important split within the Catarrhini group: that between the old-world monkeys and the apes.

A distance matrix can be used as the input to a clustering method. Algorithms for clustering work in the following way:

1 Join the closest two clusters to form a single larger cluster.

2 Recalculate distances between all the clusters.

3 Repeat steps 1 and 2 until all species have been connected to a single cluster.

Initially, each species forms a cluster on its own. The clusters get bigger during the process, until they are all connected to a single tree. We already discussed clustering methods in a different context in Chapter 2. We will discuss two clustering methods here that are particularly relevant for phylogenetics: the UPGMA (unweighted pair group method with arithmetic mean) and neighbor-joining methods.

8.3.2 The UPGMA method

This is one of a family of hierarchical clustering methods described by Sneath and Sokal (1977). It has some important limitations, which means that it is now not often used in phylogenetic research studies. However, we will discuss it first, because it is conceptually the simplest phylogenetic method.

Step 1 of the clustering algorithm is to connect the two closest species, in this case the chimpanzee and the pygmy chimpanzee. Step 2 is to calculate the distance from the chimpanzee/pygmy chimpanzee cluster to all the other species. In the UPGMA method, the distance between two clusters is simply defined as the mean of the distances between species in the two clusters. For example, the distance of

human from the chimpanzee/pygmy chimpanzee cluster is the mean of the human–chimpanzee and human–pygmy chimpanzee distances. If, at some point, we want to calculate the distance between a cluster with two species and a cluster with three species, then this will be the average of six (= 2 × 3) numbers. Defining the distances between clusters as an arithmetic mean in this way is a fairly obvious definition. However, there are many other ways of doing it. It is this definition that distinguishes the UPGMA from the other clustering methods given by Sneath and Sokal (1977).

Proceeding with our primates, it turns out that the distance from the human to the two chimpanzees is smaller than any of the other remaining distances, and therefore the next step is to connect these together. After repeating the steps of the clustering algorithm, we eventually obtain the tree in Fig. 8.6. This is a rooted tree, with branch lengths that are proportional to the inferred amount of time separating the species, i.e., the UPGMA method assumes that there is a strict molecular clock and that all species evolve at the same rate. The lengths of the branches are determined by the heights of the internal nodes in the tree above the base line. The height of the node connecting chimpanzee and pygmy chimpanzee is 0.0074. This is half the distance between the two species. The factor of a half arises because we are assuming that half of the changes separating the two species occurred on the branch leading to one species and half occurred on the other. Now consider the earliest node in Fig. 8.6, which connects the old-world monkeys to the apes.

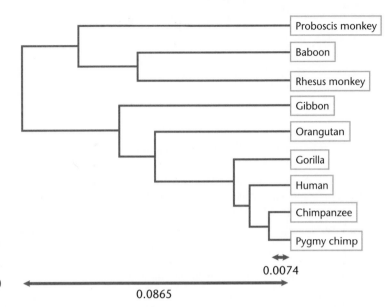

Fig. 8.6 Tree obtained using the UPGMA method with the matrix in Fig. 8.5. The PHYLIP package was used for tree construction (Felsenstein 2001) and the Treeview program (Page 2001) was used to prepare the figure.

The distance between these groups is the mean of the 18 numbers in the top right section of the matrix in Fig. 8.5. The height of the node is half this mean distance, which is 0.0865. Notice that this method tells us that the root of the tree is at the point where the last two clusters are connected – in this case the old-world monkeys and the apes. Most other methods do not give the position of the root.

If we pick any three species from the UPGMA tree, there are always two that are more closely related to each other than to the third. According to the tree, the distances from the third species to the first and second species are equal to one another, because these two distances are both determined by the same branch point. The distance from the first to the second species is smaller than this. This property is called **ultrametricity**: whichever three species we take from the UPGMA tree, the two largest of the three distances between these species will be equal. It is clear that if we consider divergence times between species, as in Fig. 8.1(a), the two largest divergence times for any three species must be equal. Thus, divergence times form an exactly ultrametric matrix. If sequences evolve according to a molecular clock, the distances between them will be approximately ultrametric, but not exactly so. This is because substitutions in gene sequences are random

events. The two lineages descending from a node will have the same number of substitutions **on average**, but in any one particular case, one lineage will by chance have slightly more substitutions than the other. Thus, even in an ideal world where the molecular clock is exactly true, the distances between sequences will only be approximately ultrametric. In our case, if the matrix were exactly ultrametric, all the 18 distances in the top right section of the matrix would be exactly equal, whereas in reality they are only approximately equal. What the UPGMA method does is to find an ultrametric tree such that the distances measured on the tree are a close approximation to the distances in the original matrix.

In the real world, it may not be just chance that makes distances violate the ultrametric rule. If species evolve at different rates, then we get trees like Fig. 8.2, rather than Fig. 8.1. In this case, there is no reason why the two largest distances of any group of three must be equal. The problem with the UPGMA method is that it forces the species to lie on an ultrametric tree, even when the original distance matrix is far from ultrametric. In this case, the resulting tree is usually wrong. In many real data sets, the molecular clock assumption may not be a good one and, for this reason, UPGMA is often not a reliable method. We have discussed UPGMA because it is

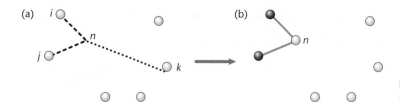

(a)

(b)

Fig. 8.7 Step one of the neighbor-joining method.

conceptually one of the simplest clustering methods. Although it would not be recommended for most serious phylogenetic studies, it is nevertheless important to understand how it works and how it differs from other methods. For the species in Fig. 8.6, the UPGMA result corresponds to what is generally believed to be true. This is not bad, considering we used the simplest possible method of calculating distances and the simplest possible clustering method. This indicates that there is quite a strong phylogenetic signal in the sequences from the Catarrhini. When the more distantly related species from Table 8.1 were included, the UPGMA did not give the correct topology, and therefore we have not shown these species in Fig. 8.6.

8.3.3 The neighbor-joining method

Neighbor joining (NJ) (Saitou and Nei 1987) is a clustering method that produces unrooted trees, rather than the rooted ones produced by UPGMA. There is a property of unrooted trees called **additivity** that we need to introduce at this point. A tree is said to be additive if the distances between the species on the tips of the tree are equal to the sum of the lengths of the branches that connect them. For example, in Fig. 8.3(a), $d_{AC} = d_{Ai} + d_{ij} + d_{jC}$. A distance matrix is said to be additive if it is possible to find a tree with specified branch lengths such that all the pairwise distances in the matrix are exactly equal to distances measured by summing along the branches of the tree. Additivity is a less restrictive condition than ultrametricity. If a distance matrix is ultrametric, it is also additive, but the reverse is not necessarily true. The NJ method constructs an additive tree such that the distances between species measured on the tree are a good approximation to the distances in the original matrix. If the original matrix

is exactly additive, NJ is guaranteed to give the correct tree, just as UPGMA is guaranteed to give the correct tree if the original distance matrix is ultrametric. However, in the real world, the distances will not be exactly additive; therefore, NJ is just one approximation.

Two species are said to be neighbors on an unrooted tree if they are connected via a single internal node. In Fig. 8.3(a), A and B are neighbors, C and D are neighbors, but A and C are not neighbors. The NJ method begins with a set of disconnected tip nodes representing the known sequences. These are shown as white circles in Fig. 8.7(a). We know the distances between all these nodes, as specified in the input distance matrix. The method chooses two nodes, i and j, that are neighbors and connects them by introducing a new internal white node, n. The original nodes, i and j, are now eliminated from the problem because they are already connected to the growing tree. They have been colored black in Fig. 8.7(b). We now have a set of disconnected white nodes that has one fewer than before. This process can be repeated until all the nodes are connected to a single tree. Details are given in Box 8.1.

Applying the NJ method to the distance matrix for the primate sequences gives the result in Fig. 8.8(a). This tree topology agrees with what we expected from the taxonomic classification (Table 8.1) in most respects. The Catarrhini and the Platyrrhini form two well-defined groups. The lemur and bushbaby cluster together, as do the two tarsiers, the two tree shrews and the two rodents. The result is easier to interpret if we place the root of the tree on the branch leading to the rodents and redraw the tree as in Fig. 8.8(b). It should be remembered, however, that the NJ method does not tell us where to place the root – we must use prior knowledge to decide where to put it.

BOX 8.1
Calculation of distances in the neighbor-joining method

When we introduce the new node, n, we need to know its distance from all the other nodes. To calculate these distances, we will assume the distances are additive. Consider one particular node, k, among the nodes still to be connected. We may write (from Fig. 8.7a):

$$d_{ij} = d_{in} + d_{jn}; \; d_{ik} = d_{in} + d_{kn}; \; d_{jk} = d_{jn} + d_{kn} \qquad (8.1)$$

We may solve these simultaneous equations to find the distance of k from the new node:

$$d_{kn} = \frac{1}{2}(d_{ik} + d_{jk} - d_{ij}) \qquad (8.2)$$

This equation can be used for each of the unconnected nodes, k. We now need the distances d_{in} and d_{jn}. From the simultaneous Eqs. (8.1), we can write:

$$d_{in} = \frac{1}{2}(d_{ij} + d_{ik} - d_{jk}) \qquad (8.3)$$

The problem with this is that it depends on one particular node, k. We could call any of the disconnected nodes k and we would get a slightly different answer for d_{in} for each one. To avoid this problem, we define the quantities

$$r_i = \frac{1}{N-2} \sum_k d_{ik}; \; r_j = \frac{1}{N-2} \sum_k d_{jk} \qquad (8.4)$$

Here, N is the total number of white nodes. The best estimate of d_{in} is to average Eq. (8.3) over all the $N-2$ disconnected nodes k. This gives:

$$d_{in} = \frac{1}{2}(d_{ij} + r_i - r_j) \qquad (8.5)$$

Obviously, because of additivity, $d_{jn} = d_{ij} - d_{in}$. We have now reached the situation in Fig. 8.7(b), where we have a reduced set of $N-1$ white nodes and therefore we can repeat the whole procedure again.

However, we did not yet say how to choose the two neighboring nodes, i and j, to connect in the first place. It might seem natural to choose the two with the smallest distance, d_{ij}, as we would with UPGMA. Unfortunately, there is a problem that is illustrated in Fig. 8.3(a). The closest two species in this tree are B and D, because there are short branches connecting them, **but** B and D are not neighbors, so it would be wrong to connect them. The NJ method has a way of getting around this. We calculate modified distances,

$$d_{ij}^* = d_{ij} - r_i - r_j \qquad (8.6)$$

and we choose the two species with the smallest d_{ij}^* rather than the smallest d_{ij}. It has been shown (Saitou and Nei 1987, see also Durbin *et al.* 1998) that if the distance matrix is additive, then the two species chosen by this rule must be neighbors.

The part of the tree covering the Catarrhini is similar to the result of the UPGMA (Fig. 8.6). The branching order within the Platyrrhini is worth noting. The marmoset and the tamarin cluster together. These are members of the Callitrichidae family in the NCBI taxonomy. In contrast, the other four new-world monkeys, classified as Cebidae, do not form a monophyletic group, and this suggests that it is not evolutionarily meaningful to group these species in the Cebidae taxon. We should beware of jumping to conclusions, because so far we have only considered a single simple phylogenetic method. However, in all the methods we tried, the same branching pattern for the new-world monkeys was found. Therefore, we

can be reasonably confident that the Cebidae are not monophyletic, according to the evidence from the rRNA gene. A more detailed study of phylogenetics of the Platyrrhini has been carried out by Horovitz, Zardoya, and Meyer (1998). Their results, using small subunit rRNA sequences, agree with our Fig. 8.8(a) in that the Saki monkeys form the earliest-branching new-world monkeys. However, results using different gene sequences and morphological evidence give different branching patterns. Thus, the Platyrrhini phylogeny is still not fully resolved.

There is another, more serious, problem with Fig. 8.8(b): the tree shrews are apparently the sister

(a)

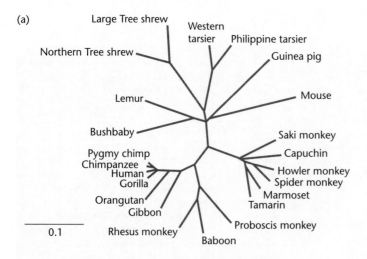

0.1

(b)

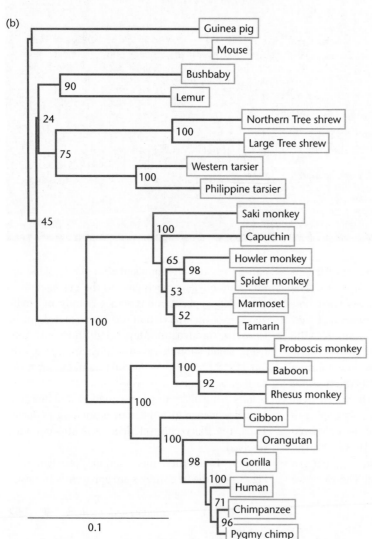

0.1

Fig. 8.8 Tree obtained with the neighbor-joining method and JC distances. (a) Unrooted. (b) Rooted with the rodents as outgroup, and with bootstrap percentages added.

group to the tarsiers. This is almost certainly wrong. It would mean that the primates were not a monophyletic group and that the order Scandentia should actually be included within the primates. We would not be willing to propose this phylogenetic arrangement, given the morphological evidence that favors placing tree shrews as a separate group outside the primates. The NJ tree must be misleading us in this case, and we will return to this point below.

To summarize: the NJ method is a practical, rapid way of getting a fairly reliable result and, as such, it is often used in real phylogenetic studies. The clustering algorithm runs in a time that is $O(N^2)$. In practice, the result can be obtained almost instantaneously, even when there are very large numbers of sequences, whereas the more complex methods described below can take large amounts of computer time. As stated above, NJ is exact if the distance matrix is additive, and it stands a good chance of giving the correct tree topology if the matrix is not far from additive. If the input distances are not close to being additive (e.g., because an inappropriate method of calculation of the pairwise distances was used, or because the sequences were not properly aligned), then NJ will give the wrong tree.

8.4 BOOTSTRAPPING

You will have noticed the numbers that appeared on the internal nodes of the tree in Fig. 8.8(b). These are called bootstrap percentages. Bootstrapping is a method of assessing the reliability of trees, introduced by Felsenstein (1985), and has since become a standard tool in phylogenetics.

Substitutions in sequences are random events. Even if the sequences are evolving in a way that is well described by the model of evolution that we use, the number of substitutions occurring on any branch of a real tree can deviate substantially from the mean value expected according to the model. This means that the distances we measure between sequences are subject to chance fluctuations. We would like to know whether these fluctuations are influencing the tree that we obtain. The bootstrapping method answers this question by deliberately constructing sequence data sets that differ by some small random fluctuations from the real sequences. The method is then repeated on the randomized sequences in order to see whether the same tree topology is obtained.

The randomized sequences are constructed by sampling columns from the original sequence alignment, as shown in Fig. 8.9. The randomized sequences are the same length as the real ones. Each column in the randomized sequences is constructed by copying one random column from the original sequence alignment. During this sampling process, some columns happen to be chosen more than once, while some columns happen not to be chosen at all (technically, this is known as "sampling with replacement"). This means that the random sequences contain slightly different information from the real ones. When we construct a tree with the randomized sequences, we are not guaranteed to get the same answer as before. If the phylogenetic signal in the data is strong, there is a lot of information about the degree of relatedness of the species spread along the whole length of the sequence. Therefore, resampling should make very little difference.

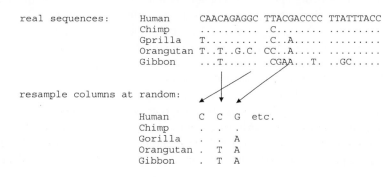

real sequences:

Human	CAACAGAGGC	TTACGACCCC	TTATTTACC
Chimp	....	.C........	....
Gprilla	T.........	.C..A.....	....
Orangutan	T..T..G.C	CC..A.....	....
Gibbon	...T......	.CGAA...T.	..GC.....

resample columns at random:

Human	C	C	G	etc.
Chimp	.	.	.	
Gorilla	.	.	A	
Orangutan	.	T	A	
Gibbon	.	T	A	

Fig. 8.9 An illustration of resampling columns for the bootstrapping method.

However, sometimes the signal indicating one tree topology over another may be rather weak. In this case, the noise that we introduce to the data when we resample may be sufficient to cause the tree-construction method to give a different result. Note that the resampling procedure is not equivalent to just reshuffling the columns of the original alignment. Phylogenetic methods treat each column independently, so a reshuffled alignment would give exactly the same answer as the original sequences, and would not tell us anything.

To carry out bootstrapping analysis, many sets of randomized sequences are constructed (usually 100 or 1,000). The tree-construction method is repeated on each set of sequences to produce a set of possible trees, some of which will be equivalent to the original tree and some of which will be different. We then look at each group of species in the original tree and we determine what percentage of the randomized trees contain this same group. This percentage gives us a measure of confidence that those species really do form a related group. The bootstrap results for the primate sequences using the NJ method are shown in Fig. 8.8(b). 1,000 replicates were done, but the numbers are shown as percentages in the figure. There are many strongly supported clades in this tree. For example, the taxa Hominidae, Cercopithecidae, Catarrhini, and Platyrrhini (as defined in Table 8.1) all have 100% bootstrap support, as do the tarsiers and the tree shrews. The prosimian group (lemur and bushbaby) also has 90% support. Other relationships in the tree are much less certain. Although the new-world monkeys, as a group, have 100% support, there are several low bootstrap values within the group, indicating that the branching order within the new-world monkeys is not well resolved by this method. The weakest point in this phylogeny is the very low value of 24% at the node linking the prosimians, tarsiers, and tree shrews. We really cannot be sure at all that these three groups form a clade. Indeed, other evidence tells us that most likely they do not, as discussed above. There is no precise rule to say how high a bootstrap percentage has to be before we can be sure that the group of species in question forms a "true" clade, although values greater than 70% are often thought to be

reasonably strong evidence. It is slightly worrying that the grouping of tarsiers and tree shrews has a bootstrap value of 75%, because this is very likely an incorrect relationship. We will see below that this apparent relationship goes away when we use a more realistic model of sequence evolution. For the moment, we note that bootstrap numbers need to be treated with caution. They are often a very useful indication of the reliability of different parts of a phylogenetic tree, but they do not "prove" anything conclusively.

Let us consider one particular point in Fig. 8.8(b) in more detail – the 71% value for the human + chimpanzee + pygmy chimpanzee clade. This value is sandwiched between two very high figures, i.e., the two chimpanzees almost always form a clade (96%), and the group gorilla + human + chimpanzee + pygmy chimpanzee always forms a clade. Thus, what is at stake is the relationship between gorilla, human, and the chimpanzees. The topology shown has the gorilla as the most distantly related of these three – this can be denoted as (gorilla,(human, chimpanzees)). The other two topologies – (human, (gorilla, chimpanzees)) and (chimpanzees,(gorilla, human)) – also turn up occasionally in the bootstrap replicates, but much less frequently than the topology in the figure. In fact, it is now generally agreed that the humans and chimpanzees are the closest of the three, although there has been considerable discussion in the past (see Hasegawa, Kishino, and Yano (1985) and references therein).

Bootstrap results are often presented in the form of consensus trees. The frequency of occurrence of each possible clade in the set of bootstrap trees is determined, and clades are ranked in descending order of frequency. The consensus tree is constructed by adding clades one at a time working from the top of the ranking list. Each clade added is the one with the highest frequency that is consistent with all the clades already added. The final consensus topology may differ slightly from the tree obtained with the original full set of data. It is then a matter of choice whether to present the original tree labeled with bootstrap percentages for the clades contained in that tree, or to present the consensus tree, which will tend to contain clades with slightly higher bootstrap values that were not present in the original tree. Well-determined clades, with high bootstrap values, will almost always occur in both the consensus and the original tree, so this issue affects only the way the results are presented for the least well-determined parts of the tree.

8.5 TREE OPTIMIZATION CRITERIA AND TREE SEARCH METHODS

8.5.1 Criteria for judging trees

As explained above, NJ constructs an additive tree such that the distances measured on the tree are approximately equal to the distances specified in the distance matrix. It is possible to define the following function, E, which measures the error made in this process:

$$E = \sum_{i,j} \frac{(d_{ij} - d_{ij}^{tree})^2}{d_{ij}^2} \qquad (8.7)$$

where d_{ij} is the distance specified in the distance matrix, d_{ij}^{tree} is the distance measured along the branches of the tree, and the sum runs over all pairs of species. This function is the sum of the squares of the relative errors made in each of the pairwise distances. E can be used as a criterion for evaluating trees. Fitch and Margoliash (1967) proposed a phylogenetic method that consists of searching for the tree that minimizes the function E. This can be considered as finding the additive tree that differs least from the input data. If the distance matrix is exactly additive, then it is possible to find a tree such that every term in the sum is exactly zero. However, usually, this is not possible, because the number of pairs of species is $N(N-1)/2$, while the number of independent branch lengths to be optimized on an unrooted tree is only $2N - 3$. Therefore, the method finds a tree for which E is as small as possible, but it cannot set E to zero exactly.

Clustering methods like NJ and UPGMA have a well-defined algorithm for constructing a tree. However, they do not have a well-defined criterion by which trees may be compared. Such a criterion is not necessary because clustering algorithms give rise automatically to a single tree as output. In contrast, the Fitch–Margoliash method has a well-defined criterion for judging trees (the best tree is the one for which E is the smallest), but there is no well-defined algorithm for constructing the best tree. It is necessary to use a program that constructs many alternative trees and then tests each one out using the criterion. This means that the Fitch–Margoliash method is much slower than NJ. In practice, if the matrix is approximately additive, NJ will give a very similar answer to Fitch–Margoliash in a shorter time. On the other hand, if the matrix is far from being additive, then neither of these methods is appropriate, and both are likely to give inaccurate tree topologies.

The idea of testing alternative trees using some criterion of optimality is a key one, because two very important phylogenetic methods, the maximum-likelihood and the parsimony methods, both use this principle. In brief, the maximum-likelihood criterion is to choose the tree on which the likelihood of observing the given sequences is highest, while the parsimony criterion is to choose the tree for which the fewest number of substitutions are required in the sequences. These criteria will be discussed fully in the following sections. In this section, we wish to consider how we might search for candidate trees. The method for tree searching is independent of the criterion used for judging the trees.

8.5.2 Moves in tree space

There is an enormous number of tree topologies, even for very small numbers of species, as shown in Table 8.2. The notation !! means the product of the odd numbers, so that $U_7 = 9!! = 9 \times 7 \times 5 \times 3 \times 1 = 945$. Only if the number of species is very small (i.e., less than about eight) will it be possible to exhaustively test every tree. More usually we will need heuristic search programs. It is useful to think of "tree space", which is the set of all possible tree topologies. Two trees are usually considered to be neighboring points in tree space if they differ by a topology change known as a nearest-neighbor interchange (NNI). To make an NNI, an internal branch of a tree is selected (a branch has been highlighted in bold in Tree 1 of Fig. 8.10). There are two

Table 8.2 How many trees topologies are there?

No. of species (N)	No. of distinct tree topologies	
	Unrooted (U_N)	Rooted (R_N)
3	1	3
4	3	15
5	15	105
6	105	945
7	945	10,395
8	10,395	135,135
9	135,135	2.0×10^6
10	2.0×10^6	3.4×10^7
N	$(2N - 5)!!$	$(2N - 3)!!$

subtrees linked to each end of this branch. A subtree may be either a single species (like A), or a branching structure that may contain any number of species (like D + E). The NNI move consists of swapping any one subtree from one end of the internal branch with any one subtree from the other end. In Fig. 8.10, Tree 2 is obtained by swapping B and C, and Tree 3 is obtained by swapping A and C. We could also swap the D + E group with either A or B, but this would also create either Tree 2 or 3. Thus, Trees 1, 2, and 3 are all neighbors of each other by a single NNI about

the highlighted branch, and there are no other trees that can be formed by making rearrangements about this same branch. Tree 4 is not a neighbor of Tree 1, as it is not accessible by a single NNI. The shortest path from 1 to 4 involves two NNIs (can you work out what these are?). It is possible to create any rearrangement of a tree by a succession of NNIs.

Moving from Tree 1 to Tree 4 is an example of another type of topology change called subtree pruning and regrafting (SPR). This consists of choosing any subtree in the original tree (in this example, just the single species D), cutting it from the tree, and then reconnecting it to any of the branches remaining on the original tree (in this example, the branch leading to B). The SPR is one type of a long-range move in tree space. Another possible long-range move, described by Swofford *et al.* (1996), is tree bisection and reconnection, in which an internal branch is selected and removed, thus forming two separate subtrees. These subtrees are reconnected by selecting one branch at random from each subtree and connecting these by a new internal branch.

We now return to the question of searching tree space to find the tree that is optimal by some criterion, such as likelihood, parsimony, or least-squares error *E*. Suppose we have an initial guess at an

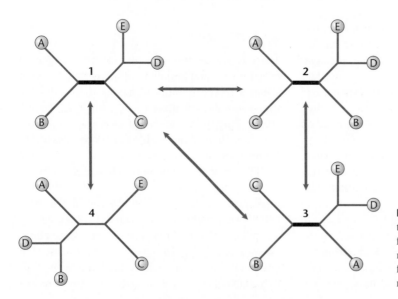

Fig. 8.10 Examples of changes in tree topology. Trees 1, 2, and 3 all differ from each other by a single nearest neighbor interchange. Tree 4 differs from tree 1 by a subtree pruning and regrafting operation.

optimum tree. This could be the tree produced by a distance-matrix method, for example. We would expect this tree to be "not too bad", according to our optimization criterion; therefore, we would expect that looking at other trees that are similar to this one might be a sensible strategy for finding the optimum tree. One way of proceeding is to test trees that are neighbors of our current tree. If we find a neighboring tree that is better, then we move to this new tree and search the neighbors of the new tree. We keep doing this until we reach a tree that is a local optimum, i.e., it has no neighbors that are better than it. This is a hill-climbing algorithm, because we always move "uphill" to a better tree at each step. We cannot guarantee that the local optimum we reach will be the global optimum tree that we are looking for. One way to check this would be to try hill climbing from lots of different starting points. A possible way to generate the starting points is by sequential addition of species. Suppose the species are listed in a random order. There is only one unrooted tree topology for the first three species in the list. We then add one species at a time to the tree. Each species is added by connecting it in the optimal way to the tree that is already present, until all species have been added. We can then try to improve the tree by hill climbing, using NNIs, until a local optimum is reached. This procedure can be repeated many times, starting with different random orders of species. We may obtain many different local optima. If we continue for long enough, then it is very likely that the best of the local optima that we find will be the global optimum. This is a heuristic search procedure, so we cannot guarantee that we have found the global optimum. All we can do is hope that if we have run the search for a long time without finding any improvement, then we have probably found the best tree.

The definition of a local optimum tree depends on the definition of what is a neighbor. If we use NNIs to define neighbors, then we may reach trees that cannot be improved by NNIs, but that could be improved by longer-range moves. In principle, long-range moves could also be included in the search strategy (and some programs do this). However, the problem with long-range moves tends to be that they are rather disruptive. If we already have a fairly good tree, then most long-range moves will create a tree that is much worse, and they will thus be a waste of computer time. The longest-range move we could imagine would be to try a completely random tree each time. We would eventually hit on the global optimum by this strategy, but it would be a very slow and inefficient search procedure. Most real search programs use a combination of NNIs and slightly longer-range moves that has been tested and found to be reasonably efficient at finding optimal trees as quickly as possible. We also note, at this point, that search strategies are not limited to moving uphill. If we occasionally allow downhill moves, then this may prevent us getting stuck on very poor local optima, and allow better optima to be found more quickly. The Markov Chain Monte Carlo method, discussed in Section 8.8, is a way of searching tree space that allows both uphill and downhill moves.

8.6 THE MAXIMUM-LIKELIHOOD CRITERION

Given a model of sequence evolution and a proposed tree structure, we can calculate the likelihood that the known sequences would have evolved on that tree. The maximum-likelihood (ML) criterion is to choose the tree that maximizes this likelihood (Felsenstein 1981). An algorithm for doing this is discussed in Box 8.2. Three qualitatively different types of parameter must be specified to calculate the likelihood of a sequence set on a tree: the tree topology; the branch lengths; and the values of the parameters in the rate matrix (base frequencies, transition/transversion ratio, etc.). It is possible to optimize all of these things at the same time. Programs are available that will search tree space making changes that alter all three types of parameter, and attempt to find the ML solution. On the other hand, it is possible to fix some things while the others are optimized. For example, we might want to find the ML tree topology and branch lengths, while keeping the rate-matrix parameters fixed at particular numerical values.

The ML criterion can be used to distinguish between a set of tree topologies specified in advance.

BOX 8.2
Calculating the likelihood of the data on a given tree

Consider the calculation of the likelihood for a given site n. In Fig. 8.11, there are two tip nodes with known sequences that happen to be A and G at this site. The likelihood that the common ancestor of these two had a base X is

$$L_n(X) = P_{XA}(t_1)P_{XG}(t_2) \qquad (8.8)$$

Here, the functions $P_{ij}(t)$ are the transition probabilities calculated using an appropriate model of sequence evolution, as described in Chapter 4. Four likelihoods are calculated using Eq. (8.8) for the four possible values of X. We may now proceed to the node labeled Y, and calculate

$$L_n(Y) = P_{YG}(t_4) \sum_X L_n(X)P_{YX}(t_3) \qquad (8.9)$$

For each of the four possible values of Y, we have to sum over the four possible values of X, weighting them by the likelihood of that value of X. For a node like W,

where both the descendant nodes are internal nodes, it is necessary to sum over the four possible values of both these descendant nodes:

$$L_n(W) = \sum_Y \sum_Z L_n(Y)P_{WY}(t_5)L_n(Z)P_{WZ}(t_6) \qquad (8.10)$$

This procedure can be repeated until we get to the root of the tree, so that we have the likelihoods for the base occurring at the root node (in this tree, W is already the root). We can now calculate the total likelihood, L_n, for site n by summing over the possible bases at the root. We assume that the prior probability of each base that could have existed at the root is given by the equilibrium frequency of the bases, π_W, according to the evolutionary model. Hence,

$$L_n = \sum_W \pi_W L_n(W) \qquad (8.11)$$

This algorithm for calculating the likelihood of a given site builds recursively from the tips towards the root. In fact, the final likelihood does not depend on the root position. When calculating the likelihood, the root can be placed at any arbitrary position on the tree, because the likelihood always turns out the same.

The likelihood is calculated for each site in this way, and it is assumed that sites evolve independently. Therefore, the likelihood for the complete sequence set, L_{tot}, is the product of the likelihoods of the sites. It therefore follows that the log of the likelihood is the sum of the logs of the site likelihoods

$$\ln L_{tot} = \sum_n \ln L_n \qquad (8.12)$$

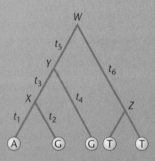

Fig. 8.11 An illustration of calculating the likelihood for a single site on a given tree. Letters A, G, and T label bases on the known sequences. Letters X, Y, Z, and W label unknown bases on internal nodes. The times (or equivalently, branch lengths) on each branch are labeled t_1, t_2, etc.

It is log-likelihood values that are usually quoted in the results of ML programs. As there are very many possible sequence sets that could evolve on any given tree, the likelihood of the real data is always very much less than 1. Therefore, $\ln L_{tot}$ is always a large negative number. The ML tree is the one with the least negative value.

For example, the bootstrapping analysis using a distance-matrix method shows where the uncertainties lie in a tree; therefore, we might wish to test a set of alternative trees that differ in their branching orders at the uncertain points in the tree. It is possible to estimate the value of the ML for each of the user-specified trees, by allowing the branch lengths and rate parameters to vary but keeping the topology fixed. This gives a ranking of the specified trees. Often the likelihoods of several different trees will differ by only a small amount. It is then important to decide whether one tree is really better than the alternatives. This can be done using a test proposed by Kishino and Hasegawa (1989). The total log likelihood is the sum of values from each site, as in Eq. (8.10). If the sites evolve independently, then we can estimate the error in the total log likelihood by calculating the standard deviation of the log likelihoods of the sites. We can then compare two different trees, and ask whether the difference in log likelihoods between the trees is significant in comparison to the error in estimation of this difference.

We would now like to see what the ML criterion says about the primate phylogeny example. We have used a search algorithm from our own phylogeny package, PHASE (Jow *et al.* 2002), to do this. There are several closely related groups of which we are already fairly certain. We are really interested in the early branch points on the tree that define the way these closely related groups are related to one another. For the ML analysis, we specified the following clades in advance:

```
1. (Mouse,Guinea Pig)
2. (Bushbaby,Lemur)
3. (Northern Tree Shrew,Large Tree
Shrew)
4. (Western Tarsier,Philippine
Tarsier)
5. (Saki Monkey,(Capuchin,((Tamarin,
Marmoset),(Howler Monkey,Spider
Monkey))))
6. ((Proboscis Monkey,(Baboon,Rhesus
Monkey)),(Gibbon,(Orangutan,(Gorilla,
(Human,(Chimpanzee,Pygmy
Chimpanzee))))))
```

Each of these clades is a rooted subtree that occurs on the NJ tree for which we are willing to assume that we know the correct branching order. There are 105 unrooted tree topologies that can be obtained by connecting these clades together. An ML algorithm was used to optimize the branch lengths and rate parameters for each of these 105 topologies separately, and to find the ML topology. Note that this involves changing the branch lengths

within the clades, as well as the branch lengths on the internal branches connecting the clades; however, rearrangement of the species within the clades is not allowed. As this algorithm is an exhaustive search of topologies, we can be sure we have found the correct ML solution (provided the clades were defined correctly at the outset). This algorithm also has the advantage of being rapid, because the number of topologies is relatively small when the number of clades is small, as in this example. We could equally well have used a standard ML search program that would not have required specifying the clades in advance, but this would have taken more computer time, and we would not have been completely sure that all the relevant trees had been searched.

Running this algorithm with the JC evolution model shows that the ML tree topology is the same as the NJ tree in Fig. 8.8(b), although the branch lengths are slightly different. There are, however, several other topologies for which the likelihood is only very slightly less than this one, and statistical tests show that these alternatives are not significantly worse than the ML topology. Hence, ML has so far not resolved any of the problems that were present with the NJ tree. One reason for this is that we are still using a very simple model of sequence evolution. The JC model makes the very restrictive assumption that the frequencies of the four bases are equal. In fact the frequencies in the sequences being studied here are 37.5% A, 24.7% C, 12.6% G, and 25.2% U, which are definitely not equal. We would also expect that the ratio of transition to transversion rates might be significantly different from 1. We therefore repeated the ML analysis using the HKY model, which allows differing base frequencies and different transition and transversion rates (see Section 4.1.3). We also know, from the alignment, that several parts of the sequence appear to be invariant across all sequences, while other sites seem to vary much more rapidly than average. To account for this, we used a model of variable rates across sites that includes a certain fraction of invariable sites, plus six categories of variable sites with a gamma distribution of rates (see Section 4.1.4). The 105 topologies for the same six clades as above were

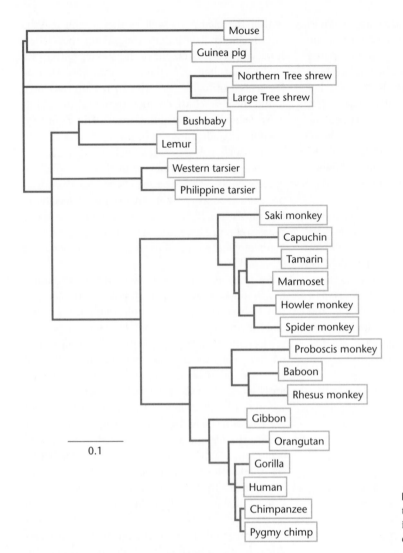

Fig. 8.12 Maximum-likelihood topology using the HKY model with invariant sites, plus six gamma-distributed rate categories.

evaluated using this model. The ML tree is shown in Fig. 8.12. The main difference from Fig. 8.8(b) is that the tree shrews are now positioned as out-groups to the primates. Thus, by using a more real-istic evolutionary model we have removed the main problem with the tree arising from the JC model. Another point to note is that there is a trifurcation between the lemurs, tarsiers, and the monkeys/apes. There are three different bifurcating topologies consistent with this, i.e., any one of these three groups could be the outgroup to the other two.

These three topologies all give exactly the same ML value, because the length of the internal branch at this point shrinks to zero, resulting in a trifurcation. There is evidence from morphological studies that it is the lemurs that are the earliest diverging of these three groups, but there is insufficient information to distinguish these possibilities in this single gene sequence. There are also several other topologies among the 105 that have lower likelihoods than the one shown, but that are not significantly less likely according to statistical tests.

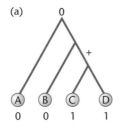 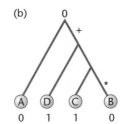 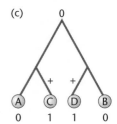

(a) 0   (b) 0   (c) 0

A 0   B 0   C 1   D 1   A 0   D 1   C 1   B 0   A 0   C 1   D 1   B 0

Fig. 8.13 The parsimony criterion used with morphological character states shows that tree (a) is preferable to trees (b) and (c).

8.7 THE PARSIMONY CRITERION

8.7.1 Parsimony with morphological characters

The parsimony criterion has a long history in phylogenetics based on morphological characters. It states that we should use the simplest possible explanation of the data, or the explanation that requires the fewest arbitrary assumptions. Often, parsimony is used with sets of binary character states, 0 and 1, where 0 represents a state thought to be ancestral, and 1 represents a state thought to be derived. For example, 1 might be the bone structure in the bird wing, and 0 might be the ancestral tetrapod forelimb bone structure. In practice, the characters used are much less obvious than this and require trained anatomists and paleontologists to identify.

Figure 8.13 shows an example where species C and D possess a derived character that is not possessed by A and B. If the species are arranged as on tree (a), the character must have evolved once on the branch marked +. If they are arranged as on tree (b), the character must have evolved once (+) and then must have been lost once (*). If the arrangement is as on tree (c), the character must have evolved independently (+) on the two branches leading to C and D. The first explanation is the simplest, or most parsimonious. The parsimony criterion says that tree (a) is to be preferred, because it has a single character-state change, whereas the other trees require two character-state changes.

If there is a single character, then it is always possible to find a tree that divides the species into two groups: the "haves" and the "have-nots". This character alone does not say anything about the branching order within each of the two groups. In order to construct a full tree, we need a large set of characters that evolved at different points in time. A shared derived character is called a **synapomorphy**. In an ideal case, we would try to find a set of characters that form synapomorphies nested one within the other, so that each character is informative about a different branch point on the tree. It would then be possible to find a tree such that there is only a single change of state in each character. In practice, there are often conflicts between different characters. For example, in Fig. 8.13, there may be a second character possessed by B and C, but not by A and D, and this character would suggest that tree (b) is preferable. Whichever tree we choose, there must be at least one of these two characters having more than one character state change. Parsimony programs take input data from a large number of characters and use a heuristic tree-search algorithm to find the tree such that the total number of changes required in all the characters is as small as possible. It is often the case that loss of a derived character can occur in species that no longer require that character – the "use it or lose it" principle. On the other hand, the independent origin of the same character twice seems much more unlikely. To account for this, it is possible to give a higher weight to a character gain than to a loss when calculating the parsimony score (see Swofford *et al.* 1996 for more details). Some characters do evolve more than once, however. This is called **homoplasy**. Characters that are subject to homoplasious changes can be misleading in parsimony, and often it is not known which characters these will be in advance.

8.7.2 Parsimony with molecular data

Parsimony can also be applied to molecular data in a straightforward way (Fitch 1971). Each column in the sequence alignment is treated as a character

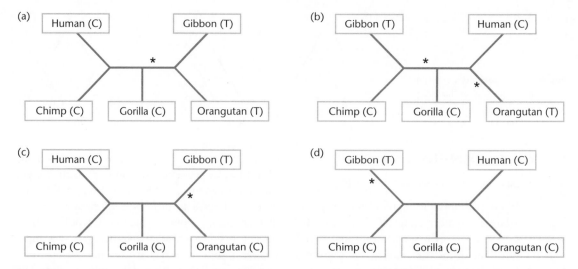

Fig. 8.14 The parsimony criterion applied to molecular data shows that tree (a) is preferable to tree (b) according to this informative site. Parsimony does not distinguish between trees (c) and (d) as this is a non-informative site.

state. A substitution event is a change in this character state, hence we simply look for the tree with the fewest substitutions. We do not know what the ancestral sequence was, hence there is no distinction between the gain and loss of a character. Molecular parsimony algorithms therefore use unrooted trees. Consider a site that is a C in human, chimpanzee, and gorilla, and is a T in orangutan and gibbon. Tree (a) in Fig. 8.14 requires a single substitution (denoted *), whereas tree (b) requires two separate substitutions. According to this site, tree (a) is preferable to (b). Not all sites are informative in the parsimony method. If a site is identical in all species, clearly this gives no information. However, the example shown in trees (c) and (d) is also non-informative. This is C in all species except for the gibbon, where it is T. Whichever tree topology we choose, it is always possible to put a single substitution on the branch leading to the gibbon, hence this site does not help us distinguish between alternative trees. The rule is that a site must have at least two of the four bases present in more than one species for it to be informative in the parsimony method.

We applied the parsimony method to the primates example. The most parsimonious tree has the same topology as obtained with both the NJ method and

the ML method with the JC model (shown in Fig. 8.8). There are several other topologies that require only one or two more substitutions than this one. The issue of distinguishing between alternative trees arises in the same way as for ML. In fact, the KH test can be used in a similar way for parsimony scores as for log likelihoods (Kishino and Hasegawa 1989). In our case, the alternatives with only a few additional substitutions are not significantly different from the most parsimonious tree. Thus, for these data, three different methods give the same answer for the best topology, and they largely agree on which parts of the tree are poorly determined. It is not always the case that parsimony gives the same answer as ML. The parsimony and ML criteria are different, and there is no guarantee that the same optimal tree will be found by the different criteria.

8.7.3 Comparison of parsimony and maximum-likelihood methods

There has been considerable debate in the literature about the relative merits of parsimony and ML. Parsimony has a longer track record in the literature, and many people still use it. Parsimony has the advantage of being fast. There is an efficient algorithm for evaluating the parsimony score on any given tree topology (Fitch 1971, Swofford *et al.* 1996), whereas with ML, for each tree topology there is still a relatively complex problem of optimizing the branch lengths and the rate-matrix parameters. Evaluation of each proposed tree therefore takes much longer for ML. Advocates of parsimony often dislike the models of sequence evolution used with ML and with distance-matrix methods, and claim that parsimony is superior because it avoids making assumptions about evolutionary models. In our example, we used the JC model with both NJ and ML methods, i.e., we assumed equal rates of all types of substitutions, and we also assumed equal rates of evolution at all sites. However, with the parsimony method, we made similar assumptions: all types of substitution contributed equally to the parsimony score, and changes at all sites were weighted equally. We saw that improving the evolutionary model by using HKY and variable rates gave a significantly different answer with ML. In principle, the parsimony score could also be improved in a similar way by weighting transitions and transversions differently, and by weighting changes at rapidly evolving sites less than those at slowly evolving sites. However, it is difficult to know what these weightings should be. In the ML framework, these effects are included as parameters in the model, and optimum values of these parameters are determined from the data. Thus, with ML, there is a systematic way of improving the evolutionary models used. There is also a rigorous way of checking whether each additional parameter gives a significantly better fit to the observed data (we will not discuss this until Section 11.2.1). We know that the models we use will never be an exact description of the true mechanism of sequence evolution, but the models we have at present do account for many of the important evolutionary factors. Therefore including these in the analysis seems preferable to ignoring them altogether. For a recent discussion of the parsimony/likelihood debate, see Steel and Penny (2000).

Parsimony methods try to minimize the number of substitutions, irrespective of the branch lengths on the tree. A substitution on a long branch counts just as much as a substitution on a short branch. ML methods allow for the fact that changes are much more likely to happen on longer branches. In fact, if a branch is long, then we expect that substitutions will have happened along this branch, and there is no reason to try to minimize the number of substitutions. Due to its neglect of branch lengths, parsimony suffers from a problem known as long-branch attraction. If two branches are long due to rapid evolutionary rates along them, then tree-construction algorithms tend to put these branches together. This can lead to the mistaken grouping of species that have little in common other than rapid evolutionary rate. This is a potential problem for many phylogenetic methods, but is particularly severe in parsimony. It is not just a problem of lack of sequence information with which to determine the correct tree: in some cases, it has been shown that parsimony will give the wrong tree, even when the length of the sequences used tends to infinity (Felsenstein 1978, Swofford *et al.* 1996).

In conclusion, parsimony is a strong method for evaluating trees based on qualitative characters for which there is no good quantitative model of evolution. It does not really mean much to have a model of the rate of evolution of birds' wings, because, as far as we know, their evolution was a one-off historical event. However, the substitutions occurring in molecular evolution occur many times in essentially the same way, and therefore it **does** make sense to have a quantitative model for the rates of these different types of change. When quantitative models are available to describe the data, we feel that likelihood-based methods are preferable to parsimony.

8.8 OTHER METHODS RELATED TO MAXIMUM LIKELIHOOD

8.8.1 Quartet puzzling

As stated previously, likelihood-based methods have several theoretical advantages over other methods but tend to be slow when dealing with large sequence sets. One way of getting around this is only to use ML to test a relatively small set of user-defined trees. It is then only necessary to optimize the parameters and branch lengths on each of the trees of interest, rather than to spend time searching through the space of tree topologies, when the vast majority of these topologies have extremely low likelihood. Predefining the known clusters, as we did in Section 8.6, is one way of restricting the search space to the region of interest.

Quartet puzzling (Strimmer and von Haeseler 1996) is another fairly rapid heuristic method of searching for the ML tree. This consists of calculating the ML tree for each quartet of four species, A, B, C, and D that can be taken from the full set. This involves testing the three unrooted trees: ((A,B),(C,D)); ((A,C),(B,D)); and ((A,D),(B,C)). The complete list of species is then considered in a randomized order. The first four are placed according to their optimal ML quartet tree. Each subsequent species from the list is then added to the growing tree one at a time in such a way that the topology of the growing tree contains as many of the ML quartet trees as possible at each step. When all the species have been added, this gives one possible candidate tree whose overall likelihood can be tested. The whole process is repeated many times, using different random orders of species. The final tree depends on the order in which the species are added, because the algorithm for species addition is a "short-sighted" or "greedy" algorithm, i.e., it does the best it can do at each step, without regard for what might happen at the next step. The set of candidate trees arising from each of the random addition orders will contain many different tree topologies. In the primates example, the most-frequently occurring topology occurred 96 times out of 1,000, and the second most-frequent topology occurred 95 times. The second most-frequent topology is the same as that given by NJ and ML with the JC model. The most frequent topology differs only in the branching order at the node linking the lemurs to the tree shrew/tarsier group. This has a very low bootstrap support of 24% in Fig. 8.8(b).

As well as using the set of trees arising from quartet puzzling as candidate trees for testing by ML, they can also be used to obtain percentage support values for the different clades. For each clade of interest we can calculate the percentage of times it occurs in the list of candidate trees. This gives percentages that can be interpreted in a similar way to bootstrap percentages, but which are not exactly equivalent because the set of trees used for the bootstrap percentages is calculated in a different way.

8.8.2 Bayesian phylogenetic methods

We have seen that often there are many trees that are only slightly worse than the optimal one according to whatever criterion we use to judge them. This tells us that there is an ensemble of possible trees consistent with the data, rather than telling us precisely what is the "correct" tree. Given this degree of uncertainty, it is rather unlikely that any one "best" tree we might choose to publish will be exactly correct in all respects. This suggests that we should consider methods that deal directly with ensembles of possible trees, rather than chasing after a single best one. This is what Bayesian methods do. The likelihoods of different trees are calculated as for ML but, rather than search for the single ML tree, a sample is taken of a large number of trees with high likelihoods. Bayesian methods calculate the posterior probabilities of different events of interest, given the information in the data and any prior information about the probabilities of these events. Bayesian methods also appear in the context of machine learning in Chapter 10. We therefore recommend that you read this section in conjunction with Section 10.2, where the ideas of prior and posterior probabilities are explained in more detail.

As with maximum likelihood, there are three types of parameter involved (the tree topology, the branch lengths, and the rate-matrix parameters), and it is necessary to specify prior probabilities for all of these. Sometimes with Bayesian methods, we have strong information about the prior values of parameters and there is relatively little information in the data. In such a case, the choice of priors would be very important. However, with phylogenetic methods, there is a large amount of information in the data, i.e., the likelihood function changes rapidly as parameters are altered. In this case, the choice of prior is not very important and it is possible to use uniform or non-informative priors. For topologies, the non-informative choice of prior is to set all possible tree topologies to have an equal prior probability. However, it would be possible to specify zero prior probability to trees that contained clades that we believed were impossible, for example, or to trees that did not contain a clade that we believed was definitely correct from prior information. For branch lengths, we have almost no idea what to expect before looking at the data; therefore, a non-informative prior again seems appropriate, such as a uniform prior distribution where all branch lengths are equally probable between zero and a specified large maximum value.

For most of the rate matrices used with nucleic acids, the base frequencies are parameters of the model (see Section 4.1). The sum of the base frequencies must equal one. A suitable non-informative choice of prior for base frequencies is therefore to set all sets of frequencies that add up to one as equally probable. A more constrained prior would be possible in this case by using peaked distributions of base frequencies. A limiting case would be to specify the base frequencies exactly and not treat them as variables at all. Substitution rate ratios also appear in models (e.g., transition/transversion parameter). If these are known, they can be specified, but it is more usual to use a non-informative prior and to allow the information in the data to determine the values.

8.8.3 The Markov Chain Monte Carlo method

In Box 8.3, we show that to calculate the posterior probabilities of different possible clades in Bayesian methods, we need to generate a large sample of trees with the property that the probability of finding a tree in the sample should be proportional to its likelihood × prior probability. The Markov Chain Monte Carlo (MCMC) method is a way of generating such a sample. If non-informative priors are used for all the parameters, then the prior probabilities cancel out of Eqs. (8.13) and (8.14), because they are constants. For this reason, we will ignore the prior in the following discussion.

The MCMC method begins with a trial tree and calculates its likelihood, L_1. A "move" is then made on this tree that changes it by a small amount, e.g., one or more branch lengths might be changed, one or more of the rate parameters might be changed, or the topology might be changed either by a nearest-

neighbor interchange or by one of the other possibilities for searching tree space discussed in Section 8.5. The likelihood of the new tree, L_2, will be slightly higher or lower than L_1. The new tree is either accepted or rejected, using a rule known as the Metropolis algorithm. If $L_2 > L_1$, tree 2 is accepted and it becomes the next tree in the sample. If $L_2 < L_1$, tree 2 is accepted with probability L_2/L_1 and rejected otherwise. If it is rejected, then the next tree in the sample is a repeat of tree 1. The Metropolis algorithm always allows moves that increase the likelihood, but it also sometimes allows moves that decrease the likelihood. It is not just a "hill-climbing" algorithm. It allows downhill moves with the correct probability, so that the equilibrium probabilities of observing the different trees are given by the likelihoods.

Suppose there were just two trees, and the MCMC moved back and forward between them. Let P_1 and P_2 be the equilibrium probabilities of observing these two trees, or, in other words, the fractions of trees in the sample that will be trees 1 and 2. At equilibrium, the probability of observing each tree must be constant, so that the probability of being in tree 1 × the rate of change from 1 to 2 must equal the probability of being in tree 2 × the rate of change from 2 to 1. This property is known as **detailed balance**. We can therefore write:

$$P_1 r_{12} = P_2 r_{21}, \text{ or equivalently, } \frac{P_1}{P_2} = \frac{r_{21}}{r_{12}} \qquad (8.16)$$

As we wish to generate trees in proportion to their likelihood, we must have $P_1/P_2 = L_1/L_2$. Combining this with Eq. (8.16) gives us

$$\frac{r_{21}}{r_{12}} = \frac{L_1}{L_2} \qquad (8.17)$$

We must therefore set the ratio of the transition rates in the simulation to be equal to the ratio of likelihoods. The Metropolis algorithm defined above does this. If $L_2 > L_1$, then $r_{12} = 1$, and $r_{21} = L_1/L_2$, and so Eq. (8.17) is satisfied. If $L_2 < L_1$, then $r_{12} = L_2/L_1$, and $r_{21} = 1$, and so Eq. (8.17) is again satisfied. It follows that if the forward and backward rates are set correctly for any pair of states, then the

BOX 8.3
Calculating posterior probabilities

The likelihood calculated in Box 8.2 is really the likelihood of the known sequences evolving on a specified tree, i.e., it is the likelihood of the data given the tree, which is written $L(data|tree)$. We have only one set of data and many possible trees, so we would like to calculate the quantity $P(tree|data)$, which is the posterior probability of the tree, given the information in the data. Bayes' theorem tells us how to do this:

$$P(tree \mid data) = \frac{L(data \mid tree)P_{prior}(tree)}{\sum_{all\ possible\ trees} L(data \mid tree)P_{prior}(tree)} \quad (8.13)$$

This says that the posterior probability of the tree is equal to the likelihood of the data, given the tree (which we already know how to calculate), multiplied by the prior probability of the tree (using any expectations that we have about the problem before looking at the data), divided by the sum of the likelihood × the prior probability over all possible trees. The problem is with "all possible trees". We already know that the number of topologies is enormous, even for a very small number of sequences, but, in addition to this sum over topologies, we also need to consider all possible branch lengths and rate-matrix parameters for each tree. Thus, in practice, the summation sign in Eq. (8.13) includes integrals over these continuous parameters, as well as sums over the discrete number of topologies, and we can never hope to calculate this exactly. Nevertheless, we can progress without doing the calculation. What we really want to know is the posterior probability that a particular clade of interest is present, given the data. This can be written as

$$P(clade \mid data) = \sum_{trees\ containing\ clade} P(tree \mid data)$$

$$= \frac{\sum_{trees\ containing\ clade} L(data \mid tree)P_{prior}(tree)}{\sum_{all\ possible\ trees} L(data \mid tree)P_{prior}(tree)}$$

$$(8.14)$$

As we cannot sum over all trees, what we need is a representative sample of a large number of trees, selected in such a way that the probability of the tree occurring in the sample is proportional to its likelihood × prior. If we have such a sample, we may write

$$P(clade \mid data) =$$

$$\frac{Number\ of\ trees\ containing\ clade}{Total\ number\ of\ trees\ in\ sample} \quad (8.15)$$

Equation (8.15) no longer contains any likelihoods, because we already used the likelihoods in generating the sample of trees; thus, it is straightforward to calculate the posterior probability from the sample of trees using this equation.

equilibrium probabilities of all the possible states will come out correctly.

The Metropolis algorithm was introduced to do simulations in statistical physics. In physical systems, the equilibrium distribution is the thermodynamic equilibrium at a fixed temperature, and the likelihoods of the states are functions of the energies of the states, $L = \exp(-E/kT)$. Application of MCMC to phylogenetics is relatively new (Larget and Simon 1999, Lewis 2001, Huelsenbeck *et al.* 2002, Jow *et al.* 2002). The art of developing a good MCMC program is in the choice of the move set. It is necessary to strike a balance between moves that alter branch lengths and those that alter topology. If changes are very small, the likelihood ratio of the states will be close to 1, and the move has a good chance of being accepted. However, a very large number of moves would be required to change the tree significantly and to obtain a broad sample of tree space. If changes are very large, then the likelihood ratio of the states will be far from 1, and the likelihood of accepting the downhill move will be very small. It is necessary to be somewhere between these extremes to have an efficient algorithm.

8.8.4 An example using MCMC

We used an MCMC program (PHASE; Jow *et al.* 2002) to analyze the primate data set. The HKY

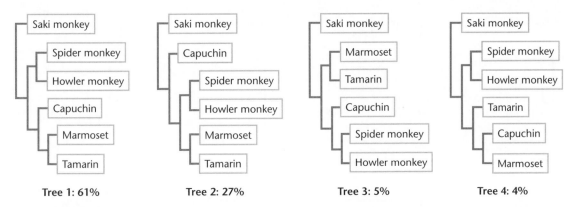

| Tree 1: 61% | Tree 2: 27% | Tree 3: 5% | Tree 4: 4% |

Fig. 8.15 The top four trees for the Platyrrhini group obtained by MCMC using the HKY model with invariant sites, plus six gamma-distributed rate categories.

model was used, allowing for invariant sites and four gamma-distributed rate categories. Several runs were performed to check consistency between the runs and to make sure that the program had converged to an equilibrium state. The tree shrews, tarsiers, lemurs, Catarrhini, and Platyrrhini were all well defined, i.e., they were all present 100% of the time during the MCMC run. However, there was substantial rearrangement of species within some of these groups and substantial variation in the positioning of these groups with respect to each other. Results for the Platyrrhini are shown in Fig. 8.15. The trees are ranked according to the frequency with which they occur in the MCMC run. Tree 1 occurs 61% of the time. In this tree, the capuchin is sister group to the marmoset and tamarin. The topology found by the NJ method (Fig. 8.8(b)), and by quartet puzzling, has the capuchin in a different position. This corresponds to Tree 2 in the MCMC method, which only occurs 27% of the time. In fact, all four top trees are identical, apart from the positioning of the capuchin. One of the strengths of the MCMC method is to illustrate which parts of a tree are well defined by the data and which parts are less well defined.

One of the hardest questions to answer with this set of sequences is the positioning of the earliest branch points in the tree. One way of presenting the MCMC results would be to give probabilities of occurrence of tree topologies for the full set of species. However, there are very many topologies that occur, and each individual tree occurs with a rather small probability. A better way to understand the results is to look at the arrangements of the high-level groups directly, rather than individual species. Figure 8.16 summarizes the results of one MCMC run in terms of the four groups: tree shrews, tarsiers, lemurs, and monkeys (= Catarrhini + Platyrrhini). The figure shows the seven most-frequently occurring topologies for these groups, ranked in order, with the percentage of the time each configuration occurred. We can present the results this way because these groups are all well defined. Rearrangements at the species level within these groups (such as the movement of the capuchin in Fig. 8.15) are ignored in Figure 8.16. Here, the most frequent tree occurs only 34% of the time. This is not much more than the trees ranked 2 and 3. So this method tells us that the information in this single gene is not sufficient to determine the relationship between these groups with any certainty.

It is interesting to compare these results with those of Ross, Williams, and Kay (1998) who constructed phylogenetic trees of both living and fossil primate species using the parsimony method applied to characters derived from tooth and skull shape. The morphological evidence indicates that tarsiers and monkeys are more closely related to each other than they are to the lemurs. This corresponds to Tree 7 in Fig. 8.16, which has only 2% support from

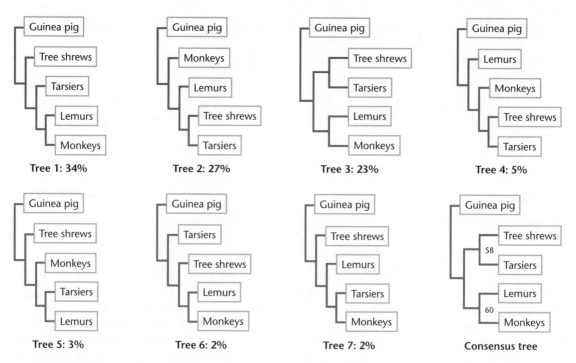

Fig. 8.16 Top seven trees for the principal groups obtained by the MCMC method. The consensus tree is also shown.

the MCMC method. Recent molecular work, using Alu sequences inserted into nuclear genes, also supports Tree 7 (Schmitz *et al.* 2001). The evidence from the small subunit rRNA gene is therefore in disagreement with both of these studies. We emphasize that the primates example discussed in this chapter uses a single gene only. It is presented as an illustration of the methods and is not intended as a serious attempt at the phylogeny of these species. Recent studies of the phylogeny of mammals with much larger sequence data sets (Murphy *et al.* 2001, Hudelot *et al.* 2003) group tarsiers with lemurs, as in Tree 5.

One point to note in Fig. 8.16 is that the top four trees differ only in the position where the guinea pig outgroup joins the rest of the tree. In other words, the guinea pig was "wandering about" on the tree during the Monte Carlo run. This may suggest that the guinea pig was a poor choice of outgroup. However, we have also performed runs with the mouse as outgroup and these did not prove any

more conclusive. This example is fairly typical in that it contains both easy and difficult parts of the tree to determine. In using MCMC, we have used much more computer time than with a simple method like NJ. It is therefore slightly disappointing that, even with a sophisticated method, we cannot be sure about some important aspects of the tree. Nevertheless, this conclusion is better than the false conclusion that we might draw by looking only at the single best tree obtained by some simpler method. We recommend the use of MCMC in phylogenetics because it is the best way of making maximum use of the information in the data. In a long MCMC run, we consider very many trees and very many sets of rate parameters. These are evaluated and averaged over in a way that has a sound statistical basis. If, at the end of the day, the information in the data is insufficient to answer some questions with confidence, then we simply need more data.

Figure 8.16 also illustrates an interesting point with respect to consensus trees. Majority-rule consensus programs are used to calculate consensus trees from sets of trees obtained via bootstrapping, as described in Section 8.4. The same method can be used to obtain consensuses of the set of trees generated in an MCMC run. The result is shown in the bottom right of the figure. The consensus tree is the same as Tree 3, and not Tree 1. The clade of tree shrews + tarsiers occurs 58% of the time. This is obtained by summing the frequencies of Trees 2, 3, 4, and several other low-frequency tree topologies that are not shown. Similarly, the clade lemurs + monkeys occurs 60% of the time. The consensus tree

has topology 3, because this is the only topology that contains both these clades. It is therefore important to realize that a consensus tree is not necessarily the most frequent tree in the set of trees from which it was derived. Drawing a consensus tree is just one way of summarizing the set of trees, and it does not give complete information about the set.

This completes our survey of phylogenetic methods. In Chapter 11, we will return to further questions related to molecular evolution and we will give several examples of interesting studies in molecular phylogenetics.

SUMMARY

There are many different ways of drawing phylogenetic trees. Sometimes, trees that appear to be different can have some of their groups interchanged, rather like "mobiles", so that it becomes clear that they are actually the same. It is important to be able to spot when this is the case. When looking at published trees, it is important to check whether the branch lengths are drawn to scale and whether the tree is intended to be rooted or unrooted. A given unrooted tree can be rooted on any of its branches, and the resulting rooted trees would all have a different evolutionary interpretation.

Distance-matrix methods in phylogenetics begin by the calculation of a matrix of pairwise distances between the sequences. Distances can be calculated using many different models of sequence evolution, and the numerical values of the distance for any pair of sequences will depend on the model used. Once the matrix of distances is calculated, many different techniques can be used to construct a tree. The simplest of these is the UPGMA, which is a hierarchical clustering method that assumes a constant rate of evolution in all the species. This assumption is often too strict for real sequences and can lead to unreliable results. The neighbor-joining method is another clustering method that works well if the input data are close to additive, i.e., if the pairwise distances between the species can be expressed as the sum of the lengths of the branches on the tree that connect the pairs of species. NJ is rapid, and useful for large data sets and for initial investigations with new sequences.

The Fitch–Margoliash method is another distance-matrix method that is based on a strict optimization criterion, rather than a clustering algorithm. This method finds an additive tree that minimizes the mean square

differences between the distances measured on the tree and the distances defined in the input distance matrix. This method requires a tree-search program to try many candidate trees and select the optimal one.

The maximum parsimony method was developed for binary characters that describe morphological features of organisms. The principle is to select the tree that requires the fewest possible changes in the character states (i.e., character gains or losses). For molecular data, the parsimony principle is to choose the tree for which the total number of substitutions required in the sequences is smallest.

The maximum-likelihood method uses another criterion for selection of the optimal tree. The likelihood of observing a given set of sequences can be calculated on any proposed tree. This is a function of the tree topology, the branch lengths, and the values of the parameters that define the evolutionary model used. The principle is to choose the tree for which the likelihood is maximized. Both likelihood and parsimony methods require a tree-search program to produce candidate trees. As calculation of the likelihood for any one tree is complex, it may require a large amount of computer time to search for the maximum-likelihood tree. Nevertheless, modern maximum-likelihood programs can be used effectively with data sets of realistic size, and likelihood-based methods are recommended for several reasons: they allow fitting of model parameters to the sequence data being used; they allow tests to be made to distinguish between models or between alternative trees; and they are less susceptible to problems of long-branch attraction than some simpler methods.

Bootstrapping, which can be applied to any of the above methods, is a way of estimating the reliability of different parts of a phylogenetic tree. A large number (at least 100) of randomized sequence data sets are generated, where each column of data is a copy of a randomly selected column in the real sequence alignment. The method is repeated on each of the randomized data sets, giving a set of slightly different trees. For each clade in the tree for the original data set, we calculate the bootstrap percentage, i.e., the percentage of time the clade appears in the set of trees from the randomized data. High bootstrap percentages (say, >70%) indicate statistical support for the presence of the clade.

Bayesian phylogenetic methods are a recent development of likelihood methods. Bayesian methods also calculate the likelihood of observing a given sequence set on a given tree, but instead of searching for a single tree that optimizes the likelihood, it takes an average over possible trees, weighting them according to their likelihood. In practice, a simulation technique known as Markov Chain Monte Carlo is used to generate a sample of possible trees, such that the probability of each tree arising in the sample is proportional to its likelihood. Quantities of interest, such as the posterior probability of formation of given clades, can then be calculated by averaging the properties of the trees in the sample. These probabilities provide similar information to bootstrap probabilities.

REFERENCES

Durbin, R., Eddy, S.E., Krogh, A., and Mitchison, G. 1998. *Biological Sequence Analysis – Probabilistic Models of Proteins and Nucleic Acids*. Cambridge, UK: Cambridge University Press.

Felsenstein, J. 1978. Cases in which parsimony or compatibility methods will be positively misleading. *Systematic Zoology*, **27**: 401–10.

Felsenstein, J. 1981. Evolutionary trees from DNA sequences: A maximum likelihood approach. *Journal of Molecular Evolution*, **17**: 368–76.

Felsenstein, J. 1985. Confidence limits on phylogenies: An approach using the bootstrap. *Evolution*, **39**: 773–81.

Felsenstein, J. 2001. PHYLIP Phylogeny Inference Package version 3.6. Available from http://evolution.genetics.washington.edu/phylip.html.

Fitch, W.M. 1971. Towards defining the course of evolution: Minimum change for a specific tree topology. *Systematic Zoology*, **20**: 406–16.

Fitch, W.M. and Margoliash, E. 1967. Construction of phylogenetic trees. A method based on mutation distances as estimated from cytochrome *c* sequences is of general applicability. *Science*, **155**: 279–84.

Hasegawa, M., Kishino, H., and Yano, T. 1985. Dating of the human–ape splitting by a molecular clock of mitochondrial DNA. *Journal of Molecular Evolution*, **22**: 160–74.

Horovitz, I., Zardoya, R., and Meyer, A. 1998. Platyrrhini systematics: A simultaneous analysis of molecular and morphological data. *American Journal of Physical Anthropology*, **106**: 261–81.

Hudelot, C., Gowri-Shankar, V., Jow, H., Rattray, M., and Higgs, P.G. 2003. RNA-based phylogenetic methods: Application to mammalian mitochondrial RNA sequences. *Molecular Phylogeny and Evolution*, **28**: 241–52.

Huelsenbeck, J.P., Larget, B., Miller, R.E., and Ronquist, F. 2002. Potential applications and pitfalls of Bayesian inference phylogeny. *Systematic Biology*, **51**: 673–88.

Jow, H., Hudelot, C., Rattray, M., and Higgs, P.G. 2002. Bayesian phylogenetics using an RNA substitution model applied to early mammalian evolution. *Molecular Biology and Evolution*, **19**: 1591–601.

Kishino, H. and Hasegawa, M. 1989. Evaluation of the maximum likelihood estimate of the evolutionary tree topologies from DNA data, and the branching order in *Hominoidea*. *Journal of Molecular Evolution*, **16**: 111–20.

Larget, B. and Simon, D.L. 1999. Markov Chain Monte Carlo algorithms for the Bayesian analysis of phylogenetic trees. *Molecular Biology and Evolution*, **16**: 750–9.

Lewis, P.O. 2001. Phylogenetic systematics turns over a new leaf. *Trends in Ecology and Evolution*, **16**: 30–7.

Murphy, W.J., Elzirik, E., Johnson, W.E., Zhang, Y.P., Ryder, O.A., and O'Brien, S.J. 2001. Molecular phylogenetics and the origins of placental mammals. *Nature*, **409**: 614–18.

Page, R. 2001. Treeview version 1.6.6. Available from http://taxonomy.zoology.gla.ac.uk/rod/treeview.html.

Ross, C., Williams, B., and Kay, R.F. 1998. Phylogenetic analysis of anthropoid relationships. *Journal of Human Evolution*, **35**: 221–306.

Saitou, N. and Nei, M. 1987. The neighbor-joining method: A new method of constructing phylogenetic trees. *Molecular Biology and Evolution*, **4**: 1406–25.

Schmitz, J., Ohme, M., and Zischler, H. 2001. SINE insertions in cladistic analyses and the phylogenetic affiliations of *Tarsius bancanus* to other primates. *Genetics*, **157**: 777–84.

Sneath, P.H.A. and Sokal, R.R. 1977. *Numerical Taxonomy*. San Francisco: W.H. Freeman.

Steel, M. and Penny, D. 2000. Parsimony, likelihood, and the role of models in molecular phylogenetics. *Molecular Biology and Evolution*, **17**: 839–50.

Strimmer, K. and von Haeseler, A. 1996. Quartet puzzling: A quartet maximum likelihood method for reconstructing tree topologies. *Molecular Biology and Evolution*, **13**: 964–9.

Swofford, D.L., Olsen, G.J., Waddell, P.J., and Hillis, D.M. 1996. Phylogenetic inference. In D.M. Hillis (ed.), *Molecular Systematics*, 2nd edition. Sunderland, MA: Sinauer.

PROBLEMS

8.1

(a) Construct the UPGMA tree for the following distance matrix:

(a)

	A	B	C	D
A	0	9	7	5
B	9	0	8	10
C	7	8	0	8
D	5	10	8	0

(b)

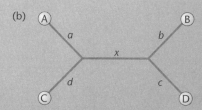

(b) In the unrooted tree above, *a*, *b*, *c*, *d*, and *x* represent the lengths of the branches. For an additive tree, the distances measured by summing along the branches must be equal to the pairwise distances, e.g.,

$$a + x + b = d_{AB}$$

Write down the six equations that must be satisfied if the data in the matrix above are to correspond to an additive tree with the topology shown. Hence, find the branch lengths and show that the data are, in fact, exactly additive.
(c) Describe qualitatively what would happen if the distance d_{BD} were equal to 12 instead of 10. What methods would you use to find an unrooted tree in that case?

8.2 The genus *Weedus* is a group of small rapidly growing green plants. You would like to obtain the phylogenetic tree for the following species, using the sequence information below. From morphological studies, the species *Pseudoweedus parsimonis* is believed to be an outgroup, and Tree A is believed to be correct.

Species	Symbol	Sites
		1 2 3 4 5 6
W. vulgaris	V	G A G C A T
W. major	M	C A G C A T
W. felsensteinii	F	T A G T G T
W. sempervivens	S	C A G T G T
P. parsimonis	P	T G G T A C

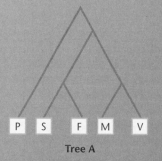

Tree A

(a) Explain the difference between informative and non-informative sites in parsimony methods. Which of the sites are non-informative in this example and why?
The algorithm for obtaining the number of required substitutions in the parsimony method works in the following way:

- Proceed upwards from the tips to the root. Consider the common ancestor of two nodes whose allowed states are known.
- If the sets of states of the two nodes are nonoverlapping, assign the union of these two sets to the ancestral node. Example: for site 1 on Tree A (shown below), the common ancestor of S and F is assigned to CT, which is the union of C and T. This means that either C or T will do at this point.
- If the sets of states of the two nodes are overlapping, assign the intersection of the two sets to the ancestral node. Example: the intersection of CT and CG is C, and therefore the common ancestor of the four *Weedus* species is assigned to C.
- Wherever a union is taken, at least one substitution is required. This is denoted * in the figure below. Tree A requires three substitutions in total at site 1.

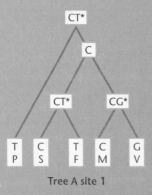
Tree A site 1

(b) Various combinations of nucleotide states are possible at the internal nodes according to the maximum-parsimony criterion. Redraw the figure for Tree A site 1, labeling only one possible state on each internal node, such that the states are one possible maximum-parsimony solution. Mark crosses on the branches where the substitutions occur.

(c) Use maximum-parsimony to compare Tree A to Trees B and C below. Calculate the total number of substitutions required at each of the informative sites for each of trees A, B, and C, and hence obtain the total number of substitutions required for the sequences on each tree. **Non-informative sites may be ignored for this calculation**. Does this analysis support the conclusions from morphology that Tree A is correct?

Tree B

Tree C

(d) From this set of sequences alone, how can we be sure that *P. parsimonis* is really an outgroup? Assuming you had access to much longer gene sequences and a computer, what other analyses would you carry out to determine whether the sequence data support the morphology-based tree?

8.3 This exercise gets you to try out some phylogeny software using the sequences in the CytBProt.txt and CytBDNA.txt files. These can be downloaded from www.blackwellpublishing.com/higgs. There are many different phylogenetic programs available for use with molecular data. We will refer to three packages below that we use frequently ourselves:
PHYLIP – available from http://evolution.genetics. washington.edu/phylip.html
Tree-Puzzle – available from http://www.tree-puzzle.de/
PHASE – available from http://www.bioinf.man.ac.uk/resources/phase/
PHYLIP has been a standard tool in phylogenetics for many years. This package has a very wide range of methods implemented. It is easy to install, has excellent documentation, and it generally does what it says it does. We find it useful for initial analyses of new data sets using quick methods like distance-matrix methods. We also make frequent use of its routines for bootstrapping,

calculating the consensus tree, and drawing trees. You will learn a lot from reading the PHYLIP documentation, even if you don't use these programs much afterwards.

Tree-Puzzle is a useful way of finding maximum-likelihood trees with many different models of sequence evolution. It also implements the quartet-puzzling method.

PHASE is our own package, which uses the MCMC method. For a comprehensive list of other software packages, see the PHYLIP site.

The species chosen in the cytochrome b example files cover the full range of vertebrate species and have been chosen to illustrate some interesting biological questions that are not fully understood at present. The table below lists the species and the principal taxonomic groups to which they belong.

(a) Create multiple alignments of the protein and DNA sequences using Clustal (see Problem 6.2), or whichever method you prefer. Are there any parts of the automatic alignments that you distrust? Are there any slight adjustments you would make? Are there any unreliable parts of the alignments that you would exclude from phylogenetic analysis? One way of viewing alignments is with GeneDoc (http://www.psc.edu/biomed/genedoc/). This can be used for manual editing – i.e., sliding a few amino acids here or there, and deleting columns with many gaps prior to running the phylogeny programs. These sequences are fairly conserved and easy to align

more or less across their full length; therefore, little editing will probably be necessary. You can also export the alignment from GeneDoc in PHYLIP format, which is the input format required by many phylogenetic programs.

(b) Try a variety of different methods using both the protein and DNA sequences. Which parts of the tree are the same with all the methods, and which parts vary? Make some bootstrap data sets and find out the bootstrap support values for the different nodes in the tree. What can you conclude from all this about both the biology and the methods used?

Some parts of the phylogeny of these species are well supported by the cytochrome b gene. For example, in our attempts with these sequences, we found that the six mammals reliably group together, as do the two birds, and the three bony fish. This is good, because most biologists are confident that these groups are true monophyletic groups. However, there are many other more problematic points:

• The platypus is a monotreme, the wombat is a marsupial, and the other four mammals are all placental mammals. It is usually assumed that the monotremes are the earliest diverging of these three groups. However, several recent studies have found that monotremes and marsupials are sister groups. We found that the relative positions of the three mammal groups varied according to the method used and according to whether DNA or protein alignment was used.

Cephalochordates	Amphioxus	*Branchiostoma lanceolatum*
Jawless fish	Lamprey	*Petromyzon marinus*
Cartilaginous fish	Shark	*Scyliorhinus canicula*
Bony fish	Eel	*Anguilla japonica*
	Goldfish	*Carassius auratus*
	Cod	*Gadus morhua*
Lungfish	Lungfish	*Protopterus dolloi*
Amphibians	Frog	*Xenopus laevis*
	Salamander	*Mertensiella luschani*
Reptiles	Iguana	*Iguana iguana*
	Turtle	*Chelonia mydas*
	Alligator	*Alligator mississippiensis*
Birds	Falcon	*Falco peregrinus*
	Ostrich	*Struthio camelus*
Mammals	Platypus	*Ornithorhynchus anatinus*
	Wombat	*Vombatus ursinus*
	Armadillo	*Dasypus novemcinctus*
	Elephant	*Loxodonta africana*
	Human	*Homo sapiens*
	Dog	*Canis familiaris*

• What is the closest relative of the birds? According to these sequences, we found it to be the alligator. This is thought to be true by most biologists. As a result of this, the reptiles are not a monophyletic group. We can, however, define a wider group called sauropsids that includes both reptiles and birds, and that is usually thought to be monophyletic. Are the sauropsids monophyletic according to the cytochrome b sequences? What are the relative positions of the turtle, iguana, and the alligator/bird clade?

• Amphibians are usually thought to be a monophyletic group. However, the salamander and frog sequences used here seem quite different from one another, and these species did not cluster together in our analyses. This points to the limitations of using a single gene for phylogenetics. Possibly one of these sequences has evolved in an unusual way that is not well described by the models of evolution used by the programs.

• The lungfish are often thought to be a sister group of the tetrapods (i.e., amphibians + reptiles + birds + mammals), although their position has been controversial. Can we say anything about this issue with these sequences? The problems mentioned above with the amphibians make this difficult.

• The most distantly related species in this set is thought to be the amphioxus, which is a member of the chordate phylum but is not a true vertebrate. The next most distant is the lamprey, which is classed as a vertebrate, but is jawless. All the other species listed here are gnathostomes (i.e., jawed vertebrates). We would therefore expect that amphioxus and lamprey are outgroups in this analysis. The point at which these sequences connect to the rest of the tree should tell us the position of the root of the gnathostome tree. Biologists have long assumed that the cartilaginous fish (sharks and rays) are the earliest branching group among the gnathostomes – hence the lamprey and shark should be connected in our example. This has been challenged recently in several studies using complete mitochondrial genome sequences. It was found that the lamprey appeared to connect to the gnathostome tree close to the tetrapods, thus making the sharks a sister group of the bony fish. What do you find with the cytochrome b sequences? It could well be that mitochondrial sequences are misleading in this respect (note that cytochrome b is one gene from the mitochondrial genome).

• If we are not confident that the lamprey and amphioxus sequences provide a reliable way of rooting the rest of the tree, we might try removing these two sequences. Repeat the analysis without these species, and assume that the shark is the root for the remaining species. What does this do to the positions of the amphibians and the lungfish?

(c) It is clear that there are many questions that cannot be answered with the cytochrome b sequences in this set. To make progress, we need to add either more species for the same gene, or more genes, or both. There are now many complete mitochondrial genomes. A convenient place from which further mitochondrial gene sequences can be downloaded is the OGRe database (http://ogre.mcmaster.ca). Several of the questions raised above can be answered more fully using the complete set of mitochondrial genes and using a more extensive set of species. However, some of these questions are still not resolved. Try looking up the latest papers on some of the points mentioned above.

SELF-TEST
Phylogenetic methods

1 The number of possible distinct unrooted trees with five species is

A: 15

B: 25

C: 5!

D: $\dfrac{5!}{3! \times 2!}$

3 If the same four trees are now taken to be rooted, which statement is correct?

A: All four trees are non-equivalent.

B: All four trees are equivalent.

C: Only trees (i) and (iii) are equivalent.

D: Only trees (iii) and (iv) are equivalent.

4 Which of these statements about phylogenetic methods is correct?

A: Bootstrapping can be used as a measure of confidence of the evolutionary model used in the phylogeny.

B: Bootstrapping cannot be done if the rate of substitution varies across sites.

C: Transition substitutions will usually saturate at a smaller divergence time than transversion substitutions.

D: Long branch attraction arises when the lengths of the sequences in the analysis vary a great deal.

2 The following four trees are unrooted and branch lengths are not drawn to scale. Which statement is correct?

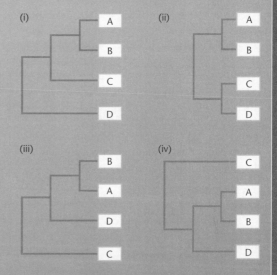

A: All four trees are non-equivalent.

B: All four trees are equivalent.

C: Only trees (i) and (iii) are equivalent.

D: Only trees (iii) and (iv) are equivalent.

5 The alignment below is thought to be part of a protein-coding gene. Which of the four conclusions would you draw?

A: This is probably part of a pseudogene.
B: This is more likely to be part of an RNA-coding gene than a protein.
C: There is evidence that transversions occur more frequently than transitions.
D: There is evidence that synonymous substitutions occur more frequently than non-synonymous substitutions.

```
Wombat      :   GTTAATGAGTGGTTATCCAGAAGTGAGATA   :   30
Opossum     :   GTTAATGAGTGGTTATTCAGAAGTGAGATA   :   30
Rhinoceros  :   GTTAATGAGTGGTTTCCAGAAGTGAAATA    :   30
Pig         :   GTTAATGAGTGGTTTCTAGAACGGAAATA    :   30
Hedgehog    :   GTGAATGAATGGCTTCCAGAAGTGAACTG    :   30
Human       :   GTTAATGAGTGGTTTCCAGAAGTGAACTG    :   30
Hare        :   GTTAACGAGTGGTTCTCCAGAAGTGAAATG   :   30
```

6 The following four phylogenetic trees have been proposed for the species in the previous sequence alignment. The standard parsimony criterion (minimization of the required number of mutations) is used to distinguish between the four trees using the above sequence data. Which conclusion would you draw?

According to the parsimony criterion:
A: Tree 1 is preferable to all other trees.
B: Trees 1 and 2 are preferable to Trees 3 and 4.
C: Trees 1 and 3 are preferable to Trees 2 ad 4.
D: Trees 1 and 4 are preferable to Trees 2 and 3.

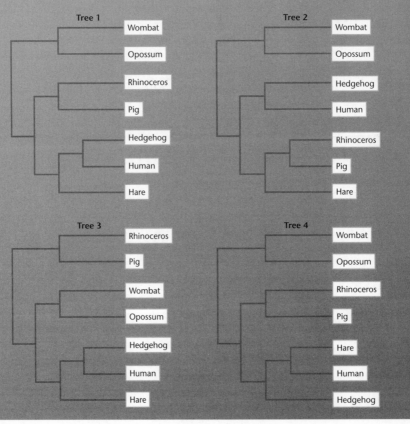

7 Which of the four trees A, B, C, or D corresponds to the result of the UPGMA algorithm applied to the distance matrix below? (The branch lengths are not drawn to scale).

	W	X	Y	Z
W	–	0.3	0.4	0.5
X	0.3	–	0.5	0.5
Y	0.4	0.5	–	0.4
Z	0.5	0.5	0.4	–

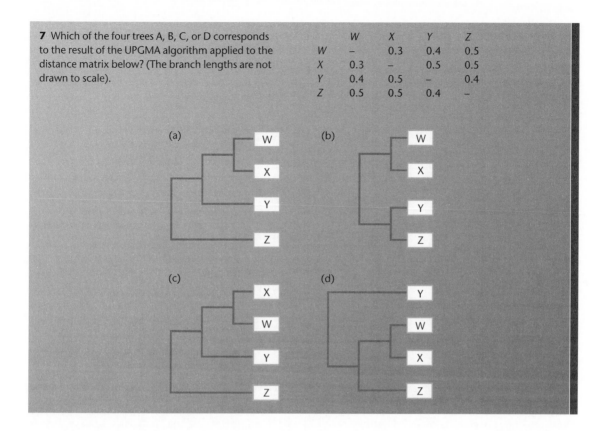

(a)

(b)

(c)

(d)

8 100 bootstrap data sets are created for a set of sequences. The neighbor-joining method is applied to these data sets to give 100 trees. The consensus tree of this set of 100 trees is given below with bootstrap percentages indicated. It has been rooted by assuming that U is the outgroup. Which of the four conclusions can be drawn?

A: The clade X+Y+Z never occurs in the set of 100 trees.
B: The clade W+X+V never occurs in the set of 100 trees.
C: The clade W+X+V could occur in up to 6 of the 100 trees.
D: The clade W+V could occur in up to 14 of the 100 trees.

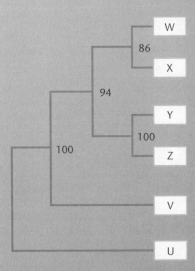

9 Which of the following statements about phylogenetic methods is correct?

A: The maximum-likelihood method determines the tree for which the likelihood of the tree given the data is largest.
B: The maximum-likelihood tree may sometimes contain branches of zero length.
C: If the trees from maximum-likelihood, parsimony and neighbor-joining methods all have the same topology, this must be the correct topology.
D: In Bayesian phylogenetic methods, if the prior probabilities of two trees are equal, then the posterior probabilities must also be equal.

(a)

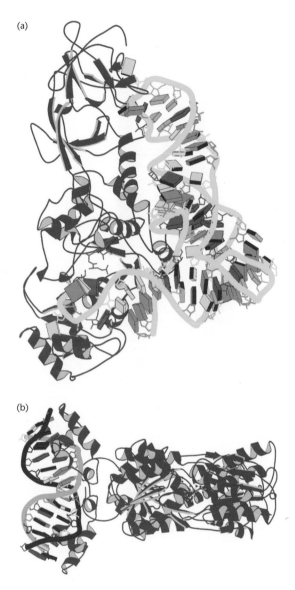

(b)

Plate 2.1 (a) Crystal structure of a complex of glutaminyl tRNA with glutaminyl-tRNA synthetase (Sherlin *et al.* 2000, *Journal of Molecular Biology*, **299**: 431–46). (b) Crystal structure of the dimeric lac repressor protein bound to DNA (Bell and Lewis 2001, *Journal of Molecular Biology*, **312**: 921–6). Image reproduced with permission from NDB (Berman *et al.* 1992). ID number PR0025 and PD0250.

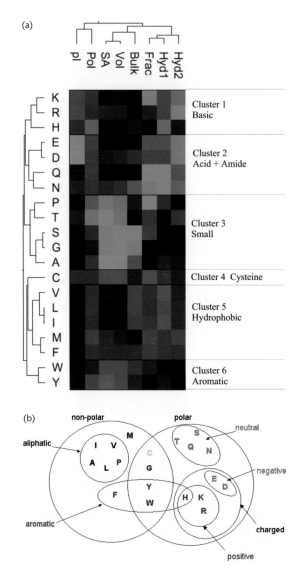

Plate 2.2 (a) Hierarchical clustering of the amino acids performed using CLUTO (Karypis 2000, University of Minnesota technical report #02–017), as described in Section 2.6.2. (b) Where clusters fail: the Venn diagram illustrates the overlapping properties of amino acids. The color scheme is the one implemented in the CINEMA sequence alignment editor (Parry-Smith *et al.* 1998, *Gene*, **221**: GC57–GC63): red for acidic; blue for basic; green for polar neutral; purple for aromatic; black for aliphatic; yellow for cysteine (disulfide potential); and brown for proline and glycine (which have unique structural properties, especially in helices).

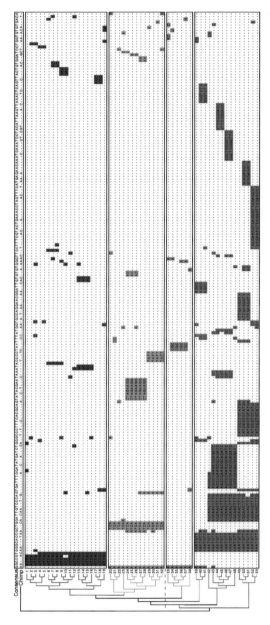

Plate 3.1 The pattern of bases observed at informative sites in human mitochondrial genomes from many different ethnic groups and the phylogenetic tree deduced from this information. Reproduced from Ingman et al. (2000. *Nature*, **408**: 708–13), with permission from Nature Publishing Group.

(a)

PSSM of IPB000817A (Prion;) 90 sequences.

(b)

OPSD_SHEEP	P Y A G V A F Y I F T H Q G S D F G P I F M T I P A F
OPSD_BOVIN	P Y A G V A F Y I F T H Q G S D F G P I F M T I P A F
OPSD_MOUSE	P Y A S V A F Y I F T H Q G S N F G P I F M T L P A F
OPSD_HUMAN	P Y A S V A F Y I F T H Q G S N F G P I F M T I P A F
OPSD_XENLA	P Y A Y V A F Y I F T H Q G S N F G P V F M T V P A F
OPSD_CHICK	P Y A S V A F Y I F T N Q G S D F G P I F M T I P A F
OPSD_LAMJA	P Y A S V A F Y I F T H Q G S D F G A T F M T L P A F
OPSG_CHICK	P Y A V V A F W I F T N K G A D F T A T L M A V P A F
OPSG_CARAU	P Y A T V A A W I F F N K G A D F S A K F M A I P A F
OPSB_GECGE	P Y A A T A I W I F T N R G A A F S V T F M T I P A F
OPSU_BRARE	P Y A G V A W Y I F T H Q G S E F G P V F M T L P A F

(c)

OPSG_GECGE	P Y A S F V S F A A A N P G Y A F H P L A A A L P A Y
OPSG_ASTFA	P Y A S F A T F S A L N P G Y A W H P L A A A L P A Y
OPSG_HUMAN	P Y A F F A C F A A A N P G Y P F H P L M A A L P A F
OPSR_HUMAN	P Y T F F A C F A A A N P G Y A F H P L M A A L P A Y
OPSR_ANOCA	P Y T V F A C F A A A N P G Y A F H P L A A A L P A Y
OPSR_CHICK	P Y T F F A C F A A A N P G Y A F H P L A A A L P A Y
OPSR_ASTFA	P Y T F F A C F A A A N P G Y A F H P L A A A M P A Y
OPSR_CARAU	P Y T F C A C F A A A N P G Y A F H P L A A A M P A Y

Plate 9.1 (a) Logo for the prion block depicted in Fig. 9.3. The completely conserved glycine and highly conserved proline residues that form the GPG anchor for this block are clearly visible toward the center of the logo. (b) Excerpts from sequence alignments of vertebrate visual receptors. The coloring of the amino acids follows the scheme shown in Plate 2.2(b). Part of the typical signature of rhodopsin (denoted OPSD) sequences is shown. (c) Part of the typical signature of red/green opsins (denoted OPSR and OPSG). Note that the red and green sequences are highly similar, such that it is almost impossible to separate them using typical sequence analysis methods – the central motif, N-P-G-Y-[AP]-[FW]-H-P-L, which includes a conserved histidine, is particularly striking. More startling is the fact that within the rhodopsin alignment are several "rogue" sequences, namely of green, blue (OPSB), and purple (OPSU) opsins, that more closely resemble the rhodopsin sequence signature than they do their own pigment sequences – in particular, the chicken (CHICK) and goldfish (CARAU) green pigments do not contain the N-P-G-Y-[AP]-[FW]-H-P-L motif characteristic of green sequences.

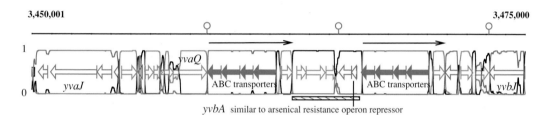

Plate 10.1 Application of a three-state M1–M2 model to the *Bacillus subtilis* genome reveals an atypical segment (3,463–3,467 kb, underlined) surrounded by ABC transporter gene duplication (thin black arrows). Filled arrows represent genes of known function, empty arrows, those of unknown function, and red hairpins represent transcriptional terminators. The colored curves show the probabilities of being in each of the three hidden states at each point in the sequence. The magenta state matches genes on the (+) strand, whereas cyan matches genes on the (–) strand. The black state fits either intergenic regions or atypical genes. Reproduced from Nicolas *et al.* (2002, *Nucleic Acids Research*, **30**: 1418–26) with permission of Oxford University Press.

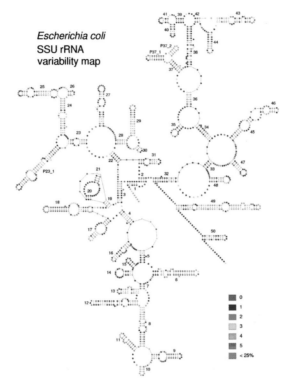

Plate 11.1 The secondary structure of SSU rRNA in *E. coli*. The color scheme shows the degree of variability of the sequence across the bacterial domain. Category 0 (purple) sites are completely conserved. Categories 1 to 5 range from very slowly evolving (blue) to rapidly evolving (red). The gray sites are present in less that 25% of the species considered, hence no measure of evolutionary rate was made. Reproduced with permission from the European Ribosomal RNA database http://oberon.fvms.ugent.be:8080/rRNA/index.html

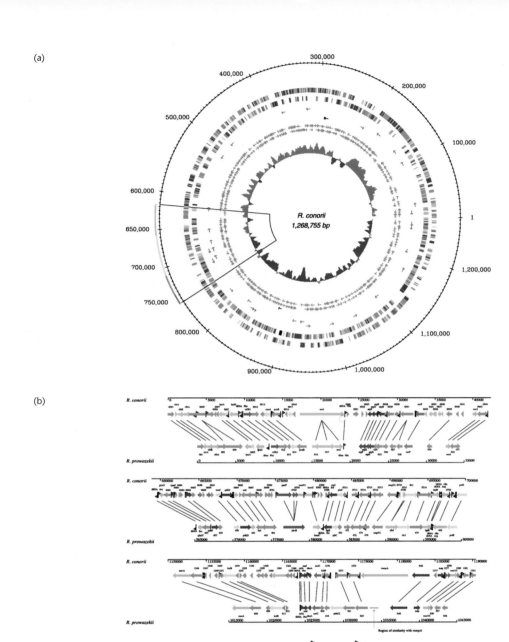

Plate 12.1 The genome of *Rickettsia conorii*, reproduced from Ogata *et al.* (2001), Copyright 2001 AAAS. (a) The outer circle gives the nucleotide positions in bases measured anticlockwise from the origin of replication. The second and third circles show the positions of ORFs on the plus and minus strands, respectively. The colors used indicate different functional classes of gene. The arrows in the fourth and fifth circles show the positions of tRNA genes, and the three black arrows show rRNAs. The sixth and seventh circles indicate short repeated sequences. The eighth circle shows G–C skew. The genome is found to be largely colinear with *R. prowazekii*, except for the shaded sector on the lower left. (b) Three distinct regions of the *R. conorii* genome aligned with homologous regions of the *R. prowazekii* genome. Most of the genes are present in the same order in both species. The top and middle sections show split genes in *R. prowazekii* and *R. conorii*, respectively. The bottom section contains a non-coding region of *R. prowazekii* that shows some sequence similarity to a functional gene (*rOmpA*) in *R. conorii*.

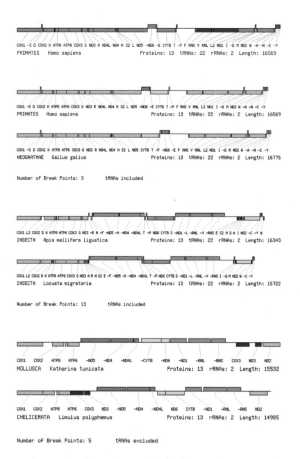

COX1 -S D COX2 K ATP8 ATP6 COX3 G ND3 R ND4L ND4 H S2 L ND5 -ND6 -E CYTB T -P F RNS V RNL L2 ND1 I -Q M ND2 W -A -N -C -Y
PRIMATES   Homo sapiens   Proteins: 13   tRNAs: 22   rRNAs: 2   Length: 16569

COX1 -S D COX2 K ATP8 ATP6 COX3 G ND3 R ND4L ND4 H S2 L ND5 -ND6 -E CYTB T -P F RNS V RNL L2 ND1 I -Q M ND2 W -A -N -C -Y
PRIMATES   Homo sapiens   Proteins: 13   tRNAs: 22   rRNAs: 2   Length: 16569

COX1 -S D COX2 K ATP8 ATP6 COX3 G ND3 R ND4L ND4 H S2 L ND5 -ND6 -E F RNS V RNL L2 ND1 I -Q M ND2 W -A -N -C -Y
NEOGNATHAE Gallus gallus   Proteins: 13   tRNAs: 22   rRNAs: 2   Length: 16775

Number of Break Points: 3   tRNAs included

COX1 L2 COX2 D K ATP6 ATP8 COX3 G ND3 -R N -F -ND5 -M -ND4 -ND4L T -P ND6 CYTB S -ND1 -L -RNL -V -RNS E S2 M Q A I ND2 -C -Y W
INSECTA   Apis mellifera ligustica   Proteins: 13   tRNAs: 22   rRNAs: 2   Length: 16343

COX1 L2 COX2 D K ATP6 ATP8 COX3 G ND3 A R N S2 E -F -ND5 -H -ND4 -ND4L T -P ND6 CYTB S -ND1 -L -RNL -V -RNS I -Q M ND2 W -C -Y
INSECTA   Locusta migratoria   Proteins: 13   tRNAs: 22   rRNAs: 2   Length: 15722

Number of Break Points: 13   tRNAs included

COX1   COX2   ATP8   ATP6   -ND5   -ND4 -ND4L   -CYTB   -ND6   -ND1   -RNL   -RNS COX3   ND3   ND2
MOLLUSCA   Katharina tunicata   Proteins: 13   rRNAs: 2   Length: 15532

COX1   COX2   ATP8   ATP6   COX3   ND3   -ND5   -ND4 -ND4L   ND6   CYTB   -ND1   -RNL   -RNS   ND2
CHELICERATA Limulus polyphemus   Proteins: 13   rRNAs: 2   Length: 14985

Number of Break Points: 5   tRNAs excluded

Plate 12.2 Comparisons of gene order in pairs of animal mitochondrial genomes constructed using the OGRe database (Jameson *et al.* 2003, *Nucleic Acids Research*, **31**: 202–6). http://ogre.mcmaster.ca

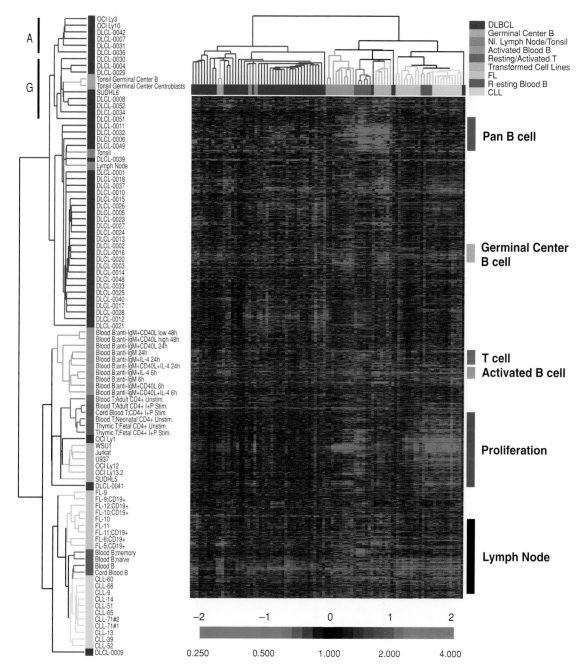

Plate 13.1 Hierarchical clustering of gene expression data depicting the relationship between 96 samples of normal and malignant lymphocytes (reproduced from Alizadeh *et al.* 2000, *Nature*, **403**: 503–11, with permission of Nature Publishing Group).

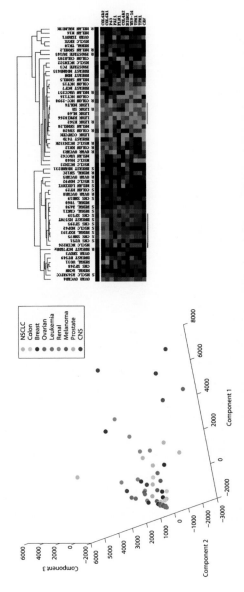

Plate 13.2 Analysis of microarray data from 60 cancer cell lines (reproduced from Slonim 2002. *Nature Genetics* supplement, **32**: 502–7, with permission of Nature Publishing Group). (a) Projection of the samples onto the first three principal component axes reveals a certain amount of clustering between some types of cancer. (b) Hierarchical clustering of the same data. S and R indicate samples that are extremely sensitive or resistant to the drug cytochalasin D.

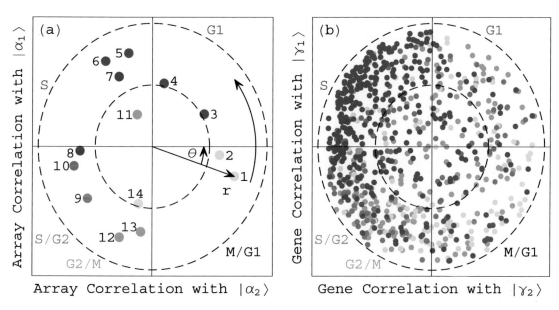

Plate 13.3 Analysis of microarray data from the yeast cell cycle using SVD (reproduced from Alter *et al.* 2000, *Proceedings of the National Academy of Sciences USA*, **97**: 10101–6, Copyright 2000 National Academy of Sciences, USA). (a) Arrays 1–12 correspond to successive time-points of the cell. The points for the 12 arrays proceed in an almost circular pattern around the space defined by the first 2 eigenarrays (denoted $|\alpha_1>$ and $|\alpha_2>$). (b) Each of 784 cell-cycle-regulated genes is shown as a point in the space of the first two eigengenes (denoted $|\gamma_1>$ and $|\gamma_2>$ and). The array points are color-coded according to the five cell-cycle stages: M/G$_1$ (yellow), G$_1$ (green), S (blue), S/G$_2$ (red), and G$_2$/M (orange). The gene points are colored according to the stage at which they are significantly up-regulated.

(a)

(b)

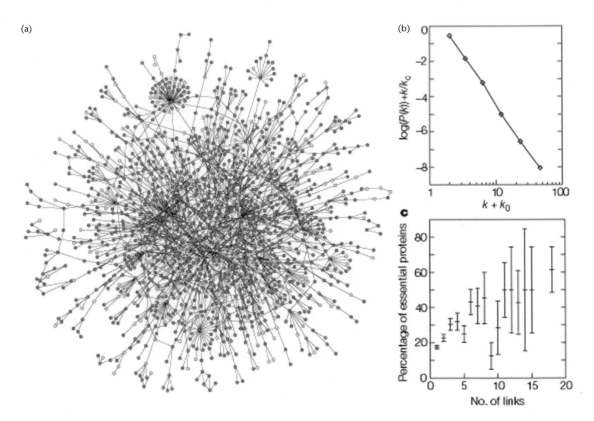

Plate 13.4 Map of protein–protein interactions in yeast (Jeong *et al.* 2001, *Nature*, **411**: 41–2, with permission of Nature Publishing Group). Each node is a protein, and each link is a physical protein–protein interaction, identified usually by Y2H experiments. The color of a node indicates the effect of deleting the corresponding protein (red, lethal; green, non-lethal; orange, slow growth; yellow, unknown).

Patterns in protein families

9

CHAPTER PREVIEW

In this chapter, we look beyond the familiar pairwise sequence alignment techniques and begin to appreciate the added value and increased diagnostic performance that accrue from the use of multiple sequence information. In particular, we focus on the different pattern-recognition methods that are used to encode conserved regions in sequence alignments (including regular expressions, fingerprints, and blocks), which are in turn used to characterize particular protein families and to build signature databases. We complete the chapter by considering how these methods can be used together to shed light on the structure and function of complex multi-gene families, with particular emphasis on the analysis of G protein-coupled receptors; we see how these diagnostic approaches offer different, but nevertheless complementary, biological insights.

9.1 GOING BEYOND PAIRWISE ALIGNMENT METHODS FOR DATABASE SEARCHES

In Chapter 7, we discussed database search methods based on pairwise alignment of a query and database sequences (e.g., BLAST (Altschul, Gish, and Miller 1990, Altschul *et al.* 1997) and FASTA (Pearson and Lipman 1988, Pearson 2000)). In many cases, such search tools are highly effective; however, they are not infallible and additional evidence may sometimes be required in order to reliably diagnose a sequence as a member of a given family and hence to make a meaningful functional assignment. There are many reasons for this. With more than 30 million sequences from over 150,000 organisms in GenBank in 2004 (Benson *et al.* 2004), search outputs can be complex and redundant: e.g.,

results can be dominated by irrelevant matches to low-complexity or highly repetitive sequences; the presence of modular or multi-domain proteins can also complicate interpretation, as it may not be clear at what level a match has been made (e.g., at the level of a single module/ domain, or several modules/ domains, or of the whole protein); multi-gene families may likewise complicate matters, as distinguishing orthologs and paralogs may be problematic; the annotation of retrieved matches may be incorrect; and, given database size and increasing levels of noise, a correct match may simply fail to achieve a higher score than a random database sequence, or sequences.

Some of these problems can now be tackled more or less automatically, for example by using masking devices in BLAST to filter out low-complexity regions from queries. However, recognizing orthology with any degree of reliability is a far more exacting task (Huynen and Bork 1998), but nevertheless an important one if we wish to establish the correct phylogenetic relationships between genes. Protein family databases offer a small step in this direction. Depending on the type of analysis method used, they may help to elucidate relationships in considerable detail, including superfamily, family, subfamily, and species-specific sequence levels. The ability to make familial distinctions in such precise ways makes these techniques useful and often powerful adjuncts to routine sequence database searches (Attwood 2000a,b).

It is often said that bioinformatics is a "knowledge-based" discipline. This means that many of the search and prediction methods that have been used to greatest effect in bioinformatics exploit information that has already been accumulated about the problem of interest, rather than working from first principles. Most of the methods discussed in this chapter adopt this kind of knowledge-based approach to sequence analysis. Typically, we have a set of known examples of sequences of a given class and we try to identify patterns in the data that characterize these sequences and distinguish them from those that are not of this class. The basic problem addressed by the methods described here is then to determine whether an unknown sequence is a member of the class or classes of sequences that we have already identified by means of such characteristic patterns.

A pattern can be more or less anything that is unique to the sequences of the class of interest, and hence that is different from other sequences. Usually, we are interested in the most conserved aspects of sequence families, so the patterns describe features that are common to all members of the class. The simplest type of pattern would be a set of specific amino acids found in every example of a protein of a given type. However, life is rarely as simple as this. Even the most conserved parts of sequences tend to vary somewhat between species. Therefore, patterns have to be sufficiently flexible to account for some degree of variation.

To make allowance for different degrees of variation, the methods discussed here use some type of scoring system, where the object is to assign a high score to sequences that are believed to be members of the class and a low score to those that are not. In Chapter 7, we defined the sensitivity and selectivity of a method (Eqs. 7.1 and 7.2). At that point, we were thinking of database search algorithms using a single query sequence. In the present context, we are thinking of scoring systems for recognizing sequences in a protein family. The same considera-

tions apply, however. A good method should have high sensitivity, i.e., it should correctly identify as many true-positive (real) members of the family as possible. It should also have high selectivity, i.e., very few false-positive sequences should be incorrectly predicted to be members of the family.

It should be clear that the more sequences we already have in a given family, the more information we have about it and the more likely we are to be able to find some pattern that accurately diagnoses its members. This is typical of knowledge-based methods, where the more we know to start with, the easier the problem becomes. An important way in which information from many sequences can be combined is by defining some kind of position-specific scoring matrix (or PSSM, usually pronounced "possum"). PSSMs are sometimes also termed "profiles", and there are several different ways of constructing them. However, the key point is that the score of matching any given amino acid at a given position depends on the set of amino acids that occurred at that position in the family of proteins we are studying. So, for example, our protein family might have a conserved leucine residue at position 10, so the PSSM would have a high positive score for aligning a leucine from the query sequence to position 10 of the family. This is in contrast to pairwise alignment algorithms, where the score for aligning a leucine with a leucine is always the same, no matter where it occurs in the sequences. Scoring systems for pairwise alignment like PAM (Dayhoff 1965, Eck and Dayhoff 1966) and BLOSUM (Henikoff and Henikoff 1992) are derived from average properties of many sequence alignments. Leucines are not always conserved, and the PAM score for an L–L match is not particularly high by comparison with some other scores, such as for W–W and C–C. A PSSM is able to capture the additional information that the matching leucine **is** very significant for **this particular** sequence family, and hence to assign this feature a correspondingly high score. Several of the methods discussed in this chapter use PSSMs.

In Chapter 5, we described a range of protein family databases and mentioned the different analytical techniques used to create them. It is import-

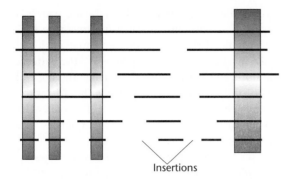

Fig. 9.1 Schematic illustration of a sequence alignment, showing how gap insertion brings equivalent parts of the alignment into the correct register, leading to the formation of conserved regions, or motifs (shaded blocks). These provide tell-tale signatures that can be used to diagnose new family members.

ant to understand the differences between the various approaches, because they have different diagnostic strengths and weaknesses and hence work best under different circumstances. The underlying principle behind all of the methods we described is that within multiple alignments can be found conserved regions, or motifs (Fig. 9.1), that are believed to reflect shared structural or functional characteristics of the constituent sequences. These regions can be used to build diagnostic signatures by means of which we can potentially recognize new family members (Attwood 2000a,b). The main methods used to derive such signatures fall essentially into three groups, based on the use of single motifs, multiple motifs, or full domain alignments (recall Fig. 5.6). Some of these techniques are outlined in the sections that follow.

9.2 REGULAR EXPRESSIONS

9.2.1 Definition of regexs

Probably the simplest pattern-recognition method to understand is the regular expression, or regex, often simply (and confusingly) referred to as a "pattern". Here, sequence data within a motif are reduced to a consensus expression that captures the conserved essence of the motif, disregarding poorly conserved regions. Regexs, then, discard sequence data, retaining only the most significant motif information, as shown in Table 9.1.

Slightly different conventions are used to depict regexs: the one we shall describe is the format specified in PROSITE (Falquet *et al.* 2002). Here,

Table 9.1 Derivation of a regex from a conserved motif from the prion protein family.

Alignment	Regex
ETDVKMMERVVEQMCVTQY	
ETDVKMMERVVEQMCITQY	
ETDVKIMERVVEQMCITQY	
ETDIKIMERVVEQMCTTQY	
ETDVKMMERVVEQMCVTQY	
ETDVKMMERVVEQMCITQY	
ETDIKMMERVVEQMCITQY	E-x-[ED]-x-K-[IVM](2)-x-[KR]-V-[IV]-x-[QE]-M-C-x(2)-Q-Y
ETDIKIMERVVEQMCITQY	
ETDIKMMERVVEQMCITQY	
ETDMKIMERVVEQMCVTQY	
ETDIKMMERVVEQMCITQY	
ETDIKIMERVVEQMCITQY	
ETDMKIMERVVEQMCVTQY	
ETDVKMIERVVEQMCITQY	
EMENKVVTKVIREMCVQQY	
EMENKVVTKVIREMCVQQY	

residues that are absolutely conserved at a particular position in a motif are denoted by single characters on their own; where different residues with similar properties occupy the same position, those residues are "allowed" within square brackets; where different residues that share no similar properties occupy the same position, a wild-card "x" denotes that any residue may occur; where specific residues cannot be tolerated at a position, those residues are "disallowed" within curly brackets; where there are consecutive positions with identical characteristics, the number of such positions is indicated in parentheses; where a numerical range is given (usually following a wild-card), the sequences defined by the regex are of different length (this notation tends to be used to describe variable loop regions).

Using this convention, in the expression derived from the 19-residue motif shown in Table 9.1, we see that positions 1, 5, 10, 14, 15, 18, and 19 are completely conserved; positions 3, 9, 11, and 13 allow one of two possible residues; positions 6 and 7 allow one of three possible residues; and positions 2, 4, 8, 12, 16, and 17 can accommodate any residue.

9.2.2 Searching with regexs

Typically, the software that makes use of regexs does not tolerate similarity, and searches are thus limited in scope to the retrieval of identical matches. For example, let us suppose that a sequence matches the expression in Table 9.1 at all but the ninth position, where it has His conservatively substituting for the prescribed Lys and Arg. Such a sequence, in spite of being virtually identical to the expression, will nevertheless be rejected as a mismatch, even though the mismatch is a biologically feasible replacement. Alternatively, a sequence matching all positions of the pattern, but with an additional residue inserted at position 16, will again fail to match, because the expression does not cater for sequences with more than two residues between the conserved Cys and Glu. Searching a database with regexs thus results either in an exact match, or no match at all.

The strict binary outcome of this type of pattern matching has severe diagnostic limitations. Creating a regex that performs well in a database search is always a compromise between the tolerance that can be built into it, and the number of matches it will make by chance: the more permissive the pattern, the noisier its results, but the greater the hope of finding distant relatives; conversely, the stricter the pattern, the cleaner its results, but the greater the chance of missing true-positive sequences not catered for within the defined expression.

A further limitation of this approach hinges on the philosophy of using single motifs to characterize entire protein families. This effectively requires us to know which is the most conserved region of a sequence alignment, not simply with the data currently at our disposal, but in all future sequences that become available. In some cases (highly conserved enzyme active sites, for example), the choice of motif may be clear-cut. However, for families where there are several conserved regions, how can we know which will remain the most conserved as the sequence databases grow, and new, possibly more divergent, family members become available?

Sometimes, part of an alignment originally used to characterize a particular protein family may change considerably, more so than neighboring regions. For PROSITE, this has three main consequences: (i) if the diagnostic performance of the original regex has fallen off with time, it will eventually require modification to more accurately reflect current family membership; (ii) if new family members are too divergent and cannot simply be captured by modification of the existing expression, either completely new (i.e., taken from different regions of conservation) or additional regexs must be derived; and (iii) if multiple regexs still cannot provide adequate diagnostic performance, it may become necessary to create a profile (Section 9.4).

9.2.3 Rules

Regexs are most effective when used to characterize protein families, where well-conserved motifs (typically 15–25 residues in length), perhaps diagnostic of some core piece of the protein scaffold, are clearly discernible. However, it is also possible to identify much shorter (3–6 residues), generic patterns of

Table 9.2 Example functional sites and the regexs used to detect them.

Functional site	Regex
Glycosaminoglycan attachment site	S-G-x-G
Protein kinase C phosphorylation site	[ST]-x-[RK]
Amidation site	x-G-[RK](2)
Prenyl group binding site (CAAX box)	C-{DENQ}-[LIVM]-x

Table 9.3 Illustration of the effects of introducing tolerance into regexs, and of motif length and database size, on the number of matches retrieved from sequence database searches.

Regex	No. of exact matches (Swiss-Prot 40.26)	No. of exact matches (TrEMBL 21.7)
A-N-N-A	373	2,522
A-N-N-[AV]	743	
A-N-N-[AVL]	1,227	
[AV]-N-N-[AV]	1,684	
[AV]-[ND]-[ND]-[AV]	7,936	
A-N-N	5,734	>10,000

residues within alignments that are not family specific, but are typical of functional sites found across all protein families. Such sequence features are believed to be the result of convergence to a common property: they may denote, for example, lipid or sugar attachment sites, phosphorylation or hydroxylation sites, and so on (see Table 9.2). Because of their often extremely short length, such residue patterns cannot provide discrimination and can only be used to suggest whether a certain type of functional site might exist in a sequence, information which must ultimately be verified by experiment.

The diagnostic limitations of short motifs are readily explained – simply, the shorter the motif (and the greater the size of the database), the greater the chance of making random matches to it. Consider, for example, the sequence motif Ala-Asn-Asn-Ala (ANNA). In Swiss-Prot there were 373 exact matches to this motif (and 2522 matches in its six-fold larger supplement, TrEMBL), and 5734 exact matches to the shorter form Ala-Asn-Asn (ANN) – see Table 9.3. Not surprisingly then, each of the regexs in Table 9.2, which are derived from real biological examples, has more than 10,000 matches in Swiss-Prot (version 40.26), begging extreme caution in their interpretation! Thus, short

motifs are diagnostically unreliable (because they are non-specific), and matches to them, in isolation, are relatively meaningless. In early versions of PROSITE, residue patterns for generic functional sites were termed "rules" to distinguish them from family-specific regexs, although this helpful distinction now seems to have been discarded.

9.2.4 Permissive regexs

As already discussed, rules and family-specific regexs are the basis of PROSITE. The diagnostic problems outlined above led to the inclusion into this database of alternative discriminators (profiles) where regexs failed to provide adequate discrimination (these are discussed in more detail in Section 9.5). Another response to the strict nature of regex pattern matching is to build an element of tolerance or "fuzziness" into the expressions. As we saw in Chapters 2 and 5, one approach is to consider amino acid residues as members of groups, as defined by shared biochemical properties: e.g., FYW are aromatic, HKR are basic, ILVM are aliphatic, and so on. Taking such biochemical properties into account, the motif depicted in Table 9.1 yields the more permissive regex shown in Table 9.4.

Table 9.4 Permissive regex derived using amino acid property groups.

Alignment	Regex
ETDVKMMERVVEQMCVTQY	
ETDVKMMERVVEQMCITQY	
ETDVKIMERVVEQMCITQY	
ETDIKIMERVVEQMCTTQY	
ETDVKMMERVVEQMCVTQY	
ETDVKMMERVVEQMCITQY	
ETDIKMMERVVEQMCITQY	E-x-[EDQN]-x-K-[LIVM](2)-x-[KRH]-[LIVM](2)-x-[EDQN]-M-C-x(2)-Q-Y
ETDIKIMERVVEQMCITQY	
ETDIKMMERVVEQMCITQY	
ETDMKIMERVVEQMCVTQY	
ETDIKMMERVVEQMCITQY	
ETDIKIMERVVEQMCITQY	
ETDMKIMERVVEQMCVTQY	
ETDVKMIERVVEQMCITQY	
EMENKVVTKVIREMCVQQY	
EMENKVVTKVIREMCVQQY	

It is evident from this example that such regexs are more relaxed, accepting a wider range of residues at particular positions. This has the potential advantage of being able to recognize more distant relatives, but has the inherent disadvantage that it will also match many more sequences simply by chance. For example, let us return to the motif ANNA, which had 373 exact matches in Swiss-Prot (Table 9.3). If we introduce one permissive position (say, the final A may belong to the group AV), then we find 743 matches; if we extend the group to include Leu (i.e., AVL), we retrieve 1227 matches; with two permissive positions, the number increases to 1684; and with slight variation at all positions, the number soars to 7936 – searching TrEMBL, the number of matches exceeds 10,000. Clearly, the more tolerant each position within a motif with regard to the types of residue allowed, the more permissive the resulting regex; and the larger the database and shorter the motif (as with regex rules), the worse the situation becomes.

Because a regex represents the minimum expression of an aligned motif, sequence information is lost and parts of it become ill-defined. The more divergent the sequences, the fuzzier the regex, and the more likely the expression is to make false matches.

Results of searching with regexs must therefore be interpreted with care – if a sequence matches an expression, the match may not be biologically meaningful; conversely, if a sequence **fails** to match an expression, it may still be a family member (it may simply deviate from the regex by a single residue). To address some of these limitations, more sophisticated approaches have been devised to improve diagnostic performance and separate out biologically meaningful matches from the sea of noise contained in large databases.

9.3 FINGERPRINTS

9.3.1 Creating fingerprints

Typically, sequence alignments contain not one, but several conserved regions. Diagnostically, it makes sense to use as many of these regions as possible to build a signature for the aligned family, so that, in a database search, there is a higher chance of identifying a distant relative, whether or not all parts of the signature are matched.

The first method to exploit this approach was "fingerprinting" (Attwood, Eliopoulos, and Findlay 1991, Parry-Smith and Attwood 1992, Attwood

Table 9.5 (a) An ungapped motif; and (b) its corresponding frequency matrix (each motif column corresponds to a row in the matrix).

```
YVTVQHKKLRTPL
YVTVQHKKLRTPL
YVTVQHKKLRTPL
AATMKFKKLRHPL
AATMKFKKLRHPL
YIFATTKSLRTPA
VATLRYKKLRQPL
YIFGGTKSLRTPA
WVFSAAKSLRTPS
WIFSTSKSLRTPS
YLFSKTKSLQTPA
YLFTKTKSLQTPA
```
(a)

T	C	A	G	N	S	P	F	L	Y	H	Q	V	K	D	E	I	W	R	M	B	X	Z
0	0	2	0	0	0	0	0	0	7	0	0	1	0	0	0	0	2	0	0	0	0	0
0	0	3	0	0	0	0	0	2	0	0	0	4	0	0	0	3	0	0	0	0	0	0
6	0	0	0	0	0	0	6	0	0	0	0	0	0	0	0	0	0	0	0	0	0	0
1	0	1	1	0	3	0	0	1	0	0	0	3	0	0	0	0	0	0	2	0	0	0
2	0	1	1	0	0	0	0	0	0	0	3	0	4	0	0	0	0	1	0	0	0	0
4	0	1	0	0	1	0	2	0	1	3	0	0	0	0	0	0	0	0	0	0	0	0
0	0	0	0	0	0	0	0	0	0	0	0	0	12	0	0	0	0	0	0	0	0	0
0	0	0	0	0	6	0	0	0	0	0	0	0	6	0	0	0	0	0	0	0	0	0
0	0	0	0	0	0	0	0	12	0	0	0	0	0	0	0	0	0	0	0	0	0	0
0	0	0	0	0	0	0	0	0	0	0	2	0	0	0	0	0	0	10	0	0	0	0
9	0	0	0	0	0	0	0	0	0	2	1	0	0	0	0	0	0	0	0	0	0	0
0	0	0	0	0	0	12	0	0	0	0	0	0	0	0	0	0	0	0	0	0	0	0
0	0	4	0	0	2	0	0	6	0	0	0	0	0	0	0	0	0	0	0	0	0	0

(b)

and Findlay 1993). Here, groups of motifs are excised from alignments, and the sequence information they contain is converted into matrices populated only by the residue frequencies observed at each position of the motifs, as illustrated in Table 9.5. This type of scoring system is said to be **unweighted**, in the sense that no additional scores (e.g., from mutation or substitution matrices) are used to enhance diagnostic performance.

In the example shown in Table 9.5(a), the motif is 13 residues long and 12 sequences deep. The maximum score in the resulting frequency matrix (Table 9.5(b)) is thus 12. Inspection of the matrix indicates that positions 7, 9, and 12 are completely conserved, corresponding to Lys, Leu, and Pro residues, respectively. Any residue not observed in the motif takes no score – the matrix is thus **sparse**, with few positions scoring and most with zero score.

The use of raw residue frequencies is not popular because their scoring potential is relatively limited and, for motifs containing small numbers of similar sequences, the ability to detect distant homologs is compromised because the matrices do not contain sufficient variation. In a protein fingerprint, however, this effect is to some extent offset by the combined use of multiple motifs and iterative database searching, which together increase the diagnostic potential of the final signature.

In practice, then, the motifs are converted to frequency matrices, which are used to scan sequences in a Swiss-Prot/TrEMBL composite database using a sliding-window approach (where the width of the window is the width of the motif). This is our first example of a position-specific scoring system in this chapter. For each window position, an absolute score is calculated and stored in a hit list, rank-ordered with the highest score at the top of the list. During the fingerprint-creation process, in order not to miss true family members, it is usual to capture the top 2000 scores, but more may be culled from the database on request (e.g., if a family contains many thousands of members – a quick rule of thumb is to collect approximately three times the number of hits as there are likely family members).

After the database-scanning and hit-list-creation processes, a hit-list comparison algorithm is used to discover which sequences match all the motifs, in the correct order with appropriate distances between them. It is unusual to include all 2000 hits in this process – normally, about 300 are top-sliced from the lists (but this value can be modified, depending on family size). Sequences that fulfil the true-match criteria are then added to the seed motifs, and the process is repeated until convergence – i.e., the point at which no further fully matching sequences are identified between successive database scans. At convergence, the comparison step is repeated one last time but with substantially fewer hits (often 100 or less, again depending on family size) to ensure that no false matches have been included in the final fingerprint.

9.3.2 The effect of mutation data matrices

As mentioned above, in creating a fingerprint, discriminating power is enhanced by iterative database scanning. The motifs therefore grow and become more mature with each database pass, as more sequences are matched and further residue information is included in the matrices – effectively, this process successively shifts the seed frequencies toward more representative values as the finger-

print incorporates additional family members. For example, after three iterations, the motif shown in Table 9.5(a) has grown to a depth of 73 sequences; as seen in Table 9.6(a), at the end of this process, only position 9 remains conserved, and more variation is observed in the matrix as a whole. Nevertheless, although clearly more densely populated than the initial frequency matrix, this matrix is still relatively sparse: there are still more zero than nonzero positions. In database searches, therefore, this matrix will perform cleanly (with little noise) and with high specificity.

This is to be contrasted with a situation where, for example, a PAM matrix is used to weight the scores, in order to allow more distant relationships to be recognized. The effect of weighting the initial matrix in this way is illustrated in Table 9.6(b). As can be seen, the PAM-weighted result, even though based on the initial sparse matrix, is highly populated. Consequently, in database searches, such a matrix will perform with high levels of noise and relatively low specificity. Thus, a distant relative may well achieve a higher score, but inevitably so will random matches, as residues that are not observed in the initial (or indeed in the **final**) motifs are given significant weights. Compare, for example, the ninth motif position, which even after three database iterations had remained conserved; in the PAM-weighted matrix, this position is no longer completely conserved, four other residues (Phe, Val, Ile, and Met), **which were not observed in 73 true family members**, being assigned large positive scores.

As a result of the poor signal-to-noise performance of weighted matrices, the technique of fingerprinting that underpins the PRINTS database has adhered to the use of residue frequencies. Diagnostic performance is enhanced through the iterative process, but the full potency of the method is gained from the mutual context of matching motif neighbors. This is important, as the method inherently implies a biological context to motifs that are matched in the correct order, with appropriate intervals between them. This allows sequence identification even when parts of a fingerprint are absent. For example, a sequence that matches only

Table 9.6 (a) Frequency matrix derived from the initial motif shown in Table 9.5(a) after three database iterations; and (b) PAM-weighted matrix derived from the same motif.

T	C	A	G	N	S	P	F	**L**	Y	H	Q	V	K	D	E	I	W	R	M
0	0	4	0	0	0	0	8	4	34	0	0	15	0	0	0	1	7	0	0
0	4	15	0	0	0	0	0	7	0	0	0	37	0	0	0	10	0	0	0
50	0	0	0	0	3	0	18	0	0	0	0	0	0	0	0	0	0	0	2
3	0	12	2	1	8	0	3	6	0	0	0	14	0	0	0	15	2	0	7
9	2	2	2	1	1	0	0	0	0	1	25	0	20	0	6	0	0	4	0
14	0	2	0	0	4	0	14	0	8	31	0	0	0	0	0	0	0	0	0
0	0	1	0	0	0	0	0	0	0	0	0	0	70	0	0	0	0	2	0
0	0	2	1	0	17	0	0	0	0	0	0	0	52	0	0	0	0	1	0
0	0	0	0	0	0	0	0	**73**	0	0	0	0	0	0	0	0	0	0	0
0	0	0	0	0	0	0	0	0	0	0	5	0	0	0	0	0	0	68	0
44	0	0	0	0	6	0	0	0	0	12	11	0	0	0	0	0	0	0	0
0	0	1	0	0	0	69	0	0	0	3	0	0	0	0	0	0	0	0	0
2	0	11	0	0	7	0	0	53	0	0	0	0	0	0	0	0	0	0	0

(a)

T	C	A	G	N	S	P	**F**	L	Y	H	Q	**V**	K	D	E	**I**	W	R	**M**
-29	-22	-29	-48	-24	-24	-46	40	-13	62	-10	-40	-22	-38	-44	-44	-15	16	-30	-22
-1	-32	-1	-18	-20	-10	-13	-9	20	-22	-21	-18	32	-23	-22	-20	32	-61	-26	19
0	-36	-18	-30	-24	-12	-30	36	0	24	-18	-36	-6	-30	-36	-30	6	-30	-30	-6
3	-29	3	-4	-10	-1	-7	-22	3	-31	-19	-15	14	-12	-15	-13	11	-52	-15	11
3	-48	-1	-8	7	1	-4	-54	-31	-46	6	14	-17	23	6	5	-20	-48	14	-9
2	-27	-7	-19	-3	-5	-13	0	-16	6	8	-10	-11	-15	-13	-11	-7	-37	-12	-15
0	-60	-12	-24	12	0	-12	-60	-36	-48	0	12	-24	60	0	0	-24	-36	36	0
6	-30	0	-6	12	12	0	-48	-36	-42	-6	0	-18	30	0	0	-18	-30	18	-12
-24	-72	-24	-48	-36	-36	-36	**24**	**72**	-12	-24	-24	**24**	-36	-48	-36	**24**	-24	-36	**48**
-12	-50	-20	-32	2	-2	0	-50	-34	-48	26	18	-24	32	-6	-6	-24	10	62	-2
24	-29	7	-5	5	6	0	-36	-24	-31	6	1	-6	1	4	4	-6	-56	-4	-14
0	-36	12	-12	-12	12	72	-60	-36	-60	0	0	-12	-12	-12	-12	-24	-72	0	-24
-6	-44	-2	-18	-16	-10	-12	-10	22	-24	-18	-14	10	-22	-24	-18	6	-40	-26	16

(b)

five of eight motifs may still be diagnosed as a true match if the motifs are matched in the right order and the distances between them are consistent with those expected of true neighboring motifs. Note, however, that for the purposes of fingerprint creation, only sequences that match all the motifs are admitted into the iterative process and hence are allowed to contribute to the evolving frequency matrices. Partial matches are nevertheless still recorded in the database in order to provide as complete a picture as possible of all fingerprint matches in a given version of Swiss-Prot/TrEMBL.

9.3.3 Searching with fingerprints

Once stored in PRINTS (Attwood *et al.* 2003), fingerprints may be used to diagnose new family members. At this point, it is important to note that the scanning

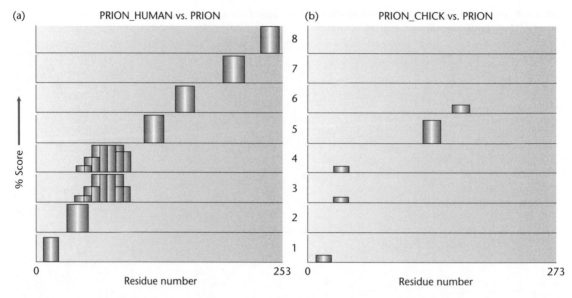

(a) PRION_HUMAN vs. PRION

(b) PRION_CHICK vs. PRION

% Score

Residue number

0   253   0   273

Fig. 9.2 Graphs used to visualize protein fingerprints. The horizontal axis represents the query sequence, the vertical axis the % score of each motif (0–100 per motif), and each block a residue-by-residue match in the sequence, its leading edge marking the first position of the match. Solid blocks appearing in a systematic order along the sequence and above the level of noise indicate matches with the constituent motifs. The graphs depict prion fingerprints of the human prion protein (a) and of its chicken homolog (b). The human prion protein is clearly a true-positive match, containing all eight motifs; the chicken homolog fails to make a complete match, but can still be identified as a family member because of the diagnostic framework provided by the five well-matched motif neighbors.

algorithms used to create fingerprints (Parry-Smith and Attwood 1992) and to search against them once deposited in the database (Scordis, Flower, and Attwood 1999) are rather different. The fingerprint-creation process is deliberately conservative, exploiting only observed residue frequencies to generate clean, highly selective discriminators; this is vital, as inadvertent inclusion of false matches would compromise diagnostic performance. Conversely, to allow a wider diagnostic net to be cast, the fingerprint-search method is more permissive; the onus is then on the user to determine whether a match is likely to be true or not. Here, then, the motifs stored in PRINTS are used to generate (ungapped) profiles, using BLOSUM matrices to boost the matrix scores. The rest of the algorithm is much the same as for fingerprint creation, in so far as motifs are required to be matched in the correct order with appropriate distances between them. However, rather than rely on the rank-ordering of absolute scores in a hit list,

probability (p) and expect (E) values are calculated both for individual-motif matches and for multiple-motif matches, giving an overall estimate of the likelihood of having matched a fingerprint by chance (Scordis, Flower, and Attwood 1999).

Probably the best way of appreciating what it means to match a fingerprint is to depict the result graphically, as illustrated in Fig. 9.2. Within the graphs shown, the horizontal axis represents the query sequence, the vertical axis the percent score of each motif (0–100 per motif). Using a sliding-window approach, for each motif, a window whose width is the width of the motif is moved one residue at a time from N to C terminus, and the profile score calculated for each position. Where a high-scoring match is made above a predefined threshold (17–20% identity), a block is drawn whose leading edge marks the first position of the match. Blocks appearing in a systematic order along the sequence and above the level of noise indicate matches with each of the constituent motifs. In Fig. 9.2, graphs are plotted for (a) a complete match of the human prion sequence against the prion fingerprint, and (b) a partial, but nevertheless true, match of the chicken prion sequence against the same fingerprint. Here, we can see that even though parts of the signature are missing (motifs 2, 7, and 8), the chicken is related to the human sequence. Inspection of the human prion database entry reveals that motifs 2–4 reside in the region of the N-terminal octapeptide repeat. When we compare the chicken sequence, not surprisingly, we find that it is characterized by a modified repeat, which explains why a match to motif 2 is missing, and why the matches to motifs 3 and 4 are relatively weak; the sequence also shows considerable divergence toward the C terminus, which is why motifs 7 and 8 are missing. Thus fingerprints not only allow the diagnosis of true family members that match them completely, but also permit the identification of partially matching relatives, yielding important insights into the differences between them.

9.4 PROFILES AND PSSMS

9.4.1 Blocks

As we have discussed, the scoring system for matching the motifs of a fingerprint uses unweighted residue frequencies and, in situations where few sequences are available, this may compromise diagnostic performance. With the signal-to-noise caveats mentioned in Section 9.3, it is possible to build alternative motif representations by applying different weighting schemes.

One such approach is embodied in the Blocks database (Henikoff *et al.* 2000, Henikoff and Henikoff 1994a). Here, conserved motifs, termed blocks, are located by searching for spaced residue triplets (e.g., Asn-x-x-Glu-x-x-x-x-x-x-x-x-Glu, where x represents any amino acid). Any redundant blocks generated by this process are removed, only the highest-scoring blocks being retained – block scores are calculated using the BLOSUM62 substitution matrix. The validity of blocks found by this method is confirmed by the application of a second motif-finding algorithm, which searches for the highest-scoring set of blocks that occur in the correct order without overlapping. Blocks found by both methods are considered reliable and are stored in the database as ungapped local alignments. These are then calibrated against Swiss-Prot to obtain a measure of the likelihood of a chance match. Two scores are noted for each block: one denotes the score above which a match is likely to be significant (the median true-positive score); the other denotes the score below which a match is probably spurious (the 99.5th percentile of true-negative scores). The ratio of these calibration scores multiplied by 1000 is referred to as the block strength, which allows the diagnostic performance of individual blocks to be meaningfully compared (Henikoff and Henikoff 1991, 1994b).

A typical block is illustrated in Fig. 9.3. Sequence segments within the block are clustered to reduce multiple contributions to residue frequencies from groups of closely related sequences. This is particularly important for deep motifs (i.e., with contributions from tens or hundreds of sequences), which can be dominated by numerous virtually identical sequences. Each cluster is then treated as a single segment, each of which is assigned a score that gives a measure of its relatedness. The higher the weight, the more dissimilar the segment is from other segments in the block; the most distant segment is given a weight of 100.

9.4.2 Searching with blocks

As with fingerprinting, blocks may be used to detect additional family members in sequence databases. In a manner reminiscent of fingerprint searches, blocks are converted to position-specific substitution matrices, which are used to make independent database searches. The results are compared and, where several blocks detect the same sequence, an *E* value is calculated for the multiple hit. Clearly, the more blocks matched, the greater the confidence that the sequence is a family member. However, as with all weighting schemes, there is a diagnostic trade-off between the ability to capture all true family members with a block or set of blocks, and the likelihood of making false matches. Unless a

```
ID     Prion; BLOCK
AC     IPB000817A; distance from previous block=(-16,28)
DE     Prion protein
BL     GPG;  width=42; seqs=90; 99.5%=1888; strength=1666
PRIO_CHICK|P27177  (  29) GKPSGGGWGAGSHRQPSYPRQPGYPHNPGYPHNPGYPHNPGY  36

PRIO_TRIVU|P51780  (  26) KPKPRPGGGWNSGGSNRYPGQPGSPGGNRYPGWGHPQGGGTN 100

PRIO_AOTTR|P40245  (  16) KKRPKPGGWNTGGSRYPGQSSPGGNRYPPQSGGWGQPHGGGW   8
PRIO_ATEPA|P51446  (  23) KKRPKPGGWNTGGSRYPGQGSPGGNRYPPQGGGWGQPHGGGW   6
PRIO_BOVIN|P10279  (  26) KRPKPGGGWNTGGSRYPGQGSPGGNRYPPQGGGGWGQPHGGG   4
PRIO_CALJA|P40247  (  23) KKRPKPGGWNTGGSRYPGQGSPGGNRYPPQGGGWGQPHGGGW   6
PRIO_CAMDR|P79141  (  26) KRPKPGGGWNTGGSRYPGQGSPGGNRYPPQGGGGWGQPHGGG   4
PRIO_CANFA|O46501  (  26) KRPKPGGGWNTGGSRYPGQGSPGGNRYPPQGGGGWGQPHGGG   4
PRIO_CAPHI|P52113  (  26) KRPKPGGGWNTGGSRYPGQGSPGGNRYPPQGGGGWGQPHGGG   4
PRIO_CEBAP|P40249  (  23) KKRPKPGGWNTGGSRYPGQGSPGGNLYPPQGGGWGQPHGGGW   9
PRIO_CEREL|P79142  (  26) KRPKPGGGWNTGGSRYPGQGSPGGNRYPPQGGGGWGQPHGGG   4
PRIO_COLGU|P40251  (  23) KKRPKPGGWNTGGSRYPGQGSPGGNRYPPQGGGWGQPHGGGW   4
PRIO_CRIGR|Q60506  (  23) KKRPKPGGWNTGGSRYPGQGSPGGNRYPPQGGGTWGQPHGGG   4
PRIO_FELCA|O18754  (  26) KRPKPGGGWNTGGSRYPGQGSPGGNRYPPQGGGGWGQPHGGG   4
PRIO_GORGO|P40252  (  23) KKRPKPGGWNTGGSRYPGQGSPGGNRYPPQGGGWGQPHGGGW   4
PRIO_HUMAN|P04156  (  23) KKRPKPGGWNTGGSRYPGQGSPGGNRYPPQGGGWGQPHGGGW   4
PRIO_MACFA|P40254  (  23) KKRPKPGGWNTGGSRYPGQGSPGGNRYPPQGGGWGQPHGGGW   4
PRIO_MANSP|P40255  (  16) KKRPKPGGWNTGGSRYPGQGSPGGNRYPPQGGGWGQPHGGGW   4
PRIO_MESAU|P04273  (  23) KKRPKPGGWNTGGSRYPGQGSPGGNRYPPQGGGTWGQPHGGG   4
PRIO_MOUSE|P04925  (  23) KKRPKPGGWNTGGSRYPGQGSPGGNRYPPQGGTWGQPHGGGW   8
PRIO_MUSPF|P52114  (  26) KRPKPGGGWNTGGSRYPGQGSPGGNRYPPQGGGGWGQPHGGG   4
PRIO_ODOHE|P47852  (  26) KRPKPGGGWNTGGSRYPGQGSPGGNRYPPQGGGGWGQPHGGG   4
PRIO_PANTR|P40253  (  23) KKRPKPGGWNTGGSRYPGQGSPGGNRYPPQGGGWGQPHGGGW   4
PRIO_PIG|P49927    (  26) KRPKPGGGWNTGGSRYPGQGSPGGNRYPPQGGGGWGQPHGGG   4
PRIO_PONPY|P40256  (  23) KKRPKPGGWNTGGSRYPGQGSPGGNRYPPQGGGWGQPHGGGW   4
PRIO_PREFR|P40257  (  23) KKRPKPGGWNTGGSRYPGQGSPGGNRYPPQGGGGWGQPHGGG   4
PRIO_RABIT|Q95211  (  24) KRPKPGGGWNTGGSRYPGQSSPGGNRYPPQGGGWGQPHGGGW   7
PRIO_RAT|P13852    (  23) KKRPKPGGWNTGGSRYPGQGSPGGNRYPPQSGGTWGQPHGGG   6
PRIO_SAISC|P40258  (  23) KKRPKPGGWNTGGSRYPGQGSPGGNRYPPQGGGWGQPHGGGW   6
PRIO_SHEEP|P23907  (  26) KRPKPGGGWNTGGSRYPGQGSPGGNRYPPQGGGGWGQPHGGG   4
PRIO_SIGHI|Q9Z0T3  (  23) KKRPKPGGWNTGGSRYPGQGNPGGNRYPPQGGGTWGQPHGGG   5
PRIO_THEGE|Q95270  (  16) KKRPKPGGWNTGGSRYPGQGSPGGNRYPPQGGGGWGQPHGGG   4
PRP1_TRAST|P40242  (  26) KRPKPGGGWNTGGSRYPGQGSPGGNRYPSQGGGWGQPHGGG   6
PRP2_BOVIN|Q01880  (  26) KRPKPGGGWNTGGSRYPGQGSPGGNRYPPQGGGGWGQPHGGG   4
PRP2_TRAST|P40243  (  26) KRPKPGGGWNTGGSRYPGQGSPGGNRYPPQEGGDWGQPHGGG  11

PRIO_ATEGE|P40246  (  16) KKRPKPGGWNTGGSRYPGQGSPGGNRYPPQGGGWGQPHGGGW   6
PRIO_CALMO|P40248  (  16) KKRPKPGGWNTGGSRYPGQGSPGGNRYPPQGGGSWGQPHGGG   7
PRIO_CERAE|P40250  (  23) KKRPKPGGWNTGGSRYPGQGSPGGNRYPPQGGGGWGQPHGGG   4
PRIO_CERAT|Q95145  (  16) KKRPKPGGWNTGGSRYPGQGSPGGNRYPPQGGGWGQPHGGGW   4
PRIO_CERMO|Q95172  (  16) KKRPKPGGWNTGGSRYPGQGSPGGNRYPPQGGGGWGQPHGGG   4
PRIO_CERPA|Q95174  (  16) KKRPKPGGWNTGGSRYPGQGSPGGNRYPPQGGGGWGQPHGGG   4
PRIO_CERTO|Q95176  (  16) KKRPKPGGWNTGGSRYPGQGSPGGNRYPPQGGGGWGQPHGGG   4
PRIO_CRIMI|Q60468  (  23) KKRPKPGGWNTGGSRYPGQGSPGGNRYPPQGGGTWGQPHGGG   4
PRIO_MUSVI|P40244  (  26) KRPKPGGGWNTGGSRYPGQGSPGGNRYPPQGGGGWGQPHGGG   4
...
//
```

Fig. 9.3 Block for the prion protein family, in which sequence segments are clustered and weighted according to their relatedness – the most distant sequence within the block scores 100 (for convenience, part of the block has been deleted, as denoted by . . .).

family is characterized by only one block, individual block matches are not usually biologically significant; multiple-block matches are much more likely to be real, provided the blocks are matched in the correct order and have appropriate distances between them.

The information content of blocks can be visualized by examination of their so-called sequence logos (Plate 9.1(a)). A logo is a graphical display of an aligned motif, consisting of color-coded stacks of letters representing the amino acid residues at each position in the motif. Letter height increases with increasing frequency of the residue, such that the most conserved positions have the tallest letters. Within stacks, the most frequently occurring residues also occupy the highest positions, so that the most prominent residue, at the top of the stack, is the one most likely to occur at that position. To reduce bias resulting from sequence redundancy within blocks, weights are calculated using a PSSM. This reduces the tendency for over-represented sequences to dominate stacks, and increases the representation of rare amino acid residues relative to common ones. The logo calculated for the prion block depicted in Fig. 9.3 is illustrated in Plate 9.1(a). The highly conserved anchor triplet, GPG, is clearly visible toward the center of the logo.

9.4.3 Profiles

By contrast with motif-based pattern-recognition techniques, an alternative approach is to distil the sequence information within **complete** alignments into scoring tables, or profiles (Gribskov, McLachlan, and Eisenberg 1987, Bucher and Bairoch 1994). A profile can be thought of as an alternating sequ-

ence of "match" and "insert" positions that contain scores reflecting the degree of conservation at each alignment position. The scoring system is intricate: as well as evolutionary weights (e.g., PAM scores), it includes variable penalties to weight against insertions and deletions occurring within core secondary structure elements; match, insertion, and deletion extension penalties; and state transition scores.

Part of a typical profile is illustrated in Fig. 9.4. Position-specific scores for insert and match states are contained in /I and /M fields respectively. These take the form:

```
/DEFAULT: MI=-26; I=-3; IM=0; MD=-26; D=-3; DM=0;
/M: SY='F';M=-2,-3,-3,-4,2,-3,-2,1,-2,0,-1,-2,-3,-3,-4,-2,-1,0,-5,2;
/M: SY='I';M=-1,-5,-2,-3,-2,-3,0,1,1,-1,1,-1,-2,-1,1,-1,0,1,-4,-4;
/M: SY='A';M=2,-3,1,0,-5,2,-2,-1,-1,-3,-2,1,1,0,-2,2,2,0,-8,-5;
/M: SY='L';M=-3,-8,-5,-4,2,-6,-2,2,-4,6,4,-3,-3,-2,-3,-3,-2,1,-3,0;
/M: SY='Y';M=-4,-2,-6,-6,9,-7,0,-1,-5,-1,-3,-3,-6,-5,-6,-4,-4,-4,-1,11;
/M: SY='D';M=1,-6,3,3,-7,0,0,-2,-1,-4,-3,2,0,1,-2,0,0,-2,-9,-6;
/M: SY='Y';M=-5,-3,-6,-6,10,-7,-1,-1,-2,-1,-2,-3,-6,-5,-5,-4,-4,-1,11;
/M: SY='K';M=-1,-6,1,1,-4,-2,0,-2,2,-3,-1,1,-1,1,1,0,0,-3,-7,-6;
/M: SY='A';M=1,-4,1,0,-5,1,-1,-1,0,-3,-1,1,0,0,0,1,1,-1,-7,-6;
/M: SY='R';M=0,-5,0,0,-5,-1,0,-1,1,-3,-1,1,0,1,1,0,0,-2,-5,-5;
/M: SY='R';M=0,-5,1,1,-6,0,1,-2,1,-4,-2,1,0,1,2,1,0,-2,-5,-5;
/M: SY='E';M=1,-6,2,2,-6,0,0,-2,-1,-4,-2,1,1,1,-1,0,0,-3,-8,-6;
/M: SY='D';M=0,-6,2,2,-6,0,1,-3,0,-5,-3,2,-1,2,-1,0,0,-4,-7,-4;
/M: SY='D';M=0,-8,4,3,-6,0,0,-2,-1,-3,-2,2,-2,2,-2,0,-1,-3,-9,-6;
/M: SY='L';M=-2,-8,-5,-5,2,-5,-3,3,-4,7,5,-4,-3,-3,-4,-3,-2,3,-4,-2;
/M: SY='S';M=1,-4,1,1,-5,1,0,-2,1,-4,-2,1,0,0,0,1,1,-2,-6,-5;
/M: SY='F';M=-3,-7,-6,-6,6,-5,-3,3,-2,5,3,-4,-5,-4,-5,-4,-3,1,-3,3;
/M: SY='Q';M=-1,-6,0,0,-3,-2,1,-1,1,-2,0,0,-1,1,1,-1,0,-1,-6,-4;
/M: SY='K';M=-1,-8,0,1,-3,-2,0,-2,3,-3,0,1,0,2,2,0,0,-3,-6,-6;
/M: SY='G';M=2,-5,1,0,-7,7,-3,-4,-2,-6,-4,1,-1,-2,-4,2,0,-2,-10,-8;
/M: SY='D';M=1,-7,5,4,-8,1,1,-3,0,-5,-3,2,-1,2,-2,0,0,-4,-10,-6;
/M: SY='I';M=0,-5,-1,-2,-2,-2,-1,2,0,0,1,-1,-2,0,0,-1,0,1,-6,-5;
/M: SY='L';M=-2,-6,-5,-5,3,-5,-3,4,-3,6,4,-4,-4,-3,-4,-3,-2,3,-5,0;
/M: SY='Q';M=-1,-5,-1,-1,-3,-2,0,0,0,-2,-1,0,-1,0,0,-1,0,-1,-6,-3;
/M: SY='V';M=0,-4,-3,-4,-1,-3,-3,5,-3,3,3,-2,-2,-2,-3,-2,0,5,-8,-4;
/M: SY='L';M=-1,-6,-3,-3,-1,-3,-2,2,-3,3,2,-2,-2,-2,-3,-2,-1,2,-5,-3;
/M: SY='D';M=0,-6,3,3,-6,0,1,-3,2,-5,-2,2,-1,2,1,0,0,-4,-7,-5;
/M: SY='K';M=-1,-6,0,0,-2,-1,0,-3,3,-4,-1,1,-1,0,1,0,0,-3,-6,-4;
/M: SY='N';M=1,-4,1,1,-5,0,0,-2,0,-3,-2,1,1,0,-1,1,1,-1,-7,-5;
   /I: MI=0; I=-1; MD=0; /M: SY='X'; M=0; D=-1;
/M: SY='G';M=1,-5,0,0,-5,1,-2,-1,-2,-3,-2,0,0,-1,-2,0,0,-1,-8,-6;
/M: SY='G';M=-1,-6,3,3,-7,3,0,-4,-1,-5,-4,2,-1,1,0,-1,0,-3,-10,-6;
/M: SY='W';M=-9,-12,-9,-11,1,-11,-4,-8,-5,-3,-6,-6,-8,-7,3,-4,-8,-9,26,0;
/M: SY='W';M=-7,-9,-9,-9,0,-9,-4,-5,-5,-1,-4,-6,-7,-6,2,-3,-6,-6,18,-1;
/M: SY='K';M=-1,-7,0,0,-3,-2,0,-2,3,-1,1,1,2,0,-1,-3,-5,-5;
/M: SY='G';M=2,-3,0,-1,-6,3,-3,-2,-3,-4,-3,0,0,-2,-3,1,0,0,-10,-6;
/M: SY='Q';M=-2,-6,0,0,-3,-3,1,-2,0,-2,-1,0,-2,1,1,-1,-1,-3,-5,-3;
   /I: MI=0; I=-2; MD=0; /M: SY='X'; M=0; D=-2;
/M: SY='T';M=0,-4,-1,-1,-4,0,-2,0,-1,-2,0,0,-1,-1,-1,0,1,0,-7,-5;
/M: SY='T';M=0,-5,0,0,-3,-1,-1,-1,-1,1,-3,-1,1,-1,0,0,1,1,-1,-6,-4;
/M: SY='G';M=0,-5,0,-1,-5,3,-2,-3,-1,-5,-3,0,-1,-1,-1,1,0,-2,-7,-6;
/M: SY='K';M=0,-6,1,1,-5,-1,1,-2,2,-4,-1,1,-1,2,2,0,0,-3,-6,-6;
/M: SY='R';M=-1,-6,-1,-1,-5,-3,1,-1,1,-3,-1,0,-1,1,3,-1,-1,-2,-2,-6;
/M: SY='G';M=1,-5,0,0,-6,6,-3,-3,-3,-5,-4,0,-1,-2,-4,1,0,-2,-10,-6;
/M: SY='W';M=-5,-5,-5,-5,2,-6,-2,-2,-4,-1,-3,-3,-6,-5,-3,-3,-4,-4,4,3;
/M: SY='F';M=-3,-5,-6,-6,6,-5,-3,4,-1,3,2,-4,-4,-5,-4,-3,-2,2,-4,3;
/M: SY='P';M=2,-4,-1,-1,-7,-1,0,-3,-2,-4,-3,-1,8,0,0,1,0,-2,-8,-7;
/M: SY='G';M=1,-3,0,0,-4,2,-1,-2,0,-3,-2,0,-1,-1,1,1,-1,-1,-6,-5;
/M: SY='N';M=1,-5,2,1,-5,0,1,-2,1,-4,-2,2,0,0,0,1,1,-2,-7,-4;
/M: SY='Y';M=-5,-1,-7,-7,10,-8,-1,-1,-5,-1,-3,-3,-7,-6,-6,-4,-4,-5,0,13;
/M: SY='V';M=0,-3,-3,-5,-2,-2,-3,5,-3,2,2,-2,-2,-3,-4,-1,0,5,-8,-5;
/M: SY='E';M=1,-6,2,3,-6,0,0,-2,1,-4,-2,1,0,2,0,0,0,-3,-8,-6;
/M: SY='P';M=0,-5,-1,-1,-2,-2,-1,-2,-1,-3,-2,0,1,-1,-2,0,-1,-2,-6,-3;
```

Fig. 9.4 Example PROSITE profile, showing position-specific scores for insert and match positions. Penalties within insert positions are highlighted bold: here, the values are more tolerant of indels by comparison with the large overall penalties set by the DEFAULT parameter line.

```
/I: [SY=char1; parameters;]
/M: [SY=char2; parameters;]
```

where `char1` is a symbol that represents an insert position in the parent alignment; `char2` is a symbol that represents a match position in the parent alignment; and `parameters` is a list of specifications assigning values to various position-specific scores (these include initiation and termination scores, state transition scores, insertion/match/deletion extension scores, and so on).

In the example shown in Fig. 9.4, the sequence can be read off from the successive /M positions: F-I-A-L-Y-D, etc. It is evident that the profile contains three conserved blocks separated by two gapped regions. Within the conserved blocks, small insertions and deletions are not totally forbidden, but are strongly impeded by large gap penalties defined in the DEFAULT data block: `MI=-26`, `I=-3`, `MD=-26`, `D=-3` (`MI` is a match-insert transition score, `I` is an insert extension score, `MD` is a match-delete transition score, and `D` is a deletion extension score). These penalties are superseded by more permissive values in the two gapped regions (e.g., in the first of these, `MI=0`, `I=-1`, `MD=0`, etc.).

The inherent complexity of profiles renders them highly potent discriminators. They are therefore used to complement some of the poorer regexs in PROSITE, and/or to provide a diagnostic alternative where extreme sequence divergence renders the use of regexs inappropriate. Before finishing this section on profile methods, we note that PSI-BLAST, which we already discussed in Chapter 7, can also be classed as a profile method, as it builds up a PSSM based on the first BLAST search, and uses this to perform further searches.

9.5 BIOLOGICAL APPLICATIONS – G PROTEIN-COUPLED RECEPTORS

9.5.1 What are G protein-coupled receptors?

The best way to try to understand the range of pattern-recognition methods described in this chapter is to consider how they have been applied to real biological examples. To this end, the following sections concern the analysis of G protein-coupled receptors, comparing and contrasting the results of using pairwise and family-based search tools, and examining the biological insights that each of these approaches affords.

G protein-coupled receptors (GPCRs) constitute a large, functionally diverse and evolutionarily triumphant group of cell-surface proteins (Teller *et al.* 2001). They are involved in an incredible range of physiological processes, adapting a common structural framework to mediate, for example, vision, olfaction, chemotaxis, stimulation and regulation of mitosis, and the opportunistic entry of viruses into cells (Lefkowitz 2000, Hall, Premont, and Lefkowitz 1999). This functional diversity is achieved via interactions with a wide variety of ligands (including peptides, glycoproteins, small molecule messenger molecules (such as adrenaline), and vitamin derivatives (such as retinal)), the effects of which stimulate a range of intracellular pathways through coupling to different guanine nucleotide-binding (G) proteins.

9.5.2 Where did GPCRs come from?

How did this extraordinary functional versatility arise? The GPCRs we see in modern organisms are part of a vast multi-gene family, which is thought to have diverged from an ancient ancestral membrane protein. Details of the origins of GPCRs are still contentious, but clues are perhaps to be found in the bacterial opsins. The bacterial opsins are retinal-binding proteins that provide light-dependent ion transport and sensory functions to a family of halophilic bacteria. They are integral membrane proteins that contain seven transmembrane (TM) domains, the last of which contains the chromophore attachment point. There are several classes of this bacterial protein, including bacteriorhodopsin and archaerhodopsin, which are light-driven proton pumps; halorhodopsin, a light-driven chloride pump; and sensory rhodopsin, which mediates both photoattractant (in the red) and photophobic (in the UV) responses.

From an evolutionary perspective, two features of bacterial opsins are particularly interesting. First, the protein has a 7TM architecture (bacteriorhodopsin was the first membrane protein whose structure was determined at atomic resolution (Henderson *et al.* 1990, Pebay-Peyroula *et al.* 1997)); second, its response to light is mediated by retinal (a vitamin A derivative), which is bound to the protein via a lysine residue in the seventh TM domain. There are striking parallels here with the protein rhodopsin, a GPCR that plays a central role in light reception in higher organisms. For a long time, the architecture of rhodopsin was believed to be like that of bacteriorhodopsin, with 7TM helices snaking back and forth across the membrane (NB, GPCRs are sometimes referred to as "serpentine" receptors). Indeed, so great was the confidence in their structural similarity that almost all of the early models of the structure of rhodopsin were based on the bacteriorhodopsin template. The crystal structure of bovine rhodopsin was solved relatively recently (Palczewski *et al.* 2000), and confirmed both the 7TM architecture and the location of its bound chromophore, retinal, which is attached via a Schiff's base to a lysine residue in the seventh TM domain.

In spite of these striking similarities, however, there are also significant differences. First, bacteriorhodopsin does not couple to G proteins. Second, there is no significant sequence similarity between rhodopsin and bacteriorhodopsin, beyond the general hydrophobic nature of the TM domains and the retinal-attachment site. Third, the rhodopsin crystal structure highlighted important differences in the packing of its helices relative to that seen in bacteriorhodopsin, rendering many of the early homology models virtually worthless. In light of the obvious similarities and differences between them, it therefore remains unclear whether the 7TM architecture was recruited by the GPCRs from the bacterial opsins and adapted to fulfil many different functional roles, or whether nature invented the 7TM framework more than once to confer transport and sensory functions on its host cells – without significant sequence similarity, we can neither confirm homology nor rule out analogy. Consequently, the optimistic researchers who modeled rhodopsin by **homology** with bacteriorhodopsin learned some painful lessons about the dangers of overinterpreting data and about the importance of understanding the evolutionary origins of the molecules we are investigating.

9.5.3 GPCR orthologs and paralogs

If the ancient origins of GPCRs are unclear, their more recent history is somewhat better understood (partly because many GPCRs share a high degree of sequence similarity and partly because, today, we have an abundance of sequence information on which to perform comparative or phylogenetic analyses). GPCRs have successfully propagated throughout the course of evolution both by speciation events (orthology) and via gene-duplication events within the same organism (paralogy). It is beyond the scope of this chapter to provide a detailed discussion of all GPCRs, but we will focus for a moment on one of the families in order to highlight the challenges that biology presents to the bioinformatician. We will do this by taking a closer look at rhodopsin, because it was one of the first GPCRs to be studied in detail and the largest GPCR superfamily is named after it – the rhodopsin-like GPCRs.

As already mentioned, rhodopsin is a light receptor. It resides in the disk membranes of retinal rod cells and functions optimally in dim light. Animals that depend on good night vision (cats, for example, which hunt by night) have large numbers of rhodopsin-harboring rods in their retinas. By contrast, the visual machinery of some animals is optimized for bright light; the retinas of these animals are packed with greater numbers of cone cells – these contain the blue-, red-, and green-pigment proteins termed opsins. Opsins and rhodopsins share ~40% sequence identity and are believed to have evolved from an ancestral visual pigment via a series of gene-duplication events. The duplication that gave rise to the distinct red- and green-sensitive pigment genes is considered to be relatively recent because the DNA sequences of the red and green genes share ~98% sequence identity, suggesting that there has been little time for change. This observation finds support in comparisons of the distribution of visual genes in new- and old-world monkeys: new-world monkeys have only a single visual-pigment gene on the X chromosome; by contrast, old-world monkeys have two X-chromosome visual pigment genes (similarly, in humans, the red- and green-sensitive genes lie on the X chromosome). Thus, the addition of the second X-chromosome gene must have occurred some time after the separation of the new and old world monkeys, around 40 million years ago (Nathans 1989).

The location of the red- and green-pigment genes on the X chromosome has long been known to underpin red–green color blindness. Males will have variant red–green discrimination if their single X chromosome carries the trait; females will be affected only if they receive a variant X chromosome from both parents. In fact, males normally have a single copy of the red-pigment gene but one, two, or three copies of the green-pigment gene. Different types of abnormal red–green color discrimination then arise from genetic exchanges that result in the loss of a pigment gene or the creation of part-red/part-green hybrid genes (Nathans 1989).

So, to recap, an ancestral visual pigment has given rise to the rhodopsin family. In vertebrates, the family includes an achromatic receptor (rhodopsin) and four chromatic paralogs (the purple, blue, red, and green opsins). In humans, the gene for the green receptor may occur in different copy numbers – distorted color perception occurs when the numbers and types of these genes present in an individual are aberrant. The opsins and rhodopsin share a high degree of sequence similarity, with greatest similarity between the red and green pigment proteins; and the proteins are also highly similar to orthologs in different species. If we are interested in sequence analysis, this seems like a fairly straightforward scenario. However, the family holds further layers of complexity and its analysis requires careful thought. Although, as we might expect, most rhodopsins have a typical "rhodopsin-like" sequence and most opsins have typical "pigment-like" sequences, there are exceptions – thus, for example, green opsins of the goldfish, chicken, and blind cave fish more closely resemble the canonical rhodopsin sequence than they do other green sequences. Likewise, blue opsins of the gecko and green chameleon are more similar to rhodopsin sequences than to other blue sequences. These unusual sequence relationships

are illustrated in Plate 9.1(b) and (c). Unraveling this kind of relationship, where a protein with particular biological attributes has the "wrong" sequence, using naïve search tools can therefore lead to erroneous conclusions. Nevertheless, it is precisely this kind of anomaly that helps to shed light on the evolutionary ancestry of protein families: in this example, the results suggest, perhaps surprisingly, that the gene for achromatic (scotopic) vision (rhodopsin) evolved out of those for color (photopic) vision (opsins) (Okano et al. 1992).

9.5.4 Why GPCRs are interesting and how bioinformatics can help to study them

As we have seen, the long evolutionary history of GPCRs has generated considerable complexity for just one of its gene families. Now consider that there are over 800 GPCR genes in the human genome. These fall into three major superfamilies (termed rhodopsin-, secretin-, and metabotropic glutamate receptor-like), which are populated by more than 50 families and 350 receptor subtypes (Takeda et al. 2002, Foord 2002, Attwood 2001); members of the different superfamilies share almost no recognizable sequence similarity, but are united by the familiar 7TM architecture. The receptors are activated by different ligands, some of which are bound in the external loop regions, others being bound within the TM scaffold – the opsins are unusual in being covalently attached to their retinal chromophore. The activated receptors then interact with different G proteins or other intracellular molecules to effect their diverse biological responses. The extent of the family, the richness of their evolutionary relationships, and the complexity of their interactions with other molecules are not only challenging to understand, but are also very difficult to analyze computationally. If we wish to derive deeper insights into the ways in which genomes encode biological functions, our traditional pairwise similarity search tools are likely to be limited. These tools have been the mainstays of genome annotation endeavors because they reduce a complex problem to a more tractable one – that of identifying and quantifying relationships between sequences. But identifying relationships between sequences is clearly not the same as identifying their functions, and failure to appreciate this fundamental point has generated numerous annotation errors in our databases.

Interest in GPCRs reaches beyond mere academic intrigue and the desire to place them into functionally related families. The receptors are also of profound medical importance, being involved in several major pathophysiological conditions, including cardiovascular disease, cancer, and pathological irregularities in body weight homeostasis. The pivotal role of GPCRs in regulating crucial cellular processes, and the diseases that arise when these processes go wrong, has placed them firmly in the pharmaceutical spotlight, to such an extent that, today, they are the targets of the majority of prescription drugs: more than 50% of marketed drugs act at GPCRs (including a quarter of the 100 top-selling drugs), yielding sales in excess of US$16 billion per annum (Flower 1999). GPCRs are therefore "big business".

Nevertheless, there is still only one crystal structure available to date (their membranous location renders GPCRs largely intractable to conventional crystallographic techniques). Moreover, the huge numbers of receptors have defied our ability to characterize them all experimentally. Consequently, many remain "orphans", with unknown ligand specificity. In the absence of experimental data, bioinformatics approaches are therefore often used to help identify, characterize, and model novel receptors (Gaulton and Attwood 2003, Attwood and Flower 2002): e.g., if a receptor is found to share significant sequence similarity with one of known function, it may be possible to extrapolate certain functional and structural characteristics to the uncharacterized protein. Bioinformatics approaches are therefore important in the ongoing analysis of GPCRs, offering complementary tools with which to shed light on the structures and functions of novel receptors, and the hope of discovering new drug targets.

9.5.5 Detecting sequence similarity

We have already seen in Chapter 7 that the fastest and most frequently used method for identifying sets of sequences related to a query is to search sequence databases using pairwise alignment tools such as BLAST and FASTA (Altschul, Gish, and Miller 1990, Altschul *et al.* 1997, Pearson and Lipman 1988, Pearson 2000). Several novel GPCRs have been identified in this way as a result of their similarity to known receptors (Takeda *et al.* 2002, Lee *et al.* 2001, Radford, Davies, and Dow 2002, Hewes and Taghert 2001, Zozulya, Echeverri, and Nguyen 2001).

However, pairwise alignment/search tools have various inherent problems, not least that different programs often return different results (to be rigorous, BLAST and FASTA should be used together, as they give different perspectives on the same query, often yielding different top hits). These programs are consequently limited when it comes to the more exacting tasks of identifying the precise families to which receptors belong and the ligands they bind. The problem is, it is difficult to determine at what level of sequence identity ligand specificity will be conserved: some GPCRs, with as little as 25% identity, share a common ligand (e.g., the histamine receptors); others, with greater levels of identity, do not (e.g., receptors for melanocortin, lysophosphatidic acid, and sphingosine 1-phosphate). Thus, it is often impossible to tell from a BLAST result whether an orphan receptor is likely to be a member of a known family, to which it is a top hit, or whether it will bind a novel ligand. Consider, for example, the recently de-orphaned urotensin II receptor: BLAST indicates that it is most similar to the type 4 somatostatin receptors, yet it is now known to bind a different (though related) ligand (Fig. 9.5). Thus, it is clearly a dangerous (though all too common) assumption that the top most statistically significant hit is also the most **biologically** relevant.

9.5.6 Diagnosing family membership

Earlier in this chapter, we discussed how search methods based on alignments of protein families have evolved to address some of the problems of pairwise approaches, and these have given rise to a range of family databases, as outlined in Chapter 5. The principle underlying such methods is that by aligning representative family members, it is

```
Query length: 389 AA
Date run: 2002-10-18 09:08:29 UTC+0100 on sib-blast.unil.ch
Taxon: Homo sapiens
Database: XXswissprot
             120,412 sequences; 45,523,583 total letters
             SWISS-PROT Release 40.29 of 10-Oct-2002

Db      AC       Description                                                  Score E-value

sp      Q9UKP6   UR2R_HUMAN Urotensin II receptor (UR-II-R) [GPR14] [Ho...     782   0.0
sp      P31391   SSR4_HUMAN Somatostatin receptor type 4 (SS4R) [SSTR4]...     167   3e-41
sp      O43603   GALS_HUMAN Galanin receptor type 2 (GAL2-R) (GALR2) [G...     147   4e-35
sp      P30872   SSR1_HUMAN Somatostatin receptor type 1 (SS1R) (SRIF-2...     144   3e-34
sp      P32745   SSR3_HUMAN Somatostatin receptor type 3 (SS3R) (SSR-28...     140   3e-33
sp      P35346   SSR5_HUMAN Somatostatin receptor type 3 (SS5R) (SSTR5)...     140   6e-33
sp_vs   P30874-01 SPLICE ISOFORM B of P30874  [SSTR2] [Homo sapiens...        134   3e-31
sp      P30874   SSR2_HUMAN Somatostatin receptor type 2 (SS2R) (SRIF-1...     134   3e-31
sp      P48145   GPR7_HUMAN Neuropeptides B/W receptor type 1 (G protei...     133   7e-31
sp      O60755   GALT_HUMAN Galanin receptor type 3 (GAL3-R) (GALR3) [G...     132   2e-30
sp      P41143   OPRD_HUMAN Delta-type opioid receptor (DOR-1) [OPRD1] ...     128   2e-29
sp_vs   P35372-01 SPLICE ISOFORM 1A of P35372  [OPRM1] [Homo sapien...        125   1e-28
sp      P35372   OPRM_HUMAN Mu-type opioid receptor (MOR-1) [OPRM1] [Ho...     125   1e-28
```

Fig. 9.5 BLAST output from a search of Swiss-Prot (release 40.29) with the human urotensin II receptor sequence (UR2R_HUMAN, Q9UKP6). Note, there is no clear cut-off between the urotensin, somatostatin, and galanin receptor matches.

possible to identify conserved regions (whether short "motifs" or complete domains) that are likely to have critical structural or functional roles. These regions may then be used to build diagnostic signatures, with enhanced selectivity and sensitivity, allowing new family members to be detected more reliably.

As we have seen, the family-based approaches differ both in terms of the extent of the alignment they use and in the methods used to encode conserved regions (recall Fig. 5.6). The main techniques use: (i) single conserved regions encoded as regexs (as in PROSITE (Falquet et al. 2002) and eMOTIF (Huang and Brutlag 2001)); (ii) multiple motifs encoded as fingerprints (as in PRINTS (Attwood et al. 2003)) or sets of blocks (in Blocks (Henikoff et al. 2000)); or (iii) complete domains encoded as profiles (Gribskov et al. 1987, Bucher and Bairoch 1994) (as in PROSITE (Falquet et al. 2002)) or HMMs (Krogh et al. 1994, Eddy 1998) (as in Pfam (Bateman et al. 2002)).

These different approaches have different diagnostic strengths and weaknesses, which must be understood when applying them. For example, regexs have to be matched exactly, which often leads to high error rates: many true relationships are missed because sequences deviate slightly from the expression, and many false matches are made because the patterns are short and non-selective. With fingerprints, although individual motifs may be relatively short, greater selectivity is achieved by exploiting the mutual context of motif neighbors – false-positive matches can then be more easily distinguished from true family members, as they usually fail to match most of the motifs within a given fingerprint. However, as no weighting scheme is used to derive fingerprints, those that encode superfamilies may be too selective and consequently may miss true family members. By contrast, the scoring methods used in the derivation of Blocks exploit weighted PSSMs, rather than the frequency matrices typical of fingerprinting – this may have the effect not only of boosting weak biological signals, but also of upweighting spurious matches. Moreover, blocks are automatically derived and their numbers and diagnostic potential often differ from equivalent fingerprints, often causing confusion among users. On the other hand, both profiles and HMMs perform well in the diagnosis of distant family members of large divergent superfamilies but, unlike fingerprints, rarely offer the means to diagnose sequences at the level of families or subfamilies (they may also be prone to profile dilution if they have been derived by purely automatic means).

To simplify sequence analysis, many of these databases have been integrated into InterPro (Mulder *et al.* 2003), which amalgamates their family documentation into a coherent whole, and allows multiple repositories to be searched simultaneously. In the following section, we will consider how InterPro and its constituent databases can be used to shed light on features of GPCR sequences. In particular, we draw attention to the different perspectives offered by their search tools, and discuss the importance this is likely to have when diagnosing novel sequences and attempting to make predictions about likely ligand specificity or G protein coupling.

9.5.7 Analysis of GPCRs

The databases described above all provide signatures for GPCRs. PROSITE contains regexs for the three major superfamilies. The expressions are relatively short, and cover N-terminal regions or parts of the TM domains. The database also contains a single family-level expression, for the opsins. The error rate associated with these signatures is high. The expression-encoding TM domain 3 of the rhodopsin-like superfamily, for example, fails to detect more than 100 known superfamily members, including the human P2Y6, UDP-glucose, NPY5, and PAR2 receptors. These sequences are annotated as false-negatives, but many additional sequences, such as the KiSS-1 receptor and the orphan SALPR (somatostatin and angiotensin-like peptide receptor), fail to match and are not recorded in the documentation. The actual error rate is therefore considerably higher than it first appears (Gaulton and Attwood 2003).

PROSITE also contains profiles encoding the 7TM regions of each of the GPCR superfamilies; these perform better, with few or no false-positives and false-negatives recorded. Similarly, Pfam contains 7TM-encoding HMMs for each of the superfamilies, which perform well and detect distant family members missed by regexs. Both databases also provide

discriminators for the N-terminal domain of the secretin receptor-like superfamily.

These signatures are useful diagnostic tools that can help to identify putative GPCRs, and many have been used to search recently sequenced genomes (Brody and Cravchik 2000, Remm and Sonnhammer 2000). However, as there is only a single signature at the family level (the PROSITE expression for opsin), searches with this sort of signature offer virtually no opportunity to identify the family to which an orphan receptor might belong, to predict the type of ligand it might bind, or to predict the type of G protein to which it might couple. In trying to characterize a putative GPCR, it is not enough to say that the sequence might be a rhodopsin-like receptor, because this does not tell us of which of the possible ~50 families it is a member (and hence its ligand-binding specificity), and which of the family subtypes it most closely resembles (and hence its likely G protein-coupling preference).

PRINTS contains the most extensive collection of GPCR signatures of any family database, with more than 270 fingerprints (Attwood, Croning, and Gaulton 2002). The nature of the analysis method it uses means that highly selective fingerprints can be created in a hierarchical manner for receptor superfamilies, families, and subtypes. Therefore, searches of this resource may not only identify the superfamily to which a receptor belongs, but may also elucidate its family and subtype relationships (see Fig. 9.6).

The ability of fingerprints to distinguish between closely related receptor families and subtypes is illustrated in Fig. 9.7. Here, the sequence of the human urotensin II receptor was searched against (a) its own fingerprint and (b) the somatostatin receptor family fingerprint (Gaulton and Attwood 2003). Recall Fig. 9.5, in which the BLAST search could not clearly distinguish the urotensin II receptor from the somatostatin receptors. Here, by contrast, the PRINTS result clearly highlights the differences between them – the receptor perfectly matches its own fingerprint (each motif matched is indicated by a block on the graph), but fails to match any of the motifs that characterize the somatostatin receptors. This is important. The result reflects the fact that although the urotensin II and somatostatin receptors

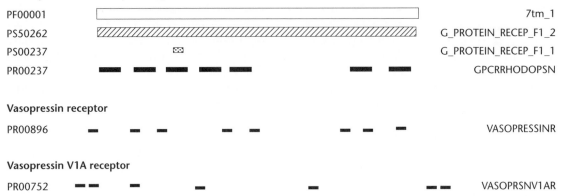

Fig. 9.6 Output from a search of InterPro with the human vasopressin 1A receptor sequence (V1AR_HUMAN). The receptor matches the Pfam HMM (PF00001 – white bar), PRINTS fingerprint (PR00237 – black bars), and PROSITE regular expression (PS00237 – spotted bar) and profile (PS50262 – striped bar) for the rhodopsin-like superfamily of GPCRs. However, only PRINTS gives family- and subtype-level diagnoses, with matches to the vasopressin receptor family (PR00896) and vasopressin V1A receptor subtype (PR00752) fingerprints.

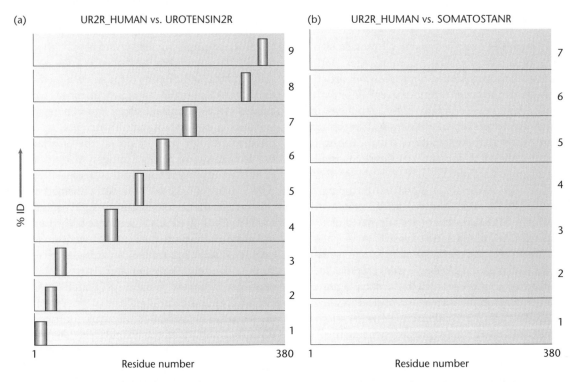

Fig. 9.7 Output from a search of the human urotensin II receptor sequence against (a) its own fingerprint and (b) the somatostatin receptor family fingerprint using PRINTS' GRAPHScan tool (Scordis, Flower, and Attwood 1999). Within each graph, the horizontal axis represents the sequence, and the vertical axis the percentage score (identity) of each fingerprint element (0–100 per motif). Filled blocks mark the positions of motif matches above a 20% threshold. Here, we see that the receptor matches all nine motifs of its own fingerprint but fails to make any significant matches to the somatostatin receptor fingerprint.

share a high degree of overall similarity (this is what BLAST "sees"), they differ in key regions that are likely to determine their functional specificity (i.e., what fingerprints "see"). Thus, BLAST, FASTA, and superfamily-level discriminators operate at a generic similarity level, while fingerprints provide much more fine-grained insights based on both the similarities and the **differences** between closely related sequences.

Several other methods have also been developed for classifying GPCRs, including those that use support vector machines, phylogenetic analysis, and analysis of receptor chemical properties. These will not be discussed further here, but interested readers are referred to Joost and Methner (2002), Karchin, Karplus, and Haussler (2002), and Lapinsh *et al.* (2002).

9.5.8 *The functional significance of signatures*

Although most protein family signatures are derived solely on the basis of amino acid conservation, without explicit consideration of function, it is evident that the conserved regions they exploit do correspond to structural and functional motifs. For example, at the superfamily level, all GPCRs share the same architecture; consequently, the superfamily signatures primarily encode the 7TM domains (which makes sense, because this is what they all have in common). If we think about individual families within a particular GPCR superfamily, however, what are the defining characteristics? The 7TM architecture is the same, isn't it, so what makes an opsin sequence different from an olfactory receptor sequence, and different again from a muscarinic receptor sequence, and so on? If the structure is the same, we can reasonably suppose that the difference between them must lie in the way the different receptors function – i.e., in the ligands they bind. Detailed analyses of the fingerprints that define the different GPCR families suggest that this is the case: motifs of family-level fingerprints are often found in parts of the TM domains and extracellular portions of the receptors, reflecting the regions in which they bind a common ligand. But what of individual receptor subtypes? These not only share the same struc-

ture, but also bind the same ligand as their sibling subtypes, so what makes them different? In this case, we conclude that the defining characteristics are in their G protein-coupling preferences (whether Gi, Go, Gq, etc.). Once again, examination of subtype fingerprints indicates that the constituent motifs are more prevalent in intracellular regions, particularly the third intracellular loop and C-terminal regions, most likely to be involved in G protein coupling.

Reassuringly, in the case of family-level fingerprints, experimental evidence is beginning to confirm their functional significance: residues shown by mutational or molecular modeling studies to be involved in ligand binding often fall within the motifs that characterize the family. In addition, recent modeling studies have predicted 3D structures for the sphingosine 1-phosphate (S1P) receptors EDG1 and EDG6, and identified residues likely to be involved in binding the ligand (Parrill *et al.* 2000, Wang *et al.* 2001, Vaidehi *et al.* 2002). Figure 9.8

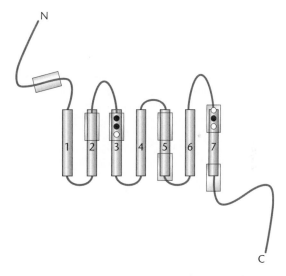

Fig. 9.8 Schematic diagram representing the endothelial differentiation gene (EDG) family of sphingosine 1-phosphate receptors. Positions of the fingerprint motifs for this family are indicated by rectangles; circles mark the positions of residues known to be important in ligand binding in the EDG1 receptor (black) (Remm and Sonnhammer 2000, Attwood *et al.* 2002) and in the EDG6 receptor (white) (Joost and Methner 2002). Models of these receptors predict that the ligand binds within the TM regions, close to domains 2, 3, 5, and 7.

shows a schematic diagram of an S1P receptor, with the positions of these critical residues indicated by asterisks, and motifs of the family fingerprint indicated by shaded boxes; all of the ligand-binding residues identified to date lie within two of the motifs (Gaulton and Attwood 2003). Moreover, the remaining motifs lie in regions that appear to be close to the ligand-binding site (TM domains 2, 3, 5, and 7), or in regions that may be involved in G protein coupling, reflecting the overlapping coupling specificity of receptors in this family. These results highlight the fact that, in addition to facilitating classification of orphan receptors, fingerprints may also provide vital information about residues and regions of receptors that are likely to have crucial functional roles.

SUMMARY

Search tools based on pairwise alignment (BLAST and FASTA) have become the methods of choice for many researchers – they are quick and easy to use. To quantify the similarity between matched sequences, the methods use different scoring schemes (PAM or BLOSUM), but matched residue pairs all carry equal weight along the sequence according to the values in the chosen scoring matrix. In searches for highly divergent sequences, such tools may lack sensitivity, failing to retrieve true matches from the "twilight zone". Conversely, in searches for closely related sequences, they may lack selectivity – it may be difficult to determine, merely on the basis of an E value, whether a matched sequence is a true functional equivalent (the distinction between orthologs and paralogs may not be clear).

To improve the sensitivity and selectivity of database searches, family-based methods were developed. The innovation was to use information from multiply aligned sequences to increase the diagnostic potential of searches. The methods still largely use scoring matrices to quantify residue matches, but benefit from the additional residue information stored in each column of the aligned sequences. This is important, because the alignment tells us which residues are the most conserved – the residues therefore carry position-specific weights, according to the degree of conservation of each alignment column. The use of multiple sequences therefore allows PSSMs to be developed to capture the evolutionary essence of a given protein family.

Different methods have been developed to encode the conserved information in sequence alignments, using this both to characterize protein families and to diagnose new family members. The methods exploit single motifs, multiple motifs, or complete domains.

Single-motif methods often use regexs to encapsulate the conserved residues or residue groups from the most conserved region of an alignment, but these tend to lack diagnostic power. Family-specific regexs were first used to create PROSITE. The database also includes non-family-specific rules, very short regexs that describe functional sites, such as glycosylation sites, phosphorylation sites, etc. Rules are not diagnostic – they only indicate that certain residues exist in a sequence, but not that they are functional. To enhance the diagnostic performance of regexs, degrees of variability may be built into expressions by tolerating substitutions according to physicochemically based residue groupings. Such regexs are the basis of eMOTIF.

Multiple-motif methods exploit the fact that most alignments contain more than one conserved region, and were first used in the fingerprint method that underpins PRINTS. Motifs are converted to (position-specific) residue frequency matrices, but no additional weighting scheme is used to enhance the scores. The method gains diagnostic potency from the biological context afforded by matching motif neighbors, so a sequence can still be diagnosed as a likely family member even if parts of the signature are absent. However, fingerprints may lack diagnostic power if derived from protein families with only few members. To overcome this problem, weighting schemes were introduced, leading to the concept of blocks, the basis of the Blocks database. The potential disadvantage of weighted blocks is that the scoring scheme may introduce noise. PRINTS is also available in Blocks format, in which PRINTS' raw motifs are converted to weighted blocks; PRINTS' motifs are also used to form permissive regexs as part of the eMOTIF resource. The fingerprint approach therefore underpins many different diagnostic procedures. A further strength of the method is that it is not confined to the diagnosis of members of divergent superfamilies: because the method is selective, it has also been used to distinguish families from their constituent superfamilies, and subfamilies from their parent families. No other method offers such a hierarchical discriminatory framework.

Profile methods encode both conserved and gapped regions, and exploit position-specific scores and substitution matrices to enhance diagnostic performance. In seeking divergent members of large superfamilies, they tend to be diagnostically more potent than regexs, fingerprints, and blocks; consequently, many are included in PROSITE. In deriving profiles, care must be taken not to include false family members, as this will lead to profile dilution.

These software tools and databases have been used in the computational analysis of many different protein families, and the GPCRs provide a good illustration of the relative merits of each. The family-based methods are particularly valuable in helping to characterize novel receptors and, in some cases, may allow identification of functional motifs responsible for binding ligands or G proteins.

Nevertheless, in studying complex gene families, it is important to appreciate the limitations of the particular computational methods being applied. No method is infallible and none can replace the need for biological experimentation and validation. Best results are obtained by using all the available resources and understanding the problems associated with each. Regexs are good at detecting short functional motifs. PROSITE offers a regex for the rhodopsin-like GPCR superfamily and another for the retinal attachment site of opsins. Profiles and HMMs are good at detecting generic similarities between divergent members of large superfamilies. Superfamily-level profiles and HMMs are available from PROSITE and Pfam. Fingerprints are good at distinguishing members of functionally related families and subfamilies. PRINTS offers over 270 fingerprints for the rhodopsin-like GPCR families and receptor subtypes; it thus offers more fine-grained diagnoses, providing the opportunity not only to determine the superfamily to which a GPCR might belong, but also to identify the particular family of which it is a member, and the specific receptor subtype it most resembles, hence offering insights into the ligand a receptor is likely to bind and the G protein(s) to which it might couple.

While experimental validation still lags behind the number of available sequences, together, pairwise and family-based search methods may therefore help to begin the process of characterizing novel proteins and to direct these much-needed future functional experiments.

REFERENCES

Altschul, S.F., Gish, W., and Miller, W. 1990. Basic local alignment search tool. *Journal of Molecular Biology*, **215**: 403–10.

Altschul, S.F., Madden, T.L., Schaffer, A.A., Zhang, J., Zhang, Z., Miller, W., and Lipman, D.J. 1997. Gapped BLAST and PSI-BLAST: A new generation of protein database search programs. *Nucleic Acids Research*, **25**: 3389–402.

Attwood, T.K. 2000a. The quest to deduce protein function from sequence: The role of pattern databases. *International Journal of Biochemistry and Cell Biology*, **32**(2): 139–55.

Attwood, T.K. 2000b. The role of pattern databases in sequence analysis. *Briefings in Bioinformatics*, **1**: 45–59.

Attwood, T.K. 2001. A compendium of specific motifs for diagnosing GPCR subtypes. *Trends in Pharmacological Sciences*, **22**(4): 162–5.

Attwood, T.K., Blythe, M.J., Flower, D.R., Gaulton, A., Mabey, J.E., Maudling, N., McGregor, L., Mitchell, A.L., Moulton, G., Paine, K., and Scordis, P. 2003. PRINTS and PRINTS-S shed light on protein ancestry. *Nucleic Acids Research*, **30**: 239–41.

Attwood, T.K., Croning, M.D., and Gaulton, A. 2002. Deriving structural and functional insights from a ligand-based hierarchical classification of G protein-coupled receptors. *Protein Engineering*, **15**: 7–12.

Attwood, T.K., Eliopoulos, E.E., and Findlay, J.B.C. 1991. Multiple sequence alignment of protein families showing low sequence homology: A methodological approach using pattern-matching discriminators for G-protein linked receptors. *Gene*, **98**: 153–9.

Attwood, T.K. and Findlay, J.B.C. 1993. Design of a discriminating fingerprint for G-protein-coupled receptors. *Protein Engineering*, **6**(2): 167–76.

Attwood, T.K. and Flower, D.R. 2002. Trawling the genome for G protein-coupled receptors: The importance of integrating bioinformatic approaches. In Darren R. Flower (ed.), *Drug Design – Cutting Edge Approaches*, pp.60–71. London: Royal Society of Chemistry.

Bateman, A., Birney, E., Cerruti, L., Durbin, R., Etwiller, L., Eddy, S.R., Griffiths-Jones, S., Howe, K.L., Marshall, M., and Sonnhammer, E.L. 2002. The Pfam protein families database. *Nucleic Acids Research*, **30**: 276–80.

Benson, D.A., Karsch-Mizrachi, I., Lipman, D.J., Ostell, J., and Wheeler, D.L. 2004. GenBank: Update. *Nucleic Acids Research*, **32**: D23–6.

Brody, T. and Cravchik, A. 2000. *Drosophila melanogaster* G protein-coupled receptors. *Journal of Cell Biology*, **150**: F83–8.

Bucher, P. and Bairoch, A. 1994. A generalized profile syntax for biomolecular sequence motifs and its function in automatic sequence interpretation. *Proceedings of the International Conference on Intelligent Systems for Molecular Biology*, **2**: 53–61.

Dayhoff, M.O. 1965. *Atlas of Protein Sequence and Structure*. Silver Spring, MD: National Biomedical Research Foundation.

Eck, R.V. and Dayhoff, M.O. 1966. *Atlas of Protein Sequence and Structure 1966*. Silver Spring, MD: National Biomedical Research Foundation.

Eddy, S.R. 1998. Profile hidden Markov models. *Bioinformatics*, **14**: 755–63.

Falquet, L., Pagni, M., Bucher, P., Hulo, N., Sigrist, C.J., Hofmann, K., and Bairoch, A. 2002. The PROSITE database, its status in 2002. *Nucleic Acids Research*, **30**: 235–8.

Flower, D.R. 1999. Modelling G-Protein-Coupled Receptors for Drug Design. *Biochimica et Biophysica Acta*, **1422**: 207–34.

Foord, S.M. 2002. Receptor classification: Post genome. *Current Opinion in Pharmacology*, **2**: 561–6.

Gaulton, A. and Attwood, T.K. 2003. Bioinformatics approaches for the classification of G protein-coupled receptors. *Current Opinion in Pharmacology*, **3**: 114–20.

Gribskov, M., McLachlan, A.D., and Eisenberg, D. 1987. Profile analysis: Detection of distantly related proteins. *Proceedings of the National Academy of Sciences USA*, **84**: 4355–8.

Henderson, R., Baldwin, J.M., Ceska, T.A., Zemlin, F., Beckmann, E., and Downing, K.H. 1990. Model for the structure of bacteriorhodopsin based on high-resolution electron cryo-microscopy. *Journal of Molecular Biology*, **213**: 899–929.

Henikoff, S. and Henikoff, J.G. 1991. Automated assembly of protein blocks for database searching. *Nucleic Acids Research*, **19**: 6565–72.

Henikoff, S. and Henikoff, J.G. 1992. Amino acid substitution matrix from protein blocks. *Proceedings of the National Academy of Sciences USA*, **89**: 10915–19.

Henikoff, S. and Henikoff, J.G. 1994a. Position-based sequence weights. *Journal of Molecular Biology*, **243**(4): 574–8.

Henikoff, S. and Henikoff, J.G. 1994b. Protein family classification based on searching a database of blocks. *Genomics*, **19**(1): 97–107.

Henikoff, J.G., Greene, E.A., Pietrokovski, S., and Henikoff, S. 2000. Increased coverage of protein families with the blocks database servers. *Nucleic Acids Research*, **28**: 228–30.

Hewes, R.S. and Taghert, P.H. 2001. Neuropeptides and neuropeptide receptors in the *Drosophila melanogaster* genome. *Genome Research*, **11**: 1126–42.

Hall, R.A., Premont, R.T., and Lefkowitz, R.J. 1999. Heptahelical receptor signaling: Beyond the G protein paradigm. *Journal of Cell Biology*, **145**: 927–32.

Huang, J.Y. and Brutlag, D.L. 2001. The eMOTIF database. *Nucleic Acids Research*, **29**: 202–4.

Huynen, M.A. and Bork, P. 1998. Measuring genome evolution. *Proceedings of the National Academy of Sciences USA*, **95**: 5849–56.

Joost, P. and Methner, A. 2002. Phylogenetic analysis of 277 human G-protein-coupled receptors as a tool for the prediction of orphan receptor ligands. *Genome Biology*, **3**: research0063.1–16.

Karchin, R., Karplus, K., and Haussler, D. 2002. Classifying G-protein-coupled receptors with support vector machines. *Bioinformatics*, **18**: 147–59.

Krogh, A., Brown, M., Mian, I.S., Sjolander, K., and Haussler, D. 1994. Hidden Markov models in computational biology. Applications to protein modeling. *Journal of Molecular Biology*, **235**: 1501–31.

Lapinsh, M., Gutcaits, A., Prusis, P., Post, C., Lundstedt, T., and Wikberg, J.E.S. 2002. Classification of G-protein coupled receptors by alignment-independent extraction of principal chemical properties of primary amino acid sequences. *Protein Science*, **11**: 795–805.

Lee, D.K., Nguyen, T., Lynch, K.R., Cheng, R., Vanti, W.B., Arkhitko, O., Lewis, T., Evans, J.F., George, S.R., and O'Dowd, B.F. 2001. Discovery and mapping of ten novel G protein-coupled receptor genes. *Gene*, **275**: 83–91.

Lefkowitz, R.J. 2000. The superfamily of heptahelical receptors. *Nature Cell Biology*, **2**: E133–6.

Mulder, N.J., Apweiler, R., Attwood, T.K., Bairoch, A., Bateman, A., Binns, D., Biswas, M., Bradley, P., Bork, P., Bucher, P., *et al.* 2002. InterPro: An integrated documentation resource for protein families, domains and functional sites. *Briefings in Bioinformatics*, **3**: 225–35.

Nathans, J. 1989. The genes for color vision. *Scientific American*, **260**(2): 28–35.

Okano, T., Kojima, D., Fukada, Y., Shichida, Y., and Yoshizawa, T. 1992. Primary structures of chicken cone visual pigments: Vertebrate rhodopsins have evolved out of cone visual pigments. *Proceedings of the National Academy of Sciences USA*, **89**: 5932–6.

Palczewski, K., Kumasaka, T., Hori, T., Behnke, C.A., Motoshima, H., Fox, B.A., Le Trong, I., Teller, D.C., Okada, T., Stenkamp, R.E., *et al.* 2000. Crystal structure of rhodopsin: A G protein-coupled receptor. *Science*, **289**: 739–45.

Parrill, A.L., Wang, D., Bautista, D.L., Van Brocklyn, J.R., Lorincz, Z., Fischer, D.J., Baker, D.L., Liliom, K., Spiegel, S., and Tigyi, G. 2000. Identification of EDG1 receptor residues that recognize sphingosine 1-phosphate. *Journal of Biological Chemistry*, **275**: 39379–84.

Parry-Smith, D.J. and Attwood, T.K. 1992. ADSP – A new package for computational sequence analysis. *Computer Applications in the Biosciences*, **8**(5): 451–9.

Pearson, W.R. 2000. Flexible sequence similarity searching with the FASTA3 program package. *Methods in Molecular Biology*, **132**: 185–219.

Pearson, W.R. and Lipman, D.J. 1988. Improved tools for biological sequence comparison. *Proceedings of the National Academy of Sciences USA*, **85**: 2444–8.

Pebay-Peyroula, E., Rummel, G., Rosenbusch, J.P., and Landau, E.M. 1997. X-ray structure of bacteriorhodopsin at 2.5-A from microcrystals grown in lipidic cubic phases. *Science*, **277**: 1676–81.

Radford, J.C., Davies, S.A., and Dow, J.A. 2002. Systematic G-protein-coupled receptor analysis in *Drosophila melanogaster* identifies a leucokinin receptor with novel roles. *Journal of Biological Chemistry*, **277**: 38810–17.

Remm, M. and Sonnhammer, E. 2000. Classification of transmembrane protein families in the *Caenorhabditis elegans* genome and identification of human orthologs. *Genome Research*, **10**: 1679–89.

Scordis, P., Flower, D.R., and Attwood, T.K. 1999. FingerPRINTScan: Intelligent searching of the PRINTS motif database. *Bioinformatics*, **15**: 523–4.

Takeda, S., Kadowaki, S., Haga, S., Takaesu, H., and Mitaku, S. 2002. Identification of G protein-coupled receptor genes from the human genome sequence. *FEBS Letters*, **520**: 97–101.

Teller, D.C., Okada, T., Behnke, C.A., Palczewski, K., and Stenkamp, R.E. 2001. Advances in determination of a high-resolution three-dimensional structure of rhodopsin, a model of G protein-coupled receptors (GPCRs). *Biochemistry*, **40**: 7761–72.

Vaidehi, N., Floriano, W.B., Trabanino, R., Hall, S.E., Freddolino, P., Choi, E.J., Zamanakos, G., and Goddard, W.A. III. 2002. Prediction of structure and function of G protein-coupled receptors. *Proceedings of the National Academy of Sciences USA*, **99**: 12622–7.

Wang, D.A., Lorincz, Z., Bautista, D.L., Liliom, K., Tigyi, G., and Parrill, A.L. 2001. A single amino acid determines lysophospholipid specificity of the $S1P_1$ (EDG1) and LPA_1 (EDG2) phospholipid growth factor receptors. *Journal of Biological Chemistry*, **276**: 49213–20.

Zozulya, S., Echeverri, F., and Nguyen, T. 2001. The human olfactory receptor repertoire. *Genome Biology*, **2**: research0018.1–12.

PROBLEMS

9.1 Discuss the diagnostic implications of the following PROSITE entry:

```
ID RCC1_2; PATTERN.
AC PS00626;
DT JUN-1992 (CREATED); DEC-1992 (DATA UPDATE); JUL-1998 (INFO UPDATE).
DE Regulator of chromosome condensation (RCC1) signature 2.
PA [LIVMFA]-[STAGC](2)-G-x(2)-H-[STAGLI]-[LIVMFA]-x-[LIVM].
NR /RELEASE=38,80000;
NR /TOTAL=68; /POSITIVE=10; /UNKNOWN=0; /FALSE_POS=58;
NR /FALSE_NEG=0; /PARTIAL=0;
CC /TAXO-RANGE=??E??;
DR P31386, ATS1_YEAST, T; P52499, RCC_CANAL , T; P21827, RCC_YEAST , T;
DR P28745, RCC_SCHPO , T; P25171, RCC_DROME , T; P18754, RCC_HUMAN , T;
DR P23800, RCC_MESAU , T; Q92834, RPGR_HUMAN, T; Q15034, Y032_HUMAN, T;
DR P25183, RCC_XENLA , T;
DR Q06077, ABRB_ABRPR, F; O26952, ADEC_METTH, F; P27875, AGS_AGRRA , F;
DR Q10334, ALAT_SCHPO, F; P08691, ARB1_ECOLI, F; P33447, ATTY_TRYCR, F;
```

```
DR  P40968, CC40_YEAST, F; P29742, CLH_DROME  , F; P52105, CSGE_ECOLI, F;
DR  P95830, DNAJ_STRPN, F; O46629, ECHB_BOVIN, F; P55084, ECHB_HUMAN, F;
DR  Q60587, ECHB_RAT   , F; P56096, FTSW_HELPY, F; P13006, GOX_ASPNG , F;
DR  P09789, GRP1_PETHY, F; P08111, L2GL_DROME, F; Q08470, L2GL_DROPS, F;
DR  P12234, MPCP_BOVIN, F; P40614, MPCP_CAEEL, F; O61703, MPCP_CHOFU, F;
DR  Q00325, MPCP_HUMAN, F; P16036, MPCP_RAT  , F; Q56111, OMS2_SALTI, F;
DR  P70682, PGH2_CAVPO, F; P35336, PGLR_ACTCH, F; Q39766, PGLR_GOSBA, F;
DR  Q39786, PGLR_GOSHI, F; P05117, PGLR_LYCES, F; P26216, PGLR_MAIZE, F;
DR  P48978, PGLR_MALDO, F; P24548, PGLR_OENOR, F; Q02096, PGLR_PERAE, F;
DR  P48979, PGLR_PRUPE, F; Q05967, PGLR_TOBAC, F; P35338, PGLS_MAIZE, F;
DR  P35339, PGLT_MAIZE, F; P26580, POLG_HPAV2, F; P26581, POLG_HPAV4, F;
DR  P26582, POLG_HPAV8, F; P06442, POLG_HPAVC, F; P08617, POLG_HPAVH, F;
DR  P06441, POLG_HPAVL, F; P13901, POLG_HPAVM, F; P14553, POLG_HPAVS, F;
DR  P31788, POLG_HPAVT, F; Q26734, PROF_TRYBB, F; P77889, PYRE_LACPL, F;
DR  Q60028, RBL2_THIDE, F; P23255, T2D2_YEAST, F; P38129, T2D4_YEAST, F;
DR  Q50709, Y09A_MYCTU, F; Q50622, Y0B2_MYCTU, F; P37969, Y246_MYCLE, F;
DR  P38352, YB9W_YEAST, F; P77433, YKGG_ECOLI, F; P45026, YRBA_HAEIN, F;
DR  P33989, YRPM_ACICA, F;
3D  1A12;
DO  PDOC00544;
//
```

This entry has been updated since it was first deposited into PROSITE. Has the situation changed? If so, in what way? What other methods are available for sequence analysis and how might their diagnostic performances differ from the above example?

9.2 A database curator has run some standard database searches on an unannotated EMBL sequence to create the following Swiss-Prot entry:

```
ID  APE_HUMAN
AC  P01130
DT  1-APR-2000 (Rel. 40, Created)
DE  APOLIPOPROTEIN E PRECURSOR (APO-E) (FRAGMENT)
GN  APOE
OS  Homo sapiens (Human)
OC  Eukaryota; Metazoa; Chordata; Craniata; Vertebrata; Euteleostomi;
OC  Mammalia; Eutheria; Primates; Catarrhini; Hominidae; Homo
CC  -!- FUNCTION: APO-E MEDIATES BINDING, INTERNALIZATION and CATABOLISM
CC  OF LIPOPROTEIN PARTICLES. IT CAN SERVE AS A LIGAND FOR THE
CC  SPECIFIC APO-E RECEPTOR (REMNANT) OF HEPATIC TISSUES
CC  -!- SUBCELLULAR LOCATION: EXTRACELLULAR
CC  -!- PTM: N-GLYCOSYLATED
CC  -!- PTM: PKC- AND CK2-PHOSPHORYLATED
CC  -!- PTM: MYRISTOYLATED
CC  -!- SIMILARITY: CONTAINS 1 LY DOMAIN
CC  -!- SIMILARITY: CONTAINS 1 EGF-LIKE DOMAIN
DR  EMBL; L29401; -; NOT_ANNOTATED_CDS. [EMBL / GenBank / DDBJ]
DR  PIR; A01383; QRHULD
DR  PFAM; PF00008; EGF
DR  PFAM; PF00058; LY DOMAIN
DR  PROSITE; PS00001; ASN_GLYCOSYLATION
```

```
DR PROSITE; PS00005; PKC_PHOSPHO_SITE
DR PROSITE; PS00006; CK2_PHOSPHO_SITE
DR PROSITE; PS00008; MYRISTYL
KW Glycoprotein; LY domain; EGF-like domain
FT DOMAIN 10 75 LY DOMAIN
FT DOMAIN 76 115 EGF-LIKE
FT CARBOHYD 75 75 POTENTIAL
SQ SEQUENCE 120 AA; 95375 MW
  DEKRLAHPFS LAVFEDKVFW TDIINEAIFS ANRLTGSDVN LLAENLLSPE DMVLFHNLTQ
  PRGVNWCERT TLSNGGCQYL CLPAPQINPH SPKFTCACPD GMLLARDMRS CLTEAEAAVA
// 
```

On the basis of results from searching PROSITE and Pfam, the curator has annotated the protein as possessing an LY domain and an EGF domain, together with glycosylation and phosphorylation sites. How reliable is the result? How or why do you think annotations of this sort arise? What additional steps could you take to confirm the various different aspects of the annotation?

This entry has been updated since it was first deposited in Swiss-Prot. Has the situation changed? If so, in what way?

9.3

(a)

```
ID OPSU_BRARE STANDARD; PRT; 354 AA.
DE ULTRAVIOLET-SENSITIVE OPSIN (ULTRAVIOLET CONE PHOTORECEPTOR PIGMENT)
DE (ZFO2).
RN [1]
RP SEQUENCE FROM N.A.
RC TISSUE=EYE;
RX MEDLINE; 93317613.
RA ROBINSON J., SCHMITT E.A., HAROSI F.I., REECE R.J., DOWLING J.E.;
RT "Zebrafish ultraviolet visual pigment: absorption spectrum, sequence,
RT and localization.";
RL Proc. Natl. Acad. Sci. U.S.A. 90:6009-6012(1993).
CC -!- FUNCTION: VISUAL PIGMENTS ARE THE LIGHT-ABSORBING MOLECULES THAT
CC MEDIATE VISION. THEY CONSIST OF AN APOPROTEIN, OPSIN, COVALENTLY
CC LINKED TO CIS-RETINAL.
CC -!- SUBCELLULAR LOCATION: INTEGRAL MEMBRANE PROTEIN.
CC -!- TISSUE SPECIFICITY: THE COLOR PIGMENTS ARE FOUND IN THE CONE
CC PHOTORECEPTOR CELLS.
CC -!- PTM: SOME OR ALL OF THE CARBOXYL-TERMINAL SER OR THR RESIDUES MAY
CC BE PHOSPHORYLATED (BY SIMILARITY).
CC -!- MISCELLANEOUS: THIS OPSIN HAS AN ABSORPTION MAXIMA AT 360 NM.
CC -!- SIMILARITY: BELONGS TO FAMILY 1 OF G-PROTEIN COUPLED RECEPTORS.
CC OPSIN SUBFAMILY.
DR EMBL; L11014; AAA85566.1; -.
DR PIR; A48191; A48191.
DR GCRDB; GCR_0754; -.
DR ZFIN; ZDB-GENE-990415-271; ZFO2.
DR PFAM; PF00001; 7tm_1; 1.
DR PROSITE; PS00237; G_PROTEIN_RECEPTOR; 1.
DR PROSITE; PS00238; OPSIN; 1.
KW Photoreceptor; Retinal protein; Transmembrane; Glycoprotein; Vision;
KW Phosphorylation; G-protein coupled receptor.
```

(b)

```
ID  OPSB_CONCO  STANDARD;  PRT;  350 AA.
DE  BLUE-SENSITIVE OPSIN (BLUE CONE PHOTORECEPTOR PIGMENT).
RN  [1]
RP  SEQUENCE FROM N.A.
RX  MEDLINE; 96288978.
RA  ARCHER S., HIRANO J.;
RT  "Absorbance spectra and molecular structure of the blue-sensitive rod
RT  visual pigment in the conger eel (Conger conger).";
RL  Proc. R. Soc. Lond., B, Biol. Sci. 263:761-767(1996).
CC  -!- FUNCTION: VISUAL PIGMENTS ARE THE LIGHT-ABSORBING MOLECULES THAT
CC  MEDIATE VISION. THEY CONSIST OF AN APOPROTEIN, OPSIN, COVALENTLY
CC  LINKED TO CIS-RETINAL.
CC  -!- SUBCELLULAR LOCATION: INTEGRAL MEMBRANE PROTEIN.
CC  -!- TISSUE SPECIFICITY: ROD SHAPED PHOTORECEPTOR CELLS WHICH MEDIATES
CC  VISION IN DIM LIGHT.
CC  -!- PTM: SOME OR ALL OF THE CARBOXYL-TERMINAL SER OR THR RESIDUES MAY
CC  BE PHOSPHORYLATED.
CC  -!- MISCELLANEOUS: THIS OPSIN HAS AN ABSORPTION MAXIMA AROUND 487 NM.
CC  -!- SIMILARITY: BELONGS TO FAMILY 1 OF G-PROTEIN COUPLED RECEPTORS.
CC  OPSIN SUBFAMILY.
DR  EMBL; S82619; AAB37721.1; -.
DR  GCRDB; GCR_1872; -.
DR  PFAM; PF00001; 7tm_1; 1.
DR  PROSITE; PS00237; G_PROTEIN_RECEPTOR; 1.
DR  PROSITE; PS00238; OPSIN; 1.
KW  Photoreceptor; Retinal protein; Transmembrane; Glycoprotein; Vision;
KW  Phosphorylation; G-protein coupled receptor.
```

(a) and (b) show excerpts from Swiss-Prot entries OPSB_BRARE and OPSB_CONCO, UV- and blue-sensitive opsins for zebra fish and conger eel, respectively. Compare the annotations for the two sequences. Do you note any differences? If so, what are they? Do you think the annotations are accurate? If not, why not? How might you verify their accuracy? How do you think the annotations were derived? Discuss the implications for this and other database entries. Are the annotations the same in Swiss-Prot today?

9.4 Examine the following excerpt from Swiss-Prot entry OPSG_CARAU, a visual receptor from the goldfish.

```
ID  OPSG_CARAU  STANDARD;  PRT;  349 AA.
AC  P32311;
DT  01-OCT-1993 (Rel. 27, Created)
DT  01-OCT-1993 (Rel. 27, Last sequence update)
DT  15-MAR-2004 (Rel. 43, Last annotation update)
DE  Green-sensitive opsin 1 (Green cone photoreceptor pigment 1).
OS  Carassius auratus (Goldfish).
RN  [1]
RP  SEQUENCE FROM N.A.
RX  MEDLINE=93120096; PubMed=8418840; [NCBI, ExPASy, EBI, Israel, Japan]
RA  Johnson R.L., Grant K.B., Zankel T.C., Boehm M.F., Merbs S.L.,
RA  Nathans J., Nakanishi K.;
RT  "Cloning and expression of goldfish opsin sequences.";
RL  Biochemistry 32:208-214(1993).
```

```
CC -!- FUNCTION: Visual pigments are the light-absorbing molecules that
CC mediate vision. They consist of an apoprotein, opsin, covalently
CC linked to cis-retinal.
CC -!- SUBCELLULAR LOCATION: Integral membrane protein.
CC -!- TISSUE SPECIFICITY: The color pigments are found in the cone
CC photoreceptor cells.
CC -!- PTM: Phosphorylated on some or all of the serine and threonine
CC residues present in the C-terminal region.
CC -!- MISCELLANEOUS: Maximal absorption at 509 nm.
CC -!- SIMILARITY: Belongs to family 1 of G-protein coupled receptors.
CC Opsin subfamily.
```

Extract the sequence from Swiss-Prot. With your knowledge of primary and secondary databases, run InterPro and PRINTS searches (i.e., use InterProScan and FingerPRINTScan). Once you have obtained the PRINTS result, examine the graphical option. By way of comparison, use PRINTS' GraphScan tool to scan the sequence against the OPSINREDGRN fingerprint.

What conclusions do you draw from these search results? How do you reconcile them with the Swiss-Prot annotation? Is the annotation wrong (clue – examine the Swiss-Prot entry for chicken green-sensitive opsin, OPSG_CHICK)? What implications might your conclusions have for automatic sequence analysis software, such as that used to annotate TrEMBL entries?

9.5 Examine the TrEMBL entry for the *C. elegans* sequence below.

```
ID Q23293 PRELIMINARY; PRT; 320 AA.
AC Q23293;
DT 01-NOV-1996 (TrEMBLrel. 01, Created)
DT 01-JUN-2000 (TrEMBLrel. 14, Last annotation update)
DE SIMILARITY TO MAMAMALIAN MU- AND KAPPA-TYPE OPIOID RECEPTORS.
GN ZC404.11.
OS Caenorhabditis elegans.
OC Eukaryota; Metazoa; Nematoda; Chromadorea; Rhabditida; Rhabditoidea;
OC Rhabditidae; Peloderinae; Caenorhabditis.
RN [1]
RP SEQUENCE FROM N.A.
RC STRAIN=BRISTOL N2;
RX MEDLINE=94150718 [NCBI, ExPASy, Israel, Japan]; PubMed=7906398;
RA Watson A., Weinstock L., Wilkinson-Sproat J., Wohldman P.;
RT "2.2 Mb of contiguous nucleotide sequence from chromosome III of C.
RT elegans.";
RL Nature 368:32-38(1994).
RN [2]
RP SEQUENCE FROM N.A.
RC STRAIN=BRISTOL N2;
RA Bentley D., Le T.T.;
RL Submitted (APR-1996) to the EMBL/GenBank/DDBJ databases.
RN [3]
RP SEQUENCE FROM N.A.
RC STRAIN=BRISTOL N2;
RA Waterston R.;
RL Submitted (APR-1996) to the EMBL/GenBank/DDBJ databases.
DR EMBL; U55363; AAA97969.1; -. [EMBL / GenBank / DDBJ] [CoDingSequence]
```

```
DR INTERPRO; IPR000276; -.
DR INTERPRO; IPR000515; -.
DR PFAM; PF00001; 7tm_1; 1.
DR PROSITE; PS00402; BPD_TRANSP_INN_MEMBR; UNKNOWN_1.
DR PRODOM [Domain structure / List of seq. sharing at least 1 domain]
DR PROTOMAP; Q23293.
DR PRESAGE; Q23293.
SQ SEQUENCE 320 AA; 36357 MW; 534DDA2EA94B1AC2 CRC64;
   MASQYNCTQF LNEFPEPKTE AAINNGIVSF MDALVKAYRP FHYYVLTFIV IFAFFANILI
   LILLTRKEMR YSGVNVTMML IAVCDLGCAI AGLSQLYLRN FSDNYSSYQT AYAQFIVYYC
   QIAFHAESLY LAVGMAFCRV ITLSSASDSR VYADGSVFLD ISPLSLANDC LFLKTSIFFS
   GSCFKILPSI LMSMFSIIIL IRIKAGKQRS NSLSHNQTDQ DAQIDRSTRF IQVVVVVFVI
   TETPQGFFSV LGSISINDYI NYYQHLSIFM NILAFFNTTT SFIIYSTLSS KFRKLFAQLF
   VPDVILEKYG RVFQKLELSF
//
```

The sequence is annotated as being similar to the mammalian μ- and κ-type opioid receptors, and includes a PROSITE signature characteristic of inner membrane binding-protein-dependent transport systems. How reliable do you think the annotation is? With your knowledge of primary and secondary databases, run BLAST, FASTA, and InterPro searches, and examine PROSITE entry PS00402 (BPD_TRANSP_INN_MEMBR).

Now find Swiss-Prot entries OPRM_HUMAN and OPRD_HUMAN. Search the sequences against the PRINTS database and examine the graphical output; repeat the search for Q23393. In light of your results, is function assignment on the basis of sequence similarity alone safe? Why? Is more information required? If so, of what sort? What factors might influence the reliability of database annotations? What could be done to improve the reliability of database annotations?

9.6 Examine the Swiss_Prot entry for the human cytomegalovirus sequence below:

```
ID UL78_HCMVA STANDARD; PRT; 431 AA.
AC P16751;
DT 01-AUG-1990 (Rel. 15, Created)
DT 01-AUG-1990 (Rel. 15, Last sequence update)
DT 16-OCT-2001 (Rel. 40, Last annotation update)
DE Hypothetical protein UL78.
GN UL78.
OS Human cytomegalovirus (strain AD169).
OX NCBI_TaxID=10360;
RN [1]
RP SEQUENCE FROM N.A.
RX MEDLINE=90269039; PubMed=2161319; [NCBI, ExPASy, EBI, Israel, Japan]
RA Chee M.S., Bankier A.T., Beck S., Bohni R., Brown C.M., Cerny R.,
RA Horsnell T., Hutchison C.A. III, Kouzarides T., Martignetti J.A.,
RA Preddie E., Satchwell S.C., Tomlinson P., Weston K.M., Barrell B.G.;
RT "Analysis of the protein-coding content of the sequence of human
RT cytomegalovirus strain AD169.";
RL Curr. Top. Microbiol. Immunol. 154:125-169(1990).
CC -!- SUBCELLULAR LOCATION: Integral membrane protein (Potential).
DR EMBL; X17403; CAA35351.1; -. [EMBL / GenBank / DDBJ] [CoDingSequence]
DR PIR; S09841; S09841.
DR InterPro; IPR000276; GPCR_Rhodpsn.
DR InterPro; Graphical view of domain structure.
DR PROSITE; PS50262; G_PROTEIN_RECEP_F1_2; UNKNOWN_1.
```

```
DR ProDom [Domain structure / List of seq. sharing at least 1 domain]
DR BLOCKS; P16751.
KW Hypothetical protein; Transmembrane.
FT TRANSMEM 42 62 Potential.
FT TRANSMEM 74 94 Potential.
FT TRANSMEM 111 131 Potential.
FT TRANSMEM 153 173 Potential.
FT TRANSMEM 202 222 Potential.
FT TRANSMEM 236 256 Potential.
FT TRANSMEM 279 299 probable.
FT CARBOHYD 105 105 N-linked (GlcNAc...) (Potential).
SQ SEQUENCE 431 AA; 47357 MW; 34668FE7F908C657 CRC64;
  MSPSVEETTS VTESIMFAIV SFKHMGPFEG YSMSADRAAS DLLIGMFGSV SLVNLLTIIG
  CLWVLRVTRP PVSVMIFTWN LVLSQFFSIL ATMLSKGIML RGALNLSLCR LVLFVDDVGL
  YSTALFFLFL ILDRLSAISY GRDLWHHETR ENAGVALYAV AFAWVLSIVA AVPTAATGSL
  DYRWLGCQIP IQYAAVDLTI KMWFLLGAPM IAVLANVVEL AYSDRRDHVW SYVGRVCTFY
  VTCLMLFVPY YCFRVLRGVL QPASAAGTGF GIMDYVELAT RTLLTMRLGI LPLFIIAFFS
  REPTKDLDDS FDYLVERCQQ SCHGHFVRRL VQALKRAMYS VELAVCYFST SVRDVAEAVK
  KSSSRCYADA TSAAVVTTT TSEKATLVEH AEGMASEMCP GTTIDVSAES SSVLCTDGEN
  TVASDATVTA L
//
```

The sequence is annotated as being similar to the rhodopsin-like GPCRs. With your knowledge of primary and secondary databases, run BLAST, FASTA, InterPro, Blocks, and Blocks-format-PRINTS searches.

Consider the implications of your results. Can you tell from these sequence analysis results whether or not this is a GPCR? How do you think annotations of this sort arise? What implications might your conclusions have for sequence analysis software?

9.7 An automatic analysis system from a Web server provided the following analysis for a researcher's query sequence:

```
LENGTH:    150
KEYWORDS: TRANSPORT PROTEIN
KEYWORDS: TRANSMEMBRANE PROTEIN
KEYWORDS: LEUCINE ZIPPER
PROSITE:  MYRISTYL_SITE
PROSITE:  CK2_PHOSPHO_SITE
PROSITE:  PKC_PHOSPHO_SITE
PROSITE:  TYR_PHOSPHO_SITE
PROSITE:  ASN_GLYCOSYLATION_SITE
TRANSMEM: 1    (POTENTIAL)
TRANSMEM: 2    (POTENTIAL)
TRANSMEM: 3    (POTENTIAL)
TRANSMEM: 4    (POTENTIAL)
TRANSMEM: 5    (POTENTIAL)
TRANSMEM: 6    (POTENTIAL)
TRANSMEM: 7    (POTENTIAL)
```

The researcher concludes that the protein is a membrane protein with seven TM domains and five PROSITE matches. How likely do you think this interpretation is to be correct, and why? What steps might you take to support your conclusions?

SELF-TEST
Protein families and databases

1 In PROSITE, the term PATTERN indicates that the entry describes:
A: a block.
B: a profile.
C: a regular expression.
D: a fuzzy regular expression.

2 In PROSITE, the NR lines indicate:
A: the comment field.
B: the list of true-positive sequences matched by the signature.
C: the statistics of the diagnostic performance of the signature.
D: the non-redundant list of sequences whose 3D structures are known.

3 When searching the Blocks and PRINTS databases, a match is judged significant if:
A: a single motif is matched.
B: two motifs are matched.
C: the *E* value is above e^{-4}.
D: a combined *E* value above a given threshold is reported for a multiple-motif match.

4 TrEMBL is:
A: an automatically annotated composite protein sequence database.
B: an automatically annotated supplement to the EMBL database.
C: an automatically annotated supplement to the InterPro database.
D: a translation of coding sequences in the EMBL database.

5 UniProt is:
A: the universal protein sequence database derived from Swiss-Prot and TrEMBL.
B: the universal protein resource derived from Swiss-Prot and PIR-PSD.
C: the universal protein family resource.
D: the universal protein structure database.

6 InterPro is:
A: an integrated protein family database.
B: an integrated protein sequence database.
C: an integrated protein structure database.
D: an integrated protein interaction database.

7 In a sequence database of a given size, which of the following expressions is likely to retrieve more matches:
A: D-A-V-I-D
B: [DE]-A-V-I-[DE]
C: [DE]-[AVILM]-X-E
D: D-A-V-E

8 Which of the regexs below is not compatible with the following motif:
```
PIFMIPAFYFTW1EMQCS
PIFMIPAFYFSWIELQGS
PIFMVPAFYFSWIQMAAS
PLMALPAFYFSWWSLVTS
PLMALPAYYFSWWHLKTS
PLVTIGAFFFSWIDLSYS
```
A: P-[IL]-[FMV]-X-[IVL]-[PG]-A-[FY]2-F-[TS]-W-[IW]-X-[ML]-X2-S
B: P-[IVL]-[FMV]-[MAT]-[IVL]-[PG]-A-[FY]2-F-[TS]-W-[IW]-X-[ML]-X2-S
C: P-[IL]-[FMV]-X-[IVL]-[PG]-A-[FY]4-[TS]-W-[IW]-X-[ML]-X2-S
D: P-[IVL]-X2-[IVL]-[PG]-A-[FY]2-F-[TS]-W-[IW]-X-[ML]-X2-S

9 Which of the following groupings is not compatible with the traditional Venn diagram of overlapping amino acid physicochemical properties:
A: FYWH, AVILMP, DEQN, KR, STCG
B: FM, YWH, AVILP, DE, KR, STQN, CG
C: FYW, AVILM, DE, KRH, STNQ, C, PG
D: FYW, AVILPST, DEQN, KRH, MCG

10 With knowledge of the physicochemical properties of the amino acids, which of the following hydropathic rankings is unlikely to be correct:
A: FILVWAMGTSYQCNPHKEDR
B: IFVLWMAGCYPTSHENQDKR
C: IVLKCMAGTSWYPHDNEQFR
D: FYILMVWCTAPSRHGKQNED

11 Which of the sequences below is not compatible with the following regex:
H-[VILM]-G(2)-S-[EDQN]-T-A-[VILM]
A: HMGGSQTAM
B: HVGGSETAV
C: HIGGSSTAL
D: HIGGSETAL

12 The midnight zone is the region of sequence similarity:
A: above 20% identity.
B: where sequence alignments are not statistically significant, as the same alignment may have arisen by chance.
C: below 20% identity.
D: where sequences fail to be detected by even the most sensitive sequence-based search algorithms.

13 The twilight zone is the region of sequence similarity:
A: above 50% identity.
B: where sequence alignments are not statistically significant, as the same alignment may have arisen by chance.
C: below 50% identity.
D: where sequences fail to be detected by even the most sensitive sequence-based search algorithms.

Probabilistic methods and machine learning

CHAPTER

10

CHAPTER PREVIEW

Here we introduce probabilistic models that can be used to recognize patterns in sequence data and to classify sequences into different types. We introduce relevant ideas from Bayesian statistics, including prior and posterior probability distributions. We discuss the idea of machine learning, where a learning algorithm is used to choose parameter values that optimize the performance of a program. We describe hidden Markov models, which are probabilistic models often used to encode families of biological sequences, and we describe neural network models, another machine-learning approach that can spot patterns in sets of sequences. We discuss some of the applications of these models in bioinformatics.

10.1 USING MACHINE LEARNING FOR PATTERN RECOGNITION IN BIOINFORMATICS

The previous chapter focused on pattern-recognition methods based on protein sequence alignments, all of which attempted to encode the conserved regions (from motifs to complete domains), the assumption being that these would provide diagnostic signatures for the aligned families of sequences. People are rather good at spotting patterns in data – our brains seem to work that way. So if we look at a multiple sequence alignment, it does not take us long to spot if there is a conserved residue. However, some types of pattern are much more difficult to spot. Suppose, for example, that a particular region of a thousand DNA bases has 55% GC content while the neighboring region has only 45% GC. Looking at the complete DNA sequence, we might not notice that the regions were different, and if we did notice, we would have a

hard job deciding where to set the boundary between the regions. A difference like this could be biologically important. It might signify that one region was a coding sequence and the other was not, or that one region had been inserted into the genome from elsewhere. This is an example of a pattern that could be spotted by a probabilistic method that works with statistical properties of different sequences. The first part of this chapter develops probabilistic models of sequences that can be applied to problems such as this.

In the example in the previous paragraph, it would be easy to write a program to calculate the frequencies of the bases in a 1000-base window of a DNA sequence that slides this window along a genome looking for regions of unusual composition. Before we could write such a program, however, we would need to have had the idea that GC frequency was an interesting thing to measure. What about patterns that we didn't even think of? How can we write a program to spot something if we don't know what it is we are looking for?

This is where machine-learning algorithms come in. The two methods discussed here that can be classed under the machine-learning heading are hidden Markov models (HMMs) and neural networks (NNs). Both these systems can be used to assign sequences, or parts of sequences, into different classes (e.g., high versus low GC content in DNA, or α helix versus β strand versus loops in proteins). These systems have many internal parameters – these are the weights of connections in NNs and the emission probabilities in HMMs, as will be described later. Changing the values of these parameters allows the system to recognize different types of pattern. Usually, we do not know how to choose the parameters in order for the system to best recognize the patterns of interest. The systems have to "learn" the best parameter values themselves.

The simplest type of problem for a machine-learning algorithm is one where it has to learn to output "yes" to sequences that are of a certain type and "no" to others. We can begin with some randomly chosen values of the internal parameters of the program. We then choose a set of known examples of true-positive and true-negative sequences, and we see how well the program performs. As the parameters are random, it will randomly output **yes** or **no** for each input sequence, and will therefore make a lot of mistakes. Methods are available for **training** the algorithm to give progressively better answers. The way this works will be described in more detail in later sections of this chapter. In essence, however, the training procedure consists of making small changes to the internal parameters of the algorithm in such a way that it gets more correct answers next time round. Training proceeds step by step, until, after many steps, the internal parameters are honed so that the maximum possible number of correct answers is obtained. If the internal details of the algorithm are sufficiently well designed and flexible, then it should be possible to train the algorithm to get almost all correct answers; however, methods like this rarely score 100% (once again, think about sensitivity and selectivity – Section 7.1.4).

The set of known examples on which the algorithm is trained is called the training set. The larger the number of examples in the training set, the more likely it is that the algorithm will be able to learn useful diagnostic features of the class of sequences of interest, and the more accurate we would expect future predictions of the algorithm to be. We might therefore be tempted to use all the known examples we have available for training. The problem with this is that we would then have no way of telling

how good the algorithm is on examples that it has not seen before. Usually, known examples are divided up into two sets. The algorithm is trained to perform as well as possible on one set, and it is then tested on the other set. The performance on the test set gives us an idea of how well it can be expected to do when we try it out on unknown sequences in the future. If the method is working well, then it should perform almost equally well on the test set as on the training set. If this is true, we say that the method is able to generalize well. This means that the algorithm has learned important general properties of the sequences in the training set that also apply to the test set. If, however, the algorithm performs well on the training set but poorly on the test set, then it must have learned very specific properties of the training set that do not apply to the test set. In this case, the algorithm is not very useful, and it will probably be necessary to change the internal design of the algorithm (not just the parameter values) in order to improve it.

At this point, it is worth emphasizing a couple of general points about machine-learning algorithms before proceeding to the methods themselves. When we choose the training set, we are obviously relying on knowing which examples are positive and which negative (i.e., this is a knowledge-based method). This information usually comes from experimental studies. However, it may still be true that we don't know what the pattern is that we are looking for. For example, if we wish to predict the position of α helices in proteins, we may have many examples of protein sequences with known structure on which to train a pattern-recognition algorithm. Even so, it may not be obvious to us what features of the sequences determine the positions of the helices. Therefore, we hope that the system can learn useful features to diagnose the helices without us having to tell it what it should look for.

10.2 PROBABILISTIC MODELS OF SEQUENCES – BASIC INGREDIENTS

10.2.1 Likelihood ratios

Sections 10.2–10.4 will cover HMMs, a powerful method used to describe the statistical properties of individual sequences and families of sequences. Before we can define HMMs, however, we need to consider several aspects of the theory separately. We will build up from a very simple example.

Suppose we are interested in membrane proteins and we wish to predict which parts of the sequences are transmembrane (TM) helices and which parts are loops either side of the membrane. One important feature of the helices is that they have a high content of hydrophobic amino acid residues. This should provide some information with which to distinguish potential helical regions in a sequence of unknown structure. Let p_a and q_a be the frequencies of amino acid a in the helices and the loops, respectively. These frequencies can be measured in a set of proteins of known structure. Now consider a section of N amino acids taken from an unknown protein, and let x_i be the amino acid at position i in this sequence. The likelihoods L_1 and L_0 that the sequence occurs in a helical or a non-helical region are just the products of the frequencies of the amino acids in the sequence:

$$L_1 = \prod_{i=1}^{N} p_{x_i}, \qquad L_0 = \prod_{i=1}^{N} q_{x_i} \qquad (10.1)$$

These two likelihoods are both very small numbers: if all amino acids had equal frequency, we would get $L = (1/20)^N$. The values of L_1 and L_0 do not mean much on their own, but the likelihood ratio L_1/L_0 is more informative: it tells us whether the sequence is more or less likely to be a helical or a non-helical region.

It is common to deal with logarithms of likelihood ratios rather than the ratios themselves because these are more manageable in computers (they do not lead to overflow and underflow errors) and because they give rise to additive scoring systems. This point was also discussed in Section 4.3 on log-odds amino acid scoring matrices. In the present case, the log-odds score is

$$S = \ln\left(\frac{L_1}{L_0}\right) = \sum_{i=1}^{N} \ln\left(\frac{p_{x_i}}{q_{x_i}}\right) \qquad (10.2)$$

A positive S means the sequence is more likely to be a helical region than a non-helical region. In order to use this method for prediction, we could chop our

sequences of unknown structure into segments of a given length (say $N = 30$) and calculate the log-odds score for each segment. Segments with positive scores would be predicted to be helical regions.

10.2.2 Prior and posterior probabilities

In Bayesian statistics, prior probabilities are probabilities that we associate with alternative hypotheses, based on our previous experience of similar problems. They describe our expectations of the probabilities of the different hypotheses being true for a new example before we look at the data of the new example. The case of the TM region prediction is an example where we have alternative models of the data: models 0 and 1, defined by the two sets of amino acid frequencies, q_a and p_a. When we examine a new sequence, we thus have two alternative hypotheses: that the sequence is described by model 0 or by model 1.

Let us be slightly more general. Suppose we have two or more models to describe some type of data. We will call these models M_k, where $k = 0, 1, 2 \ldots$ etc. We have a particular new example of data of this type (call this D), and we want to know the probability that this new example is described by each of our models. From our previous experience, we assign prior probabilities $P^{prior}(M_k)$ to each model. The sum of these probabilities over all the models must add up to one. There is no theory about how to choose the prior probabilities. We have to use our common sense. If we know nothing about the problem, then we can assign equal prior probabilities to all the alternatives. If we know that one of the alternatives is unlikely to be true (e.g., this model might describe a rare type of sequence that we are not expecting to find in our new data), then we can assign it a low prior probability.

The next step is to calculate the likelihood of the data according to each of the models. This is written $L(D \mid M_k)$ and is read as "the likelihood of the data, given model k". These likelihoods are defined as the probability distributions over all possible sets of data that could be described by the model. If we sum the likelihoods over all possible data for any given model, we must get one (i.e., the likelihood is normalized). For example, in Eq. (10.1), L_0 is really $L(D \mid M_0)$, where D is the sequence $x_1 \ldots x_N$ and M_0 is the model defined by the frequencies q_a. You may like to check that the sum of L_0 over all possible sequences of length N is equal to one. In general, when we think of a model, we also know how to calculate the likelihood of any given data according to the model. This might involve some complicated equations and we may need to write a program to calculate it for us. Nevertheless, we know how to calculate it. The problem is that what we really want to know is the probability that the new data are described by model k, not the probability that the data would arise if model k were true. What we want to calculate is called the posterior probability of model k, given the data, which is written $P^{post}(M_k \mid D)$. This is proportional to the prior probability of model k times the likelihood of generating the data according to the model, i.e.,

$$P^{post}(M_k \mid D) \propto L(D \mid M_k)\, P^{prior}(M_k).$$

Written in this way, the posterior probability is not properly normalized. We are interested in the relative probabilities of the models for one set of data, so we need to normalize by summing over all the models (in contrast to the likelihood, which is normalized by summing over all the possible data for one model). Thus, we may write:

$$P^{post}(M_k \mid D) = \frac{L(D \mid M_k) P^{prior}(M_k)}{\sum\limits_{k} L(D \mid M_k) P^{prior}(M_k)} \qquad (10.3)$$

The basic principle of Bayesian statistical methods is that we should make our inferences using posterior probabilities. If the posterior probability of one model is very much higher than the rest, then we can be confident that this is the best model to describe the data. If the posterior probabilities of different models are similar to one another, then we cannot rule out the alternatives.

In the simple case, where there are just two models, M_0 and M_1, we can simplify the notation by writing the two prior probabilities as P_0^{prior} and P_1^{prior}. Hence, from Eq. (10.3), we may write the posterior probability of model 1 as

$$P_1^{post} = \frac{L_1 P_1^{prior}}{L_0 P_0^{prior} + L_1 P_1^{prior}} \qquad (10.4)$$

It is useful to define a log-odds score in a similar way to Eq. (10.2):

$$S' = \ln\left(\frac{L_1 P_1^{prior}}{L_0 P_0^{prior}}\right) = S + \ln\left(\frac{P_1^{prior}}{P_0^{prior}}\right) \qquad (10.5)$$

The difference between S' and S is simply the addition of a constant term, which is the log of the ratio of the priors. Thus, if we considered a set of potential sequences under our model for TM helices, the ranking of these sequences would be the same using S as S'; however, the conclusions as to which sequences to predict as helices would be different. If we expected TM helices to be rare in the set of sequences to be examined, we would set P_1^{prior} to be less than P_0^{prior}. This would mean that S' would be shifted down with respect to S, so that the likelihood ratio term in Eq. (10.2) would have to be correspondingly higher in order for S' to be positive and for us to predict the sequence to be a helix. If the two prior probabilities are equal, the second term in Eq. (10.5) is zero, and $S' = S$. Thus, if we use the original score, S, in Eq. (10.2) for the prediction of a helix, we are implicitly making the assumption that any sequence we look at has an equal prior probability of being a helix or a non-helix.

The posterior probability can be written in terms of the log-odds score, as follows:

$$P_1^{post} = \frac{1}{1 + \dfrac{L_0 P_0^{prior}}{L_1 P_1^{prior}}} = \frac{1}{1 + \exp(-S')} \qquad (10.6)$$

If $S' = 0$, $P_1^{post} = 1/2$. If S' is large and positive, P_1^{post} tends to one. If S' is large and negative, P_1^{post} tends to zero.

10.2.3 Choosing model parameters

So far, we have talked about distinguishing between a discrete set of alternative models. Often, however, the models themselves have continuous parameters and we want to estimate these parameters from the data. For example, our model of the helical regions was defined by the set of frequencies of the amino acids p_a. How should we choose the values of the frequencies?

In our set of known TM helices, let the observed number of amino acids of type a be n_a, and let the total number of amino acids be n_{tot}. The likelihood of this occurring is

$$L = \prod_{a=1}^{20} (p_a)^{n_a} \qquad (10.7)$$

The simplest way to choose the frequencies is via the principle of maximum likelihood (ML). We choose parameter values so that the likelihood of observing the data is greatest. It can be shown that the above function, L, is maximized when the frequency parameter for each amino acid is equal to the observed frequency of that amino acid in the data:

$$p_a = n_a/n_{tot} \qquad (10.8)$$

This result seems like common sense. The proof is Problem 10.2 at the end of this chapter.

We can immediately generalize this to give a probabilistic model for a profile (i.e., a set of aligned proteins). Suppose we have an ungapped alignment of a protein motif of length N sites containing K sequences. We want to develop a position-specific score system that reflects the fact that the frequencies of the amino acids are different at each site. The ML values of the frequencies at site i are $p_{ia} = n_{ia}/K$, where n_{ia} is the observed number of times amino acid a occurs at site i. As before, let x_i be the amino acid at site i in a new sequence. The likelihood of this sequence, according to the profile model, is

$$L_{prof} = \prod_{i=1}^{N} p_{ix_i} \qquad (10.9)$$

As usual, this likelihood is only meaningful if we compare it to something. We can define a model 0, with likelihood L_0 given by Eq. (10.1), where the frequencies, q_a, in the model are to be interpreted as average amino acid frequencies in general proteins other than the particular sites in the profile we are interested in. By analogy with Eq. (10.2), the log-odds score for the profile model is

$$S = \ln \left(\frac{L_{prof}}{L_0} \right) = \sum_{i=1}^{N} \ln \left(\frac{p_{ix_i}}{q_{x_i}} \right) \qquad (10.10)$$

We could also treat this model in a Bayesian way, by introducing a prior probability term, as we did in Eq. (10.5). However, there is another, more important, way in which a Bayesian approach is useful in this model – this relates to the way the p_{ia} parameters are estimated. When building a profile model, we may only have a few tens of sequences in the alignment. There will thus be considerable fluctuation in the observed frequencies at each site, because we have only sampled a small number of sequences. In particular, it is quite likely that some amino acid, a, will not occur at all at a given site in our alignment. If $n_{ia} = 0$, the ML frequency of a is $p_{ia} = 0$. This means that the likelihood L_{prof} will be set to zero for any future sequence found to have a at site i. Clearly, the model is too strict if it completely forbids the future occurrence of an amino acid at a site simply because we have not yet observed it in the limited amount of data we have so far.

The Bayesian method of dealing with this problem is to use **pseudocounts**. Instead of basing the frequencies solely on the observed number of occurrences of each amino acid, we add a pseudocount to each total, based on our prior expectations. Before looking at the alignment, our best guess is that the amino acid frequencies are the same as the average frequencies in other proteins: q_a. If we included A additional sequences in our alignment with these prior frequencies, then the expected number of additional occurrences of amino acid a in each site would be Aq_a. Our estimate of the frequencies at the site, including the pseudocounts, is then

$$p_{ia} = \frac{n_{ia} + Aq_a}{K + A} \qquad (10.11)$$

When K is small compared to A, the estimated frequencies are strongly influenced by the prior frequencies q_a. However, as the amount of real data accumulates, K becomes much larger than A, and the estimated frequencies are governed only by the observed data at the site. This is what always happens in Bayesian methods: when there are few data, the prior is important, but when we have a lot of data, the result becomes independent of the prior. The value of the parameter A thus controls the amount of weight we put on the prior, and the number of observations of real data that we need in order to "over-rule" the effect of the prior. More information on pseudocounts is given by Durbin *et al.* (1998).

The posterior probability Eq. (10.3) can also be written down for continuous models. We will use θ to represent the continuous parameters of the model (it could be a single variable or more than one, but for simplicity we will just write one). The posterior probability of θ then becomes

$$P^{post}(\theta \mid D) = \frac{L(D \mid \theta)P^{prior}(\theta)}{\int L(D \mid \theta)P^{prior}(\theta)d\theta} \qquad (10.12)$$

The sum has now become an integral because we are dealing with continuous parameters. The best estimate of the parameters of the model, according to Bayesian analysis, is to take the average of the parameter over the posterior distribution. In Box 10.1, we consider the derivation of Eq. (10.11) more carefully in terms of explicit prior and posterior distributions. We show that the frequencies given by the pseudocount argument above correspond to mean values obtained from the posterior probability distribution.

In general, there will be many parameters to integrate over, and so it may not be possible to do the integral exactly. However, it may be possible to determine model parameters without doing this integration. One approximation is to use the value of

θ that maximizes $L(D \mid \theta)P^{prior}(\theta)$. This is called the maximum posterior probability solution. This may be compared to the maximum likelihood solution, which simply chooses the θ that maximizes $L(D \mid \theta)$, ignoring the prior. Another approach is to use the Markov Chain Monte Carlo (MCMC) method, which allows us to sample possible values from the posterior distribution of continuous parameter models when the integral in Eq. (10.12) is too hard to do exactly. MCMC was discussed in the context of molecular phylogenetics in Chapter 8. The reader may wish to review Section 8.8 after reading this section. In phylogenetics, we have both continuous parameters to determine (branch lengths and rate matrix parameters) and a discrete set of topologies that we are trying to distinguish. MCMC can be used for any type of problem where we have a likelihood function. Usually, we are interested in the value of the model parameters, or some function of these parameters. As MCMC generates samples with the correct posterior probability distribution, we simply need to take the average of the quantity of interest from the samples generated. When there are large quantities of data available, the likelihood functions become sharply peaked about the optimum parameter values. This means that the mean value of a parameter from the posterior distribution, the value that maximizes the posterior probability, and the maximum likelihood value should all be very similar.

BOX 10.1
Dirichlet prior distributions

When we introduced pseudocounts, we said that q_a was the average frequency of amino acid a in proteins other than the particular sites of interest. To be more precise, we need to define a prior distribution, $P^{prior}(\theta_a)$, over the amino acid frequencies, θ_a, such that the average frequency of amino acid a is $\bar{\theta}_a = q_a$. The most commonly used prior distribution is the Dirichlet distribution:

$$P^{prior}(\theta_a) = \frac{1}{Z} \prod_{a=1}^{20} \theta_a^{\alpha_a-1} \delta\left(\sum_{a=1}^{20} \theta_a - 1\right) \qquad (10.13)$$

In the above equation, the delta function expresses the constraint that the sum of the frequencies must add up to one, and Z is a normalization factor that ensures that the total probability, when integrating over all possible values of the θ_a, is equal to one. The parameters α_a control the shape of the distribution. It can be shown that the mean values of the frequencies are

$$\bar{\theta}_a = \frac{\alpha_a}{\sum_a \alpha_a} \qquad (10.14)$$

If we let $\alpha_a = Aq_a$, then we have a prior distribution such that $\bar{\theta}_a = q_a$ as required, as A cancels out and

$\sum_a q_a = 1$ for any set of frequencies. The Dirichlet distribution is useful as a prior whenever we have a set of parameters constrained to add up to one, e.g., the emission probabilities of the symbols in an HMM or the transition probabilities between the states. The parameter A does not affect the average frequencies, but it does affect the shape of the distribution. The larger the value of A, the more constrained the distribution is to be close to its average. To see this, let us consider only one type of amino acid with a frequency θ and mean frequency q (we have dropped the subscripts for convenience). Equation (10.13) looks much simpler:

$$P^{prior}(\theta) = \frac{1}{Z}\theta^{Aq-1}(1-\theta)^{A(1-q)-1} \qquad (10.15)$$

The factor $(1-\theta)$ is the combined frequency of all the other types of amino acids, apart from the one of interest, and the mean frequency of the sum of these other types is $1-q$. Let us suppose that our prior expectation is that the mean frequency of this amino acid should be $q = 0.05$. Figure 10.1(a) shows the shape of the prior distribution for several different values of A: when $A = 200$, the distribution is sharply peaked close to $\theta = q$; for $A = 40$, the peak is much broader; and for $A = 20$, the distribution is continuously decreasing. All these distributions have the same mean. The larger the value of A we choose, the more confident we are in specifying that θ must be close to q.

Now consider the posterior distribution after observing data. Let there be K observed amino acids, of which n are of the type we are considering. The likelihood of this is a binomial distribution $L(n \mid \theta) = C_n^K \theta^n (1-\theta)^{K-n}$. The posterior distribution is proportional to the likelihood times the prior:

$$P^{post}(\theta \mid n) \propto C_n^K \theta^n (1-\theta)^{K-n} \frac{1}{Z}\theta^{Aq-1}(1-\theta)^{A(1-q)-1}$$

Writing this as a properly normalized distribution, we obtain:

$$P^{post}(\theta \mid n) = \frac{1}{Z'}\theta^{n+Aq-1}(1-\theta)^{K-n+A(1-q)-1} \qquad (10.16)$$

where Z' is a new normalization constant obtained from calculating the integral on the bottom of Eq. (10.12). Now we can see why the Dirichlet distribution was chosen for the prior in the first place: with this choice, the posterior distribution is a member of the same family of curves as the prior. As Eq. (10.16) is a Dirichlet, we

can use Eq. (10.14) to determine the mean posterior value of θ:

$$\bar{\theta} = \frac{n + Aq}{n + Aq + K - n + A(1-q)} = \frac{n + Aq}{K + A} \qquad (10.17)$$

This is equivalent to Eq. (10.11), which we derived originally using the pseudocount argument. Figure 10.1(b) shows the shape of the posterior distribution (10.16) in the case where the prior has $q = 0.05$, but the frequency in the data is 0.5.

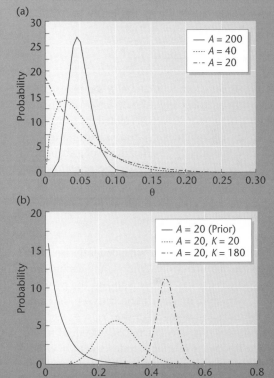

Fig. 10.1 (a) Dirichlet prior distributions for a single variable θ (see Eq. (10.15)), with mean value $q = 0.05$ and different values of A. Larger values of A give distributions more closely peaked about the mean. (b) The effect of increasing the amount of data N on the posterior probability distribution. The prior has $q = 0.05$ and $A = 20$. The data are chosen with frequency 0.5. As N increases, the posterior distribution moves towards a peak centered at $\theta = 0.5$.

10.3 INTRODUCING HIDDEN MARKOV MODELS

10.3.1 Markov models – Correlations in sequences

So far, we have considered the frequencies of amino acids in sequences but we have not considered the possibility of correlations between the residues at neighboring sites. Consider a very long sequence containing n_a amino acids of type a. Let there be n_{ab} occasions where an amino acid of type b immediately follows an a. We can define r_{ab} to be the conditional probability that the amino acid at site i is b, given that the one at site $i-1$ is a. The ML value of r_{ab} from the data is

$$r_{ab} = P(x_i = b \mid x_{i-1} = a) = n_{ab}/n_a \qquad (10.18)$$

A first-order Markov model of a sequence is defined by an alphabet (e.g., the 20-letter amino acid alphabet, or the four-letter DNA alphabet), a matrix of conditional probabilities, r_{ab}, and a set of frequencies for the initial state, q_a. The likelihood of a sequence, $x_1 \ldots x_N$, occurring according to a first-order model is

$$L = q_{x_1} \prod_{i=2}^{N} r_{x_{i-1} x_i} \qquad (10.19)$$

The first state is chosen independently from the amino acid frequency distribution, and each subsequent state is chosen conditionally on the previous one. If there were no correlation between one state and the next, we would find that $r_{ab} = q_b$, i.e., the probability of state b occurring is just equal to the average frequency of state b, independent of which state went before it. In that case, Eq. (10.19) reduces to a simple product of amino acid frequencies, as in Eq. (10.1). A model with independent frequencies at each site is called a zero-order Markov model. We can also model longer range correlations in sequences. In general, a model is called k^{th}-order if the probability of a state occurring at a site depends on the states at the k previous sites. So for a second-order model, we would need to define conditional probabilities $r_{abc} = P(x_i = c \mid x_{i-1} = b, x_{i-2} = a) = n_{abc}/n_{ab}$, in a similar way to Eq. (10.18).

Let us return to the problem of distinguishing TM helix positions in proteins. Rather than use zero-order models for the helix and non-helix regions, we could define two first-order models, as in Eq. (10.19), and define a log-odds score based on these – this would probably provide a better means of prediction than the zero-order models, because the first-order models contain more information about real sequences. Real protein sequences certainly do not have independent residues at neighboring sites. Nevertheless, even if we include first-order correlations (or higher order ones as well), we are still missing some important features of the problem. So far, we have envisaged chopping our test protein into segments, and testing each segment separately. But how long should the segments be? All the helical regions might not be the same length. How do we know where to chop the sequence? The boundary between helix and loop regions might fall in the middle of the segment we are testing. Instead of having two separate models for the two parts of the sequence, it would be nice if we had a model to describe the whole of a sequence, which would tell us where the beginning and end of a helix were. It is time to introduce our first HMM.

10.3.2 A simple HMM with two hidden states

In a Markov model, the probability of any given letter appearing in the sequence may depend on the letter(s) that went before. In an HMM, the probability of a letter may depend on the letter(s) that went before, but it also depends on which hidden state the model is in. Hidden states are so called because they are not visible when looking only at the sequence. For our model, we need two hidden states: state 0 (the loops) and state 1 (the helices). It is also useful to define begin (B) and end (E) states. An allowed path through the HMM starts from the begin state, passes any number of times through states 0 and 1, and ends at the end state. Each time we enter a 0 or 1 state, the model emits one letter chosen from the allowed alphabet of the model. In our case, this means one amino acid is added to the sequence. The begin and end states do not emit letters to the sequence. We can define the emission probabilities of the amino acids in states 0 and 1 to be the same amino acid frequencies that we used at the beginning of Section 10.2:

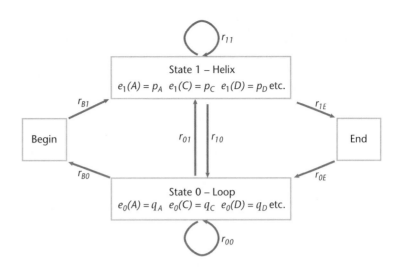

Fig. 10.2 A simple HMM with two hidden states for distinguishing helical and loop regions of proteins.

$$e_0(a) = q_a, \qquad e_1(a) = p_a \qquad (10.20)$$

This means that we are supposing the amino acids occur independently (zero-order model), as long as we remain in either state 0 or 1. However, we will use a first-order model to describe the transitions between the hidden states. The transition probabilities are shown in Fig. 10.2. The probabilities r_{B1} and r_{B0} are the probabilities that the first residue in the sequence will be in either a helix or a loop. If we enter the helix state, then there is a probability r_{11} of remaining in the helix, a probability r_{10} of moving into a loop, and a probability r_{1E} of reaching the end of the sequence. Similar probabilities control the transitions from the loop state. The values of these probabilities control the relative frequency and relative lengths of the regions of types 0 and 1. For example, if r_{01} is very small, it will be very difficult to initiate a new helical region; hence the model will predict very few helices. Also, we expect that both helical and loop regions in a protein will be fairly long (tens of amino acids, not just two or three). This means that the probabilities of remaining in a state (r_{11} and r_{00}) will be larger than the probabilities of switching between states (r_{10} and r_{01}).

This model is able to describe a sequence of any length containing any number of helical and loop regions. A path through the model for a given sequence might look like this:

Sequence x_i GHMESSAGEQLLKQCYTINSIDEWHLNT
Path π_i B0011111110000000011111100000E

The path variables, π_i, describe the hidden states at each point in the sequence. The likelihood of the path given above can be written down in terms of the product of the emission probabilities and the transition probabilities for each site:

$$L = (r_{B0}e_0(G)) \times (r_{00}e_0(H)) \times (r_{01}e_1(M))$$
$$\times \ .....$$
$$(r_{00}e_0(T) \times r_{0E}) \qquad (10.21)$$

The model can be used to determine the most likely positions of helices and loops within the sequence. This is known as **decoding**. The first way of doing this is to find the most probable path through the model, i.e., we find the sequence of hidden states for which the likelihood, calculated as in Eq. (10.21), is the highest. This gives a straightforward prediction that each site is either helix or loop. The second way is to consider all possible paths through the model, weighted according to their likelihood, and to calculate the probability that each site is in each of the hidden states. We would expect some sites where $P(\pi_i = 1)$ is very close to one, and some where $P(\pi_i = 1)$ is very close to zero; these will be confidently predicted to be either helical or loop regions. There will probably be other sites for which the probability is intermediate; for these, the model is unable to predict the state of the site with confidence. Such ambiguous sites might well occur at the boundaries between regions that are strongly predicted to be helix and loop; they might also occur if a short section of the sequence in a loop region had amino acid frequencies that were more typical of helices, for example. The algorithms for these two decoding methods are known as the Viterbi and the forward/backward algorithms. These are described in Boxes 10.2 and 10.3.

10.3.3 Choosing HMM parameters

The most probable paths through the model for a given sequence depend on the emission and transition probabilities. These parameters can be determined from the information in sequences with known structure. We can write down the paths (0s and 1s) below each known example. From these, we can calculate the ML values of the probabilities. As in Eq. (10.8) above, the ML values of the emission frequencies will be given by the observed frequencies in the data. If amino acid a occurs n_{ka} times in regions of state k, and the total number of residues in state k is n_k^{tot}, then

$$e_k(a) = n_{ka}/n_k^{tot} \tag{10.22}$$

Similarly, if we observe m_{kj} occasions where hidden state j follows hidden state k, then the ML transition probabilities are

$$r_{kj} = m_{kj}/n_k^{tot} \tag{10.23}$$

This is equivalent to Eq. (10.18) for the first-order Markov model. If some of the transitions occur very rarely, or not at all, in the known examples, then we are best to use prior information in choosing the frequencies. The simplest way to do this would be to add pseudocounts to the number of observed transitions, just as we added pseudocounts to the number of observed amino acids of each type in Eq. (10.11).

At the beginning of this chapter, we introduced the idea of machine learning. A machine-learning method is one where the algorithm learns to recog-

nize patterns in the data as well as possible by tuning the free parameters in the model. What we described in the previous paragraph does not seem much like learning at first sight. This is because the model that we have is simple enough for us to be able to write down the ML parameter values. So the algorithm can be set immediately to the optimal choice of parameters. In a more complicated case, we might not be able to calculate the optimal parameters analytically. We could then use a step-by-step numerical method to optimize the parameters. One such method is called gradient ascent or gradient descent, depending on whether we are trying to maximize or minimize a function. Here, we want to maximize the likelihood. We can imagine the likelihood function as a mountain surface that we are trying to climb. If we are able to calculate the derivative of the likelihood with respect to each of the parameters, then we know which is the quickest way up. The gradient-ascent method begins at some initial guess of parameter values, calculates the derivatives, and moves a small distance in the direction of the steepest increase. It then recalculates the gradients from the new point, and repeats the process until it reaches the ML point where the derivatives are zero. This step-by-step process seems a bit more like a gradual learning process. In any case, in this context, "learning" is just another word for optimization. In Section 10.5, we will discuss neural networks, where the analogy between optimization and learning may seem more intuitive.

If the model parameters are chosen based on a set of known examples, as when we used a set of protein sequences with known helix positions for optimizing the likelihood above, this is referred to as **supervised learning**. In other words, we are telling the algorithm what it should learn. However, it is also possible to use **unsupervised learning**, where the algorithm can maximize the likelihood within the framework of the model, but without being told what the patterns are to learn. For example, we could define a model with two hidden states, defined by two sets of frequencies, without specifying the meaning of the two states. The learning process would then determine the best way of partitioning the sequence into two different types of subsequence. It would highlight the fact that the properties of the sequence were not uniform along its length, and it would show in what way they were non-uniform.

The simplest way by which an unsupervised-learning process can be implemented in HMMs is called Viterbi training. This begins with an initial guess as to the model parameters. The Viterbi algorithm (Box 10.2) is used to calculate the values of the hidden states on the most probable path for each sequence in the training set. From this, we can calculate n_{ka}, the number of times that amino acid a occurs where the most probable path is in state k, and m_{kj}, the number of times that there is a transition in the most probable path from state k to state j. We can then use Eqs. (10.22) and (10.23) to calculate new values of the emission and transition probabilities for the HMM, and the whole process can be repeated with the new model parameters. After repeating this several times, we will reach a self-consistent state, where the observed values on the most probable paths are consistent with the input model parameters, and so the new parameter values are the same as the old ones. This model is then trained as well as possible on the data.

A slightly more complex training method is called the Baum–Welch (or expectation-maximization) algorithm. This relies on the forward and backward algorithms (Box 10.3) instead of the Viterbi algorithm. Beginning with some initial guess as to the model parameters, we can calculate the probability $P(\pi_i = k)$ that point i on the sequence is in state k (Eq. (10.35) in Box 10.3). Hence, the expected number of times that letter a appears in state k, averaged over all possible paths, is

$$\overline{n_{ka}} = \sum_{\substack{sites \\ where \\ x_i = a}} P(\pi_i = k) \qquad (10.24)$$

Similarly, the probability that point i is in state k, while $i + 1$ is in state j, is given by Eq. (10.36). Hence, the expected number of times that a transition from state k to state j occurs is

$$\overline{m_{kj}} = \sum_i P(\pi_i = k, \pi_{i+1} = j) \qquad (10.25)$$

Calculating the expected values of these quantities is the "expectation" part of the algorithm. The "maximization" step consists of calculating new model parameters by inserting $\overline{n_{ka}}$ and $\overline{m_{kj}}$ into Eqs. (10.22) and (10.23). As we said above, the values of the emission and transition probabilities given by these equations are the ones that maximize the likelihood of seeing the observed numbers of amino acids and transitions.

The two steps of the algorithm can be repeated several times until the change in parameters is negligible and the algorithm converges. Both the Viterbi and Baum–Welch training algorithms converge towards local maxima, where the fit of the model to the data cannot be improved. However, there is no guarantee that the global maximum set of model parameters has been found. It is therefore necessary to run the algorithms from several different starting points to check for trapping in suboptimal local maxima. This is less of a problem with supervised-learning methods, where the parameters are initialized with values thought to be reasonable, based on the patterns we expect to see in the data.

10.3.4 Examples

We have been using the example of membrane proteins to motivate this section, as it is easy to appreciate that the helical and loop regions in the structure are likely to give rise to sequences with regions of different amino acid compositions. However, the model discussed is a very general way of distinguishing heterogeneous regions within sequences. It is convenient to refer to this model as the M1–M0 model, meaning that the transitions between the hidden states are determined by a first-order Markov process, but that the symbols in the sequence are independent (zero order). Recall that, in the above discussion, the emission probabilities, $e_k(a)$, have been assumed to be dependent on the hidden state k, but not on the previous symbol in the sequence. Durbin *et al.* (1998) present an example of the M1–M0 model under the heading "the occasionally dishonest casino". In this case, a sequence of die rolls is observed that has been generated either by a fair die, with equal probability of each number, or a

BOX 10.2
The Viterbi algorithm

The Viterbi algorithm calculates the most probable path of hidden states for a given sequence, $x_1 \ldots x_N$. Consider an HMM with emission probabilities $e_k(a)$ for emitting letter a when in hidden state k, and with transition probabilities r_{hk} between hidden states h and k. Let $v_k(i)$ be the likelihood of the most probable path for the first i letters in the sequence, given that the ith letter is in state k. The algorithm is a type of dynamic programming routine, similar to those used for sequence alignment (see Chapter 6). We can initialize the algorithm by setting

$$v_k(1) = r_{Bk}e_k(x_1) \tag{10.26}$$

as there is only one path to each of the hidden states for the first letter of the sequence. For subsequent letters, $i = 2$ to N, we have the recursion relation

$$v_k(i) = \max_h(v_h(i-1)r_{hk}e_k(x_i)) \tag{10.27}$$

Finally, v_E, the likelihood of the best total path ending in the end state, is:

$$v_E = \max_h(v_h(N)r_{hE}) \tag{10.28}$$

Just as with sequence alignment, we need to store an array of pointers for back-tracking, i.e., we can define variables $B_k(i)$ that are set equal to the hidden state h that gave the maximum term in step (10.27). In this way, we can trace the path through the model back from the end to the beginning.

The time taken by the Viterbi algorithm is $O(KN)$, where K is the number of hidden states summed over at each step of the recursion. Thus, it is a polynomial time algorithm, even though the number of possible paths through the model scales as K^N.

loaded die, with unequal probabilities. The casino occasionally switches between the two dice, and the object of the HMM is to distinguish the points at which switches were made from observation of the sequence of numbers rolled.

We have not tried the M1–M0 model on TM helices, so we do not know how well it would work in practice. However, this model has been used for real biological sequence analysis. In one of the first applications of HMMs in bioinformatics, Churchill (1989) used a model with only two symbols, 0 (representing A or T) and 1 (representing G or C), to look for heterogeneities in DNA sequences. The hidden states in this model represent regions of high and low probability of occurrence of G or C. This model was sufficient to detect heterogeneities in mitochondrial genomes and in the bacteriophage λ sequence. Natural generalizations of the M1–M0 model are the M1–M1 and M1–M2 models, meaning that the transitions between the hidden states are still determined by a first-order process, but the emission probabilities are determined by first- or second-order processes. Thus, for M1–M1, we would require emission probabilities of the form $e_k(x_i = a \mid x_{i-1} = b)$. Nicolas

et al. (2002) analyzed the genome of *Bacillus subtilis*. They found that while the M1–M0 model did not reveal any interesting structure, the M1–M2 model did. They used a four-letter alphabet (A, C, G, T) and had three hidden states. The model was trained in an unsupervised manner using the expectation–maximization algorithm. After training, the first two hidden states matched the coding regions on the plus and minus strands of the genome (which have different sequence statistics), and the third usually matched non-coding regions between the genes. However, some coding regions with unusual composition also matched the third state, suggesting that these regions must have arisen by prophage insertion or by some kind of horizontal transfer. Plate 10.1 shows the probabilities of sites being in each of the states, as a function of position on the genome. The illustrated region contains sections that match strongly to each of the states, and the probabilities switch fairly sharply at the boundaries between them.

It is possible to build up HMMs with much more complex structures that can recognize more specific types of pattern in sequence data. The architecture of an HMM (i.e., the number of states and the way in

BOX 10.3
The forward and backward algorithms

The forward and backward algorithms allow us to calculate the probability $P(\pi_i = k)$ that the ith letter in the sequence is in hidden state k. For the forward algorithm, let $f_k(i)$ be the sum of the likelihoods of all the paths for the part of the sequence from x_1 to x_i, given that x_i is in state k. To initialize the forward algorithm,

$$f_k(1) = r_{Bk}e_k(x_1) \tag{10.29}$$

The recursion step is

$$f_k(i) = e_k(x_i) \sum_h f_h(i-1)r_{hk} \tag{10.30}$$

which is analogous to the Viterbi algorithm (Eq. (10.27)), except that we are summing all contributions instead of taking only the maximum term. The final step is

$$L_{tot} = \sum_h f_h(N)r_{hE} \tag{10.31}$$

Here, L_{tot} is the total likelihood of all the paths for the given sequence.

For the backward algorithm, let $b_k(i)$ be the sum of the likelihoods of all the paths for the part of the sequence from x_{i+1} to x_N, given that x_i is in state k. This algorithm is initialized with the end state:

$$b_k(N) = r_{Bk} \tag{10.32}$$

The recursion runs downwards from N to 1:

$$b_k(i) = \sum_h r_{kh}e_h(x_{i+1})b_h(i+1) \tag{10.33}$$

The final step is

$$L_{tot} = \sum_h r_{Bh}e_h(x_1)b_h(1) \tag{10.34}$$

Note that Eqs. (10.31) and (10.34) should give the same numerical result, otherwise there must be a bug in the program!

From these quantities, we can now determine the probability that point i in the sequence is in state k. This is just the sum of the likelihoods of all paths, with $\pi_i = k$ divided by the total likelihood of all paths:

$$P(\pi_i = k) = \frac{f_k(i)b_k(i)}{L_{tot}} \tag{10.35}$$

It is worth noting that the emission probability, $e_k(x_i)$, has been included in the definition of $f_k(i)$ but not in the definition of $b_k(i)$. This slight asymmetry means that this term is not counted twice at the top of Eq. (10.35).

A second useful quantity to calculate is the probability that point i is in state k, while $i + 1$ is in state j:

$$P(\pi_i = k, \pi_{i+1} = j) = \frac{f_k(i)r_{kj}e_j(x_{i+1})b_j(i+1)}{L_{tot}} \tag{10.36}$$

This result comes in useful for the Baum–Welch training algorithm.

---

which they are connected) will be different for each application. As an example of a very specific type of HMM architecture, let us consider a model designed to spot coiled-coil domains in proteins. Coiled coils are associations of two or more α helices that wrap around each another. They occur in many different proteins, including tropomyosin, hemagglutinin (from the influenza virus), and the DNA-binding domains of transcription factors. There are approximately 3.5 residues per turn of an α helix. This leads to a repeating pattern of seven residues (a heptad) corresponding to two turns. Figure 10.3(a) shows a schematic diagram looking down such helices, with

residue sites labeled *a* to *g*. The two helices attract one another owing to the presence of hydrophobic residues at sites *a* and *d*. The frequencies of the amino acids vary at the seven positions, and this provides a strong signal for identification of coiled-coil domains (in fact, there are slightly more than 3.5 residues per turn, which is why the two helices spiral round each other, rather than running in parallel).

A straightforward way to detect this signal was given by Lupas, Vandyke, and Stock (1991). They developed a profile score system dependent on the relative amino acid frequencies at each site (cf. Eqs. (10.9) and (10.10) above), and calculated this score

(a)

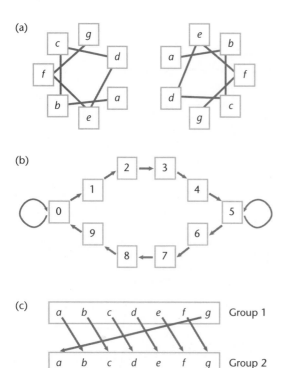

(b)

(c)

Fig. 10.3 (a) The helical wheel representation of a two-helix coiled-coil domain, showing sites *a* to *g*. (b) The model contains nine groups of states, plus an initiation/end group 0. (c) Each of the nine groups contains seven states representing the seven possible positions in the helix. States in one group are linked to the state at the following helix position in the next group. Redrawn from Delorenzi and Speed (2002).

within a sliding window of 28 residues (four heptads). A treatment of this problem using HMMs has been developed by Delorenzi and Speed (2002). This model is shown in Fig. 10.3(b). State 0 represents residues outside the helical regions, and emits amino acids with a background frequency distribution (analogous to state 0 in our simple model in Fig. 10.2). Boxes 1 to 9 in Fig. 10.3(b) are groups of seven states, representing the seven positions in the coiled-coil pattern. The emission probabilities of amino acids from *a* to *g* are determined by seven different site-specific frequency distributions. A path through the model may remain in state 0 any number of times, before beginning a helix by entering one of the group 1 states. As shown in Fig. 10.3(c), each state is con-

nected with a high probability to the state at the next helix position in the next group of states, i.e., $1a \rightarrow 2b \rightarrow 3c \ldots$ or $1f \rightarrow 2g \rightarrow 3a$. This means that once a helix begins, the path usually follows the correct set of positions in the helix. Transitions that disrupt the helix are permitted with small probabilities (e.g., $1a \rightarrow 2c$). A path must pass through each of groups 1 to 9 before returning to state 0. However, group 5 states, which represent the middle region of the helix, are linked to other group 5 states ($5a \rightarrow 5b$, $5b \rightarrow 5c$, etc.). This means that the helical region can be any number of states long. A slightly simpler version of this model would only include the group 5 states and state 0. Addition of groups 1–4 and 6–9 allows the model to account for properties of helices that might be different at the ends from the middles. It also means that every helix must be at least nine residues long. The parameters in the model were obtained using a version of the Baum–Welch training algorithm, using known examples of coiled-coil proteins.

We will finish this section on HMMs with the same problem that we started with: prediction of the position of helices in TM proteins. Krogh *et al.* (2001) constructed the complex HMM shown in Fig. 10.4. This model contains 25 states representing the helix core. The connections between these states are such that the path must pass through at least five states. This constrains the helix core to be between five and 25 residues long. The cap regions, at the end of the helices, consist of five states that the path must pass through. This allows amino acid frequencies at the ends of helices to be different from those in the interior. Loops of up to 20 amino acids are treated by specialized states in the model. Longer non-helical regions are treated by a "globular" state that connects to itself, thus allowing any length of protein. The model distinguishes between the cytoplasmic and non-cytoplasmic sides of the membrane, so that it is also able to predict the orientation of proteins within the membrane, as well as the helix position. This is a good example of the way an HMM can be designed to incorporate realistic features of particular biological problems. Although HMM designs can reach a high degree of complexity, the basic algorithms for decoding and training remain the same. HMMs are thus extremely flexible. How well a model

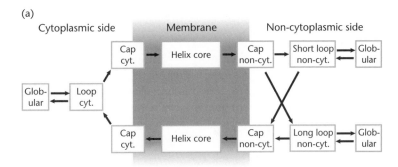

(a)

Cytoplasmic side | Membrane | Non-cytoplasmic side

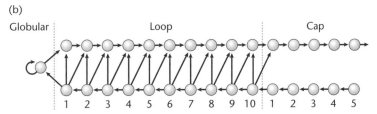

(b)

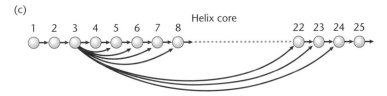

(c)

Fig. 10.4 A detailed HMM for prediction of the positions of TM helices in membrane proteins. The model contains specialized states, representing the helix core, the cap regions at the end of the helices, loop regions connecting helices, and globular regions both inside and outside the membrane. From Krogh *et al.* (2001). Copyright (2001) with permission from Elsevier.

works in practice will depend on whether there is really a distinguishable signal present in the data, and also on whether the structure of the model is properly designed so that it is compatible with the structure of the patterns that are present.

10.4 PROFILE HIDDEN MARKOV MODELS

Chapter 9 described techniques based on profiles, where position-specific scores were used to describe aligned families of protein sequences. A drawback of profile methods is their reliance on *ad hoc* scoring schemes – a coherent statistical theory has been developed to describe ungapped sequence alignments, but scoring gapped alignments depends on empirical estimates. To address this limitation, and replace the fixed gap penalties implemented in profiles, probabilistic approaches using HMMs have been developed.

A particular type of HMM known as a profile HMM is used to capture the information in an aligned set of proteins. The alignment below could be represented by the profile HMM shown in Fig. 10.5.

```
1 2       3
W H  .  . E n
W H  .  . Y .
W -  .  . E .
S H  .  . E .
T H  e  . Y .
W H  e  r E .
```

Columns 1, 2, and 3 of the alignment are "well-aligned" sites that contain an amino acid in almost all sequences. These columns are represented by the match states M1, M2, and M3. In addition, there are insert states, labeled I, between each of the match states, and delete states, labeled D, below each match state. The match and insert states emit amino

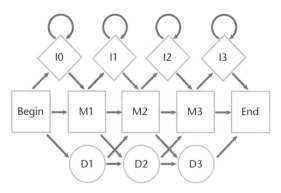

Fig. 10.5 A profile HMM representation of a protein sequence alignment, showing match (M), insert (I), and delete (D) states.

acid symbols to the sequence. The emission probabilities will be position specific, and will reflect the frequencies of the residues observed in the alignment used to train the model. The delete states do not emit symbols. The "-" character in the alignment signifies that the path of the sequence WE passes through the state D2 instead of M2. Residues emitted by insert sequences have been shown in lower case in the above alignment: e.g., the "n" in WHEn is emitted by the I3 state. The insertion states have connections back to themselves. Hence, any number of residues can be inserted between two match states: e.g., both the "e" and "r" in WHErE are emitted by the I2 state. A "." has been used in the alignment to fill in the space above and below inserted characters; it thus indicates that the path of that sequence does **not** pass through an insert state. This is distinct from "-", which indicates that the path **does** pass through a delete state.

To specify the model completely, we need to define emission probabilities for all the match and insert states, and transition probabilities between all states. These can be calculated from the observed number of amino acids in each column, and from the observed number of transitions of each type, as with Eqs. (10.22) and (10.23) above. As we have a large number of parameters to estimate, and as we may not have many sequences in the alignment, it is useful to include prior information for the para-

meters. This can be done by introducing pseudocounts or Dirichlet priors, as explained in Section 10.2.3.

Once we have all the model parameters, we can use the model to align new sequences to it. The most probable alignment of a sequence to the model can be calculated using the Viterbi algorithm (Box 10.2). This provides a convenient way to build up large alignments starting from a relatively small set of sequences in a seed alignment. We can also use the Viterbi algorithm to train the model. We can estimate an initial model from an initial alignment, then calculate the most probable paths for each sequence in the original alignment. This gives us a new version of the alignment, which gives us a new model, and so on, until the alignment no longer changes. As described in Section 10.3.3, the Baum–Welch method can also be used for training, where the expectation values of the parameters are calculated over all paths, rather than just the most probable paths. In principle, it is possible to use completely unsupervised learning with profile models, i.e., to start with unaligned sequences. However, this is not often done, because it can lead to trapping of the learning algorithm in suboptimal alignments. It is more usual to begin with a fairly good first estimate of the alignment, and allow the training method to refine the alignment by making small changes to the model parameters. There are now several software packages designed to build profile HMMs and train them based on user-supplied alignments, in particular HMMER (Eddy 2001) and SAM (Hughey, Karplus, and Krogh 2001).

In general, the probabilities of transitions to insert and delete states will be smaller than the probabilities of transitions to match states. This means that there is a cost associated with putting gaps in alignments. Profile scoring systems, discussed in Section 9.4, use a rather *ad hoc* choice of gap penalties. An advantage of profile HMMs is that the transition probabilities are derived from the alignment itself, thereby providing more relevant scores for the insertions and deletions. Insertions will not be equally likely at all positions, and this information will be captured in the HMM training process.

During training, it may be necessary to change the number of match states in the model. For example, if more than half the sequences in the alignment turn out to have an inserted residue at a given position, it makes more sense to add an extra match state to represent this position. Similarly, if many sequences have a delete state in a given position, it is more sensible to remove the match state from this position and treat the few residues in this position as insertions. Training routines can be designed that make such changes, where necessary, at each iteration. Another refinement of the training process is the possibility of giving different weights to different sequences. This is important when the alignment used contains groups of closely related sequences and a few outliers. If sequences are weighted equally, the model may specialize to describe the majority very well but the outliers very poorly. A more general model, describing all the sequences in the alignment reason-ably well, would be preferable. This can be achieved by downweighting closely related sequence groups (see Durbin *et al.* 1998 for more details).

There are several points about the architecture of the profile model worth noting. The model shown in Fig. 10.4(a) has no connections between insert and delete states, which means that it is not possible to have an inserted residue in the alignment immediately followed by a deletion. Connections of this type can be added if desired (e.g., I0 → D1, D1 → I1 etc.), and they were in fact included in the original profile model used by Krogh *et al.* (1994). However, the probabilities associated with these transitions should be small, because if an insertion occurred next to a gap, it would usually be more natural to move the inserted residue into the match state, thus eliminating both insertion and gap.

The layout shown has insertion states at the beginning and end; thus, when a sequence is aligned to a model, any number of unaligned residues can be included before and after the aligned part. Essentially, the alignments created will be local with respect to the sequence (because unaligned sections can occur before and after the match states), but global with respect to the model (because once the match states are entered, the path must proceed through the whole model). The Plan 7 architecture used in HMMER (Eddy 2001) makes this more explicit, by

including states representing non-aligned residues before and after the main profile model. A loop from the end to the beginning is also added containing a state for unaligned residues. This means that the model can represent multiple occurrences of the same domain in a protein sequence, or multiple occurrences of the same gene on a genome. Another feature that can be included in the Plan 7 model is a transition from the begin state of the main model to each of the internal match states, and transitions from the internal match states to the end of the main model. This means that paths can pass through only part of the aligned region, making the model a local alignment with respect to both the model and the sequence.

If we have an HMM based on a given alignment, it is possible to search databases for other sequences that match the model. The principle is to calculate a log-odds score that tells us which sequences in the database are more likely to be described by the HMM model parameters than by a general null model of random sequences. High-scoring sequences can be aligned with the model, and a new model can then be built, based on the extended sequence alignment. The whole process can then be iterated. The Pfam database (Bateman *et al.* 2002) is a library of profile HMMs built up in this way. The library can be used to assign new proteins to families according to which HMM gives the highest score for the new sequence.

In general, HMMs are extremely powerful for discriminating between protein families but, as with all methods, they have certain limitations. Perhaps the biggest concern is that an HMM will do exactly what it is trained to do. In Pfam, models are trained by means of largely handcrafted seed alignments, which generally provide reliable sources of data. However, if dissimilar sequences are fed into the modeling process, the model parameters will adapt to give higher likelihoods to these new sequences. This makes it more likely that the model will match additional incorrect sequences in the future. The addition of new sequences to the model therefore needs to be done with care in order to avoid progressive corruption and loss of information that was in the original seed alignment (a process known as profile dilution). The same point was also made with regard to PSI-BLAST in Section 7.1.3.

10.5 NEURAL NETWORKS

10.5.1 The basic idea

Neuroscientists have been impressed by the complex system of interconnected nerve cells in the brain. It is clear that the information processing and storage capabilities of the brain are distributed throughout a network of cells, and that each single neuron has a relatively simple behavior. Computer scientists have been inspired by this idea to create artificial neural networks (ANNs) that can be trained to solve particular problems, many of which fall into the category of pattern recognition. ANNs have been used very widely and successfully in many different research disciplines and have generated a huge literature. In fact, most scientists in the field omit the word "artificial". They are motivated to use NNs as a practical means of solving their own specific problems and are not too worried whether the network model has any relationship to real neural systems. The purpose of this section is to discuss how NNs can be used to solve pattern-recognition problems in bioinformatics.

A typical NN might look like Fig. 10.6. Circles represent individual neurons, each of which has several inputs and one output. The output signal of each neuron is a real number between 0 and 1. It is convenient to think of the neuron as being "on" when its output is close to 1, and "off" when its output is close to 0. The output of each neuron is a function

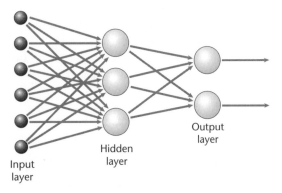

Fig. 10.6 Architecture of a typical feed-forward NN with an input layer, a hidden layer, and an output layer.

of its inputs. The network in Fig. 10.6 is termed feed-forward, because the outputs from one layer of neurons feed forward as inputs to the next layer (indicated by the arrows). The signals at the input layer (shown in black) are determined directly from the data shown to the network.

In bioinformatics applications, the data are usually short segments of protein or DNA sequence (say, 10 or 20 residues or nucleotides). A sequence can be encoded into inputs in several ways. The input signals determine the signals produced by the hidden layer, which in turn determine the signals of the output layer. The output signal represents the answer to the problem for the particular set of input data. The simplest type of problem is a "yes" or "no" classification. For this, we only need one output neuron, which outputs either 1 or 0. For example, we may want to know whether the sequence shown to the network is a member of a class of sequences of interest or not. If we have n output neurons, the output of the network can be in 2^n states, each of which could be associated with a different meaning. The output is often a predicted property of a single point in the sequence, such as the central residue in an input window, rather than a property of the whole window (e.g., it could represent whether the central residue is predicted to lie in an α helix). By sliding an input window along the sequence, the network can be used to make predictions for each residue.

The behavior of the network is determined by a set of internal parameters that control the output of each neuron. NNs are usually used in a supervised-learning framework. We have a training set of known examples of data, each with a known answer that we want the network to give. Training a network consists of choosing the internal parameters so that the network gives the correct output for each of the training examples (or for as many of the examples as possible). If the network is sufficiently complicated, it will usually be possible to train it to give almost all correct answers on the training set. We can then evaluate the performance of the network by using a test set, also of known examples. If the network gets a high fraction of correct answers in the test set, then we can say that it has learned useful general features of the training data, and that it is able to generalize to give reliable predictions on other sequences.

A useful review of early bioinformatics applications of NNs has been given by Hirst and Sternberg (1992). Problems tackled include the prediction of promoter regions in DNA and of mRNA splicing sites, and the location of signal peptide cleavage sites and of protein secondary structural elements, like α helices and β turns. NNs can be used for many different problems of this kind, provided we have sufficient data on which to train the model.

10.5.2 Data input

Usually in bioinformatics, the input to an NN will be a sequence. We wish to present a window of a DNA or protein sequence to the network and have it classify the input sequence in some way. To convert a DNA sequence to input values, it is usual to have four inputs, one for each base in the sequence (A, C, G, and T). To represent a specific input sequence, a binary code is used, where one of each group of four inputs will be set to 1, while the other three are 0. So a window of six DNA bases with sequence TTCGAA could be represented by 24 inputs with values

$$0\ 0\ 0\ 1\ 0\ 0\ 0\ 1\ 0\ 1\ 0\ 0\ 0\ 0\ 1\ 0\ 1\ 0\ 0\ 0\ 1\ 0\ 0\ 0$$

Protein sequences can also be encoded in this way, using 20 inputs per residue. This results in large networks, with large numbers of parameters to optimize. The number of inputs can be reduced by using quantitative properties of the amino acids, rather than a binary code: e.g., Schneider and Wrede (1993) used physicochemical properties of amino acids, like hydrophobicity, polarity, and surface area.

When real values are used as inputs, they are usually scaled into the range −1 to 1, or 0 to 1. As well as making the network simpler, using real amino acid properties gives the network a helping hand in pattern recognition: amino acids with similar properties will have similar inputs, whereas, when binary coding is used, all the amino acids appear equally different and the network itself has to learn any relevant similarities from the training data.

In these examples, each residue in the sequence is represented by one or more specific neurons. However, it is also possible to use global properties of the sequence as inputs, such as the frequencies of the residues within the input window, or the frequencies of codons in DNA sequences. For some problems in bioinformatics, the input is not a property of the sequence data at all: e.g., Murvai *et al.* (2001) used a network for classifying proteins into different families, where the inputs were BLAST scores and associated statistical parameters for matching the query sequence with the families. For further information, the book by Wu and McLarty (2000) contains a useful section on encoding data for input to NNs.

10.5.3 The single neuron

The model used for a single neuron is shown in Fig. 10.7. The inputs to the neuron are denoted y_i (where $i = 1 \ldots n$). If the neuron is on the first layer of a network, then the inputs represent the data. If the neuron is in the interior of a network, then the inputs to this neuron are the outputs of other neurons. Each connection from an input to the neuron j has a weight w_{ij} associated with it. These weights may be positive or negative in sign. The

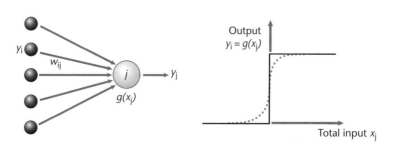

Fig. 10.7 A single artificial neuron, with several inputs and one output. The output is a function of the total input, and can either be a step function (solid line) or a sigmoidal function (dotted line), as in Eq. (10.38).

total input, x_j, to neuron j is a weighted sum of the individual inputs:

$$x_j = \sum_{i=1}^{n} w_{ij}\, y_i \qquad (10.37)$$

Neuron j has an output y_j that is a function of its total input: $y_j = g(x_j)$. The simplest possible output function is a step function, where the output is 0 if the input is negative, and the output is 1 if the input is positive. It is more usual, however, to use a sigmoidal function for $g(x_j)$, as shown in Fig. 10.7, the most common one being

$$g(x_j) = \frac{1}{1 + \exp(-x_j)} \qquad (10.38)$$

This function changes smoothly from 0 to 1, when the input changes from negative to positive. The advantage of this is that the derivative dg/dx_j is well defined for the sigmoidal function, and we will need the derivative later in the NN learning algorithm. At the moment, the neuron "switches on" when the input is above zero. However, we may want to introduce a non-zero threshold to the neuron, so that it only switches on when the input is above some value, θ_j. This can be done by subtracting θ_j from the total input, and using the same function $g(x_j)$ for the output. It is convenient to write

$$x_j = \sum_{i=1}^{n} w_{ij}\, y_i - \theta_j = \sum_{i=0}^{n} w_{ij}\, y_i \qquad (10.39)$$

This shows that the threshold can be included in the sum as an extra input, y_0, that is set permanently to 1, with a weight $w_{0j} = -\theta_j$. This extra input is known as a bias.

10.5.4 The perceptron

A network consisting of a single output neuron with multiple inputs, as shown in Fig. 10.7, is known as a perceptron. A perceptron can be trained to solve certain problems of the yes or no type: it outputs either 1 or 0 for any given set of input data (when we use the sigmoidal output function, we will interpret everything with $y_j > 1/2$ as a 1 and with $y_j < 1/2$ as a 0).

Consider a perceptron with only two inputs. To train the perceptron, we need a training set with examples of input data that we know belong to class 1 and to class 0. Data are in the form of pairs of numbers (y_1^n, y_2^n), where n labels the nth example in the training set. We will denote the correct output (or "target") for the training examples as t^n, which is either 0 or 1.

From Eq. (10.39), the total input, including the bias, is $x^n = w_0 + w_1 y_1^n + w_2 y_2^n$. We have dropped the subscript j for the output neuron because there is only one. The output for each of the training examples is $y^n = g(x^n)$. Training consists of choosing the weights so that $y^n = t^n$ for every n, if possible. Consider the following training set:

$$
\begin{array}{lll}
(y_1^n, y_2^n) & \rightarrow & t^n \\
1. \quad (0, 1/2) & & 1 \\
2. \quad (1, 1) & & 1 \\
3. \quad (1, 1/2) & & 0 \\
4. \quad (0, 0) & & 0
\end{array}
\qquad (10.40)
$$

To begin, suppose we are using a step function output, where the output depends only on the sign of the input. In order for the output to be correct, we require $x^n > 0$ for the examples where $t^n = 1$, and $x^n < 0$ for the examples where $t^n = 0$. This gives us four inequalities:

$$w_0 + \frac{1}{2} w_2 > 0$$

$$w_0 + w_1 + w_2 > 0$$

$$w_0 + w_1 + \frac{1}{2} w_2 < 0 \qquad (10.41)$$

$$w_0 < 0$$

We need to choose some values of w so that all these inequalities are true. The diagram in Fig. 10.8 allows us to spot a solution immediately. The dotted line has the equation

$$y_2 = \frac{1}{4} + \frac{1}{2} y_1 \qquad (10.42)$$

All points (y_1, y_2) above the line have

$$-\frac{1}{4} - \frac{1}{2} y_1 + y_2 > 0 \qquad (10.43)$$

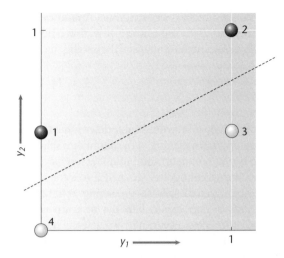

Fig. 10.8 Illustration of the perceptron problem defined by Eq. (10.40). The two black points must be mapped to output 1, and the two white points must be mapped to output 0. Finding a solution consists of choosing a line that separates the black and white points.

while the same function is less than 0 for all points below the line. In other words, a solution to the problem is to set

$$w_0 = -\frac{1}{4}; \quad w_1 = -\frac{1}{2}; \quad w_2 = 1 \qquad (10.44)$$

Finding the network weights is equivalent to finding a line that separates points 1 and 2 from 3 and 4. We now know how the network will behave in future: if a new example of data (y_1, y_2) is put into the network, it will output 1 if the point is above the dotted line, and 0 if it is below.

From Figure 10.8, we see that there is some freedom to move the line up and down and to change its slope. Any line that separates the black and white points will give the same correct outputs on the training data, but the outputs given to future inputs will depend on where we put the line. If we had more examples in the training set, this would give more constraints on the choice of the weights, and the reliability of future predictions would be better. This example shows why we need the bias term w_0. Without this, the dotted line would have to pass through the origin in the figure, which would not be a correct solution.

In a general perceptron problem with N inputs, we can consider the training examples as points in N-dimensional space. Choosing the weights consists of finding a surface in $N - 1$ dimensions that separates the two sets of points. Not all problems of the **yes–no** type can be correctly solved by perceptrons. For example, if we change the targets in Eq. (10.40) so that points 1 and 3 should give output 1, and points 2 and 4 should give output 0, then there is no line that will correctly separate the points, and hence there is no choice of weights such that the network will get the correct output for every example in the training set. If a line (or hyperplane in $N - 1$ dimensions) does exist, then the problem in known as **linearly separable**. Most interesting problems, however, are not linearly separable. This means that a perceptron cannot solve the problem, and we need more complex networks with multiple layers.

Before leaving the perceptron, let us consider what happens if we use the sigmoidal output function $y^n = g(x^n)$ in Eq. (10.38), instead of the step function. In this case, the targets, t^n, are still 0 or 1, but the outputs, y^n, are continuous numbers between 0 and 1 that we would like to be as close as possible to 0 or 1. We can define an error function, E, that measures how close the outputs are to the targets:

$$E = \sum_n (y^n - t^n)^2 \qquad (10.45)$$

The logical way to choose the weights is to fix them so that E is as small as possible. To minimize E, we need to find weights such that $\dfrac{\partial E}{\partial w_i} = 0$, for $i = 0$, 1, and 2. The choice of weights given in Eq. (10.44) is actually the one that minimizes E in this example. This is intuitively obvious from the symmetry of the problem: the line shown in Fig. 10.8 goes "midway" between the black and white points. The key point here is that when we have a sigmoidal output function, we have an unambiguous way to define the best choice of weights that gives a particular solution. We are not free to move the line about, as we were with the step function.

10.5.5 Multi-layer networks

A feed-forward NN of the type shown in Fig. 10.6 can be used to solve many types of problem that are not linearly separable and cannot be solved by perceptrons. There can be more than one hidden layer between the input and output layers in some cases, although a single layer is often sufficient for many realistic problems. Let us suppose that there are several output neurons, and that y_k^n is the output of the kth output neuron, when the nth example from the training set is put into the network. We now define the target value of the kth neuron as t_k^n. We are free to choose the target patterns in any way that seems sensible. As an example, a network with three outputs can be used for protein structure prediction with three categories: α helix, β strand, and loop. The target patterns $(0,0,1)$, $(0,1,0)$, and $(1,0,0)$ can be used to represent the three categories. The same problem could also be tackled using two output neurons, with $(0,1)$ and $(1,0)$ representing α helix and β strand, and $(0,0)$ representing loop. The first way is probably easier to interpret, because when the outputs have continuous values, we can check which of the three neurons has the highest output value in order to predict the state.

Using sigmoidal output functions for the neurons, the error made by the network when shown the nth input is a sum over the output neurons

$$E^n = \sum_k (y_k^n - t_k^n)^2 \qquad (10.46)$$

and the total error for the whole training set is

$$E = \sum_n E^n \qquad (10.47)$$

We wish to find a set of weights, w_{ij}, that minimizes E. There is one weight for every connection in the network. This means that E is a complex function of many variables, and we need a numerical method to find the minimum. A general way of minimizing a function is called gradient descent. Beginning with some initial guess at the set of weights, w_{ij}, we calculate E and also the partial derivatives $\dfrac{\partial E}{\partial w_{ij}}$ with respect to each of the weights; these are the slopes of the function we are trying to minimize. The best way to find the minimum is to go down hill. We therefore set the new estimates of the weights, w_{ij}', to be

$$w_{ij}' = w_{uj} - \eta \frac{\partial E}{\partial w_{ij}} \qquad (10.48)$$

Here, η is a constant whose value determines the size of the step we are making in the parameter space: its value is not too important, but if it is too large, the algorithm will not converge, and if it is too small, it will be unnecessarily slow. Having updated the weights, we can recalculate the derivatives and repeat the process. Eventually, we will reach a set of weights where all the derivatives are zero. This is a minimum in E. We cannot guarantee that this is a global minimum, however, so it is necessary to run the minimization procedure from several starting points and to make sure that we end up in the same place. If so, we can be reasonably confident that we have found the global minimum.

The gradient-descent method is a general way of minimizing a function. The trick to applying it with NNs is to calculate the partial derivatives. The algorithm for doing this is called back-propagation, and is explained in Box 10.4. This algorithm is due to Rumelhart, Hinton, and Williams (1986). Although some algebra is required to derive the result, the final formula for the derivatives is simple, and the method is easy to implement on a computer. Back-propagation algorithms and related methods are often used in practice. More information is given by Baldi (1995) and Baldi and Brunak (2002).

10.5.6 How many neurons are necessary?

The number of neurons necessary on each layer will depend on the complexity of the problem. Increasing the number gives the system greater flexibility and makes it more likely to be able to give correct answers to each example in the test set. However, this increases the number of free parameters to be estimated from the training data (i.e., the weights, w_{ij}). When there are too many neurons, the problem of overfitting may arise. For a network to be useful, it must be able to generalize. In other words, it must be able to learn

BOX 10.4
The back-propagation algorithm

Our object is to calculate the partial derivatives for use in the gradient-descent Eq. (10.48). These can be written as sums over the examples in the training set:

$$\frac{\partial E}{\partial w_{ij}} = \sum_n \frac{\partial E^n}{\partial w_{ij}} \qquad (10.49)$$

so we can consider just one training example at a time, and sum the contributions at the end. When we change the weight w_{ij}, we change the input x_j at neuron j. From Eq. (10.37) we know

$$\frac{\partial x_j}{\partial w_{ij}} = y_i^n \qquad (10.50)$$

Using the chain rule, we can write

$$\frac{\partial E^n}{\partial w_{ij}} = \frac{\partial x_j}{\partial w_{ij}} \frac{\partial E^n}{\partial x_j} = y_i^n \delta_j^n \qquad (10.51)$$

where we have defined δ_j^n to be $\frac{\partial E^n}{\partial x_j}$. But we still need to calculate δ_j^n. Suppose, initially, that j is an output neuron. This neuron therefore contributes directly to the error function in Eq. (10.46); differentiating this, we have

$$\delta_j^n = \frac{\partial E^n}{\partial x_j} = 2(y_j^n - t_j^n) \frac{dy_j^n}{dx_j} \qquad (10.52)$$

Note that changing x_j affects y_j, but does not affect the other outputs, y_k, so only one term in the sum (10.46) remains after we take the derivative. Also, we know that y_j is a direct function $g(x_j)$ of its input. For the sigmoidal function (10.38), we have

$$\frac{dy_j^n}{dx_j} = \frac{d}{dx_j}\left(\frac{1}{1 + \exp(-x_j)}\right) = \frac{\exp(-x_j)}{(1 + \exp(-x_j))^2}$$
$$= y_j^n(1 - y_j^n) \qquad (10.53)$$

Hence, we have δ_j^n for each output neuron in terms of the output signal y_j^n that we know already:

$$\delta_j^n = 2(y_j^n - t_j^n)y_j^n(1 - y_j^n) \qquad (10.54)$$

Now suppose that neuron j is in the hidden layer, and therefore does not contribute directly to the error function (10.46). The output of j nevertheless goes into the neurons in the output layer, and therefore neuron j affects the error indirectly via all the connections from j to the output layer. Therefore, we may write

$$\delta_j^n = \frac{\partial E^n}{\partial x_j} = \frac{dy_j^n}{dx_j} \frac{\partial E^n}{\partial y_j^n} = y_j^n(1 - y_j^n) \sum_k \frac{\partial x_k}{\partial y_j^n} \frac{\partial E^n}{\partial x_k} \qquad (10.55)$$

The sum above is over all the neurons k to which j is connected. We can now write both the derivatives in this sum in terms of things we already know. This gives a closed form equation for the δ_j^n in the hidden layer:

$$\delta_j^n = y_j^n(1 - y_j^n) \sum_k w_{jk} \delta_k^n \qquad (10.56)$$

As the neurons k are in the output layer, we can calculate δ_k^n for these neurons from Eq. (10.54). This is the origin of the term back-propagation. We first calculate the δ values for the output neurons, then we work back to the hidden layer using Eq. (10.56) – in networks with more than one hidden layer, we can also use this equation to work back from one hidden layer to the previous hidden layer. Hence, with Eqs. (10.54) and (10.56) we have all the information needed to calculate the δ values and the partial derivatives for a network of any size. Combining everything together, the gradient-descent rule (10.48) can be written:

$$w_{ij}' = w_{uj} - \eta \sum_n y_i^n \delta_j^n \qquad (10.57)$$

To use the method, we go through the network from input layer to output layer in order to calculate the y_j^n values for each neuron. The δ_j^n are then calculated from the y_i^n by working back from the output to the input.

patterns in the training data, and then be able to spot these patterns when they occur in other examples. If a network is overfitting the data, it may very precisely distinguish all the examples in the training set, but give very poor predictions on other examples. It is very important that neural networks should be evaluated using a test set that is distinct from the data used for training. Good performance on the training set is not necessarily a predictor of good performance on the test set. There is no good theory to tell us how to design a network for a particular problem, and it is often not known how large a network will need to be until tests are made with real data.

The work of Schneider and Wrede (1993) provides an instructive example. Signal peptides are short sections at the N-terminal end of certain proteins that are responsible for targeting these proteins to the secretory pathway. The signal peptides are later cut from the protein by a signal peptidase. The problem of Schneider and Wrede (1993) was to predict the position of signal-peptidase cleavage sites. The amino acid sequences in the known examples were not very well conserved, but it was hoped that an NN would be able to spot conserved patterns in the amino acid properties. They used a window of 13 residues from positions -10 to $+3$ in the sequence relative to the cleavage site. There was a single output neuron that signaled 1 if a cleavage site was predicted to be between residues -1 and 1. Inputs were real numbers representing physico-chemical properties of the amino acids. Four different network designs were compared: a perceptron with one input per residue, representing any single amino acid property; a network with one hidden layer and one input property per residue; a network with one hidden layer and several input properties per residue; and a network with two hidden layers and several input properties per residue.

In this study, even the simplest perceptron model was able to make a reasonable attempt at the problem. The single, most useful amino acid property used as input was found to be hydrophobicity, for which the network scored 91% correct predictions on the test set. This increased to 97% for the network with a single hidden layer and one input property. The networks with two hidden layers performed well on the training set, but poorly on the test set, i.e., they were overfitting the training data and could not generalize to the test set. The number of neurons in the hidden layers was also varied, and sometimes overfitting occurred if this number was too large.

Signal-peptide recognition is a well-studied problem in bioinformatics. The SignalP Web server is now available for predicting signal-peptide cleavage sites (Nielsen *et al.* 1997). Two networks are used, one to distinguish between signal peptides and non-signal peptides, and another to predict the position of the cleavage site. The server also offers an HMM method for the same problem. Another NN method for signal peptides (Jagla and Schuchhardt 2000) is worth noting because of the way the sequence is encoded. Each amino acid is represented by two real numbers in the same way as when physicochemical properties are used. However, these numbers are not assigned in advance. Instead, they are treated as parameters of the model that can be learned by the network. As well as learning to recognize the cleavage sites, the network determines some "pseudo properties" of the amino acids that are most useful for predicting the cleavage sites. It was found that two pseudo properties generated by the network had some correlation with real properties (hydrophobicity and chromotographic retention coefficient), although this correlation was not particularly strong.

A general point about NN design is that the larger the set of training examples used, the more complex the network can be without suffering from overfitting. This is because the training examples put constraints on the model parameters, and provide extra information with which to estimate the parameters. When there are large numbers of training examples, the model cannot specialize to exactly match any particular sequence, without reducing its ability to match other sequences. It is therefore forced to do its best to match average consensus patterns in the whole data set, which is what we want. With smaller training sets, it is better to stick to simpler models.

10.6 NEURAL NETWORKS AND PROTEIN SECONDARY STRUCTURE PREDICTION

Protein secondary structure is usually represented by three states, α helix, β strand, and loop. Note that a β strand is a segment of a chain that may be hydrogen-bonded with other strands to form a β sheet, and a loop is (by default) anything that is not an α helix or a β strand (see Section 2.2 for some examples of protein structures). It has been known for a long time that different residues have different frequencies in the three structural states. Scoring systems that measure the propensity of each amino acid for each of the states have been devised, and used as the basis of secondary structure prediction methods (Chou and Fasman 1978, Garnier, Osguthorpe, and Robson 1978). These methods have had only moderate success (largely because 25 years ago the data sets available for training were too small – only 15 proteins were used to train Chou and Fasman's first method, and 64 proteins were used in a later update!). The predicted output state for each residue is better than a random guess, but is not sufficiently reliable to draw useful conclusions about the structure of an unknown protein. Protein structure prediction is a notoriously difficult problem. We expect the secondary structure of any given residue to be strongly influenced by the sequence in which it finds itself. One reason to hope that NNs will be useful here is that they can combine information from the whole of the input sequence window (typically each hidden layer neuron is connected to every input neuron). NNs should, in principle, be able to spot patterns involving correlations in residues at different points in the input window. Although NN training is supervised (we tell the program what the answer is for each training example), the learning algorithm is free to assign the weights any way it likes; so it may be able to spot patterns that were not obvious features visible to the human eye.

The most straightforward approach to secondary structure prediction is to represent each input amino acid by 20 binary inputs, and to use three output neurons, with target patterns $(0,0,1)$, $(0,1,0)$, and $(1,0,0)$ to represent the three structural states possible for the central residue in the window, as discussed previously. There have been many variants of this idea; Wu and McLarty (2000) have given a good review of these. One of the earliest and most informative papers using this method is that of Qian and Sejnowski (1988). An important additional feature introduced by these authors is the use of a second network to filter the output of the first. The second network uses the set of three output numbers for each residue from the first network as its inputs. Once again, there are three outputs for the second stage that predict the structural state of the central

residue. The second stage is able to correct some of the mistakes made by the first stage; in particular, it is able to learn that when the first stage predicts small numbers of residues of one type in the middle of a sequence of a second type (e.g., a couple of helical residues in the middle of a β strand) this should really be an unbroken sequence of the second type. An accuracy of 64% correctly predicted residues in the training set was obtained in this study.

Qian and Sejnowski (1988) show how the average magnitude of the weights in their network changes during the training process. Large weight magnitudes (of either positive or negative sign) indicate that the feature on the corresponding input has an important role in predicting the target output. The magnitude of the important weights therefore grows during training. Input features that are essentially independent of the target output merely put noise into the network. We expect training to reduce the magnitudes of the corresponding weights. In this study, it was found that weights corresponding to the few residues in the center of the input window grew more rapidly during training than those corresponding to the outer residues in the window. This is intuitive, because we would expect the state of the central residue to be more sensitive to its immediate neighbors than to residues further away along the sequence.

In a binary encoding, one input of each group of 20 corresponds to each amino acid. Qian and Sejnowski (1988) also show how the weights for each amino acid vary with the position in the window. Some amino acids are stronger predictors of secondary structure than others: e.g., for proline, the weights corresponding to the α-helix output are strongly negative, and those corresponding to the loop output are positive, indicating that proline rarely occurs in helices and usually occurs in loops (proline has an unusual chemical structure that does not usually fit into helical geometry (Fig. 2.6), although TM domains in membrane proteins are an exception to this); other amino acids, like alanine, leucine, and methionine, have positive weights for the α-helix output; and some of the weights show patterns that are not symmetric about the central residue, meaning that the influences of upstream and downstream residues on the central residue are not necessarily equivalent. Examination of the trained values of the weights reveals the information the network is able to learn about the statistics of protein sequences. This study is also illustrative of the problem of overfitting. A network with 40 neurons in the hidden layer can be trained to predict the structures in the training set much more accurately than a network without a hidden layer. Unfortunately, the more complex network performs no better than the simpler one on the test set.

A considerable improvement in prediction accuracy was achieved by Rost and Sander (1993). Their network, known as PHDsec, uses profiles as input rather than single sequences. The network is trained using alignments of proteins with known consensus secondary structures. Twenty inputs per residue are used for the 20 amino acid frequencies at each position in the alignment. Additional inputs detail the fraction of insertions and deletions. When a new sequence is submitted to the network, an automatic multiple alignment is made by searching Swiss-Prot, and a profile is made from this alignment for input to the network. For each position in the window, there is an extra spacer input whose value is set to 1, if this position is beyond the end of the protein. This will occur if the central residue for which the prediction is made is near the end of the sequence. Global information about the whole of the protein is also used as input: this includes the total length of the protein, the distance of the central residue of the window from each end of the protein, and the frequencies of the 20 amino acids in the whole protein. Rost and Sander (1993) also use the idea of a jury of networks. Networks trained in slightly different ways may give different predictions. It is therefore possible to combine the outputs of several different networks by some sort of a voting procedure to give a single final output. This hopefully reduces problems of noise and overfitting in the outputs of any one of the networks used.

For a long time, PHDsec has been one of the best available methods for protein secondary structure prediction, with a claim of over 70% accuracy in assignment of residues to structural states. Sequences can be automatically submitted to PHDsec via the PredictProtein server (Rost 2000). Two other groups have recently produced NN methods that claim small but noticeable improvements in accuracy. Jones (1999) obtained 76–8% correctly predicted residues, and Petersen et al. (2000) reached 77–80%; both methods use profiles as input that are constructed automatically by PSI-BLAST.

Percentage accuracies of different methods are directly influenced by the choice of sequences used to test them. In principle, there should be no similarity between sequences in the training and test sets. If the test sequences are homologous to the training sequences, then we expect the performance on the test set to be falsely high, and to overestimate the performance to be expected in future. Most researchers recognize this point and go to lengths to be fair in assessing their results. However, it is far from clear what criterion should be used to consider sequences as being independent of one another. Real sequences will be found with a continuous range of similarities. At what level do we make the cut-off? Should we use a cut-off based on sequence similarity or structural similarity? There is no single, easy answer to this.

The protein structure community has become very organized in assessing methods of structure prediction, both at the secondary and tertiary levels. In view of the general difficulty of the problem, the large number of alternative approaches, and the disappointingly poor performance of many methods, a series of large-scale experiments, called CASP (Critical Assessment of Structure Prediction), has been run allowing different research groups to test their methods on the same sequences. The CASP organizers release the sequences of proteins whose structures are about to be determined experimentally. Structure prediction groups submit their predictions to the organizers prior to publication of the experimental structure. Results of different groups can then be compared when the true structures are published. The latest two experiments are CASP4 (Moult et al. 2001) and CASP5 (Moult et al. 2003).

Although we touch on structure prediction at several points in this book, a thorough review of all possible methods is beyond our current scope. In the context of this chapter on pattern recognition, a general point about prediction methods is that the most successful are knowledge based – i.e., they look for similarities to sequences of known structure, or they use training sets of known examples (as with machine-learning approaches). *Ab initio* methods, beginning only with a single sequence and fundamentals, such as interatomic forces, tend to be less successful. Thus, pattern-recognition techniques are providing an answer to the practical question of structure prediction (what is the structure of this protein?) and gradual improvements are being made. However, Nature does not have access to sets of training data or profiles of PSI-BLAST hits. Even if a method, such as an NN, can be developed with 100% accuracy, we will still not really understand the fundamental question of how a protein folds, unless we also understand the problem from an *ab initio* point of view. A somewhat dissatisfying aspect of NNs is that it is not always possible to say exactly **why** they work, even when they do.

SUMMARY

A common task in bioinformatics is to classify sequences into a number of different categories. The simplest case would be where there are just two categories – "yes" the sequence belongs to a group of sequences of interest, and "no" the sequence does not belong to it. A more general case is when there are several categories, and it is necessary to decide into which category the new sequence best fits. Probabilistic models can be developed to address questions of this type. It is necessary to define a model for each category, from which we can calculate the likelihood of the sequence data according to the model. We may also define prior probabilities for each model. These are our expected probabilities of finding a new sequence in each category, before we look at the new sequence itself. Within a Bayesian framework, we want to determine the posterior probability of the model, given the data, which is proportional to the prior probability of the model multiplied by the likelihood of the data given the model. Using these principles, we can make inferences about sequences that are based on sound statistics. A useful way of expressing the posterior probabilities is to convert them into log-odds scoring systems.

In a Markov model of a protein sequence, the probability that any particular amino acid occurs at a given position depends on which amino acid lies immediately before it, or on which combination of amino acids occurs at several positions before it. In an HMM, the probability of an amino acid occurring depends on the values of hidden variables determining which state the sequence is in at that point (e.g., a helix or loop region of a membrane protein). HMMs can be trained to represent families of sequences. When a given sequence is compared to the model, the optimal path of the sequence through the states of the model can be calculated. This allows prediction of which parts of a sequence corres- pond to which states in the model. A particular case of this is profile HMMs, where the model describes a sequence alignment, and calculating the optimal path through the model corresponds to aligning the new sequence with the profile.

A machine-learning algorithm is one that learns to solve a given problem by optimization of the internal parameters of the program. Often, the learning is supervised, i.e., the program is given a set of right and wrong examples called the training set. The parameters of the algorithm are adjusted so that the method performs as well as possible on the training set. By contrast, an unsupervised-learning algorithm proceeds by optimization of a pre-specified function, but is not supplied with a set of examples. Supervised-learning algorithms are evaluated by using a second set of known examples (the test set). A good algorithm should also be able to perform well on the test set, not just the training set. If this is the case, we say that the algorithm is able to generalize, and we expect predictions made on unknown data to be reliable.

NNs are machine-learning algorithms particularly suited to classification and pattern-recognition problems. Networks are usually composed of several layers of neurons. Each neuron has several inputs and one output. The output depends on the weighted sum of the inputs. Outputs of neurons in one layer become the inputs to neurons in the next layer. The input data (e.g., a sequence) are coded in the form usually of binary inputs to a set of neurons in the first layer of the network. The output of the final layer is the answer specified by the network (e.g., whether or not the sequence belongs to the class of sequences of interest). The network learns by optimizing the weights associated with the inputs to each neuron using the back-propagation algorithm. Some of the most successful methods of protein secondary structure prediction use NNs.

REFERENCES

Baldi, P. 1995. Gradient-descent learning algorithm overview: A general dynamical systems perspective. *IEEE Transactions on Neural Networks*, **6**: 182–95.

Baldi, P. and Brunak, S. 2002. *Bioinformatics: The Machine Learning Approach* (2nd Edition). Cambridge, MA: MIT Press.

Bateman, A., Birney, E., Cerruti, L., Durbin, R., Etwiller, L., Eddy, S.R., Griffiths-Jones, S., Howe, K.L., Marshall, M., and Sonnhammer, E.L.L. 2002. The Pfam protein families database. *Nucleic Acids Research*, **30**: 276–80.

Chou, P.Y. and Fasman, G.D. 1978. Prediction of the secondary structure of proteins from their amino acid sequence. *Advances in Enzymology*, **47**: 45–148.

Churchill, G.A. 1989. Stochastic models for heterogeneous DNA sequences. *Bulletin of Mathematical Biology*, **51**: 79–94.

Delorenzi, M. and Speed, T. 2002. An HMM model for coiled-coil domains and a comparison with PSSM-based predictions. *Bioinformatics*, **18**(4): 195–202.

Durbin, R., Eddy, S.R., Krogh, A., and Mitchison, G. 1998. *Biological Sequence Analysis: Probabilistic Models of Proteins and Nucleic Acids*. Cambridge, UK: Cambridge University Press.

Eddy, S. 2001. HMMER: Profile hidden Markov models for biological sequence analysis. http://hmmer.wustl.edu/.

Garnier, J., Osguthorpe, D.J., and Robson, B. 1978. Analysis of the accuracy and implications of simple methods for predicting the secondary structure of globular proteins. *Journal of Molecular Biology*, **120**: 97–120.

Hirst, J.D. and Sternberg, M.J.E. 1992. Prediction of structural and functional features of protein and nucleic acid sequences by artificial neural networks. *Biochemistry*, **31**: 7211–18.

Hughey, R., Karplus, K., and Krogh, A. 2001. SAM: Sequence alignment and modelling software system. http://www.cse.ucsc.edu/research/compbio/sam.html.

Krogh, A., Larsson, B., von Heijne, G., and Sonnhammer, E.L.L. 2001. Predicting transmembrane protein topology with a hidden Markov model: Application to complete genomes. *Journal of Molecular Biology*, **305**: 567–80.

Jagla, B. and Schuchhardt, J. 2000. Adaptive encoding neural networks for the recognition of human signal peptide cleavage sites. *Bioinformatics*, **16**: 245–50.

Jones, D.T. 1999. Protein secondary structure prediction based on position-specific scoring matrices. *Journal of Molecular Biology*, **292**: 195–202.

Lupas, A., Vandyke, M., and Stock, J. 1991. Predicting coiled coils from protein sequences. *Science*, **252**: 1162–4.

Moult, J., Fidelis, K., Zemla, A., and Hubbard, T. 2001. Critical assessment of methods of protein structure prediction (CASP): Round IV. *Proteins*, Suppl. 5, 2–7.

Moult, J., Fidelis, K., Zemla, A., and Hubbard, T. 2003. Critical assessment of methods of protein structure prediction (CASP): Round V. *Proteins*, Suppl. 6, 334–9.

Murvai, J., Vlahovicek, K., Szepesvari, C., and Pongor, S. 2001. Prediction of protein functional domains from sequences using artificial neural networks. *Genome Research*, **11**: 1410–17.

Nicolas, P., Bize, L., Muri, F., Hoebeke, M., Rodolphe, F., Ehrlich, S.D., Prum, B., and Bessieres, P. 2002. Mining *Bacillus subtilis* chromosome heterogeneities using hidden Markov models. *Nucleic Acids Research*, **30**: 1418–26.

Nielsen, H., Engelbrecht, J., Brunak, S., and von Heijne, G. 1997. Identification of prokaryotic and eukaryotic signal peptides and prediction of their cleavage sites. *Protein Engineering*, **10**: 1–6. http://www.cbs.dtu.dk/services/SignalP-2.0/

Qian, N. and Sejnowski, T.J. 1988. Predicting the secondary structure of globular proteins using neural network models. *Journal of Molecular Biology*, **202**: 865–84.

Rost, B. 2000. The predict protein server. http://www.embl-heidelberg.de/predictprotein/predictprotein.html.

Rost, B. and Sander, C.J. 1993. Improved prediction of protein secondary structure by use of sequence profiles and neural networks. *Proceedings of the National Academy of Sciences USA*, **90**: 7558–62.

Rumelhart, D.E., Hinton, G.E., and Williams, R.J. 1986. Learning representations by back-propagating errors. *Nature*, **323**: 533–6.

Schneider, G. and Wrede, P. 1993. Development of artificial neural filters for pattern recognition in protein sequences. *Journal of Molecular Evolution*, **36**: 586–95.

Wu, C.H. and McLarty, J.W. 2000. *Neural Networks and Genome Informatics*. Amsterdam: Elsevier.

PROBLEMS

10.1

(a) A promoter protein, P, is discovered to bind to sites containing the sequence AGCCGA in genomic DNA. A newly sequenced organism has a genome that contains 30% A and T, 20% G and C. Assuming that the bases in the genome are independent, how many binding sites for P would you expect to find by chance in a 1 Mb region of genome?

(b) How sensible is the assumption that DNA bases are independent? How could the predictions in part (a) be made more accurate if correlations between DNA bases were included?

10.2 In Section 10.2.3, we considered a model for protein sequences where the frequency of amino acid a was q_a, and the amino acid at each site was assumed to be chosen independently. In a particular sequence, let the observed number of amino acids of type a be n_a, and let the total number of amino acids be n_{tot}. The likelihood of this sequence, according to the model, is

$$L = \prod_{a=1}^{20} (q_a)^{n_a} \tag{1}$$

The problem is to determine the values of q_a that maximize L. We will give the proof – make sure you follow each step.

First, it is easier to maximize $\ln L$, rather than L:

$$\ln L = \sum_{a=1}^{20} n_a \ln q_a + \lambda \left(1 - \sum_{a=1}^{20} q_a \right) \tag{2}$$

The first term above is the log of (1). The second term has been added because we are going to use the method of Lagrange multipliers, which allows us to maximize a function, subject to a constraint (check this in a mathematics book if you need to). In this case, the constraint is that all the frequencies must add up to one. λ is called the Lagrange multiplier, and the term in brackets must

be zero, i.e., we have just added zero to $\ln L$. Differentiating (2) with respect to one particular frequency, q_1, we obtain

$$\frac{\partial \ln L}{\partial q_1} = \frac{n_1}{q_1} - \lambda$$

This must be zero if the function is maximized. Hence, $q_1 = \lambda n_1$. Similarly, for each amino acid, $q_a = \lambda n_a$. Summing this over all amino acids, we obtain

$$\sum_a q_a = \lambda \sum_a n_a$$

from which $\lambda = 1/n_{tot}$. The final result is therefore that $q_a = n_a/n_{tot}$, which means that the ML value of the frequency is equal to its observed frequency in the data. This is what we wanted to prove in Section 10.2.3.

Try the same method for a DNA sequence where the observed numbers of bases are n_A, n_C, n_G, and n_T, and the frequencies in the model are π_A, π_C, π_G, and π_T. Obviously, the calculation is exactly the same.

Now try it for a DNA sequence model where it is assumed that the frequencies of A and T are equal, and the frequencies of C and G are equal. There are only two parameters to optimize this time. The likelihood is:

$$L = \pi_A^{n_A+n_T} \pi_C^{n_C+n_T}$$

10.3

(a) Explain how a profile HMM, such as the one shown in the below figure, could be used to model an aligned set of sequences. In particular, explain the role of each different type of state in the model and explain the meaning of the arrows between the states.

(b) Describe a possible path through the model for sequences 1 and 4 in the alignment given above.

(c) Draw a table containing the number of transitions between each type of state at each position in the model.

(d) How can the transition counts calculated in part (c) be used to determine ML values for the transition probabilities? Describe one disadvantage of using ML para-

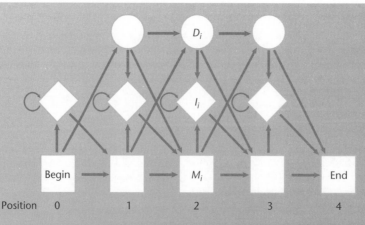

Position   0   1   2   3   4

meters and suggest an alternative method for setting the transition probabilities.

(e) Which algorithm can be used to find the optimal alignment of a new sequence to the model?

(f) In practice, it may be necessary to create a profile HMM from unaligned data. Why is this generally much harder than learning from aligned data?

(g) Given a library of HMMs modeling a number of protein families, how would you attempt to classify a novel sequence?

10.4

T^1_{XY}	A	C	T	G
A	564	848	1,381	369
C	1,193	2,502	2,566	556
T	1,231	2,637	2,808	990
G	174	830	911	439

t^1_{XY}	A	C	T	G
A	0.1784	0.2682	0.4367	0.1167
C	0.1750	?	?	0.0816
T	?	0.3440	0.3663	0.1291
G	0.0739	?	0.3870	0.1865

The top table above gives the number of neighboring nucleotide pairs found in DNA sequences selected from specific regions of the human genome (call these class 1 sequences). The lower table gives the transition probabilities of a Markov chain model estimated from these data.

(a) Explain how the transition probabilities were estimated and calculate the four missing values in the lower table. How would you justify this method of parameter estimation?

(b) Calculate the probability of generating the following sequence from a Markov chain model, with the transition probabilities estimated above,

TGCGCGCCGTG

(c) The transition probabilities below were calculated from a different set of sequences, known to have different characteristics (call these class 2 sequences).

t^2_{XY}	A	C	T	G
A	0.3001	0.2049	0.2850	0.2100
C	0.3220	0.2983	0.0776	0.3021
T	0.2481	0.2460	0.2982	0.2077
G	0.1770	0.2390	0.2924	0.2916

These transition probabilities are used to construct a second Markov chain model. Define a log-odds score that can be used to detect to which class a new sequence belongs. Calculate this score for the sequence given in part (b) and determine to which class it is most likely to belong, assuming equal prior probabilities for each class. How could the log-odds score be adjusted to deal with unequal prior probabilities for each class?

(d) Given that you have large numbers of classified sequences, how would you attach a level of confidence to the score derived in part (c)?

(e) Describe a probabilistic model of a sequence containing segments of DNA from both classes. Suggest two methods that could be used to estimate the model parameters. Which algorithm could be used to detect contiguous segments coming from a particular class?

Further topics in molecular evolution and phylogenetics

CHAPTER

11

CHAPTER PREVIEW

This chapter is a mixed bag of topics that follow on from the evolutionary models and phylogenetic methods discussed in Chapters 4 and 8. We begin with RNA sequence evolution, as rRNA phylogenies have a central role in evolutionary studies. We discuss the comparative method for determining RNA secondary structure, and also consider RNA structure prediction using free energy minimization programs. We consider how thermodynamics influences sequence evolution in RNA. Changing topic, we move on to discuss ways to estimate the parameters in evolutionary models by fitting to sequence data, and the use of likelihood ratio tests to select appropriate models for data. We then discuss a range of important biological problems that have been studied with molecular phylogenetics to illustrate the types of question that can be asked and the kinds of information that can be obtained.

11.1 RNA STRUCTURE AND EVOLUTION

11.1.1 Conservation of RNA secondary structures during evolution

Some of the most important genes for evolutionary studies are the small and large subunit rRNAs. These genes occur in all organisms, including both prokaryotic and eukaryotic genomes, and also in mitochondrial and chloroplast genomes. A typical example of a small subunit rRNA structure is shown in Plate 11.1. Each dot represents one nucleotide, and the paired, ladder-like, regions are actually double helices in the 3D configuration (as discussed in Chapter 2). The secondary-structure diagram shows which bases are paired, but does not clearly show the relative positions of the different parts of the molecule in 3D. The color scheme indicates the

degree of variability of the base at this position in a large set of bacterial sequences. Although some sites are completely conserved, other parts are quite variable in sequence. What is striking, however, is that the secondary structure changes very little between species. For almost all the helices shown in the *E. coli* structure, it is possible to find helices at the same molecular position in other species, even if their base sequences differ.

The typical length of this gene in bacteria is 1,500. In eukaryotes, it is significantly longer (up to 2,000), and in mitochondria it is significantly shorter (~950). With respect to the structure shown in Plate 11.1, there are significant numbers of insertions in the eukaryotic structure and deletions in the mitochondria. Nevertheless, the core part of the structure contains many helices that are clearly homologous between all three types of sequence. It is instructive to compare the structure diagrams from different organisms and organelles. These can be downloaded from either the Comparative RNA Web site (http://www.rna.icmb.utexas.edu/) or the European rRNA database (http://oberon.fvms. ugent.be:8080/rRNA/index.html). Owing to the length variation, rRNA genes have several different names. Small subunit rRNA is called 16S in bacteria, 18S in eukaryotes, and 12S in mitochondria.

Large subunit rRNA is called 23S in bacteria, 28S in eukaryotes, and (unfortunately!) 16S in mitochondria. The S numbers are sedimentation coefficients, and are essentially measurements of the size of the molecule. Despite the different numbers, there are only two types of gene, which we will call the small subunit and large subunit rRNAs – henceforward, SSU and LSU, respectively.

SSU rRNA has been sequenced in a very wide range of organisms, because it is extremely informative in phylogenetic studies. It is an essential gene under strong stabilizing selection, and therefore it evolves rather slowly (at least, the core parts of the molecule evolve slowly). This means that sequences from quite divergent organisms can be aligned with reasonable confidence. As we have seen in previous chapters, it is often necessary to refine automatically produced sequence alignments manually. The conserved secondary structure in rRNA is an important source of biological information that can be used when constructing RNA alignments.

SSU rRNA became very popular in phylogenetics owing to the remarkable discovery, made very early on, of archaea, the third domain of life. Woese and Fox (1977) compared SSU rRNA from a range of species using electrophoresis, before full sequences of these genes were available. They used a ribonuclease enzyme to break the gene into oligonucleotide fragments, and separated the fragments by electrophoresis to produce a characteristic fingerprint for each species. They defined a similarity score for two species that was a measure of the fraction of oligonucleotide fragments shared by the two species. This study showed that, rather than there being two fundamental types of organism, prokaryotes and eukaryotes, as had previously been supposed, there were in fact three. The prokaryotes were divided into two domains, now called bacteria and archaea, which were as different from each other as they were from eukaryotes. This division stood the test of time, and is clearly shown in phylogenetic trees using full SSU rRNA sequences (Woese, Kandler, and Wheelis 1990). The phylogenetic tree of SSU rRNA is frequently used for studying the relationship between the major groups of bacteria and archaea, and also for the early branching groups of eukaryotes. Information is now also available from some protein-coding genes for these ancient parts of the tree of life, but few protein-coding genes are as reliable and informative as rRNA for these questions. Thus, rRNA gets to the parts of the tree that other genes cannot reach, and the rRNA phylogeny is often treated as a standard to which the phylogenies of other genes are compared (see also Section 12.1).

11.1.2 Compensatory substitutions and the comparative method

The structures of SSU and LSU rRNA are obtained using what is referred to as "the comparative method". This involves examining sequence alignments of RNA genes to find pairs of sites that covary in a non-random fashion. To see how this works, we will consider tRNA. The cloverleaf structure of a typical tRNA molecule was shown in Fig. 2.4. Figure 11.1(a) shows an alignment of a set of tRNA genes (these happen to be mitochondrial tRNA-Leu genes). The gray shading illustrates conservation of sequence between species. The three bases marked X denote the position of the anticodon. The secondary-structure pattern is denoted with bracket notation. The left and right sides of brackets should be paired up in a "nested" fashion, in the same way as a mathematical expression or a computer program. The secondary structure is virtually identical in all these species. The structure is essential to the function of the molecule, and is thus conserved during evolution. Nevertheless the sequence is able to evolve while the secondary structure remains fixed, as we described above for rRNA.

Figure 11.1(b) shows the part of the alignment from the two ends of the sequence that form the seven base-pair acceptor stem of the tRNA. The Ts from the DNA gene alignment have been changed to Us, to illustrate the pairing in the RNA molecule. The rabbit sequence has matching Watson–Crick pairs in all seven positions (AU or CG pairs). Pairs in the other sequences that differ from the rabbit sequence have been shaded. For example, in Pair 1, most of the species have a GC pair, but the loach has an AU pair, and the shark a GU. Note that GU pairs are less stable than Watson–Crick pairs, but they

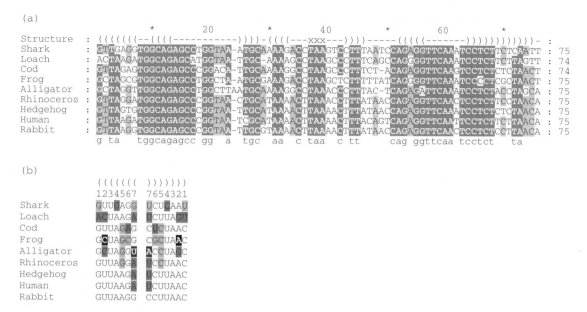

Fig. 11.1 (a) Alignment of tRNA-Leu genes from mitochondrial genomes, with conserved secondary structure illustrated using bracket notation. Gray-scale shading illustrates sequence conservation. (b) Alignment of the two halves of the aminoacyl acceptor stem of the tRNA, with shading added to illustrate compensatory substitutions.

occur with fairly high frequency (a few percent) in helical regions of RNA structures, and they are usually treated as matching pairs. At Pair 2, most species have a UA pair, but there are two species with a CG pair, and the frog has a mismatch CA. Mismatches like this are rare in regions where the secondary structure is strongly conserved. When comparing RNA sequences from different species, we often find that substitutions have occurred at both sides of a pair in such a way that the pair can still form. These are referred to as compensatory substitutions. There are many examples of this in Fig. 11.1(b). We know that mutation rates are extremely small, so it is unlikely that both mutations occurred at the same time in the same copy of the gene. More likely, a mutation in one side of the pair occurred some time prior to the second. The first mutation would have disrupted the structure and would probably have been slightly deleterious. The second mutation would then have compensated for the error made by the first. When we see compensatory mutations in an alignment, this is fairly strong evidence that the two sites are paired in the secondary structure.

In passing, we note two interesting things about tRNA-Leu genes. Mutations in human mitochondrial tRNAs result in a large number of pathological conditions. The mitomap database catalogs human mitochondrial mutations, and includes a diagram of tRNA-Leu (http://www.mitomap.org/mitomap/tRNAleu.pdf) showing the cloverleaf secondary structure, together with many point mutations that are known to give rise to disease. The story here is very similar to the example of the *BRCA1* gene discussed in Section 3.3. Mutations occur essentially randomly all over a gene, and most of these will be deleterious. We see evidence for this when we sequence lots of rare variants of the gene within the human population. When we look at an alignment of genes from different species, as in Fig. 11.1(a), we only see those changes that have survived the action of natural selection; hence, we see conservative changes that do not disrupt the structure and function of the molecule. These principles apply to RNA

Further topics in molecular evolution and phylogenetics • **259**

genes and protein-coding genes in much the same way. The second interesting thing about tRNA-Leu is that there are two distinct versions of these genes in almost all animal mitochondrial genomes: one, with anticodon UAG, translates the four-codon family CUN; the other, with anticodon UAA, translates the two-codon family UUR. These two genes usually differ at several points in the sequence, indicating that they have been independent for a long time. However, in some species, the two genes are almost identical, except for the anticodon itself. Thus, it is possible for one type of tRNA to evolve into the other by an anticodon mutation. Using a combination of phylogenetics and analysis of mitochondrial gene order, we have shown that this has occurred at least five times within the known set of complete mitochondrial genomes (Higgs *et al.* 2003).

When sequences are available for an RNA gene from a number of different species, the comparative method provides an excellent way of deducing the secondary structure (Woese and Pace 1993, Gutell *et al.* 1992, Gutell 1996). Programs are available that compare pairs of sites in an alignment to see if changes in one site are accompanied by changes in another more often than would be expected by chance. If so, we say that these sites covary. If the covariation is such that the base-pairing ability is maintained, we consider these sites to form part of a helix.

The first comparative models of rRNA structure were made in the early 1980s, when only a few sequences were available. These have been gradually modified and improved, and the current ones use information from around 7,000 species for SSU and 1,000 species for LSU. Gutell, Lee, and Cannone (2002) have compared the models from 1980 with those from 1999. For SSU rRNA, the current model contains 478 pairs, while the original contained only 353. This is because the method only makes predictions where there is significant covariation of sites. Trends of covariation are less significant and more difficult to spot when there are fewer sequences. Hence, there were areas left as unstructured in the original model, in which paired regions can now be identified with the larger data set. Of the 353 pairs originally predicted, 284 (80%) are still present in the current model. This shows that a large number of helical regions can be identified with only a small number of sequences.

For a long time, the secondary-structure models were known but tertiary structures were not. This has changed recently with the crystallization of the small and large ribosomal subunits (Wimberly *et al.* 2000, Ban *et al.* 2000). Gutell *et al.* (2002) compared the current secondary structures deduced by the comparative method with the positions of the base pairs in the crystal structures. They found that 97% of the base pairs predicted in the model were present in the crystal structures. This is a remarkable vindication of the method, and demonstrates how powerful sequence analysis techniques can be. The other 3% are false positives, i.e., they are predicted to be present but are not actually present. Although this number is small, the false-negative rate is somewhat larger. Approximately 25% of the base pairs in the crystal structures are not present in the model. The false-negative pairs show no covariation (e.g., some are completely conserved) and often consist of non-canonical base pairs, or of single pairs that are not part of standard helices. Thus, the comparative method does just about as well as it is possible to do in identifying the standard helices. In some ways, this comes as no surprise, because the comparative structures have long been taken as the reference by which to judge other structure-prediction techniques (such as the minimum free energy folding programs discussed below), because it was assumed they were good approximations to the true structure.

11.1.3 Secondary structure basics

In the next few sections, we will turn to the prediction of RNA secondary structure from sequences, using thermodynamic methods that look for the minimum free energy structure. Thermodynamic structure-prediction methods cannot really be classed as a "topic in molecular evolution"; however, they find their way into this chapter because we wish to keep the whole discussion of RNA structure in one place. The main algorithm in this category uses the dynamic programming (or recursion

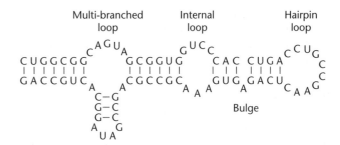

Fig. 11.2 Secondary structure of a short RNA molecule illustrating the different types of loop structure.

relation) method that we already saw for sequence alignment in Chapter 6. Thus, this topic certainly merits inclusion somewhere in a bioinformatics textbook. We also wish to highlight how thermodynamic factors influence the way a sequence evolves, so it will be useful to consider thermodynamic and evolutionary aspects of RNA together.

Figure 11.2 shows an example of a secondary structure illustrating the nomenclature for the different types of loop that can occur: hairpins (connecting the two sides of a single helix); bulges (where unpaired bases are inserted into one strand of a double helix); internal loops (connecting two helices); and multi-branched loops (connecting three or more helices).

A secondary structure can be thought of as a list of the base pairs present in the structure. To form a valid secondary structure, base pairs must satisfy several constraints. Let the bases in a sequence be numbered 1 to N. Suppose that bases i and j in the sequence are complementary, and that $j > i$. There must usually be at least three unpaired bases in a hairpin loop, so the $i - j$ pair can form if $j - i \geq 4$. Let bases k and l form another allowed pair. We want to know whether the two pairs are compatible with each other, i.e., whether the two pairs can form at the same time. A base cannot pair with more than one other base at a time, hence for the pairs to be compatible, i, j, k, and l must all be different bases. Pairs are compatible if they are in a side-by-side arrangement, as in Fig. 11.3(a), where $i < j < k < l$, or if one pair is nested within the other, as in Fig. 11.3(b), where $i < k < l < j$. The third case, Fig. 11.3(c), where the pairs are interlocking so that $i < k < j < l$, is known as a pseudoknot. Such pairs are usually assumed to be incompatible, because

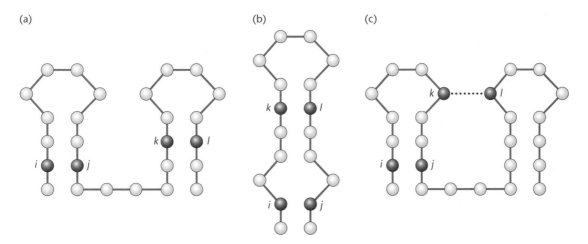

Fig. 11.3 Three possible relative positions of two RNA base pairs $i - j$ and $k - l$. Diagrams (a) and (b) show compatible pairs. Diagram (c) is a type of pseudoknot, and is usually excluded in secondary-structure prediction programs.

pseudoknot pairs are observed to be relatively infrequent in real structures and the most commonly used dynamic programming routines cannot deal with them. Thus, for our purposes, an allowed secondary structure is a set of base pairs that are compatible with each other according to case (a) or (b).

Although secondary-structure diagrams give us no information about the positions of parts of the molecule in 3D, they are quite informative about the accessibility of different parts of a molecule and the positioning of sites of interaction with other molecules. It is usually argued that the secondary structure forms first during RNA folding, that the helices and loops are well established before they are arranged into a tertiary structure (for more details see Tinoco and Bustamante 1999, and Higgs 2000). Therefore, the computational algorithms that try to find secondary structures can (hopefully) ignore tertiary structure. Tertiary structure will not be considered further in this book.

11.1.4 Maximizing base pairing

As a general rule, we expect macromolecules to fold into low-energy, thermodynamically stable structures. For RNAs, the energy is lowered when base pairs are formed. Therefore, the simplest rule for secondary-structure prediction in RNAs is to determine the structure with the maximum possible number of base pairs. The algorithm for doing this (Nussinov and Jacobson 1980) will be described here in detail, because it is the basis on which more accurate folding routines are built. For this algorithm, it is useful to think of each base pair as contributing an energy of -1 unit to the total energy of the structure. This means that finding the minimum energy structure is the same as finding the structure with the most base pairs.

Let ε_{ij} be the energy of the bond between bases i and j, which is set to -1 if the bases are complement-

ary, and to $+\infty$ if they are not complementary, to prevent the formation of disallowed base pairs. Let $E(i,j)$ be the minimum energy of the part of the sequence from bases i to j inclusive. We wish to calculate the minimum energy of the whole sequence, $E(1,N)$. It can be shown that the number of possible secondary structures for a given RNA sequence increases exponentially with the length of the molecule. However, calculation of the minimum energy structure can be done in polynomial time using dynamic programming. As with sequence alignment, the method works by writing a recursion relation that breaks down a large problem into a set of smaller problems. In Fig. 11.4, the dotted line from i to j represents the minimum energy structure of this part of the molecule, which has energy $E(i,j)$. To break this down, we consider the two possibilities illustrated on the right of Fig. 11.4. Either the final base j is unpaired (first diagram on the right), or it is paired to some other base k (second diagram on the right). In the first case, the energy is $E(i,j-1)$, because base j contributes nothing to the energy. In the second case, the pair $k-j$ divides the molecule into a section from i to $k-1$, and a section in the loop from $k+1$ to $j-1$. As pseudoknots are forbidden, it is not possible for pairs to form between these two sections of the molecule. The minimum possible energy, given that k and j are paired, is therefore the sum of the minimum energies of the two sections, plus the energy ε_{kj} of the new pair. The recursion is therefore:

$$E(i,j) = \min \begin{cases} E(i,j-1) \\ \min_{k} E(i,k-1) + E(k+1,j-1) + \varepsilon_{kj} \end{cases}$$

$$(11.1)$$

where the first min means that we are to take the minimum of the structures, where j is either unpaired or paired, and the min over k means that we are to calculate the minimum possible structure,

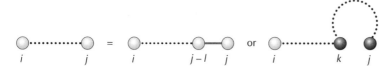

Fig. 11.4 Illustration of the dynamic programming algorithm for maximizing the number of base pairs in an RNA secondary structure (see Eq. (11.1)).

with j paired, by considering all possible bases, k, with which it could pair. The allowed range is $i \leq k \leq j - 4$. This means that the minimum energy of any chain segment can always be expressed in terms of the minimum energies of smaller segments. Before we can start using the recursion, we need to define the initial conditions. We can set $E(i,j) = 0$ for all pairs, where $j - i < 4$, because no pairs are possible in these very short sequences. Also, to make the formula work in the case where $k = i$, we need to define $E(i, i - 1) = 0$ for convenience, for each value of i. This is now sufficient to build up recursively to the minimum energy of the whole sequence.

Of course, we also want to know the minimum energy structure, not just the value of the minimum energy. This is analogous to the alignment problem, where we are more interested in the alignment itself than the score of the optimal alignment. To work out the structure, we need to use a pointer array, $B(i,j)$, that can be filled in at the same time $E(i,j)$ is calculated. If the best option is to pair j with some particular base h, then $B(i,j)$ is set to h, while if the best option is to leave j unpaired, then $B(i,j)$ is set to 0. After completing the recursion, we can work back through the B array to reconstruct the optimal structure.

11.1.5 More realistic folding routines

The structure with the maximum number of base pairs is usually quite different from the structure of a real RNA. To obtain more realistic structure predictions, we need to know more about the energies involved in RNA structure formation. When an RNA helix forms, the energy of the molecule is lowered, owing to the attractive interactions between the two strands. The entropy of the molecule is also reduced, because the single strands lose their freedom of motion and flexibility when they are bound to each other in a double helix. The **free energy** change, ΔG, of helix formation is the combination of the energy and entropy changes. At low temperatures, ΔG is negative, i.e., the helix is stable. At higher temperatures, ΔG is positive, i.e., the helix is unstable. Most RNA helices of three or more base pairs are stable at body temperature, but if an RNA

molecule is heated by a few tens of degrees, then the secondary structure will melt, i.e., the helical elements will turn back into single strands.

The free energy of helix formation can be measured in experiments with short nucleotide sequences (Freier *et al.* 1986, SantaLucia and Turner 1998), using either calorimetry or optical absorption methods. There are hydrogen bonds between the two bases in each pair, bonding the helix together. However, helix formation is **cooperative**: the base pairs are not independent of one another. Much of the stability of the helix comes from attractive stacking interactions between successive base pairs, which are in roughly parallel planes. When calculating the free energy of the helix, there is a free energy term for each two successive base pairs. In the example below, we have an AU stacked with a CG, a CG with a CG, and a CG with a GC.

```
5′ –A–C–C–G– 3′
3′ –U–G–G–C– 5′
```

Generally, GC pairs are more stable (i.e., ΔG is more negative) than AU pairs, but it should be remembered that, because of stacking interactions, the free energy of a helix is dependent on the sequence of base pairs in the helix and not just on the numbers of base pairs of each type. GU pairs are usually less stable than either AU or GC pairs. Other types of pair occurring in the middle of helices are usually unstable (i.e., they give a positive contribution to ΔG).

There are also free energy penalties associated with loops, owing to the loss in entropy of the chain when the loop ends are constrained. Some loop free energies have been measured experimentally. In general, loop parameters are known with lower accuracy than helix parameters (SantaLucia and Turner 1998), and there are some aspects, such as multi-branched loops, about which there are no thermodynamic data. It is usually assumed that the loop free energies depend on the number of unpaired bases in the loop, but not on the base sequence. Tetraloops are exceptions to this. These are particular sequences of four single-stranded bases (e.g., GNRA, where N is any base, and R is a purine) that occur frequently in length-four hairpin loops, and

	G+C (whole sequence)	G+C (helical regions)
tRNA (mitochondria)	34%	45%
tRNA (bacteria and eukaryotes)	53%	68%
tRNA (archaea)	64%	83%
rRNA (bacteria)	54%	67%

Table 11.1 Comparison of frequencies of G and C bases averaged over the whole RNA sequence with frequencies averaged only over helical regions. Data from Higgs (2000).

that have increased thermodynamic stability because of interactions between the unpaired bases.

By adding the contributions from all the helices and loops, we can assign a free energy to any given secondary structure. If the molecule is in equilibrium, we expect it to be in its most stable state most of the time, i.e., the minimum free energy structure. Dynamic programming methods have been developed to determine the minimum free energy structure using the measured free energy parameters (Waterman and Smith 1986, Zuker 1989) and have been implemented in a number of freely available software packages, such as mfold (Zuker 1998) and the Vienna RNA package (Hofacker *et al.* 1994). The recursion relations used in these programs are considerably more complicated, because they have to account for penalties for the formation of loops of different types, and there are many special cases to be considered. Nevertheless, the algorithms remain efficient, and still scale as N^3 for the full energy parameters.

11.1.6 *The influence of thermodynamics on RNA sequence evolution*

For molecules like tRNA or rRNA to function correctly, they must adopt structures that are thermodynamically stable, rather than continually fluctuating between many, very different structures. We would expect different RNA sequences to differ with respect to the stabilities of their folded states. If thermodynamic stability is important, sequences with high stability should be selected during evolution. This means that the sequences of real and random molecules may differ in their thermodynamic properties, and should show evidence of the action of selection.

One of the most important factors influencing the stability of RNA secondary structure is the fraction of G and C bases in the sequence, because helices with larger numbers of GC base pairs usually have more negative free energies. The percentage of G+C in different sequences, and in different organisms, varies considerably, owing to the different mutational biases in different organisms. If there were no selection acting, the G+C content would tend to an equilibrium value determined only by the rates of mutation to and from the different bases. One piece of evidence that suggests selection is acting on G+C content, as well as mutation, is shown in Table 11.1. Here, we compare the mean frequencies of G and C bases in large sets of RNA genes. The frequency of G+C, averaged over the whole sequences, varies considerably between the different sequence sets. However, in each case, the frequency of G+C in the helical regions is substantially larger than the average over the whole molecule, suggesting that selection prefers G and C bases in helices, and that this trend is consistent, despite the different mutational biases in the different groups.

In Section 11.1.2, we discussed the compensatory substitutions that occur in the helical regions of RNA sequences with conserved secondary structure, like tRNAs and rRNAs. A substitution in one side of a helix is often accompanied by a compensatory substitution in the other. This means that the two sites evolved in a correlated way. Several evolutionary models have been proposed to describe the evolution of such paired sites. Savill, Hoyle, and Higgs (2001) carried out a systematic comparison of these models, using likelihood ratio tests to distinguish the ability of the models to describe sequence data (see Section 11.2.1 for a description of likelihood ratio tests). One of the most useful models was found to be the general reversible seven-state model. The seven states allowed are the six common pairs in RNA helices

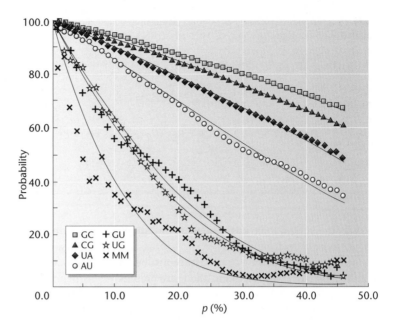

Fig. 11.5 The probability, $P_{ii}(p)$, that a base pair in state i in one sequence is also in state i in a related sequence, shown as a function of the percentage, p, of base-pair changes. Data points are measured in a large set of SSU rRNA genes from bacteria (more details in Higgs 2000). Curves are the best-fit lines for the general reversible seven-state model.

(AU, GU, GC, UA, UG, and CG), plus a mismatch state, MM, that lumps the 10 possible non-matching states together. The model is specified in a way that is directly analogous to the general reversible four-state model in Section 4.1.3, except that there are now seven frequency parameters and 21 rate parameters. Phylogenetic programs that include models of paired sites in RNA have been implemented in the PHASE package by Jow *et al.* (2002).

Figure 11.5 shows an example of fitting sequence data with this model. An alignment of 455 SSU rRNA sequences was used, spanning the full range of bacterial species (Higgs 2000). For each pair of sequences in the alignment, the fraction of paired sites, p, that differ from each other was calculated. Sequence pairs were assigned to bins representing small intervals of p. The number of times a base pair in state i aligns with a base pair in state j was calculated, and summed over all pairs of sequences in each interval of p. From this, the probability, P_{ij}, that a base pair in one sequence is in state j, given that the other sequence is in state i, was obtained as a function of p. This gives 49 different sets of data points. The data points for the diagonal elements, P_{ii}, are shown in Fig. 11.5. For any given parameter values in the seven-state model, the corresponding 49

functions can be calculated. The model parameters were optimized by using a simple least-squares fitting criterion between the theoretical curves and the data points for all 49 curves simultaneously. As we are fitting all the data points at the same time, we are choosing the model that best represents evolution over the full range of distances. Figure 11.5 shows that GC and CG pairs change the most slowly (low mutability), whereas GU, UG, and MM pairs change the most rapidly (high mutability). The mutability of a base pair is defined as the rate of substitution of a given pair to any other pair, measured relative to the average rate of substitution of the whole sequence. Table 11.2 shows the frequencies and mutabilities of the different base pairs. The first thing to note is that base pairs with higher thermodynamic stability have higher frequencies, again suggesting that natural selection prefers strong base pairs in helical regions. The second point is that high frequency base pairs have low mutability. Mutations from a high stability pair, like GC, to a low stability pair, like GU, tend to be deleterious. Selection therefore acts against them, and reduces the rate at which these changes are fixed in the population. The mutability of GC is therefore low. By a similar argument, the mutability of GU is high, because changes

Table 11.2 Comparison of base-pair frequencies and mutabilities illustrating the effects of thermodynamics on sequence evolution. Figures are given for bacterial rRNAs (set rRNA-1 in Higgs 2000).

Base pair	Frequency (%)	Mutability	Thermodynamic stability
GC	35.2	0.55	Strong
CG	29.8	0.66	Strong
AU	12.2	1.40	Moderate
UA	17.3	0.93	Moderate
GU	2.0	3.92	Weak
UG	2.1	4.36	Weak
MM	1.4	7.84	Unstable

away from GU tend to be advantageous mutations that are more likely to be fixed.

In the above discussion, it should be remembered that the base pairs we are talking about are present in the structure of the RNA molecule and not the DNA gene from which it is transcribed. The G and C in an RNA pair are at two separate sites in the gene, and these are each paired with a complementary C and G on the opposite strand of the DNA. GU pairs and mismatches are not present in the DNA!

We already discussed how compensatory mutations occur in pairs of sites that are separate in the gene but paired in the RNA. This means that substitutions at different sites in the gene are correlated with one another. This has important consequences for molecular phylogenetics. Most phylogeny programs assume that each column of a sequence alignment evolves independently. In a likelihood method, the total likelihood is taken to be the product of the likelihoods of the sites. However, in RNA sequences, the sites on opposite sides of pairs are not independent. For this reason, we have developed phylogenetic methods that use models of evolution of paired sites, such as the seven-state model described above. If these correlations are ignored, the calculations for the relative likelihoods of different phylogenetic trees can be seriously wrong (Jow *et al.* 2002).

Before leaving RNA, let us consider one further illustration of the influence of thermodynamics on evolution. As real sequences have been through the process of natural selection, they may differ from random sequences in a systematic way. One case where this can be observed is with tRNA molecules.

It is found that the minimum free energy state of real tRNAs (i.e., the cloverleaf structure) is significantly less than the minimum free energy of random sequences of the same length and the same base composition (Higgs 1993, 1995), again suggesting selection for stable secondary structures. It is also found that there are fewer alternative structures for the tRNAs within a small energy interval above the minimum free energy state than there are for random sequences. This means that the equilibrium probability of being in the minimum free energy state is larger for real than for random sequences, and that the melting of the secondary structure occurs at a higher temperature for real sequences.

This concludes our discussion of RNA structure and thermodynamics. We will examine some examples where RNA sequences have been useful in phylogenetics in Section 11.3.

11.2 FITTING EVOLUTIONARY MODELS TO SEQUENCE DATA

11.2.1 Model selection: How many parameters do we need?

In Section 4.1.3, we discussed a series of models for substitution-rate matrices in DNA sequences. These ranged from the simplest, Jukes–Cantor, in which all the substitution rates are equal, to the most general reversible model, which has four parameters for base frequencies and six parameters to control the substitution rates. These parameters can differ considerably between different sequences; therefore, it does not make sense to assign particular values to these parameters and try to apply them in every case. It is usual to estimate the values of the parameters that best describe each individual data set. We discussed maximum likelihood (ML) methods for phylogenetics in Chapter 8. If we already know the topology of the phylogenetic tree for a set of species, we can use an ML program to optimize the branch lengths and the model parameters. If we do not know the tree topology, then the phylogeny program can be set to search for the optimal topology at the same time as the branch lengths and model parameters. In either case, we will end up with the set of model parameters for which the likelihood of the observed sequences evolving is greatest. Usually, the ML values of base frequencies will come out to be very close to the average frequencies of the bases in the sequence set. The values of the rate parameters (like α and β in the K2P and HKY models) are not immediately obvious from the sequences without doing an ML calculation.

Thus, the ML principle provides a clear way of selecting the values of the parameters for a given model, but how do we choose which model to use? Models with a larger number of variable parameters can fit data more precisely than those with fewer parameters, but does this make them better models? How many parameters do we actually need? This is a general problem in fitting data. As a simple example, suppose we have a set of measurements that we plot as points on a graph. Our first thought is to fit a straight line, $y = a + bx$, through the points. This has just two parameters, a and b, and their values could be obtained by linear regression. We now observe that the best-fit line does not go exactly through the points and we suspect the data may lie on a curve rather than a straight line. We therefore add a parameter to the model, giving $y = a + bx + cx^2$, and we fit this parabolic curve through the data. It turns out that this looks much better and we believe that this function describes a real trend in the data that is not described by the straight line. However, the curve still doesn't go exactly through the points. So, should we add more parameters? If we add as many parameters as there are points on the graph, we can make the curve go exactly through all the points.

However, this curve will have a lot of wiggles in it that are probably meaningless. If we have too many parameters, we end up just fitting the noise in the data and losing sight of the important trends. Therefore, there must be a happy medium somewhere.

One statistical test that is useful for model selection is known as the likelihood ratio test. Suppose we have two models to describe a given data set: Model 1 is the simpler (with few parameters), and Model 2 is the more complex model (with more parameters). The test applies when the models are nested, i.e., when Model 1 is a special case of Model 2. For example, we might obtain Model 1 by fixing some of the parameters in Model 2 to be equal, or by setting some of the parameters to zero. The models of DNA evolution in Section 4.1.3 form a series of nested models: JC is a special case of K2P, obtained by setting $\alpha = \beta$; K2P is a special case of HKY, obtained by setting all the base frequencies to $1/4$; and HKY is a special case of GR. When two models are nested, we know that Model 2 must fit the data at least as well as Model 1. This means that the likelihood, L_2, of the data according to the more complex model must be higher than the likelihood, L_1, according to the simpler model. The likelihood ratio L_2/L_1 is therefore always greater than one. If the likelihood ratio is sufficiently large, it can be concluded that Model 2 fits the data significantly better than Model 1, while if the ratio is only slightly larger than 1, then Model 2 is not significantly better, and Model 1 cannot be rejected.

To carry out the test, we calculate the quantity δ, which is equal to twice the log of the likelihood ratio.

$$\delta = 2 \ln(L_2/L_1) = 2(\ln L_2 - \ln L_1) \qquad (11.2)$$

It can be shown that if Model 1 is true, the expected distribution of δ is a chi-squared distribution, with a number of degrees of freedom equal to the difference in the number of free parameters between the two models. We can calculate the value of δ for the real data and compare it to the distribution. If the probability of obtaining a value greater than or equal to the real value is small, we can say that Model 2 is a significantly better fit than Model 1, and we can reject Model 1.

Let us consider a simple example, where the likelihoods can easily be calculated without using a computer program. Suppose we have a single DNA sequence of total length N, where the numbers of bases of each type are N_A, N_T, N_C, and N_G. We will consider a series of three nested models whose parameters are the frequencies of the four bases.

Model 1 assumes that all four bases have frequency $1/4$. The log-likelihood of generating a sequence according to this model is

$$\ln L_1 = N \ln(1/4) \tag{11.3}$$

The number of free parameters in this model is $k_1 = 0$, because there are no parameters to optimize.

Model 2 assumes that the frequencies of A and T are equal and that the frequencies of C and G are equal. The likelihood therefore depends on the two frequencies π_A and π_C:

$$\ln L_2 = (N_A + N_T) \ln \pi_A + (N_C + N_G) \ln \pi_C \tag{11.4}$$

The values of the frequencies that maximize Eq. (11.4) are $\pi_A = (N_A + N_T)/2N$ and $\pi_C = (N_C + N_G)/2N$. The method discussed in Problem 10.2 in the previous chapter can be used for proof of this. Hence, the maximum of the log-likelihood according to Model 2 is

$$\ln L_2 = (N_A + N_T) \ln\left(\frac{N_A + N_T}{2N}\right) \\ + (N_C + N_G) \ln\left(\frac{N_C + N_G}{2N}\right) \tag{11.5}$$

Although there are two frequencies, these are related by the constraint $\pi_C = 1/2 - \pi_A$, so there is only one free parameter (i.e., $k_2 = 1$).

Model 3 allows all bases to have different frequencies. Hence, the log likelihood is

$$\ln L_3 = N_A \ln \pi_A + N_T \ln \pi_T + N_C \ln \pi_C + N_G \ln \pi_G \tag{11.6}$$

The optimal values of the frequencies are equal to their observed frequencies $\pi_A = (N_A/N$, etc.), hence the maximum of the log likelihood is

$$\ln L_3 = N_A \ln\left(\frac{N_A}{N}\right) + N_T \ln\left(\frac{N_T}{N}\right) \\ + N_C \ln\left(\frac{N_C}{N}\right) + N_G \ln\left(\frac{N_G}{N}\right) \tag{11.7}$$

There are four frequencies, but there is a constraint that they sum to 1; therefore, the number of free parameters is $k_3 = 3$.

Now we consider one particular sequence with $N = 1,000$, $N_A = 240$, $N_T = 220$, $N_C = 275$, and $N_G = 265$. The numbers of the different bases are not exactly equal, but this could be by chance. Even if we generate a sequence with all frequencies equal to $1/4$ (Model 1), we will not get exactly 250 of each base. We will now use the likelihood ratio test to ask whether Model 2 is a significantly better fit to the data than Model 1. From Eq. (11.3), $\ln L_1 = -1386.294$; from Eq. (11.5), $\ln L_2 = -1383.091$; and hence $\delta = 6.407$. Remember that the likelihood of any particular data set is always much less than 1, and hence the log likelihood is always a negative number. Model 2 has a higher likelihood than Model 1, because $\ln L_2$ is **less negative** than $\ln L_1$.

The probability distribution $P(\delta)$, if Model 1 were true, would be a chi-squared distribution, with number of degrees of freedom $k_{test} = k_2 - k_1 = 1$. The formulae for the chi-squared distributions are given in the Appendix (Section M.12). Figure 11.6(a) shows the chi-squared distribution as a smooth curve, and the real value of δ as a dashed line. The real value is right out in the tail of the distribution. The probability of getting a value greater than or equal to this (i.e., the p value of the statistical test) is approximately 1%. Thus, δ is significantly larger than we would expect by chance. Model 2 fits the data significantly better than Model 1, and we can reject the hypothesis that all four bases have equal frequency.

There is another detail about the likelihood ratio test that we need to point out at this stage. Strictly speaking, the expected distribution of δ is a chi-squared distribution only in the asymptotic limit, i.e., only when N tends to infinity. For finite values of N, the distribution is slightly different, and this

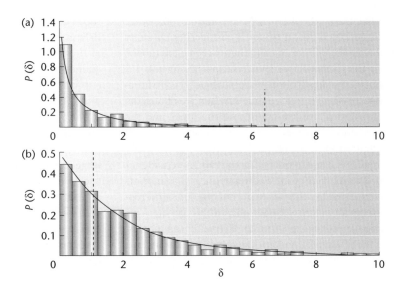

Fig. 11.6 Distributions, P(δ), for the likelihood ratio tests described in Section 11.2.1.

may alter the significance of the test if N is too small. If we are worried that we may not be close enough to the asymptotic limit, it is possible to perform an exact statistical test for finite N, without using the chi-squared distribution. In the example above, our initial hypothesis is that Model 1 is correct. Therefore, we can generate many sequences of length $N = 1,000$ by choosing each base randomly with equal frequency. For each random sequence, we can calculate δ. The distribution, P(δ), obtained in this way is shown as a histogram in Fig. 11.6(a). The p value for this test is simply the fraction of the δ values for the random sequences that are greater than or equal to the real δ. It can be seen from the figure that the histogram is very close to the curve for the chi-squared distribution. The conclusion we draw is therefore the same, whether we use the chi-squared distribution or the histogram.

So far, we showed that Model 2 is significantly better than Model 1 for fitting our data. Now let us ask whether Model 3 is significantly better than Model 2. Using the same sequence as before, $\ln L_3 = -1382.563$, from Eq. (11.7), and $\delta = 2(\ln L_3 - \ln L_2) = 1.055$. The number of degrees of freedom for the likelihood ratio test is $k_{test} = k_3 - k_2 = 2$. Figure 11.6(b) shows the chi-squared distribution with two degrees of freedom as a smooth curve. The real value of δ is in the middle of this distribution – dashed

line – and the p value is 59%. Thus, it is quite likely that δ values of this size would arise by chance if Model 2 were correct. Hence, Model 3 does not provide a significantly better fit than Model 2 and we cannot reject Model 2. Figure 11.6(b) also shows the histogram generated by the exact method for finite N. To do this, we note that the ML fit of Model 2 to the data is $\pi_A = 0.23$ and $\pi_C = 0.27$. We therefore need to generate a set of random sequences with these base frequencies, and calculate δ for each sequence by fitting both Model 2 and Model 3. As in the previous example, the histogram comes out very close to the chi-squared curve and the conclusion we draw is the same.

The likelihood ratio test can be used to select evolutionary models in phylogenetics in essentially the same way as in the cases discussed above. For examples, see Whelan and Goldman (1999), Whelan, Lio, and Goldman (2001), and Savill, Hoyle, and Higgs (2002). The ML phylogeny is obtained for the same set of sequences using two different nested models, and the corresponding δ value is compared to the chi-squared distribution. Exact tests for finite N are also often performed, but for the lengths of sequences typically used in molecular phylogenetics, we are usually close enough to the asymptotic limit for the conclusions drawn from the chi-squared test to be valid.

For the series of four DNA evolution models we considered above, it is worth considering carefully how many free parameters there are in each model. For the JC model, we showed explicitly in Box 4.1 that all the substitution probabilities, $P_{ij}(t)$, are functions of a rate α times a time t. When we obtain the ML solution, we are optimizing the times and the rates simultaneously. However, the solution is not unique, as only the product αt determines the likelihood. It is usual to normalize the rate matrix so that the average rate of substitution is one. This amounts to choosing the time scale so that one time unit represents an average of one substitution per site. This average rate constraint is applied to all the models. In the JC model, there is a single rate parameter, but the average rate constraint means that $k = 0$ for this model. For the K2P model, there are two rates, but the constraint means that it is only the ratio of these rates that is important; hence, $k = 1$. For the HKY model, there are two rates and four frequencies, and there is a constraint on the average rate and a constraint that the frequencies sum to one; hence, $k = 4$. For the GR model, there are six rates, four frequencies, and two constraints; hence, $k = 8$.

11.2.2 Estimating parameters in amino acid substitution models

The number of parameters to be optimized in DNA models is relatively small. For protein models, the number is much larger. For a general reversible 20×20 matrix, we have 20 amino acid frequencies and 190 rate parameters. This means that it is computationally much more difficult to estimate the parameters for protein models. First, the calculations are much more time consuming, and second, large amounts of sequence data are necessary to get accurate estimates of so many parameters. In principle, it is possible to estimate the rate matrix and phylogenetic tree simultaneously by ML in the same way as for the DNA models. One important case where this has been done is the mtREV model (Adachi and Hasegawa 1996). These authors used an alignment of mitochondrial protein sequences from 20 vertebrate species. The amino acid frequencies in mitochondrial proteins are significantly different from

those in nuclear proteins, and there are also some differences in the genetic code; the numerical values of the substitution rates estimated from the mitochondrial sequences are therefore noticeably different from those in nuclear protein substitution rate matrices.

More usually, however, phylogeny programs for protein sequences use matrices with specified numerical values that have been pre-calculated on large sequence databases. When these matrices are used for phylogenetics, the rates are kept fixed and are not optimized for the sequences being studied. There are several ways of estimating these matrices. We already discussed in Chapter 4 how the PAM matrices were derived, by counting substitutions between closely related sequences. This method has been criticized because it assumes that it is possible to extrapolate from closely related sequences to sequences at much larger distances. Benner, Cohen, and Gonnet (1994) estimated rate models using a method that takes account of the distance between the sequences used to derive it. They estimated a series of substitution matrices from pairs of sequences within specified distance ranges. They then extrapolated each of these to a common distance of 250 PAM units. If the same evolutionary model were applicable at all distances, then the resulting matrices should have been the same. However, significant differences were found. They argued that the genetic code has more of an influence on substitutions between closely related proteins, and that the chemical properties of the amino acids are the dominant effect for more distantly related sequences. Variability of substitution rates between sites could also lead to similar effects, particularly if the frequencies of amino acids at rapidly evolving sites are different from those at more slowly evolving sites.

If we wish to derive an evolutionary model that best describes the behavior of sequences over the full range of distances, then it seems sensible to use sequences from a wide range of distances in its estimation procedure. In principle, the ML method does this, because the likelihood calculation works for any evolutionary distance. We could take a database containing alignments of many families of proteins and calculate the ML tree for each family, using the same rate matrix for each family. The ML rate matrix obtained in this way would be the one that best describes the whole range of sequences in the database. The only problem is the large computation time required. Nevertheless, there have been several attempts to do this in an approximate way. Müller and Vingron (2000) calculated the variable time (VT) evolutionary model from a database of multiple alignments, using sequences over a wide range of distances. They chose only two sequences from each alignment. Hence, there is no tree to calculate, just a single distance for each independent pair of sequences. They also proposed a new method of optimizing the rate-matrix parameters that approximates to the ML solution, but is easier to calculate (Müller, Spang and Vingron 2002). Whelan and Goldman (2001) also used an approximate ML method. They calculated a tree for each family of proteins using a rapid distance-matrix method, and then optimized the rate-matrix parameters, while keeping the tree topologies fixed.

11.2.3 Synonymous and non-synonymous substitutions

As we mentioned in Section 3.2, synonymous substitutions in DNA sequences are likely to be neutral (or very nearly so), because they do not change the amino acid, and therefore selection acting on the protein sequence cannot intervene. In contrast, non-synonymous substitutions do change the amino acid, and stabilizing selection may prevent many of these substitutions from occurring. Most synonymous changes occur at third-codon positions, and most non-synonymous changes occur at first- and second-codon positions. The result is that, even though we assume that mutations are equally likely at all three positions, the rate of substitutions observed at first and second positions is less than that at third positions, because selection is weaker at the third position.

It is useful to define the quantity d_S as the number of synonymous substitutions per synonymous site, and d_N as the number of non-synonymous substitutions per non-synonymous site. These quantities are generalizations of the usual evolutionary distance d,

which is simply the average number of substitutions per site, irrespective of whether or not they are synonymous. (Another notation that is often found uses K_S for d_S, and K_A for d_N.) In most genes, the ratio $\omega = d_N/d_S$ is less than one, owing to the action of stabilizing selection on the amino acid sequence. This ratio is a property of a gene that indicates how strongly selection is acting on it. A highly conserved protein will have very low ω, and a more variable protein will have high ω. In the large set of mammalian proteins analyzed by Yang and Nielsen (1998), the lowest ω was 0.017, for the ATP synthase β sequence, and the highest ω was 0.838, for interleukin 6.

One of the simplest ways of estimating d_S and d_N for a pair of aligned sequences is due to Nei and Gojobori (1986). This involves counting the number of synonymous sites, S, and the number of synonymous differences between two sequences, S_d. The ratio of these quantities is the fraction of synonymous sites observed to differ. In order to correct for the possibility of multiple substitutions occurring at a site, the JC distance formula is used – Eq. (4.4):

$$ d_S = -\frac{3}{4} \ln\left(1 - \frac{4}{3} \frac{S_d}{S} \right) \tag{11.8} $$

In a similar way, the number of non-synonymous sites N, and non-synonymous differences N_d, are calculated, and the distance formula is used to obtain d_N.

There are several complications associated with this calculation. Most first- and second-position sites are clearly non-synonymous, because changes occurring here will change the amino acid. Third-position sites that are four-fold degenerate (e.g., the final site in a Val codon – see the genetic code in Table 2.1) are also clearly synonymous. However, the third positions of two-fold degenerate sites can mutate either synonymously or non-synonymously (e.g., the U in the Asp codon GAU could mutate synonymously to C, or non-synonymously to A or G). Such sites need be counted as part synonymous and part non-synonymous. Also, when a codon differs in more than one position, there are several routes by which the mutations from one to the other could have occurred. This must also be accounted for when summing the differences. A further complication is that the method does not properly account for differences in transition and transversion rates.

As a result of problems such as these with methods of counting synonymous and non-synonymous changes, Yang and Nielsen (1998) used a codon-based evolutionary model in which the rates of non-synonymous changes are multiplied by a factor ω with respect to synonymous ones. As ω is now an explicit parameter of the model, it is possible to use ML methods to estimate the value of ω that best fits the data, and also to account for other parameters, like transition and transversion rates, and differences in frequencies of the four bases.

One of the reasons for using models like this is to identify genes that seem to be under positive directional selection. This would imply that adaptation was driving the divergence of these sequences. It is actually rather difficult to find such genes – even the highest value of ω, 0.838 mentioned above, is still less than one, which implies weak stabilizing selection. Part of the problem lies with the fact that we are averaging synonymous and non-synonymous changes over the whole of the sequence. If an adaptive change occurred, it might only affect a few sites in the protein, while stabilizing selection would be acting on the other sites. Recent studies have used evolutionary models in which a distribution of ω values is possible across sites. This allows for the identification of a small number of sites with $\omega > 1$, when the majority of sites have $\omega < 1$. Yang *et al.* (2000) give several examples of genes where this seems to have occurred.

As a general conclusion from this section on fitting models to data, we note that models are being developed to describe a number of increasingly complex and more realistic situations. This involves increasing the number of parameters in the model, and inevitably increases the complexity of the computations. However, as long as this is done in a principled way, we can make sure that the additional parameters are justified and that they allow the model to reflect real trends in the data. ML methods provide a good way of estimating parameter values that are applicable in many situations. Comparison of parameter values for different sets of sequences can tell us important information about how mutation and selection influence sequence evolution.

11.3 APPLICATIONS OF MOLECULAR PHYLOGENETICS

11.3.1 The radiation of the mammals

This section presents some examples of recent results using molecular phylogenetics, illustrating the range of questions that can be addressed, and showing some of the problems that arise. The most fundamental use of molecular phylogenetics is to determine the evolutionary tree for the corresponding species. In many cases, the species have already been classified using morphological evidence. Sometimes, molecular studies confirm what is already known from morphologies, but surprises quite often arise, leading to "molecules versus morphology" disputes. Where recognizable shared derived characters are available, morphological studies provide reliable phylogenetic trees. However, where few informative characters are available, morphological studies can be inconclusive or misleading.

A good example is the evolution of the mammalian orders. Over the last decade, the use of molecular phylogenetic techniques has led to important changes in our understanding of the evolutionary tree of mammals. The number of species for which appropriate sequence information is available has been increasing rapidly, and the methods used have become increasingly sophisticated. This has led to a large interest in mammalian phylogenetics and to an increasing degree of confidence in many features of the mammalian tree that were not appreciated only a few years ago.

"Orders" are medium-level taxonomic groups that are divided up into families, genera, and species. The orders of mammals were defined from morphological evidence a long time ago. Familiar examples of mammalian orders include rodents, carnivores, primates, and bats. Most orders have well-defined morphological features that distinguish them. In most cases, when molecular sequences were examined, it was found that the groupings defined from morphology were supported (although we will highlight a few cases where this was not so). However, the relationship between the mammalian orders was not clear from morphological studies. Molecular data have allowed us to answer questions like: "Are rodents more closely related to primates or carnivores?" It is thus the earliest branch points in the mammalian tree that have been of most interest in recent work.

One important feature that now appears to have strong support is that the mammalian orders can be divided into four principal supra-ordinal groups. This has been shown convincingly using large data sets, derived mainly from nuclear genes (Madsen *et al.* 2001, Murphy *et al.* 2001a, 2001b). It has also been found using the complete set of RNAs from mitochondrial genomes (Hudelot *et al.* 2003); the tree from this latter study is shown in Fig. 11.7. The results of these studies are largely in agreement. The data sets used are mostly independent (although the mitochondrial SSU rRNA gene is included in all of them), the alignments were done independently, the sets of species used are different, and the details of the phylogenetic methods used are also different. The consensus reached is therefore quite convincing. We will number these groups following Murphy *et al.* (2001a) as:

• Group I: Afrotheria, containing Proboscidea (elephants), Sirenia (sea cows), Hyracoidea (hyraxes), Tubulidentata (aardvarks), Macroscelidea (elephant shrews), and Afrosoricida (African insectivores).

• Group II: Xenarthra (armadillos, sloths, and anteaters).

• Group III: Supraprimates, containing Primates, Dermoptera (flying lemurs), Scandentia (tree shrews), Rodentia, and Lagomorpha (rabbits).

• Group IV: Laurasiatheria, containing Cetartiodactyla (even-toed ungulates and whales), Perissodactyla (odd-toed ungulates), Carnivora, Pholidota (pangolins), Chiroptera (bats), and Eulipotyphla (Eurasian insectivores).

It has taken some time for this picture to emerge. The identification of the Afrotherian group (Stanhope *et al.* 1998) was a surprise, as it contains orders that are morphologically diverse, and superficially have little in common, apart from their presumed African origin. In Section 3.3, we already mentioned the deletion in the *BRCA1* gene that is a piece of evidence supporting the Afrotherian group. The Xenarthran group is South American and had been recognized as an early diverging group on morphological grounds. Owing to the general difficulty of determining the root of the eutherian tree using molecular phylogenetics, the positioning of Xenarthra has been unstable, sometimes appearing quite deep within the eutherian radiation (Waddell *et al.* 1999, Cao *et al.* 2000). The present picture (Delsuc *et al.* 2002) puts Xenarthra back as an early branching group, but it is not possible to distinguish whether it branches before or after Afrotheria or as a sister group to Afrotheria. In Fig. 11.7, Xenarthra and Afrotheria are sister groups, but this is only supported at 52%, and hence cannot be relied on.

Supraprimates and Laurasiatheria are both diverse groups for which a lot of sequence data have been available for some time. These groups contain several species that have presented particular problems to molecular phylogenetics. The mouse-like rodents and the hedgehogs have been repeatedly found to branch at the base of the eutherian tree in analyses of mitochondrial proteins (Penny *et al.* 1999, Arnason *et al.* 2002). This would mean that rodents would not be a monophyletic group, because the mice would be separated from other rodents, like squirrels and guinea pigs, and that hedgehogs would be separated from other insectivores, like moles and shrews. It is now generally accepted that the positioning of these problem species was an artifact arising because some lineages contradict the assumption of a homogeneous and stationary substitution model (Lin, Waddell, and Penny 2002). Our own approach, using RNA sequences from mitochondrial genomes, appears to suffer less from these problems than most other methods (Jow *et al.* 2002, Hudelot *et al.* 2003).

If we accept, for the moment, that most parts of the tree in Fig. 11.7 are probably correct, it tells us that there has been a large amount of convergent evolution in the morphology and ecological specializations of the mammals. For example, sea-living groups have evolved independently from land-living ancestors three times: whales (Cetartiodactyla,

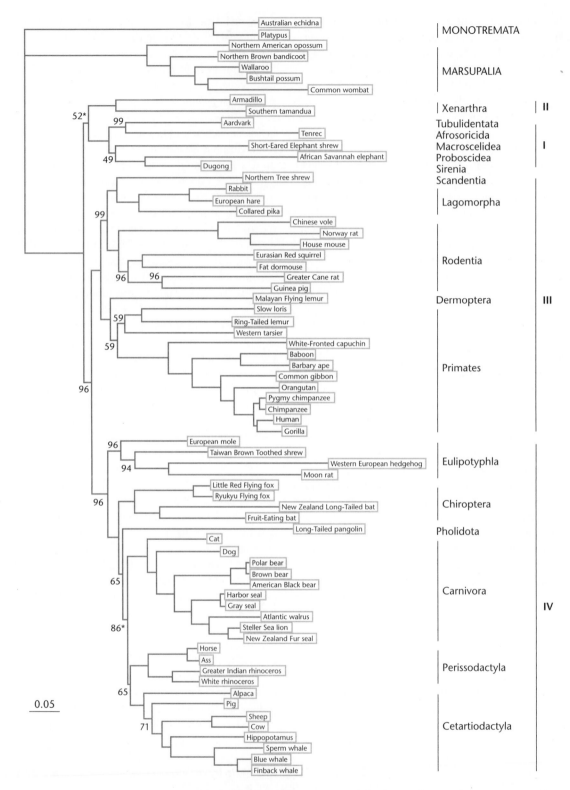

Fig. 11.7 Phylogeny of the mammalian orders obtained using the complete set of rRNA and tRNA genes from mitochondrial genomes (Hudelot *et al.* 2003). Two models of evolution were combined in this analysis: one for the paired regions of the secondary structure, and one for the unpaired regions. Posterior probabilities given on the nodes are obtained using the MCMC method. Where no percentage is given, the node has 100% support.

Group IV), seals (Carnivora, Group IV), and sea cows (Sirenia, Group I). Specialized ant-eating animals have also evolved independently three times: anteaters (Xenarthra, Group II), pangolins (Pholidota, Group IV), and aardvarks (Tubulidentata, Group I). Another amusing example is the elephant shrew, a small shrew-like African mammal, so-named because of its rather long and flexible proboscis. However, molecular evidence shows that elephant shrews are more closely related to elephants than they are to shrews, which would be hard to deduce simply by looking at them!

Convergent morphological evolution has long been recognized between marsupials (pouched mammals) and placental mammals, i.e., there are marsupial rats, marsupial wolves, etc. These groups have been separated geographically for well over 100 million years because of the isolation of Australia. Continental drift also seems to have played an important part in the separation and independent evolution of the four principal placental mammal clades. In Fig. 11.7, marsupials and monotremes (egg-laying mammals) have been included as outgroups with which to determine the root of the placental mammals. The relationship of monotremes, marsupials, and placentals has also been an issue recently, with some studies placing monotremes as a sister group to marsupials. The most recent evidence places monotremes as branching slightly before the split of marsupials and placentals (Phillips and Penny 2003).

Several groupings proposed from morphological evidence now seem unlikely to be true. From morphological evidence, the anteaters and pangolins are often placed as sister groups (e.g., Liu and Miyamoto 1999); however, molecular evidence places them far apart. Another traditional grouping based on morphology, known as Archonta, places the bats with primates, tree shrews, and flying lemurs; however, the bats seem to be widely separated from the other three orders according to molecular evidence. The tree also highlights problems with classification of insectivores. The "true" insectivore group (shrews, hedgehogs, and moles) now appear in group IV; however, other morphologically similar groups like tenrecs, golden moles,

and elephant shrews that were traditionally classed as insectivores now appear to be members of Afrotheria.

A final group worth mentioning is Cetartiodactyla, which is a combination of the traditional orders Cetacea (whales) and Artiodactyla (cows, pigs, camels, hippos, etc.). There is now general agreement from molecular, morphological, and palaeontological evidence that these two groups are closely related. Molecular trees suggest that whales are more closely related to hippos than hippos are to other artiodactyls. This means that the group formerly defined as Artiodactyla is not monophyletic: hence the introduction of the more broadly defined group Cetartiodactyla, which **is** monophyletic. The details of whale origins are beginning to become clearer from the discovery of a number of fossil taxa that represent intermediate steps, such as the "walking whale", *Ambulocetus* (Gatesy and O'Leary 2001).

Another area of dispute between molecular phylogeneticists and palaeontologists is in the dating of evolutionary events. There are few eutherian mammal fossils before the end of the Cretaceous period, 65 million years ago. This led to the belief that the radiation of mammalian orders occurred when ecological niches became vacant after the extinction of the dinosaurs. Molecular phylogenies also allow estimates of divergence times, if a date is known from fossil evidence for one branch point on the tree that can be used as a calibration. If a molecular clock is assumed, then the times of other branch points on the tree can be estimated from the relative lengths of the different branches. The molecular estimates for the timing of the radiation of mammals usually suggest that it was considerably before the end of the Cretaceous. Dates for the primate–rodent split are around 100 million years ago (Bromham, Phillips, and Penny 1999). However, there are large confidence intervals on these estimates, and different phylogenetic methods give varying answers. Estimating accurate times from phylogenies is a more difficult problem than simply obtaining the tree topology. Hedges and Kumar (2003) have reviewed methods of dating from molecular sequences. They argue that molecular dates can be accurate if derived from combined data sets of many genes, and if care is taken to exclude genes that are unusually variable in rate between species.

11.3.2 The metazoan phyla

In the above discussion of mammals, we saw that: (i) the species had already been separated into fairly well-defined groups (in this case, orders) on the basis of morphology; (ii) apart from a few surprises, the molecular evidence confirmed the existence of these groups; (iii) the groups had arisen in an apparent radiation in a fairly short period, hence the relationship between the groups was unclear from morphological evidence; (iv) molecular studies were able to shed light on some of the branching patterns between groups during the radiation; and (v) dating of events from molecular studies is still controversial. All these points are also true when we consider the evolution of the major phyla of animals (metazoa).

Most animal phyla that were previously identified from morphological studies are confirmed to be monophyletic groups using sequence comparisons. However, the relationship between these groups was poorly understood, and molecular evidence is beginning to shed light on the earliest branches in the animal tree (Valentine, Jablonski, and Erwin 1999, Adoutte et al. 2000). With the exception of some early-branching phyla, like sponges (Poriferans) and jellyfish (Cnidarians), animal phyla are classed as bilateria, i.e., having a bilateral symmetry. Figure 11.8 compares the phylogeny of the bilaterian phyla according to morphology and according to molecular phylogeny using rRNA. There is now fairly strong molecular evidence that the major split in the bilateria is between the Deuterostomes and the Protostomes, and that the latter group can be divided into two groups, known as Ecdysozoans and Lophotrochozoans. The Deuterostomes include Vertebrates, Echinoderms (i.e., starfish and sea urchins) and related taxa. The Ecdysozoans are a group of animals that molt at some point during their life cycle: these include Arthropods (insects, spiders, etc.), Nematodes (roundworms) and several other phyla not previously known to be related. The Lophotrochozoans are another diverse group, named after the lophophore (a type of feeding struc-

ture) and the trochophore (a type of larval structure) that some of them possess. The Lophotrochozoa contain some relatively complex, or "advanced", invertebrate phyla, like Molluscs, as well as some anatomically much simpler phyla, like Platyhelminths (flatworms), previously thought to have arisen near the base of the bilaterian tree. Observations like this have forced a major rethink in our understanding of developmental biology and the evolution of animal body plans. An important conclusion that had been drawn from morphological studies was the supposed relationship between Annelids (segmented worms, like earthworms) and Arthropods, both of which have a segmented body plan. It is now clear that this was misleading, because Arthropods are within Ecdysozoa and Annelids within Lophotrochozoa.

The relationships between the phyla within the two Protostome groups is still unclear, and these are left as multifurcations in Fig. 11.8. Mallatt and Winchell (2002) have made some progress in resolving the Protostome phyla using combined LSU and SSU rRNA, but there are many remaining questions. This problem is more difficult than the mammal one, because there are fewer suitable genes that can be reliably aligned over the whole range of Metazoa. However, more sequence data are continually becoming available, and we are therefore relatively optimistic that molecular phylogenies will be able to reach the parts of trees that morphological evidence has thus far been unable to reach. The dating of events is still a major problem. Fossil evidence dates the appearance of most modern animal groups to the "Cambrian explosion", around 530 million years ago. Some early metazoan fossils can be traced as far back as ~600 million years ago. Molecular dates for the divergences between these phyla are usually more than 600, and sometimes as much as 1,500, million years ago. One of the most important unresolved dates is that of the Protostome–Deuterostome divergence.

11.3.3 The evolution of eukaryotes

Moving still further backwards in evolution, another major issue is the relationship between the different

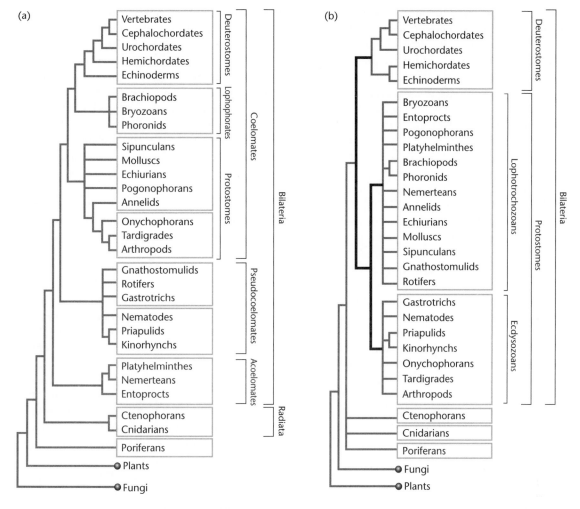

Fig. 11.8 Phylogeny of the metazoan taxa. Reproduced from Adoutte *et al.* (2000). Copyright 2000 National Academy of Sciences, USA. (a) Traditional phylogeny based on morphology and embryology. (b) Molecular phylogeny based on rRNA.

groups of eukaryotes. This is a tricky problem owing to the large variability of evolutionary rates between different taxa. Taxa with high rates appear on long branches in the tree. Phylogenetic methods often place long-branch taxa incorrectly – a problem known as long-branch attraction (Philippe and Laurent 1998). This can result in the rapidly evolving taxa being artificially placed at the base of a tree, leaving a "crown" of slowly evolving taxa clustered together. It can also result in the artificial clustering of a number of unrelated rapidly evolving taxa.

The problem is known to be severe for parsimony methods, which would give inconsistent results in such cases (Felsenstein 1978), and for UPGMA, which would force rapidly evolving taxa to the base of the tree. However, long branches present a potential problem with other methods too. Longer branches usually have larger errors in their estimated lengths, especially if the variability of rates between different sites in sequences is not accounted for.

The eukaryote tree derived from SSU rRNA sequences possesses a crown of "higher" taxa (including animals, fungi, plants, and algae) and some early-branching divergent groups (including slime-molds, euglenozoa, microsporidia, and diplomonads). Van de Peer *et al.* (2000) carried out a large-scale phylogeny of eukaryotes using over 2,500 SSU rRNA sequences, calculating distances using a method that corrects for the estimated variability of substitution rates across sites. They argue that some of the previously early-branching taxa should actually be placed within the crown, and that the positioning of these taxa near the base of the tree was an artifact of long-branch attraction. Similar conclusions were reached by Philippe and Adoutte (1998), who compared the phylogeny of eukaryotes deduced from rRNA, to that deduced from actin and tubulin protein sequences. They found that the three molecules give trees that are inconsistent with each other, and that different taxa appear to branch off at the base of the tree in each case. They conclude that there are serious problems with the positioning of these divergent groups, and that the three molecules used are beginning to reach saturation of mutations, and are hence becoming unreliable for estimating distances. Philippe and Adoutte (1998) also give an excellent discussion of the problems and pitfalls that can arise in molecular phylogenetics.

11.3.4 Miscellaneous examples

As well as tackling the deepest points in the tree of life, phylogenetic methods are often applied to groups of closely related species. Here, the studies can reveal interesting details about the ecology and biology of the species, as well as providing a taxonomic classification. A good example is the study by Losos *et al.* (1998) on the Anolis lizards of the Caribbean. Related species live on the four islands, Cuba, Hispaniola, Jamaica, and Puerto Rico. There are recognizable "ecomorphs" (known as e.g., Crown Giant, Trunk Ground, Twig) that are specialized for living in different habitats. These differ in size and color, presumably for camouflage. Most of the ecomorphs are found on all the islands. When species are clustered according to the degree of similarity in measurements of their shape and size, those in the same ecomorph cluster together, showing

that there are objectively measurable differences between them. However, when molecular phylogenies are constructed, it is found that none of the ecomorphs consist of monophyletic groups of species. There is a strong tendency for species of different ecomorphs on the same island to be more closely related to one another than to species of the same ecomorph on other islands. This indicates that ecological specialization has occurred in parallel several times, producing species that are morphologically similar as a result of the similarity of selection pressures on the different islands.

Even for closely related species, we are still used to the idea that molecular evolution occurs over periods of millions of years. However, viruses such as the Human Immunodeficiency Virus (HIV) can evolve measurably in just decades. Phylogenetic methods have established that HIV is related to Simian Immunodeficiency Viruses (SIV) that infect other primates (Hahn *et al.* 2000). HIV-1, the most common form of the virus in humans, is thought to have arisen via cross-species infections from chimpanzees, the evidence suggesting that there have been three separate cross-species transmissions. HIV-2 is another form of the virus in humans that has been shown to be related to SIV sequences from the sooty mangabey, an African monkey. Hence, this represents another independent cross-species transmission. Molecular phylogenetics is yielding important information about the likely dates of these events. The latest common ancestor of the main group of HIV-1 sequences is estimated to have existed in the period 1915–41 (Korber *et al.* 2000).

Throughout this section, we have been assuming that the sequences being studied have evolved on a tree, in other words that genes have been inherited vertically from parent to offspring in each lineage of organisms. We have ignored the possibility of horizontal transfer of genes between unrelated species. For multicellular organisms, this is perfectly reasonable, but for bacteria, many cases of horizontal gene transfer are now known. The relevance of horizontal transfer in bacterial evolution will be discussed in Chapter 12.

As the number of sequences in biological databases has rocketed over the past few years, so has the number of people wishing to construct molecular phylogenies, and the number of different programs for doing so. An excellent reference list of available programs is maintained by J. Felsenstein at http://evolution.genetics.washington.edu/phylip/software.html. There are still many challenging phylogeny problems, and still many groups of organisms for which few sequence data are available. This is an active research area, both in terms of the development of new methods and the application of phylogenetic techniques to new sequences.

SUMMARY

RNA sequences, like tRNA and rRNA, have secondary structures that are conserved during evolution. Comparison of aligned sequences of these molecules across many species has allowed the positions of secondary structural elements to be identified. The paired regions of RNA structures evolve via compensatory substitutions: often substitutions occur on both sides of a pair in such a way that the base-pairing ability is maintained. Compensatory changes cause correlations in the substitutions at different sites in the molecule. Identification of these correlations is the principal clue used in the comparative method to deduce the structure. The correlations are also important in molecular phylogenetics, as they break the usual assumption that all sites in a molecule evolve independently.

RNA secondary structure can also be predicted using a dynamic programming algorithm that minimizes the free energy. Programs of this type use energy parameters measured in experiments with RNA oligonucleotides, e.g., stacking interactions for the different combinations of base pairs, and free energy penalties for formation of hairpin loops, bulges, and internal loops. Another thermodynamics-based approach to RNA structure prediction is to follow the kinetics of the folding process. Analysis of known secondary structures shows that thermodynamic parameters have an influence on evolutionary parameters. More stable base pairs tend to have higher frequency of occurrence in helical regions than less stable pairs, and the mutability of more stable pairs is also lower. Thus, there is a general trend in tRNA and rRNA molecules to select for stable secondary structures, and this trend is apparent despite the wide variability of G+C content between different organisms.

Many different models of sequence evolution are now available for use with phylogenetic programs. These models vary in their complexity and number of free parameters. Models with more parameters will be able to fit the data more precisely that those with fewer parameters. However, the more complex model is not necessarily the better one. We need to be sure that the additional parameters in a more complex model are necessary to explain the real trends in the data, and that they are not simply fitting noise in the data. The likelihood ratio test is a standard procedure for comparing the likelihoods for a given set of data from two different models. As a null hypothesis, the test assumes that the simpler model is correct. If the null hypothesis is rejected by the test, we can say that the more complex model gives a significantly better fit to the data than the simpler one.

Increasingly more complex models of protein sequence evolution are now being developed, and various techniques based on ML are used to estimate the parameters of these models. One factor of interest to evolutionary biologists is the relative rate of synonymous and nonsynonymous substitutions in a gene. Variations in this ratio between genes tell us about the varying nature of selection on different genes, and can, in principle, reveal genes where adaptive selection has played a role in causing the divergence of sequences between species. Codon-based models are now available that can account for this.

The phylogeny of the mammalian orders and the phylogeny of the metazoan phyla are both important cases where considerable advances have been made in our understanding as a result of the use of molecular sequences. In both cases, a set of clearly defined groups had been identified from morphological studies, but there was insufficient evidence to determine the relationships between them. In the mammal case, the establishment of four major clades of eutherian mammals makes sense from the point of view of the geographical origins of the groups, and also illustrates the surprising power of convergent evolution to generate similar morphologies from different genetic stocks. In the metazoan case, the establishment of major groupings of phyla, such as Deuterostomes, Ecdysozoa, and Lophotrochozoa, is of fundamental importance in understanding developmental biology and the evolution of animal body plans. In many cases, dating of events from molecular phylogenetics is controversial, and there are still important differences to be resolved between molecular and fossil studies.

REFERENCES

Adachi, J. and Hasegawa, M. 1996. A model of amino acid substitution in proteins encoded by mitochondrial DNA. *Journal of Molecular Evolution*, **42**: 459–68.

Adoutte, A., Balavoine, G., Lartillot, N., Lespinet, O., Prud'homme, B., and de Rosa, R. 2000. The new animal phylogeny: Reliability and implications. *Proceedings of the National Academy of Sciences USA*, **97**: 4453–6.

Arnason, U., Adegoke, J.A., Bodin, K., Born, E.W., Yuzine, B.E., Gullberg, A., Nilsson, M., Short, V.S., Xu, X., and Janke, A. 2002. Mammalian mitogenomic relationships and the root of the eutherian tree. *Proceedings of the National Academy of Sciences USA*, **99**: 8151–6.

Ban, N., Nissen, P., Hansen, J., Moore, P.B., and Steitz, T.A. 2000. The complete atomic structure of the large ribosomal subunit at 2.4 Å resolution. *Science*, **289**: 905–20.

Benner, S.A., Cohen, M.A., and Gonnet, G.H. 1994. Amino acid substitution during functionally constrained divergent evolution of protein sequences. *Protein Engineering*, **7**: 1323–32.

Bromham, L., Phillips, M.J., and Penny, D. 1999. Growing up with dinosaurs: Molecular dates and the mammalian radiation. *Trends in Ecology and Evolution*, **14**: 113–18.

Cao, Y., Fujiwara, M., Nikaido, M., Okada, N., and Hasegawa, M. 2000. Interordinal relationships and timescale of eutherian evolution as inferred from mitochondrial genome data. *Gene*, **259**: 149–58.

Delsuc, F., Scally, M., Madsen, O., Stanhope, M.J., de Jong, W.W., Catzeflis, M., Springer, M.S., and Douzery, J.P. 2002. Molecular phylogeny of living Xenarthrans and the impact of character and taxon sampling on the placental tree rooting. *Molecular Biology and Evolution*, **19**: 1656–71.

Felsenstein, J. 1978. Cases in which parsimony or compatibility methods will be positively misleading. *Systematic Zoology*, **27**: 401–10.

Freier, S.M., Kierzek, R., Jaeger, J.A., Sugimoto, N., Caruthers, M.H., Nielson, T., and Turner, D.H. 1986. Improved free energy parameters for prediction of RNA duplex stability. *Proceedings of the National Academy of Sciences USA*, **83**: 9373–7.

Gatesy, J. and O'Leary, M.A. 2001. Deciphering whale origins with molecues and fossils. *Trends in Ecology and Evolution*, **16**: 562–70.

Gutell, R.R. 1996. Comparative sequence analysis and the structure of 16S and 23S RNA. In R.A. Zimmermann and A.E. Dahlberg (eds.), *Ribosomal RNA: Structure, Evolution, Processing, and Function in Protein Biosynthesis*. Boca Raton: CRC Press.

Gutell, R.R., Power, A., Hertz, G.Z., Putz, E.J., and Stormo, G.D. 1992. Identifying constraints on the higher order structure of RNA: Continued development and application of comparative sequence analysis methods. *Nucleic Acids Research*, **20**: 5785–95.

Gutell, R.R., Lee, J.C., and Cannone, J.J. 2002. The accuracy of ribosomal RNA comparative structure models. *Current Opinion in Structural Biology*, **12**(3): 301–10.

Hahn, B.H., Shaw, G.M., De Cock, K.M., and Sharp, P.M. 2000. AIDS as a zoonosis: Scientific and public health implications. *Science*, **287**: 607–14.

Hedges, S.B. and Kumar, S. 2003. Genomic clocks and evolutionary timescales. *Trends in Genetics*, **19**: 200–6.

Higgs, P.G. 1993. RNA secondary structure: A comparison of real and random sequences. *Journal of Physics I*, **3**: 43–59.

Higgs, P.G. 1995. Thermodynamic properties of transfer RNA: A computational study. *Journal of the Chemical Society Faraday Transactions*, **91**: 2531–40.

Higgs, P.G. 2000. RNA secondary structure: Physical and computational aspects. *Quarterly Review of Biophysics*, **33**: 199–253.

Higgs, P.G., Jameson, D., Jow, H., and Rattray, M. 2003. The evolution of tRNA-leucine genes in animal mitochondrial genomes. *Journal of Molecular Evolution*, **57**: 435–45.

Hofacker, I.L., Fontana, W., Stadler, P.F., Bonhoeffer, L.S., Tacker, M., and Schuster, P. 1994. *Monatshefte für Chemie*, **125**: 167. The Vienna RNA software package is available at http://www.tbi.univie.ac.at/~ivo/RNA.

Hudelot, C., Gowri-Shankar, V., Jow, H., Rattray, M., and Higgs, P.G. 2003. RNA-based phylogenetic methods: Application to mammalian mitochondrial RNA sequences. *Molecular Phylogeny and Evolution*, **28**: 241–52.

Jow, H., Hudelot, C., Rattray, M., and Higgs, P.G. 2002. Bayesian phylogenetics using an RNA substitution model applied to early mammalian evolution. *Molecular Biology and Evolution*, **19**: 1591–601.

Korber, B., Muldoon, M., Theiler, J., Gao, F., Gupta, R., Lapedes, A., Hahn, B.H., Wolinsky, S., and Bhattacharya, T. 2000. Timing the ancestor of the HIV-1 pandemic strains. *Science*, **288**: 1789–96.

Lin, Y-H., Waddell, P.J., and Penny, D. 2002. Pika and vole mitochondrial genomes increase support for both rodent monophyly and glires. *Gene*, **294**: 119–29.

Liu, F.R. and Miyamoto, M.M. 1999. Phylogenetic assessment of molecular and morphological data for eutherian mammals. *Systematic Biology*, **48**: 54–64.

Losos, J.B., Jackman, T.R., Larson, A., de Queiroz, K., and Rodriguez-Schettino, L. 1998. Contingency and determinism in replicated adaptive radiations of island lizards. *Science*, **279**: 2115–18.

Madsen, O., Scally, M., Douady, C.J., Kao, D.J., DeBry, R.W., Adkins, R., *et al.* 2001. Parallel adaptive radiations in two major clades of placental mammals. *Nature*, **409**: 610–14.

Mallatt, J. and Winchell, C.J. 2002. Testing the new animal phylogeny: First use of combined large subunit and small subunit rRNA gene sequences to classify the protostomes. *Molecular Biology and Evolution*, **19**: 289–301.

Müller, T. and Vingron, M. 2000. Modeling amino acid replacement. *Journal of Computational Biology*, **7**: 761–76.

Müller, T., Spang. R., and Vingron, M. 2002. Estimating amino acid substitution models: A comparison of Dayhoff's estimator, the resolvent approach, and a maximum likelihood method. *Molecular Biology and Evolution*, **19**: 8–13.

Murphy, W.J., Elzirik, E., Johnson, W.E., Zhang, Y.P., Ryder, O.A., and O'Brien, S.J. 2001a. Molecular phylogenetics and the origins of placental mammals. *Nature*, **409**: 614–18.

Murphy, W.J., Elzirik, E., O'Brien, S.J., Madsen, O., Scally, M., Douady, C.J., Teeling, E., Ryder, O.A., Stanhope, M.J., de Jong, W.W., and Springer, M.S. 2001b. Resolution of the early placental mammal radiation using Bayesian phylogenetics. *Science*, **294**: 2348–51.

Nei, M. and Gojobori, T. 1986. Simple methods for estimating the number of synonymous and nonsynonymous nucleotide substitutions. *Molecular Biology and Evolution*, **3**: 418–26.

Nussinov, R. and Jacobson, A.B. 1980. Fast algorithm for predicting the secondary structure of single-stranded RNA. *Proceedings of the National Academy of Sciences USA*, **77**: 6309–13.

Penny, D., Hasegawa, M., Waddell, P.J., and Hendy, M.D. 1999. Mammalian evolution: Timing and implications from using the LogDeterminant transform for proteins of differing amino acid composition. *Systematic Biology*, **48**: 76–93.

Philippe, H. and Adoutte, A. 1998. The molecular phylogeny of Eukaryota: Solid facts and uncertainties. In G.H Coombs, K. Vickerman, M.A. Sleigh, and A. Warren (eds.), *Evolutionary Relationships among Protozoa*. London: Chapman and Hall.

Philippe, H. and Laurent, J. 1998. How good are deep phylogenetic trees? *Current Opinion in Genetics and Development*, **8**: 616–23.

Phillips, M.J. and Penny, D. 2003. The root of the mammalian tree inferred from whole mitochondrial genomes. *Molecular Phylogeny and Evolution*, **28**: 171–85.

SantaLucia, J. Jr. and Turner, D.H. 1997. Measuring the thermodynamics of RNA secondary structure formation. *Biopolymers*, **44**: 309–19.

Savill, N.J., Hoyle, D.C., and Higgs, P.G. 2001. RNA sequence evolution with secondary structure constraints: Comparison of substitution rate models using maximum likelihood methods. *Genetics*, **157**: 399–411.

Stanhope, M.J., Waddell, V.G., Madsen, O., de Jong, W., Hedges, S.B., Cleven, G.C., Kao, D., and Springer, M.S. 1998. Molecular evidence for multiple origins of Insectivora and for a new order of endemic African insectivore mammals. *Proceedings of the National Academy of Sciences USA*, **95**: 9967–72.

Tinoco, I. and Bustamente, C. 1999. How RNA folds. *Journal of Molecular Biology*, **293**(2): 271–81.

Valentine, J.W., Jablonski, D., and Erwin, D.H. 1999. Fossils, molecules and embryos: New perspectives on the Cambrian explosion. *Development*, **126**: 851–9.

Van de Peer, Y., Baldauf, S.L., Doolittle, W.F., and Meyer, A. 2000. An updated and comprehensive rRNA phylogeny of crown eukaryotes based on rate-calibrated evolutionary distances. *Journal of Molecular Evolution*, **51**: 565–76.

Waddell, P.J., Cao, Y., Hauf, J., and Hasegawa, M. 1999. Using novel phylogenetic methods to evaluate mammalian mtDNA, including amino acid-invariant sites-LogDet plus site stripping, to detect internal conflicts in the data, with special reference to the positions of hedgehog, armadillo and elephant. *Systematic Biology*, **48**: 31–53.

Waterman, M.S. and Smith, T.F. 1986. Rapid dynamic programming algorithms for RNA secondary structure. *Advances in Applied Mathematics*, **7**, 455–64.

Whelan, S. and Goldman, N. 1999. Distributions of statistics used for the comparison of models of sequence evolution in phylogenetics. *Molecular Biology and Evolution*, **16**: 1292–9.

Whelan, S., Lio, P., and Goldman, N. 2001. Molecular phylogenetics: State of the art methods for looking into the past. *Trends in Genetics*, **17**: 262–72.

Wimberly, B.T., Brodersen, D.E., Clemons, W.M. Jr., Morgan-Warren, R.J., Carter, A.P., Vonhein, C., Hartsch, T., and Ramakrishnan, V. 2000. Structure of the 30S ribosomal subunit. *Nature*, **407**: 327–39.

Woese, C.R. and Fox, G.E. 1977. Phylogenetic structure of the prokaryote domain: The primary kingdoms.

Proceedings of the National Academy of Sciences USA, **74**: 5088–90.

Woese, C.R., Kandler, O., and Wheelis, M.L. 1990. Towards a natural system of organisms: Proposal for the domains archaea, bacteria and eucarya. *Proceedings of the National Academy of Sciences USA*, **87**: 4576–9.

Woese, C.R. and Pace, N.R. 1993. Probing RNA structure, function and history by comparative analysis. In R.F. Gesteland and J.F. Atkins (eds.), *The RNA World*, pp. 91–117. Cold Spring Harbor Laboratory Press.

Yang, Z. and Nielsen, R. 1998. Synonymous and non-synonymous rate variation in nuclear genes of mammals. *Journal of Molecular Evolution*, **46**: 409–18.

Yang, Z., Nielsen, R., Goldman, N., and Krabbe Pedersen, A. 2000. Codon-substitution models for heterogeneous selection pressure at amino acid sites. *Genetics*, **155**: 431–49.

Zuker, M. 1989. Finding all sub-optimal foldings of an RNA molecule. *Science*, **244**: 48–52.

Zuker, M. 1998. RNA web page – including lecture notes on RNA structure prediction and mfold software package. http://www.ibc.wustl.edu/~zuker/rna.

Genome evolution

CHAPTER

12

CHAPTER PREVIEW

Now that many complete genome sequences are available, it is possible to compare whole sets of genes between species and to study how organisms evolve at the whole genome level. This chapter describes several prokaryotic genomes, discussing how genes are gained and lost via duplication and deletion, and considering the evidence for horizontal gene transfer. We describe the COG system for clustering orthologous genes from different genomes. We also consider chloroplast and mitochondrial genomes, discussing the evidence that these organelles have arisen by endosymbiosis, and looking at the trend for reduction in the size of organelle genomes resulting from the transfer of genes to the nucleus. Finally, we discuss the mechanisms by which gene order can change during evolution, and examine methods for the reconstruction of phylogenetic trees using both gene content and gene order of complete genomes.

all the open reading frames (ORFs), predicts whether these are likely to be real genes, and, as far as possible, gives names and functions to them. We thus have a large amount of data, ripe for study using bioinformatics methods, and interesting trends are now emerging regarding how genomes evolve.

Before considering examples of prokaryotic genomes, we will briefly discuss the relationships between the major divisions of prokaryotes. Until a few decades ago, biologists had classified organisms as either eukaryotes (cells possessing a nucleus surrounded by a nuclear membrane, and usually possessing many other internal structures, like the cytoskeleton, endoplasmic reticulum, and mitochondria) or prokaryotes (cells not possessing a nucleus, and having a much simpler internal structure). At that time, the term "bacteria" was more or less synonymous with "prokaryotes". Our current understanding is that life is split into three domains: archaea, bacteria, and eukaryotes. There are thus two domains within the prokaryotes, which are as different from one another as they are from eukaryotes, in terms of the sequences of their genes, the set of genes they possess, and the metabolic processes occurring in their cells. This was discovered initially by comparative analysis of rRNA sequences (Woese and Fox 1977). Now that

12.1 PROKARYOTIC GENOMES

12.1.1 Comparing prokaryotic genomes

When we look at complete genomes, many questions spring to mind. Which genes are present? How did they get there? Are the genes present in more than one copy? Which genes are **not** present that we would expect to be present? What order are the genes in, and does this have any significance? How similar is the genome of one organism to that of another? These are questions that we could not ask before complete genomes were available. The GOLD database (see Box 12.1) lists 128 bacterial and 17 archaeal genomes complete by December 2003, and there will no doubt be more by the time you read this. The amount of information available for these genomes varies, but in many cases we have reasonably good annotation that tells us the positions of many complete rRNA sequences are available, phylogenies based on rRNA have become a standard way of classifying organisms (Woese 1987, Olsen, Woese, and Overbeek 1994, Ludwig *et al.* 1998). A recent "backbone tree", showing the relationship between representative groups of bacteria and archaea, is available from the Ribosomal Database Project (Cole *et al.* 2003).

One of the principal groups of bacteria recognized from the rRNA sequences was the Proteobacteria. This is a diverse group, containing many species of interest, for which we have a lot of complete genomes. We will use genomes from this group as examples in this section. Table 12.1 lists the species of Proteobacteria for which we have complete genomes, together with some of their genome features. The Proteobacteria fall into five subdivisions, labeled α, β, γ, δ, and ε. When material for this chapter was prepared, there were no complete genomes available from the δ subdivision, but several have appeared in the last few months.

As a first example, we will consider the *Rickettsia* group, which belong to the α subdivision. These bacteria live inside the cells of arthropod hosts as obligate parasites. *R. conorii* infects ticks, and can be transmitted to humans by tick bite, leading to Mediterranean spotted fever. *R. prowazekii* is transmitted by lice, causing typhus in humans. Plate 12.1 shows the genome of *R. conorii* (Ogata *et al.* 2001). It is a circular genome of approximately 1.2 million bases (1.2 Mb or 1,200 kb), which is smaller than average for bacteria, as can be seen from Table 12.1. The second and third circles in Plate 12.1(a) show the positions of genes on the two strands of the circular genome. We immediately get the impression that the genome is "full of genes", with little space between them.

An expanded version of three sections of the genome is shown in Plate 12.1(b). In each case, the genome of *R. conorii* is compared with the corresponding region in *R. prowazekii* (the similarities and differences between these genomes are discussed in more detail below). Plate 12.1(b) confirms our impression that most of the genome consists of protein-coding genes, and that gaps between genes are short. In fact, 81% of the *R. conorii* genome is estimated to be coding regions. These features are typical of prokaryotes, in contrast to eukaryotes, where there are often large non-coding intergenic regions. The sixth and seventh circles in Plate 12.1(a) show the positions of repeat sequences in *R. conorii*. Repeat sequence families were identified,

Table 12.1 Statistics of completely sequenced genomes from Proteobacteria.

		Length (kb)	No. genes	%G+C
Agrobacterium tumefaciens C58 Cereon	α	4,915	5,299	59.3
Brucella melitensis 16M	α	3,294	3,197	57.2
Buchnera aphidicola Ap	γ	640	564	26.3
Buchnera aphidicola Bp	γ	618	504	25.3
Buchnera aphidicola Sg	γ	641	545	25.3
Campylobacter jejuni	ε	1,641	1,654	30.6
Caulobacter crescentus	α	4,016	3,737	67.2
Coxiella burnetii	γ	2,100	2,095	42.7
Escherichia coli CFT073	γ	5,231	5,533	50.5
Escherichia coli K12	γ	4,639	4,289	50.8
Escherichia coli O157:H7 EDL933	γ	4,100	5,283	50.5
Haemophilus influenzae	γ	1,830	1,850	38.1
Helicobacter pylori 26695	ε	1,667	1,590	38.9
Helicobacter pylori J99	ε	1,643	1,495	39.2
Mesorhizobium loti	α	7,596	6,752	62.6
Neisseria meningitidis MC58	β	2,272	2,158	51.5
Neisseria meningitidis A1	β	2,184	2,121	51.8
Pasteurella multocida	γ	2,250	2,014	40.4
Pseudomonas aeruginosa	γ	6,264	5,570	66.6
Pseudomonas syringae	γ	6,397	5,615	58.0
Rastolnia solanacearum	β	5,810	5,120	67.0
Rickettsia conorii	α	1,268	1,374	32.4
Rickettsia prowazekii	α	1,111	834	29.0
Salmonella enterica Typhi Ty2	γ	4,791	4,646	52.0
Salmonella typhimurium	γ	4,857	4,597	52.2
Shewanella oneidensis	γ	4,969	4,931	46.0
Shigella flexneri 2a 301	γ	4,607	4,434	50.9
Sinorhizobium meliloti	α	6,690	6,205	62.2
Vibrio cholerae	γ	4,000	3,885	47.5
Vibrio parahaemolyticus	γ	5,165	4,832	45.4
Vibrio vulnificus	γ	5,211	5,028	46.6
Xanthomonas axonopodis	γ	5,273	4,384	64.8
Xanthomonas campestris	γ	5,076	4,182	65.1
Xylella fastidiosa	γ	2,679	2,904	52.7
Xylella fastidiosa Temecula1	γ	2,519	2,066	51.8
Yersinia pestis C0-92	γ	4,653	4,012	47.6
Yersinia pestis KIM P12	γ	4,600	4,198	47.6

with lengths between 19 and 172 bases (Ogata *et al.* 2001), that seem to be distributed randomly all over the genome. However, the size of the arrows is misleading. The repeat sequences constitute only 3.2% of the *R. conorii* genome, and the figure is even lower for *R. prowazekii*. Again, we can contrast this situation with eukaryotes, where there are many families of repetitive sequences that take up large fractions of the genome, and are probably "junk" as far as the organism is concerned. In bacteria, there seems to be selection to economize on genome length, keeping the accumulation of junk DNA to a minimum.

The arrows in Plate 12.1(b) show that transcription of neighboring genes is often in opposite

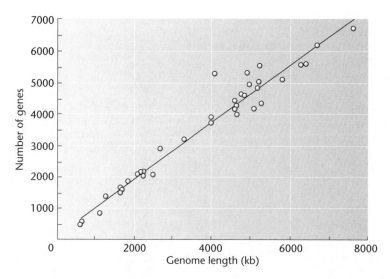

Fig. 12.1 Correlation between the total genome length and the estimated number of genes on bacterial genomes. Each point corresponds to one of the species of Proteobacteria listed in Table 12.1. The correlation coefficient for the linear regression is 0.98, i.e., there is an extremely good straight-line fit.

directions, i.e., the genes interchange between the strands. Color is used to classify genes according to their function, where this is known. There are several genes of unknown function (in green), which are simply labeled as numbered ORFs. An ORF is a continuous region of DNA, beginning with a start codon and ending with a stop codon in the same frame, with no in-frame stop codons in between. The presence of a long ORF is usually indicative of a coding region, since in random, non-coding DNA there would be a relatively high frequency of chance occurrence of stop codons. It is difficult to predict from sequence analysis alone whether an ORF really corresponds to an expressed gene, so it is possible that some of these ORFs may not represent real genes. However, in the case shown in Plate 12.1(b), many of the unknown ORFs are present in the same relative position in the genome of both *Rickettsia* species, suggesting that they are real genes that have been conserved during evolution. This is typical of the state of the art in genome annotation: most genomes contain substantial numbers of genes of unknown function.

Table 12.1 shows the lengths of complete genomes from the Proteobacteria and the estimated number of genes in each; Fig. 12.1 shows that there is a very strong correlation between these numbers. The linear regression passes almost through the origin, and the slope is just below one, i.e., there is

just less than one gene every kb, or slightly more than 1,000 bases per gene on average. The presence of this strong correlation reflects the fact that the typical length of a gene varies little between bacterial species, and that most of the space in the genome is occupied by genes. Figure 12.1 makes it clear how much genome size and numbers of genes vary between species. The largest genome in this set, that of *Mesorhizobium loti*, is 12 times larger than the smallest, that of *Buchnera aphidicola*.

12.1.2 Gene loss and gene rearrangement

Comparing genome sizes between related species gives us some insight into the mechanisms of genome evolution. Figure 12.2 shows a phylogenetic tree of Proteobacteria that we obtained from a concatenated set of 26 tRNA genes present in all species. There is not yet a consensus on the phylogeny of Proteobacteria, and some details of the tree differ, according to which genes and which phylogenetic method are used. The tree shown is likely to be correct in most aspects and is sufficient to illustrate the points we make regarding genome evolution in this section. The main observations from Fig. 12.2 are that both genome length and base composition can change quite rapidly, and that closely related species are not always similar in either length or base composition.

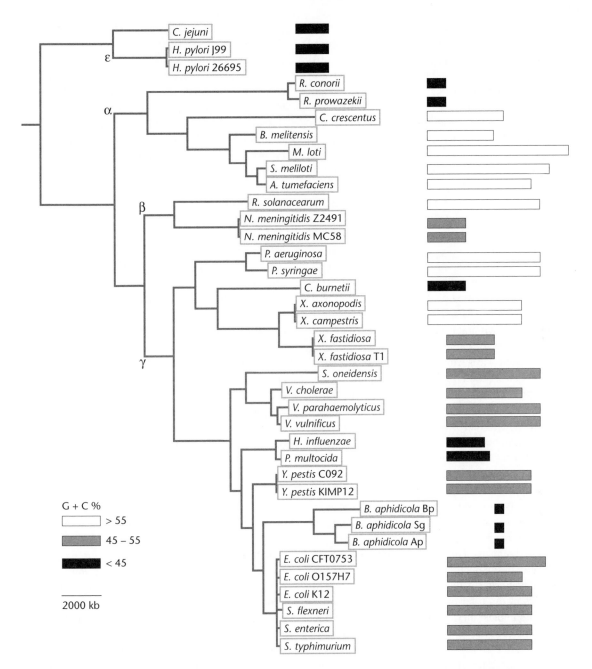

Fig. 12.2 Phylogenetic tree of Proteobacteria obtained using a set of 26 concatenated tRNA genes that are present in every genome. The bars indicate the genome length (scale bar shows 2,000 kb). Color indicates percentage of G+C bases.

Species with short genomes appear to have evolved from ancestors with much longer genomes on several occasions. The *Rickettsia* species discussed above are reasonably close to *M. loti*, the longest genome in this set, and to *Agrobacterium tumefaciens* and *Sinorhizobium meliloti*, which also have large genomes. These species are members of a group called Rhizobia, which have close association with plants. *A. tumefaciens* is a plant pathogen that causes the growth of galls on many types of plant. It is able to insert parts of its own DNA into plant genomes. Scientists have made use of this to insert specific genes into genetically engineered plants. *A. tumefaciens* has an unusual genome composed of one circular chromosome, one linear chromosome, and two smaller circular plasmids (Wood *et al.* 2001). *M. loti* and *S. meliloti* are both nitrogen-fixing bacteria that live symbiotically in the root nodules of leguminous plants. *M. loti* has a single circular chromosome and two small plasmids, while *S. meliloti* has a 3.65 Mb circular chromosome, together with 1.35 Mb and 1.68 Mb "megaplasmids", both of which are larger than the *Rickettsia* genome.

The other group with very short genomes in Fig. 12.2 is the *Buchnera aphidicola* species. These bacteria are symbionts inside the cells of aphid hosts: the abbreviations *Bp*, *Ap*, and *Sg* in Table 12.1 stand for *Baizongia pistaciae*, *Acyrthosiphon pisum*, and *Schizaphis gramium*, three species of aphid that feed by sucking the sap from different host plants. Each aphid is, in turn, host to its own variety of *Buchnera*. These are transmitted through the maternal cell line, and have little opportunity to be transmitted between organisms. Therefore, it is thought that the phylogeny of the bacteria follows the phylogeny of the aphids. *Buchnera* appear to benefit their hosts by synthesis of essential amino acids that are deficient in the aphids' diet. These species are members of the Enterobacteria group, which also includes *Escherichia coli* (a well-studied model organism that is usually a harmless resident in the human gut, but can sometimes be pathogenic), and several other human pathogens, like *Salmonella typhimurium* (a cause of gastroenteritis), and *Yersinia pestis* (responsible for bubonic plague). These latter species all have much larger genomes than *Buchnera*.

What *Buchnera* and *Rickettsia* have in common is that they live inside other cells, either symbiotically or as parasites. It seems that this specialized lifestyle tends to allow species to survive with a much-reduced set of genes. Bacteria in a host cell have a very stable environment and may be able to rely on the host to provide substances that a free-living bacterium would have to synthesize for itself. We can expect that there is a selective pressure acting to shorten genomes, because this leads to more rapid genome replication and cell division. We also expect that short deletions and unfavorable point mutations will be occurring in genes all the time. If unfavorable mutations occur in a redundant gene that is no longer necessary to the bacterium, then there is no selection against them. Consequently, a redundant gene can be rapidly lost from a genome.

Almost all the genes in *Buchnera* are also present in *E. coli* K12. Thus, the major factor in the evolution of the endosymbiont genome has been the large-scale loss of unnecessary genes. The establishment of the symbiosis with aphids is estimated to have occurred approximately 200 million years ago. Since then, the process of gene loss has slowed down, but is still detectable. When the three *Buchnera* species were compared (van Ham *et al.* 2003), it was found that 499 genes are present in all three, and a further 139 are present in only one or two of them. The interpretation is that the common ancestor of these species had a genome with at least 638 genes. As this number is larger than the current gene content of the individual species, all three genomes must have been subject to further gene loss since the establishment of symbiosis.

It is also interesting to compare the gene order in related genomes and to ask to what extent this has changed. An inversion is the reversal of a section of DNA such that any genes present in that section switch strands and have their direction of transcription reversed. A translocation is the deletion of a section of DNA from one place in a genome and its reinsertion in another place. Inversions and translocations can lead to complete reshuffling of gene order when comparing distantly related species. However, with closely related genomes, it is often possible to find regions of the genome where colinearity is maintained. The order of the genes in the three *Buchnera* species was found to be almost identical, except for two small inversions and two small translocations. For 200 million years, this represents rather little change; van Ham *et al.* (2003) therefore call this a "gene-order fossil".

Although both the gene content and gene order in the *Buchnera* group seem to have evolved rather slowly, this is not true with gene sequence. Van Ham *et al.* (2003) observed an accelerated rate of non-synonymous substitutions with respect to synonymous substitutions (see Section 11.2.3) in these species relative to free-living bacteria. This is indicative of the increased rate of fixation of mildly deleterious mutations. In Fig. 12.2, the branches leading to *Buchnera* are rather long, indicating that the tRNA genes have also evolved faster in these species than in any of the other γ-Proteobacteria. It is known that small, isolated asexual populations are subject to accumulation of deleterious mutations by a process known as Muller's ratchet. The effect of genetic drift is larger in small populations, and the efficiency of selection is reduced. It appears that the *Buchnera* genome is gradually degenerating, and that the aphids may have to find a new symbiotic bacterium at some point in the future.

A similar picture is seen in the *Rickettsia*. The number of genes in *R. conorii* is substantially larger than in *R. prowazekii*: 767 orthologous gene pairs were found between them (Ogata *et al.* 2001); 552 *R. conorii* genes have no functional counterpart in *R. prowazekii*; but only 30 *R. prowazekii* genes have no functional counterpart in *R. conorii*. The process of gene loss seems to have continued at a faster rate in *R. prowazekii* than *R. conorii* since their divergence. We already compared regions of the *R. conorii* and *R. prowazekii* genomes in Plate 12.1. This illustrates that some long genes in one species have become split into several consecutive ORFs in the other when stop codons arose by mutations in the middle of the gene. The transcription patterns of the split genes vary: in some cases, all the ORFs are still transcribed, but in others only one or other of them. So, some of the split genes may have retained their function, but we can probably conclude that gene splitting is an example of fixation of deleterious

mutations in isolated populations. The occurrence of a stop codon in the middle of a gene would normally be strongly deleterious, and would be eliminated by selection in most species. The bottom panel of Plate 12.1(b) illustrates a gene in *R. conorii* that has almost disappeared from *R. prowazekii* owing to deletions, but where some vestigial sequence similarity can still be detected, showing that the gene was once there.

As with *Buchnera*, there is very strong conservation of gene order between the two *Rickettsia*. The genomes are approximately colinear, apart from a few translocations that have occurred in the shaded sector of Plate 12.1(a). Gene order comparisons become more interesting for pairs of species that are slightly more divergent. One graphical way to represent such comparisons is by means of a dot-plot, such as that shown in Fig. 12.3 for two of the Rhizobia (Wood *et al.* 2001). Proteins from each genome were queried against the other genome using BLASTP. Each dot in Fig. 12.3 shows the position of a bidirectional best hit. There is a substantial degree of colinearity between the circular chromosome of *A. tumefaciens* and the *S. meliloti* chromosome, indicating that these chromosomes evolved from a common chromosome in the ancestor of the Rhizobia. The linear chromosome in *A. tumefaciens* seems to have arisen at a later date. It contains genes that have matches in the main *S. meliloti* chromosome that seem to have got there by individual translocations; hence, there is no pattern of conserved gene order in the linear chromosome. Suyama and Bork (2001) show dot-plots of this type for many pairs of bacterial genomes, and compare the rate at which gene order is disrupted by rearrangements to the rate of sequence evolution.

In Fig. 12.3, we see that reshuffling of genes appears to be largest around the origin and terminus of replication of the circular chromosomes. This illustrates the influence of genome replication on the occurrence of inversions and translocations. The same effect is also apparent in *Rickettsia* (Plate 12.1). The shaded region, where genome rearrangement has occurred, is next to the terminus of replication. There is one more strange feature apparent in Plate 12.1 that is also connected with genome replication.

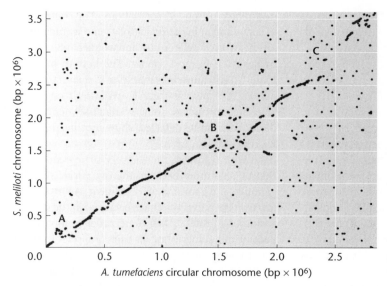

Fig. 12.3 A dot-plot illustrating regions of colinearity between the circular chromosomes of *A. tumefaciens* and *S. meliloti* (reproduced from Wood *et al.* 2001. Copyright 2001 AAAS). Each dot represents a bidirectional best hit between protein sequences using BLASTP. A and B indicate putative origin and terminus of replication. C indicates another sizeable region lacking colinearity.

The inner circle shows the G–C skew, defined as (%G − %C)/(%G + %C). One half of the genome has excess G and one half has excess C on the strand analyzed. If the other strand were analyzed, the G–C skew would be opposite, because a C on one strand is complementary to a G on the other. This shows that the base composition is not equal on the two strands. The effect arises as a result of the asymmetrical nature of DNA replication (see Section 2.3.5), where one strand is leading and one lagging. There appear to be different mutational biases on the leading and lagging strands. Bacterial chromosome replication begins at an origin of replication, proceeds around the circle in both directions, and finishes at a terminus on the opposite side of the circle. Each strand of the chromosome is leading in one half of the genome and lagging in the other half, which leads to the pattern seen in Plate 12.1. The same pattern is also seen in many other bacteria (Lobry 1996) and can be used to locate the position of the origin and terminus in newly sequenced genomes (Chambaud *et al.* 2001).

12.1.3 Gene duplication and horizontal gene transfer

So far, we have painted a dismal picture of genomes degenerating via deleterious mutations and dele-

tions. How did the big genomes arise from which the small ones have degenerated? To create a new gene in a genome, the first way we might think of would be to create it *de novo* by mutations in a DNA sequence. However, this is hard! Mutations in a sequence would have to occur in just the right way to create a coding sequence that could be translated into a protein that could do something useful. Mutations in the upstream and downstream regions of the gene would have to occur to ensure that the proper signals were present that ensure the new gene would be transcribed and translated. It seems very unlikely that a new gene would arise like this in a modern genome. At some point in the very distant past, life must have arisen, and part of the problem of the origin of life is to explain the *de novo* origin of sets of genes that work together to control the functioning of an organism. However, there must have been some last universal common ancestor (LUCA) of all currently existing species (i.e., of archaea, bacteria, and eukaryotes). We don't know much about what type of organism the LUCA was, but we can be fairly sure that it had some type of genome with lots of genes, many of which would have functions similar to the functions of genes in present-day organisms. If we consider how genomes evolved after the establishment of cellular life forms, like present-day prokaryotes, then the most obvious ways for genomes to acquire new genes are gene duplication (i.e., making a second copy of a gene you already have) and horizontal gene transfer (i.e., pinching an existing gene from somewhere else).

Duplications of regions of DNA can occur during replication. Sometimes, these will contain a whole gene, in which case the new genome now has two copies of that gene. We can also suppose that, occasionally, a larger chunk of DNA is copied, leading to the simultaneous duplication of a large number of genes. When two copies of a gene are present, sequence evolution proceeds in the normal way. Substitutions will accumulate in both sequences such that the proteins they produce will no longer be identical. Sometimes, one of the genes may accumulate deleterious mutations and become non-functional, in which case it is likely to disappear from the genome. Occasionally, however, one of the copies might acquire a new function, slightly different from the old, in which case selection would cause the retention of both copies and would lead to diversification and specialization of each of the genes for different functions.

Horizontal (or lateral) gene transfer means that a piece of DNA from an unrelated organism has been incorporated into a genome. This is in contrast to the normal vertical transmission of genes passed down by DNA replication and cell division. There are several ways in which foreign DNA can get into cells. Many bacteria have the ability to take in fragments of DNA from their surroundings through special receptor sites on the cell surface – such cells are known as "competent" cells. The foreign DNA can sometimes form a heteroduplex with the host DNA and replace the original complementary strand. Descendants of this cell then possess the foreign DNA sequence as an integral part of their genome – this is known as transformation. Some bacteria also deliberately transfer DNA to their fellow cells via conjugation. Genes controlling the formation of the conjugative bridge structure are often found on plasmids: this is how plasmids can be transferred to other cells, along with any genes they happen to contain. Another method of transfer is via bacteriophages: DNA packed in the capsids of bacteriophages is horizontally transferred when the virus invades a new cell. Some DNA elements act as transposons, which have the ability to insert and excise themselves in and out of bacterial genomes; integrons are specific locations in bacterial DNA where new DNA sections can be incorporated. Bacteria often come into contact with eukaryotic cells too, as many are parasites or live as commensals inside eukaryotic organisms. Transfer of genes from eukaryotes to prokaryotes or vice versa has also been claimed to occur. Some bacteria are **intracellular** parasites, which puts them in even closer contact with host DNA and makes horizontal transfer seem one step more likely.

The subject of horizontal transfer is controversial, owing to its medical importance and possible relevance to genetically modified organisms. The spread of antibiotic-resistance genes between bacteria is largely thought to result from horizontal transfer of genes carried on plasmids. It is also known that pathogenic and non-pathogenic strains of bacteria often differ from one another by the presence or absence of DNA elements known as pathogenicity islands. These sections contain one or more genes whose function is directly linked to the pathogenic lifestyle of the host. Therefore, it seems that the horizontal gain of new genes is a major mode of evolution and adaptation in bacteria. The possibility of horizontal transmission of genes for pesticide resistance is also a major concern in relation to testing genetically modified crops. Until we fully understand the processes involved, it is difficult to be certain whether such concerns are justified.

Let us consider a further example from the Proteobacteria: *Escherichia coli*. Most *E. coli* strains are harmless inhabitants of our digestive tract: the K12 strain is one of these. However, strain EDL933 is enterohemorrhagic (i.e., it is a pathogen inside the digestive system), while strain CFT073 causes infections of the urinary tract. Welch *et al.* (2002) compared the genomes of these three strains. They defined genes as orthologs if their sequence identity was greater than 90% and the alignments covered at least 90% of the gene. They found 2,996 genes present in all three strains. However, the total numbers of genes present in at least one of the

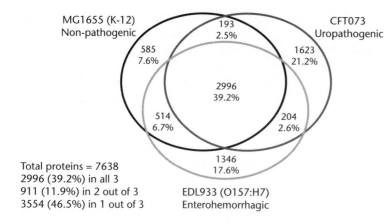

MG1655 (K-12)
Non-pathogenic

CFT073
Uropathogenic

193
2.5%

585
7.6%

1623
21.2%

2996
39.2%

514
6.7%

204
2.6%

Total proteins = 7638
2996 (39.2%) in all 3
911 (11.9%) in 2 out of 3
3554 (46.5%) in 1 out of 3

1346
17.6%

EDL933 (O157:H7)
Enterohemorrhagic

Fig. 12.4 Comparison of the gene content in three strains of *E. coli* (reproduced from Welch *et al.* 2002. Copyright 2002 National Academy of Sciences USA). The regions of the diagram illustrate the numbers of predicted proteins present in one, two, or all three strains. Remarkably few are shared between all three.

genomes is 7,638, implying that only 39% of genes are shared between all three strains. The results are summarized in Fig. 12.4. One might argue that this result is sensitive to the method by which the orthologs were detected and that it is failing to find all relevant matches; however, it is clear that there are very large differences between these genomes, and that each contains genes that have been acquired horizontally.

When CFT073 was compared to K12, 247 regions of length at least 50 bases were found to have been inserted in CFT073. These regions are CFT073-specific "islands" in the genome, totaling 1.3 Mb, while the remaining "backbone" regions of the genome, totaling 3.9 Mb, have clear homology between the strains (Welch *et al.* 2002). Genes in the backbone regions are usually very similar in sequence between the strains. The strains have thus arisen relatively recently. Evidence that the island regions consist of horizontally transferred DNA comes from the fact that most of the genes without orthologs in the other strains are on the islands, and also from analysis of codon usage. Sequence-specific patterns of mutation and selection lead to preferred usage of certain codons in a given organism (Section 3.6). Genes transferred from outside that organism will have different codon usage frequencies, and it will take some time until this signal disappears as a result of mutation and selection acting on the genes in their new environment. Codon frequencies were found to differ significantly between genes in the island and backbone regions, while codon frequencies in the backbone regions of the different strains did not differ.

Many of the genes in the islands are known to be essential for the pathogenic lifestyle of the bacteria – hence the term "pathogenicity islands". The different strains of *E. coli* have thus evolved new lifestyles by acquiring genes for new functions. This is in contrast to the three *Buchnera* species, which have diverged from each other "passively", being carried along as their host species diverged.

12.1.4 Detecting and quantifying horizontal transfer

Horizontal transfer is also controversial because it flies in the face of the traditional tree-like view of evolution. As we said above, our understanding of the relationship between bacterial species derives principally from phylogenies based on rRNA sequences. The rRNA phylogeny has become so central that it is often referred to as the "universal tree" (Woese 2000). As sequence information has accumulated from an increasing number of genes, it has become clear that not all genes give the same tree. It is then tough to decide if this is the result of artifacts of phylogenetic reconstruction, or if genes have really followed different evolutionary paths. Some authors have argued that the amount of horizontal transfer is so large that we should really think of bacterial evolution as a tangled network rather than a tree (Doolittle 2000, Boucher, Nesbo, and Doolittle

2001). Others argue that it still makes sense to look for a core set of genes that are passed down vertically as a coherent group. Some of the most fundamental genes possessed by all domains of life are those connected with genome replication and expression, and these usually give trees consistent with the rRNA tree. Woese (2000) argues that this core set of genes is giving us a valid picture of the organismal genealogy and defends the idea of a universal tree.

One family of genes for which there is agreement that horizontal transfer is common is the aminoacyl-tRNA synthetases (aaRSs), which are responsible for charging tRNAs with their appropriate amino acid. Most organisms have a set of 20 of these genes. Because of their essential role in the translation process, they occur in all three domains of life, and have been widely studied. When phylogenetic trees for aaRS genes are constructed, numerous anomalies are found, many of which can be explained by horizontal gene transfer (Doolittle and Handy 1998, Koonin, Makarova, and Aravind 2001). One interesting case is that of GluRS in *E. coli*. Most eukaryotes have a GluRS and a GlnRS that are related in sequence and probably arose via a gene duplication event. On the other hand, most bacteria have only a GluRS that is responsible for correctly charging tRNA^{Glu} and for mischarging tRNA^{Gln} with Glu. Correctly charged Gln-tRNA^{Gln} is then made from the mischarged Glu-tRNA^{Gln} by a transamidation reaction that takes place on the tRNA. It is therefore unusual to find that *E. coli* and some other proteobacteria do possess a GlnRS gene, and do not use the transamidation route. This indicates horizontal transfer of a eukaryotic gene into this group of bacteria. Many other cases of apparent horizontal transfer are known in the aaRSs, making them unusual in this respect.

For a horizontally transferred gene to be detectable, it needs to survive as a functioning gene in its new host genome. The new gene might replace an existing ortholog that is subsequently deleted, or it might have a useful function not possessed by any of the host genes. If the horizontally transferred gene were a non-functional pseudogene in its new environment, then it would rapidly accumulate mutations, and sequence analysis techniques would be

unlikely to spot it. So, if a family of genes, like aaRSs, is frequently horizontally transferred, this suggests that the corresponding proteins can function more effectively in a foreign cell than would the products of most foreign genes, rather than that the DNA of these genes is more likely to be transferred than others. The aaRSs have to interact with tRNAs, which are very conservative in their structure; other proteins might be involved in large-scale complexes with other fast-evolving proteins, so a horizontally transferred sequence might be unable to function effectively and to replace an existing gene. A contrast is sometimes drawn between aaRSs and ribosomal proteins. The latter must interact with many other proteins and RNAs inside the ribosome, so there is large scope for coevolution of these genes. Ribosomal proteins have rarely been found to be horizontally transferred.

As a final note of caution, we should remember that even rRNA genes, beloved of tree-builders, have occasionally been transferred horizontally. Yap, Zhang, and Wang (1999) found that, in the bacterium *Thermomonospora chromogena*, there are six copies of the ribosomal RNA operon, containing both SSU and LSU rRNA genes. One of these copies is substantially different from the other five, but bears a close relationship to the sequence in the related species, *T. bispora*, indicating a likely horizontal transfer from this species.

Given that horizontal transfer between bacteria occurs to some degree, an important aim is to locate particular cases and to give a quantitative estimate of the frequency of occurrence. One way to do this is to look for regions of a genome that appear to have different properties from the surrounding regions: the codon-usage patterns in islands of the *E. coli* CFT073 strain discussed above is an example of this. The frequencies of bases in different genomes differ substantially, so if a foreign piece of DNA is inserted into a genome, it will often stand out by this alone. In their study of the *E. coli* genome, Lawrence and Ochman (1998) measured the GC frequency at the first and third positions in each gene. The second position was excluded, because it is most constrained by the genetic code and therefore least variable. They defined genes as atypical if the GC frequency was more than two standard deviations away from the mean for all genes. They then looked for genes that appeared to have biased codon usage with respect to typical *E. coli* genes. They also took account of the possibility of horizontal transfer of operons. If several genes are translationally coupled, but not all of them appear unusual by the statistics above, then it is nevertheless likely that the genes were transferred together as an operon. One factor that can influence the base composition of genes is the amino acid composition of the corresponding protein. For example, Lawrence and Ochman found some proteins with large numbers of lysine residues. As the corresponding codons are AAA and AAG, a high AT content is apparent in these genes, which does not indicate horizontal transfer.

Combining all these factors, it was estimated that 755 of 4,288 ORFs in *E. coli* had been horizontally transferred, which is a huge 17.6%. What happens to these genes when they arrive in a new genome? First, they are subject to the mutational biases in the host organism. This means that any usual base frequencies and codon-usage patterns will gradually decay away until the gene looks just like its surroundings. Lawrence and Ochman (1998) had estimates of mutation rates and hence were able to obtain estimates of how long the horizontally transferred genes had been present, by looking at their base frequencies. They estimated a mean age of 14.4 million years, which translates into an estimated rate of 16 kb of DNA transferred per million years. In the 100 million years since the divergence of *E. coli* and *Salmonella*, this means a total of 1.6 Mb has been added to the *E. coli* genome – this should be compared to the total genome length of 4.6 Mb. The large rate of gain by horizontal transfer must be balanced by a significant amount of random deletion, thus creating a turnover in gene content.

These numbers should be taken with a pinch of salt. There is a degree of uncertainty in assigning which genes are horizontally transferred, and there is a large amount of educated guesswork in calculating the rates. Nevertheless, the bottom line is that horizontal transfer seems to be much more important than anyone expected before complete genomes were available. A more recent estimate (Ochman, Lawrence, and Groisman 2000) gives a horizontal transfer figure of 12.8% of the *E. coli* genome, which is larger than for all but one of the species considered. The figures vary considerably between species: e.g., 11.6% of foreign DNA was found in *Mycoplasma pneumoniae*, but none at all in *Mycoplasma genitalium*.

12.1.5 Clusters of orthologous groups (COGs)

When comparing the sets of genes in different genomes, a first question is which genes are orthologous. Orthologs are genes in different species that diverged from a common ancestral sequence as a result of speciation of the organisms. If different species had exactly the same genes, then it would be possible to take each gene in one species and find an ortholog for it in every other species. However, there are many genes that are present in some, but not all, of the complete genomes we have available, either because they have arisen for the first time in certain species, or because they have been deleted from genomes that originally possessed them. Another complication is gene duplication. Related genes in one genome that have arisen from a common ancestral sequence by gene duplication are called paralogs. Many genes in eukaryotes consist of families of paralogs. Sometimes, one gene in a bacterial genome is orthologous to a family of genes in a eukaryotic genome. Duplications can also occur independently in different lineages, leading to situations where different species each possess families of paralogs, but there is no correspondence between individual pairs of genes in the different organisms. In other words, relationships between genes may be one-to-one, one-to-many, or many-to-many. An example of complex gene duplications was given for hexokinase genes in Fig. 6.5. Here, there are four pairs of orthologs between human and rat, which can be identified because the gene duplications occurred before the divergence of the human and rat species. However, if we compare human with yeast, we can only say that there is a family of paralogs in humans and a family of paralogs in yeast, but there is no one-to-one relationship between individual genes. In this case, we might nevertheless say that the two **families** of genes are orthologous, because the genes that gave rise to the families diverged as a result of speciation at some point long ago.

The COG (Clusters of Orthologous Groups) database was set up to identify related groups of genes in complete genomes (Tatusov, Koonin, and Lipman 1997, Tatusov *et al.* 2001). Each COG is a set of genes composed of individual orthologous genes, or of orthologous groups of paralogs from three or more species. For each gene in each of the complete genomes analyzed, BLAST was used to find the best hitting gene in each of the other genomes. The clustering procedure began by looking for triangular relationships of genes in three species, such that each is the best hit of the other two. Whenever two triangles of genes shared a side, these sets of genes were combined into one cluster. Although this clustering scheme is relatively simple and easy to implement automatically, it avoids some of the dangers that might arise in some other automated schemes. For example, best hits according to BLAST (and other search algorithms) are not necessarily the most closely related when more careful phylogenetic analysis is carried out. In the COG system, each member of a cluster is a best hit of at least two other members of the group, which prevents connecting two clusters that are only related by spurious single links. Also, the COG system avoids defining a similarity threshold. It is easy to envisage a system in which all sequences are clustered if their similarity score is above some threshold; but such clusters would be very sensitive to the threshold chosen. As some orthologous groups are much more divergent than others, it would be difficult to choose a threshold appropriate to all genes.

The COG database began with seven genomes and found 720 clusters. The current version has expanded to 43 genomes, including *Saccharomyces cerevisiae* as the only eukaryote, and a range of bacteria and archaea. There are now 3,307 clusters. Of the 104,000 genes in the full set of genomes, 74,000 of them (71%) have been clustered into COGs. The percentage of genes that are clustered tends to be larger for the smaller bacterial genomes: e.g., in *Buchnera*, 568 out of 575 genes are clustered (99%). As we saw above, *Buchnera* has retained a relatively small fraction of the genes it once had, and those

that are retained appear to be essential genes for which it is possible to find orthologs in many other species. On the other hand, in *E. coli* K12, which has a much larger genome, 3,414 genes of 4,275 (80%) were clustered. For the considerably larger *S. cerevisiae* genome, only 2,290 genes of 5,955 (38%) could be placed in COGs. This indicates that a substantial number of genes in eukaryotes do not have orthologs in prokaryotes. As more complete eukaryotic genomes become available, it will presumably be possible to cluster many of these genes into COGs containing only eukaryotic species.

If we consider only the average figure of 71%, this still means that there are substantial numbers of genes for which no reliable orthologs can be found. How do we interpret this? Some of these genes could really be "unique" to a given species or closely related group of species. As more species are added, the chances of finding related species that share genes will become larger, hence the number of COGs should increase and the fraction of genes assigned to them should increase. This has indeed been happening, as can be seen by comparing the figures above, from the Web site in December 2003 (http://www.ncbi.nlm.nih.gov/COG/), with those published two years ago (Tatusov *et al.* 2001). Another interpretation is that our methods for detecting orthologous sequences are simply not good enough to detect distant matches (this must be at least partly true, but it is difficult to know how many matches are being missed). A third possibility is that some of the ORFs annotated as genes in the published genomes are not real genes. By studying COGs, Tatusov *et al.* (2001) gave evidence that many of the ORFs originally predicted in *Aeropyrum pernix* (an archaeon not closely related to previously sequenced species) were not real genes. COG analysis can also potentially help in genome annotation by detecting missing genes. If a gene is absent in a species but present in several related ones, this might suggest that it has been overlooked, and that a more careful search of the genome is warranted. Another potential use for COGs is that when a gene has been reliably assigned to a cluster, one can infer its function from that of others in the cluster. However, there are still many clusters for which the function is unknown, or at best, only partially understood.

There are several take-home messages here. Complete genomes are an important resource, but are just a beginning for future studies. Our current knowledge is insufficient to say with certainty where all the genes are in a genome. Knowing the complete set of genes does not tell us their functions. Bioinformatics methods are important for genome annotation, and methods that can compare new genomes with many previously sequenced genomes can lead to a substantial improvement in the reliability of annotation.

12.1.6 Phylogenies based on shared gene content

Sometimes, phylogenies based on different genes are not consistent with one another. This may result from inaccuracies or lack of resolution of the methods, or from horizontal gene transfer. Either way, it is interesting to try to build phylogenies based on the degree of similarity of the gene content of genomes rather than on the similarity of individual gene sequences. Arguably, this gives a better picture of evolution of the whole organism than could be obtained from any single gene.

The first step in this process is to measure the number of shared genes between genomes. Snel, Bork, and Huynen (1999) used the Smith–Waterman algorithm to compare all the genes in a set of complete genomes. Genes from two different genomes were defined to be orthologs if they were each the other's best hit in the other genome, and if the E value for the alignment was less than 0.01. The more closely related the two genomes, the larger the number of shared genes we would expect to find. However, genomes differ considerably in the number of genes they possess. Larger genomes are likely to share larger numbers of genes than smaller genomes. Also, genome size can vary considerably in closely related organisms, particularly in parasitic and endosymbiotic bacteria, which can lose large numbers of genes very quickly. To account for genome size bias, the number of shared genes was divided by the number of genes in the smaller of the two genomes. This gives a measure of similarity, S, that varies between 0 (for no shared genes) and 1 (where the genes on the smaller genome are a subset of the genes on the larger). If there were a continual process of genome evolution via gain and loss of genes, we would expect the similarity between two genomes to decay roughly exponentially with time since they diverged. We can therefore define a measure of distance between genomes as $d = -\ln S$. In this way, a matrix of distances between a set of genomes can be calculated, and this distance matrix can be used as input to distance-matrix phylogenetic programs, such as neighbor-joining and Fitch–Margoliash methods (Sections 8.3 and 8.5).

Initial attempts at doing this (Snel, Bork, and Huynen 1999, Fitz-Gibbon and House 1999) used relatively small numbers of genomes. Now that substantial numbers of genomes are available, the results become quite interesting, and Korbel *et al.* (2002) have set up a Web-based resource, called SHOT, for doing whole-genome phylogenies of this type. An example of the output of SHOT is shown in Fig. 12.5. Many aspects of this phylogeny confirm what was previously deduced from rRNA sequence phylogenies. The division into the 3 domains, archaea, bacteria, and eukaryotes, is clearly shown in the figure. The eukaryotes are divided into animals, fungi, and plants (animals and fungi being the closest two of the three groups). This point has also been shown using rRNA phylogeny (van de Peer *et al.* 2000). Unfortunately, there is, as yet, no complete genome information on the diverse array of protists and other early-branching eukaryotes. Several of the major groups of bacteria known from rRNA phylogenies are also found in the whole-genome phylogeny. But there are also aspects of the tree that are not well resolved (e.g., branching within the bacteria is rather star-like), or that conflict with sequence-based phylogenies (e.g., *D. melanogaster* appears closer to human than to *C. elegans*, but sequence-based phylogenies place *D. melanogaster* and *C. elegans* in the Ecdysozoa group within the Protostomes, with humans as Deuterostomes (Section 11.3.2)).

It might be hoped that whole-genome phylogenies would be free of some of the sensitivity of sequence-based phylogenies to the gene used, the

Fig. 12.5 Phylogeny based on shared gene content of completely sequenced genomes, calculated using the default options of the SHOT program (Korbel *et al.* 2002).

alignment, the choice of species included, and the method of tree construction. Nevertheless, there are many factors that enter into whole-genome phylogenies too. There are several ways to calculate similarity between genomes. By default, Korbel *et al.* (2002) now use a weighted average of genome sizes to normalize the number of shared genes, rather than dividing by the smaller of the two genomes. The number of shared genes between two genomes depends on the definition of orthologs and the method used to detect them. The number of genes on

each genome is also open to question. One might wish to exclude genes that have no ortholog in any of the other species. The way distance is calculated from similarity is rather crude, and is not based on a well-defined model of genome evolution. Once we have a distance matrix, there are many ways to calculate a phylogeny from it. Unfortunately, all these details affect the resulting tree to some extent. We are still in the early days of whole-genome phylogenetics, and although methods like this are unlikely to solve all the problems, they promise to give increasingly useful information as the number of complete genomes continues to rise.

An interesting study by Snel, Bork, and Huynen (2002) compares gene content among Proteobacteria and Archaea. Rather than use gene content to deduce the phylogeny, it is assumed that the tree is already known, and phylogenetic methods are used to study the mechanisms by which gene content evolves on this tree. They allow for change in gene content via gene loss, gene duplication, horizontal gene transfer, and *de novo* origin of genes. A cost is assigned for changes of each type. The gene content is known for each species on the tips of the tree. The gene content of each of the ancestral nodes of the tree is estimated by a type of maximum parsimony program that minimizes the total cost of changes in gene content across the tree. The method can be used to predict the gene content of ancestral genomes, and also to give sensible bounds to costs associated with each type of change.

The main factor influencing the results of this study is the relative cost assigned to horizontal transfer and gene deletion. Whenever a gene is present in some species but not others, it is possible to explain this entirely in terms of gene deletions or entirely in terms of horizontal transfer, as shown in the example of Fig. 12.6. If the cost of horizontal transfer is very high, the maximum parsimony solu-

tion will use only deletions. As a result, the program will predict extremely large ancestral genomes containing every gene that is present in at least one of the species. If the cost of horizontal transfer is low, the solution will not use deletions. Genes will originate internally within the tree and will be horizontally transferred many times. Hence, ancestral genomes will be predicted to be very small. Snel, Bork, and Huynen (2002) concluded that vertical descent of genes was the leading factor in genome evolution (i.e., most genes were transmitted vertically from the preceding node on the tree most of the time), followed by gene loss and gene genesis. Horizontal transfer was important in explaining the patterns of presence and absence of genes, but was less frequent than gene loss and genesis on the basis of the number of events that occurred in the most parsimonious solution.

12.2 ORGANELLAR GENOMES

12.2.1 Origin of mitochondria and chloroplasts

Both mitochondria and chloroplasts are organelles in eukaryotic cells that are contained in their own membranes and that possess their own genomes. The idea that these organelles are descended from endosymbiotic bacteria has been around for a long time (Margulis 1981). We have already seen examples of bacteria that live obligately inside larger cells, either as parasites or as symbionts. It seems a relatively small step from such bacteria to organelles that are an integral part of the cell. We now have solid evidence, from analysis of gene sequences of organellar genomes, that both mitochondria and chloroplasts arose by endosymbiosis. This has been shown by many researchers and has been reviewed by Gray and Doolittle (1982) and Gray (1992).

One study that shows this clearly is that of Cedergren *et al.* (1988). These authors aligned both SSU and LSU rRNA from representative species of archaea, bacteria, eukaryotes, mitochondria, and chloroplasts. An alignment over such a wide range is difficult, because of the large variability in length between the sequences from the different domains. Nevertheless, the secondary structure of these

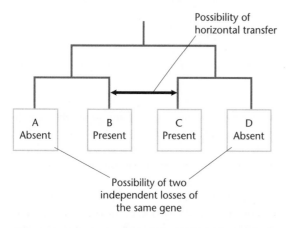

Fig. 12.6 The pattern of presence and absence of a gene in the four genomes shown here can be explained either by horizontal transfer from B to C, or vice versa, or by two independent losses of the gene from species A and D.

molecules is known and there are some regions that are very well conserved across all the domains. Cedergren *et al.* (1988) chose sequence regions from the most conserved parts of the secondary structure for their phylogenetic analysis. They found that the division of life into the three domains – archaea, bacteria, and eukaryotes – was strongly supported, and that the sequences of mitochondria and chloroplasts branched within the eubacterial domain. The chloroplast sequences were closely related to cyanobacteria, a group of normally free-living photosynthetic bacteria. This means that photosynthesis in eukaryotes (i.e., plants and algae) was invented in prokaryotes and acquired by endosymbiosis. The mitochondrial sequences were similar to those of the α-Proteobacteria, a group that includes intracellular bacteria, such as *Rickettsia*. The chief role of present-day mitochondria is to carry out aerobic respiration. This means that aerobic respiration was also invented by prokaryotes and acquired by eukaryotes by endosymbiosis.

Several of the genes involved in the electron transport system of the respiratory process are coded on mitochondrial genomes. Two of the most conserved of these are cytochrome *c* oxidase subunit 1 and cytochrome *b*. Sicheritz-Ponten, Kurland, and Andersson (1998) compared the sequences of these two genes from mitochondrial and bacterial genomes. They found that the mitochondrial genes were related to *Rickettsia* genes, and they estimated that mitochondria and *Rickettsia* diverged from one another 1,500–2,000 million years ago. This is consistent with estimates of the time of origin of eukaryotic cells from other sources. Another group of sequences that provide evidence on the origin of organelles are the Hsp60 heat-shock proteins (known as GroEL in bacteria). These are chaperonins that assist unfolded or misfolded proteins in finding their correct structure. Phylogenetic analysis of these sequences also confirms the relationships between mitochondria and *Rickettsia*, and between chloroplasts and cyanobacteria (Viale and Arakaki 1994). Although the heat-shock proteins are found in the mitochondria, the genes for these proteins are actually in the eukaryotic nucleus. The interpretation is that the Hsp60 genes entered eukaryotes in the bacterium that became the mitochondrion, and that the gene was later transferred from the mitochondrial genome to the nuclear genome. Many other genes from organelle genomes have also been transferred to the nucleus, as we shall see below.

Most mitochondrial genomes are very much smaller than bacterial genomes. For example, the human mitochondrial genome, which is typical of animal mitochondrial genomes, is only 16 kb in length and possesses 37 genes: 13 protein-coding genes associated with the electron transport chain, two rRNAs, and 22 tRNAs. The gap between mitochondria and bacteria has been reduced recently by the sequencing of several protist mitochondrial genomes (Gray *et al.* 1998). The most bacteria-like of the available mitochondrial genomes is that of *Reclinomonas americana* (Lang *et al.* 1997). This is a single-celled freshwater protist that belongs to a group known as the jakobids. The genome is of length 69 kb and contains 97 genes. The proteins present include 24 from the electron transport chain, 28 ribosomal proteins (i.e., components of the mitochondrial ribosome), four that form an RNA polymerase, six others of various functions, and five of unknown function. There are three rRNA genes, as a gene for 5S rRNA is present in addition to the usual SSU and LSU rRNAs. There is also a gene for ribonuclease P RNA, not usually found in mitochondrial genomes, and there is an expanded set of 26 tRNAs.

In addition to its having a larger repertoire of genes, there are other reasons for concluding that the *R. americana* mitochondrion more closely resembles the ancestral bacterial genome than any of the other known mitochondrial genomes. The four-component RNA polymerase encoded by this genome is similar to those found in bacteria, and it is substantially different from the nuclear encoded single-component polymerase that functions in most mitochondria. This suggests that the single-component polymerase replaced the bacteria-like polymerase genes fairly early in mitochondrial evolution. For several genes, Lang *et al.* (1997) also located sequence motifs upstream of the start codons that appear to be Shine–Dalgarno regions, and suggest that translation is initiated in the same way as

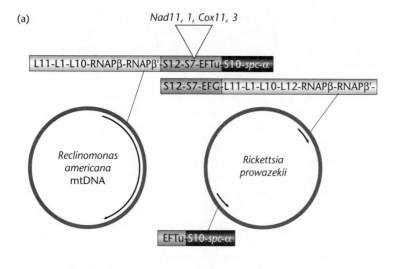

(a)

Nad11, 1, Cox11, 3

L11-L1-L10-RNAPβ-RNAPβ'-S12-S7-EFTu-S10-*spc*-α

S12-S7-EFG-L11-L1-L10-L12-RNAPβ-RNAPβ'-

Reclinomonas americana mtDNA

Rickettsia prowazekii

EFTu-S10-*spc*-α

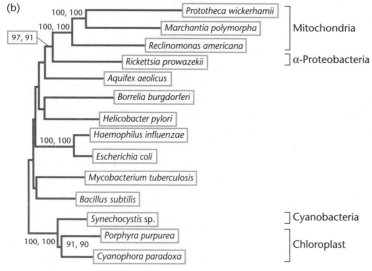

(b)

100, 100 — Prototheca wickerhamii
100, 100 — Marchantia polymorpha } Mitochondria
97, 91 — Reclinomonas americana
] Rickettsia prowazekii] α-Proteobacteria
Aquifex aeolicus
Borrelia burgdorferi
Helicobacter pylori
100, 100 — Haemophilus influenzae
Escherichia coli
Mycobacterium tuberculosis
Bacillus subtilis
Synechocystis sp.] Cyanobacteria
100, 100 — Porphyra purpurea
91, 90 — Cyanophora paradoxa } Chloroplast

Fig. 12.7 Comparison of a bacteria-like mitochondrial genome from *Reclinomonas americana* with a mitochondria-like bacterial genome from *Rickettsia prowazekii* (reproduced from Andersson *et al.* 1998, with permission of Nature Publishing Group). (a) Several regions of conserved gene order are illustrated. S10, *spc*, and α are each operons composed of several consecutive genes. (b) The phylogeny, constructed using ribosomal protein genes from bacteria, mitochondria, and chloroplasts, demonstrates that the mitochondria are related to *Rickettsia* and the chloroplasts to cyanobacteria.

in bacteria. There are also several operons (i.e., groups of consecutive, co-transcribed genes) present in both the *R. americana* and bacterial genomes. Andersson *et al.* (1998) compared the mitochondrial genome of *R. americana* with the most mitochondria-like of known bacterial genomes, *Rickettsia prowazekii*. Figure 12.7 illustrates several regions of conserved gene order between these genomes, which are still present despite over 1,500 million years of separate evolution. They also carried out phylogenetic analysis with ribosomal pro-

tein genes from both mitochondria and chloroplasts, and confirmed the expected relationships of the organelle sequences to bacteria.

12.2.2 The evolution of eukaryotic cells

Eukaryotic cells are defined by the presence of a nucleus with its own nuclear membrane. Where did the genes in the nucleus come from? Molecular phylogenies showing the three domains of life cannot tell us whether eukaryotes are more closely related to bacteria or to archaea, because trees like this are unrooted. As we are talking about the whole of life, there is no outgroup species that can be used to provide a root. This problem was overcome by using pairs of genes that were duplicated prior to the divergence of the three domains. Species from each domain therefore possess both genes. Phylogenies constructed from these gene pairs have two halves, each half providing a root for the other half. An example is the two-elongation factor genes, part of the translation apparatus. Iwabe *et al.* (1989) and Gogarten *et al.* (1989) concluded that the root lies on the branch leading to bacteria, and hence that eukaryotes are more closely related to archaea. The implication is therefore that eukaryotes arose from an ancestor that was something like an archaeon. However, some protein phylogenies suggest that eukaryotes are closer to some taxa within the bacteria. Golding and Gupta (1995) argued for a chimeric origin of the eukaryotic nucleus, possibly by fusion of an archaeon and a bacterium. However, if horizontal gene transfer was frequent in the early stages of cellular evolution, then it may be that a eukaryotic genome acquired genes from many different sources at different times. Philippe and Forterre (1999) reanalyzed the data on the rooting of trees using duplicate genes. They found that trees derived from different gene pairs were often inconsistent, and they also emphasized the problems of mutational saturation and long-branch attraction that arise in phylogenies with very divergent sequences. They concluded that the rooting of the tree of life in the eubacterial branch cannot be relied on. Thus, the origin of the nuclear genome of eukaryotes is much less clear than the origin of the organellar genomes.

At what point did the primitive eukaryote acquire mitochondria? Our understanding of this is linked to our interpretation of the phylogeny of the early-branching eukaryotes (we touched on this in Section 11.3.3). Most eukaryotes contain mitochondria, but there are several groups that do not. In some studies, these amitochondriate groups appeared at the base of the tree, and were hence referred to as Archezoa, implying that these groups were ancient taxa that branched off the main

eukaryotic line prior to the endosymbiosis event. In other words, it was assumed that the Archezoa never possessed mitochondria. However, several groups formerly called Archezoa have been found to be closely related to crown eukaryote groups in more recent trees (Embley and Hirt 1998, Roger 1999). There have also been cases where genes thought to have originated from mitochondrial genomes have been found in nuclear genomes of species that do not possess mitochondria. This suggests that these groups must have once had mitochondria and then lost them at a later stage. Thus, it is not clear whether any of the current species lacking mitochondria are actually descended entirely from organisms that never had mitochondria.

It is generally assumed that the endosymbiosis event leading to the origin of chloroplasts happened after that leading to mitochondria, as there are many eukaryotes lacking chloroplasts but very few lacking mitochondria. There are several types of chloroplast-containing algae that derived from the initial incorporation of the cyanobacterium. Some of these later evolved into land plants. It is fairly certain that the chloroplasts in all these organisms derive from a single endosymbiosis event (Gray 1992, Douglas 1998). However, there are many other chloroplast-containing eukaryotes that are thought to derive from secondary endosymbiosis events, in which an alga arising from the primary event is incorporated inside another eukaryotic cell. Secondary endosymbiosis seems to have occurred many times (Douglas 1998). Some of these cells still contain nucleomorphs (which are the remnants of the nucleus of the algal cell incorporated in the secondary event), in addition to a full eukaryotic nucleus deriving from the new host cell.

12.2.3 *Transfer of organellar genes to the nucleus*

The number of genes in chloroplast and mitochondrial genomes has reduced tremendously since their integration into the eukaryotic cell. We already saw a tendency to reduction of genome size in intracellular bacteria, like *Buchnera* and *Rickettsia*, resulting from loss of genes that were no longer necessary to the bacteria because the host cell provided substances that would otherwise need to be made by a free-living species. However, organelles have gone a lot further than this. The majority of proteins necessary for the function and replication of the organelles have been lost from the organellar genome and transferred to the nucleus.

In well-studied species like humans and yeast, most nuclear genes for mitochondrial proteins have been identified. There is a specialized database of human mitochondrial proteins (http://bioinfo. nist.gov:8080/examples/servlets/index.html) that gives details of both nuclear and mitochondria-encoded proteins in the mitochondrion. We therefore know that transfer of genes to the nucleus has been frequent in the past – but is it still going on? One way to answer this is to search for sequence similarity between mitochondrial and nuclear genomes of the same species. Blanchard and Schmidt (1996) developed a computational strategy for doing this and found a variety of identifiable mitochondrial sequences in humans, yeast, and other species. Many of these inserted sequences are just fragments that do not encode whole genes, and therefore are not important evolutionarily. However, the longest section found in the human genome was over 3,000 bp and contained tRNA and rRNA sequences.

The location of mitochondrial DNA fragments in the nucleus demonstrates that transfer may still be possible, but it does not demonstrate that the transferred genes can be established as functional replacements of the gene in the mitochondrion. In fact, in animal (i.e., metazoan) mitochondria, transfer seems to have stopped some time ago. Animal mitochondrial genomes have a very reduced numbers of genes, but almost exactly the same set of genes is found in all animal phyla. Thus, the gene content of these genomes was established before the divergence of the animal phyla and has remained constant for over 500 million years. In the flowering plants, however, many cases of recent (i.e., in the last few hundred million years) transfer from both chloroplast and mitochondrial genomes to the nucleus are known. For example, Adams *et al.* (2000) showed that there have been 26 independent losses of the ribosomal protein gene *rps10* from the mitochondrion in a set of 277 plant species.

Similarly, Millen *et al.* (2001) demonstrated the loss of the translation initiation factor 1 gene *infA* from chloroplast genomes of 24 separate lineages in a set of 300 plant species. The number of times these genes have been independently established in the nucleus is not known, but is presumed to be large.

The above studies were done using Southern blot hybridizations to probe for the presence of a known gene. When complete organelle genomes are known, a much more complete analysis is possible. Martin *et al.* (1998) studied a set of nine complete chloroplast genomes, with a cyanobacterial genome as an outgroup. They identified 205 genes present in at least one of the chloroplasts that must have been present in the common ancestral chloroplast genome. Of these, 46 were still found in all genomes. These gene sequences were used to construct a molecular phylogeny for the species. With this phylogeny, it was possible to study the process of loss that had occurred in the remaining 159 genes, using a parsimony principle. It was found that 58 genes had been lost once only, but 101 genes had been lost more than once (including a few that had been lost four times independently from the set of only nine species). Thus, parallel losses of the same gene occurred very frequently.

Transfer of an organelle gene to the nucleus involves several steps. First, a copy of a piece of organelle DNA has to find its way into the nucleus. The fragment then has to insert itself into the nuclear genome in a position that does not interrupt an existing gene. It also has to acquire a promoter, so that it gets expressed in the nucleus. Then it has to find a way of getting its protein product back to the organelle. Proteins destined to be targeted to organelles usually have signal peptides attached to the ends of their sequences. The cell has mechanisms to recognize such labeled proteins and transport them. One way in which a new organelle-targeted gene might be established is to insert itself next to a duplicate copy of the promoter and signal peptide region of an existing mitochondrial gene.

All these steps cannot be as unlikely as they seem, given the fact that transfers do occur frequently. If the process goes to the stage described above, it means that there is now a functioning copy of the gene in both the organelle and the nucleus (note that the gene in the organelle cannot be deleted unless there is a functioning copy in the nucleus, assuming that it is an essential gene for the functioning of the organelle). From this point, it is necessary to think about population genetics. Why should the gene from the organelle be deleted in preference to the copy in the nucleus? The nuclear genome usually reproduces sexually and undergoes recombination, whereas the organelle multiplies asexually. This means that the organelle genome is subject to the accumulation of deleterious mutations via Muller's ratchet. It might also be argued that the presence of recombination makes it more likely for a favorable mutation to become fixed in the sequence of the nuclear copy. Either way, if the nuclear copy becomes fitter than the organelle copy after a certain time, it is more likely that the organelle copy will be eliminated. Another relevant factor is the difference in rate of mutation in the different genomes. Animal mitochondrial genes accumulate substitutions at a far greater rate than nuclear genes or typical bacterial genes – see, for example, the rRNA trees of Cedergren *et al.* (1988), in which the animal mitochondrial sequences appear on extremely long branches. However, in plants, mitochondrial substitution rates are low in comparison to both nuclear and chloroplast rates (Wolfe, Li, and Sharp 1987). A large part of the variation in substitution rates can be attributed to variation in mutation rates. The mutation rate clearly affects the likelihood of accumulation of both favorable and unfavorable mutations in the two gene copies. Another factor is that organelle genomes from which a gene has been deleted may also have a selective advantage in speed of replication over longer genomes, and this might tend to favor the fixation of deletions into the population. For further discussion of these issues, see Blanchard and Lynch (2000) and Berg and Kurland (2000).

We might wonder why the organelle retains a genome at all. The purpose of the mitochondrial genome seems to be to code for the proteins of the electron transport chain. However, some of these have already been transferred to the nucleus, so why

not the rest? The system seems inefficient: if some genes are maintained in the organelle, then it is also necessary to maintain genes involved in the translation of those genes, like the rRNAs and tRNAs and (in some organelles) the ribosomal proteins. It has been suggested that the genes that remain are those that could not easily be transported to the organelle, because they are large hydrophobic proteins with several membrane-spanning helices. It has also been suggested that these proteins might be harmful to the cell if expressed in the cytoplasm. However, there is no clear answer to this question yet.

On the theme of the inefficiency of organellar genomes, consider the 35 kb organellar genome of the malaria parasite *Plasmodium falciparum* (Wilson *et al.* 1996). This appears to resemble chloroplast genomes from green algae, although *P. falciparum* is not photosynthetic. The genome contains rRNAs, tRNAs, ribosomal protein genes, and RNA polymerase genes, all of which are necessary purely for the transcription and translation of just three other genes. Despite the degenerate nature of this genome, it has been maintained in the group of apicomplexans, suggesting that it has a key function for the organism that is still not understood.

Despite the many losses of genes from organelles, there are relatively few cases of genes found to be inserted **into** organelle genomes. The probable simple explanation of this is that, for many types of organelle, it is physically difficult for DNA to get itself inside the organelle membrane. Plant mitochondrial genomes appear to be an exception to this. They are much larger than animal mitochondrial genomes owing to the presence of large amounts of non-coding sequence. Due to extensive variability in the lengths of the non-coding sequences, the number of genes does not correlate with the genome size. In the model flowering plant *Arabidopsis thaliana*, there are 58 genes in a length of 367 kb; in the liverwort *Marchantia polymorpha*, there are 66 genes in 184 kb; and in the green alga *Prototheca wickerhamii*, there are 63 genes in 58 kb (Marienfeld, Unseld, and Brennicke 1999). There has thus been a large expansion of genome size in the plant lineage. There is evidence for the insertion of nuclear DNA into the *A. thaliana* mitochondrial genome.

However, this does not consist of nuclear genes, but rather of copies of retrotransposons, which are mobile elements able to copy and insert themselves, usually within the nuclear genome itself. The *A. thaliana* mitochondrial genome does not contain a full set of tRNA genes, which means that some tRNAs must be imported into the organelle. Of the tRNA genes that are found in the mitochondrial genome, some appear to be of chloroplast rather than mitochondrial origin. This is the only known case of chloroplast genes being transferred to mitochondrial genomes.

12.2.4 Gene rearrangement mechanisms

Animal mitochondrial genomes provide good examples for looking at the processes leading to change in gene order. More than 470 of these genomes have been completely sequenced, and almost all contain the same set of 37 genes, but these occur in many different orders in different species. The OGRe (Organellar Genome Retrieval) database stores information on complete animal mitochondrial genomes and allows visual comparison of any pair of genomes (Jameson *et al.* 2003). The examples shown in Plate 12.2 depict mitochondrial genomes linearly – it should be remembered that they are actually circular and that the two ends are connected. Genes on one strand, transcribed from left to right, are drawn as boxes below the central line, and genes on the other strand, transcribed from right to left, are drawn as boxes above the line.

Plate 12.2(a) shows the human mitochondrial genome. The 13 protein-coding genes are shown in green, the two rRNAs in blue, and the 22 tRNAs in green. The genome has almost no space between the genes, with the exception of one large non-coding region, which is the origin of replication for the genome. The same gene order occurs in all placental mammals, and many other vertebrates too. Birds, however, differ from mammals. Plate 12.2(b) compares the human and chicken mitochondrial genomes. Regions of the genomes in which the gene order is the same in both species are shown in the same color. Points on the genome where the gene order changes between the two species are called breakpoints. There are three breakpoints in the human–chicken example. In this case, the yellow block, containing one protein and two tRNAs, has switched positions with the red block, containing one protein and one tRNA.

We can represent this process as

$$1,2,3,4 \rightarrow 1,3,2,4$$

where each number represents a gene or a block of consecutive genes. There are several ways in which the order of the two blocks can be reversed. One is by a translocation, in which one of the blocks is cut out and reinserted in its new position. A second possibility is an inversion, in which a segment of the genome is reversed and reconnected in the same position, thus:

$$1,2,3,4 \rightarrow 1,-3,-2,4$$

Here, the minus signs in front of 2 and 3 indicate that these genes (or gene blocks) are now on the opposite strand and transcribed in the opposite direction. Note that this is **not** what happened in the human–chicken case, because the red and yellow blocks are on the same strand in both species. Switching two blocks while keeping them on the same strand can be achieved with three separate inversions, like this:

$$1,2,3,4 \rightarrow 1,-2,3,4 \rightarrow 1,-2,-3,4 \rightarrow 1,3,2,4$$

However, this would mean that three inversions had occurred in the same place on the genome in a short space of time, without leaving any surviving species having the intermediate gene orders. This does not seem like a parsimonious explanation.

There is another, quite likely way in which the interchange in human and chicken could have occurred. This is via duplication followed by deletion.

$$1,2,3,4 \rightarrow 1,2,3,2,3,4 \rightarrow 1,\cancel{2},3,2,\cancel{3},4 \rightarrow 1,3,2,4$$

In this case, a tandem duplication of the region 2,3 has occurred (e.g., because of slippage during

DNA replication). There are now redundant copies of these two genes and we expect one or other copy to disappear rapidly as a result of accumulation of deleterious mutations and deletions, and possibly selection for minimizing the genome length. After the duplicate copies have disappeared, we are left with genes 2 and 3 in reverse order. However, if the central 3,2 pair had been deleted instead, the original gene order would have been maintained.

This mechanism seems likely whenever we see switching of order within two short neighboring regions. A case where the same mechanism seems extremely likely (Boore and Brown 1998) is in the region containing the consecutive tRNAs W,–A, –N,–C,–Y (the letters label tRNA genes according to their associated amino acid). This is shown on the right of the human genome diagram in Plate 12.2(a). This order occurs in the platypus, echidna, and in placental mammals, but in all the marsupials sequenced so far, the order is –A,–C,W,–N,–Y. This can be explained by a duplication and deletion of four genes:

W,–A,–N,–C,–Y →
W,–A,–N,–C,W,–A,–N,–C,–Y →–A,–C,W,–N,–Y

Note that all the genes remain on the same strand, so it is unlikely that this occurred by inversions. Interestingly, the change is not reversible: if the four genes A,–C,W,–N in the marsupials are duplicated, it is not possible to return to the order in the placentals.

The mobility of tRNA genes within the mitochondrial genome is illustrated in Plate 12.2(c). This compares two insect genomes (honey bee and locust). There are some long expanses of unchanged gene order, but there are also areas with complex reshufflings of tRNAs, including long-distance jumps of E and A. Ignoring the tRNAs, the order of the proteins and rRNAs is identical.

When genomes from different animal phyla are compared, complex rearrangement patterns are often seen. However, if tRNAs are excluded, a simple pattern sometimes emerges for the rearrangement of the proteins and rRNAs. In the comparison of a mollusc and an arthropod shown in Plate 12.2(d),

the blue gene block has been translocated and the yellow gene block has been inverted. This can be compared with the much more complex pattern of movement when tRNAs are included (see the OGRe Web site, Jameson *et al.* 2003).

12.2.5 *Phylogenies based on gene order*

Gene order has the potential to be a strong phylogenetic marker. There are so many different possible orders, even for a small genome like the mitochondrial genome, that when conserved patterns of gene order are seen, they are unlikely to have arisen twice independently. For example, the tRNA rearrangement shared by the marsupials would be a strong argument that these species form a related group, had we not already known this from a wealth of other evidence. In some cases, however, gene order can provide a strong argument to resolve issues for which there is no conclusive evidence from other sources.

One such case is the relationship between the four major groups of arthropods: chelicerates (e.g., spiders, scorpions); myriapods (centipedes and millipedes); crustaceans; and insects. Taxonomists have disagreed about this for decades, but one hypothesis that had generally been accepted is that myriapods and insects were sister groups. However, Boore, Lavrov, and Brown (1998) compared the mitochondrial gene order of these groups and found that insects and crustaceans shared a genome rearrangement not present in myriapods and chelicerates. By considering gene orders from outside the arthropods, they were able to show that the chelicerates and myriapods had retained features of gene order present in other phyla, and hence that the rearrangement in insects and crustaceans is a shared derived feature that links these as sister groups. Roehrdanz, Degrugillier, and Black (2002) have continued this approach, looking at the relationships between subgroups of arthropods at a more detailed level. Scouras and Smith (2001) give another good example of how gene order information can be linked to molecular phylogenies for several groups of echinoderms.

In cases such as those above, where clear shared derived features of gene order are located, this gives important phylogenetic information in a qualitative way. However, there have also been attempts to develop more general computational algorithms to deduce phylogeny from gene order. As with sequence-based methods, the simplest to understand are distance-matrix methods. First, we need to define a measure of distance between two gene orders. We can then determine a matrix of distances for each pair of genomes in the set considered, and input this matrix to any of the standard methods, like neighbor-joining and Fitch–Margoliash (see Chapter 8).

The issue, then, is how to measure distances between gene orders. One possibility is simply to count the number of breakpoints (Blanchette, Kunisawa, and Sankoff 1999). This is an intuitively easy-to-understand measure of how much disruption of the gene order there has been. It is also very simple to calculate. However, it is not directly related to any particular mechanism of genome rearrangement. Other possible types of distance are edit distances (Sankoff et al. 1992), which measure the minimum number of editing operations necessary to transform one gene order into another. The simplest of these would be the inversion distance, in which the only editing operations allowed are inversions of a consecutive block of genes. Distances can also be calculated that allow a combination of translocations and inversions (Blanchette, Kunisawa, and Sankoff 1996).

Although the principle of these edit distances is simple to state, the algorithms for calculating them are complex (see Pevzner 2000, and references therein). If the genes in a genome are numbered from 1 to N, then a possible gene order can be represented as a permutation of these numbers. Some algorithms deal with unsigned permutations, i.e., the strand a gene is on and its direction of transcription are ignored, and only the position of the gene matters. However, it is much more realistic to treat gene orders as signed permutations, as in the examples in Section 12.2.4, in which case a minus sign indicates a gene on the opposite strand. It turns out that there is an exact algorithm for the inversion distance (i.e., sorting signed permutations using only inversions) that runs in polynomial time. However, no such algorithm is known for edit distances that use translocations as well; hence, combined

inversion–translocation distances have to be calculated heuristically by trying many possible pathways of intermediates between the two genomes.

Somewhat surprisingly, it has been shown that, in problems with real genomes, the simple breakpoint distance is very close to being linearly proportional to both the inversion distance and the combined inversion–translocation distance (Blanchette, Kunisawa, and Sankoff 1999, Cosner et al. 2000). This means that if the distance matrices calculated with the different measures of distance are put into phylogenetic algorithms, the same shaped tree will emerge. This is an argument for using the simplest measure of distance, i.e., the breakpoint distance.

There are also several methods for constructing gene-order trees that can be thought of as maximum parsimony methods. With "normal" parsimony, using character states coded as 0 or 1 (Section 8.7.1), the parsimony cost for a given tree is the sum of the number of character-state changes over the whole tree, and the tree with the minimum cost is selected. To calculate the cost for any given tree, it is necessary to determine the optimal values of the character states on each internal node, and then to sum the changes necessary along each branch of the tree. This is directly analogous to the minimal-breakpoint tree method of Blanchette, Kunisawa, and Sankoff (1999). Here, a gene order is assigned to each internal node of a tree in such a way that the sum of the breakpoint distances along all branches of the tree is minimized. The tree is then selected that minimizes this minimum value. Bourque and Pevzner (2002) have also developed a similar method, where gene orders on the internal nodes are calculated so as to minimize the sum of the inversion distances along all the branches.

Another parsimony method for gene order (Cosner et al. 2000) creates a set of binary characters that represent each gene order and inputs the binary characters into a normal parsimony program. Each character represents the presence or absence of a consecutive gene pair in the order. Consider the four gene orders below:

(i) 1,2,3,4

(ii) 1,−3,−2,4

(iii) 1,2,−3,4

(iv) 1,3,2,4

The character for the pair 2,3 would be assigned to 1 in (i) and also in (ii), because these genes are still consecutive and run in the same direction relative to one another, whereas it would be assigned to 0 in (iii) and (iv). A parsimony program will be able to reconstruct the most parsimonious assignments of the character states on all the internal nodes. The method has the drawback that not every possible set of 0s and 1s that might arise on the internal nodes represents a valid gene order. This is because the characters are not independent. For example, if the pair 2,3 is present, then the pair 2,4 cannot be present in the same genome. Parsimony programs treat each character as being independent. Despite this drawback, Cosner *et al.* (2000) found this method useful. They use it to generate trial trees that they then screen using other criteria.

Another gene-order phylogeny method worth mentioning is the Bayesian approach of Larget *et al.* (2002). This is the only method to date that has an explicit probabilistic model of genome rearrangements. The only events permitted are inversions. The number of inversions that occur on a branch has a Poisson distribution, with a mean proportional to the branch length. A trial configuration consists of a tree with specified branch lengths and specified positions of inversions. It is possible to calculate the likelihood of any given configuration according to the probabilistic model. A Markov Chain Monte Carlo (MCMC) method is then used to obtain a representative sample of configurations weighted according to the model.

There are as yet rather few data sets that can be used with these methods, and it remains to be seen just how useful the algorithms will be. Sankoff *et al.* (1992) studied mitochondrial genomes of fungi and a few animals. Chloroplast genomes have been studied for the Campanulaceae family of flowering plants (Cosner *et al.* 2000) and for the photosynthetic protists (Sankoff *et al.* 2000). The animal mitochondrial genomes have been studied several times, but the results are not as good as might be hoped. Distance-matrix methods using breakpoint distance (Blanchette, Kunisawa, and Sankoff 1999) failed to place the Deuterostomes (i.e., Chordates + Echinoderms) together in a clade, and failed to place

the Molluscs together. These are things that sequence-based phylogenies would have no problem with. The minimal-breakpoint tree method was considerably better – it did find both of these clades, but placed the Arthropods as sister group to the Deuterostomes, rather than as part of the Ecdysozoa, as expected from sequence-based phylogeny (Section 11.3.2). With the method of Bourque and Pevzner (2002), neither the Arthropods nor the Molluscs appeared as monophyletic groups. With the method of Larget *et al.* (2002), a wide variety of alternative tree topologies arose in MCMC simulations, and the posterior probabilities did not strongly discriminate between alternatives. There was also a tendency for the Molluscs to be polyphyletic and, for some reason, the human tended to cluster with echinoderms rather than the chicken. The phylogeny of the animal phyla is still not well understood, and additional evidence deriving from gene order would be welcome. However, the results from these studies of mitochondrial gene order do not seem sufficiently reliable to draw firm conclusions.

The reason for this may lie in the gene orders themselves, rather than with problems in the methods. Several of the animal phyla contain both conservative and highly derived gene orders. There is a tendency for all the methods to group the more conservative genomes from different phyla together, and to form another separate group of all the highly derived genomes. This is particularly a problem with the distance-matrix methods, but it probably also underlies the difficulties encountered by the other methods. Table 12.2 shows breakpoint distances between mitochondrial genomes, calculated both including and excluding the tRNA genes. As we cannot show the whole matrix, we simply give distances from one example species, *Limulus polyphemus*, the horseshoe crab. *Limulus* is thought to be a fairly primitive arthropod, and is phylogenetically related to the arachnids. Examination of the gene order suggests that it has retained the order possessed by the ancestral arthropod, or is at least very close. Measuring distances from *Limulus* to the other arthropods therefore tells us how much change there has been within the arthropod phylum, while distances from *Limulus* to non-arthropod species are

Table 12.2 Breakpoint distances between mitochondrial genomes obtained from OGRe (Jameson *et al.* 2003). Distances are measured from the horseshoe crab, *Limulus polyphemus*, to the species listed.

Taxon	Species	Common name	Distance exc. tRNAs	Distance inc. tRNAs
Arachnids	*Ixodes persulcatus*	Tick	0	0
	Rhipicephalus sanguineus	Tick	3	7
Other Arthropods	*Drosophila melanogaster*	Fly	0	3
	Locusta migratoria	Locust	0	6
	Artemia franciscana	Shrimp	0	7
	Pagurus longicarpus	Hermit crab	5	18
	Thrips imaginis	Thrips	11	31
Nematodes	*Trichinella spiralis*	Worm	8	21
	Caenorhabditis elegans	Worm	14	37
Molluscs	*Katharina tunicata*	Chiton	5	20
	Loligo bleekeri	Squid	4	27
	Cepaea nemoralis	Snail	13	35
	Crassostrea gigas	Oyster	14	37
Brachiopods	*Terebratulina retusa*	Brachiopod	5	26
	Terebratalia transversa	Brachiopod	13	37
Deuterostomes	*Homo sapiens*	Human	5	20
	Paracentrotus lividus	Sea urchin	19	32

a good indication of how much change there has been between phyla.

The species in each of the boxes corresponding to the taxa in Table 12.2 should be equidistant from *Limulus* in terms of the time since the common ancestor. However, the breakpoint distances differ tremendously for the species in each box. For the two closely related tick species, the genome of *I. persulcatus* has remained unchanged, while that of *R. sanguineus* is highly derived. In the other arthropods (i.e., non-arachnids), there are many examples of insects and crustaceans, like *D. melanogaster* and *A. franciscana*, where there has been very little change, but there are also examples of unusual species, like *T. imaginis* and *P. longicarpus*, whose genomes appear to have been completely shredded and reassembled. In these latter two cases, we know that a lot of genome rearrangement has occurred in an isolated lineage of organisms in a short time. We have no

idea why these particular genomes should have been so unstable.

When comparing genomes between phyla, the distances that exclude the tRNAs are probably more informative – some of the distances with tRNAs are as large as 37, meaning that there is complete randomization, with a breakpoint after every gene. Each phylum is seen to contain conservative gene orders (e.g., *K. tunicata*, *T. retusa*, *H. sapiens*) and divergent ones (e.g., *C. gigas*, *T. transversa*, *P. lividus*). All nematode species sequenced to date seem fairly divergent. Thus, there is little support for the idea of grouping arthropods and nematodes together as Ecdysozoa. However, the nematodes are not similar to any other group either, so this just represents lack of evidence for Ecdysozoa, rather than positive evidence for an alternative. Another interesting point is that the Deuterostomes are rather less divergent than most phyla, and hence appear closer to the arthropods than many of the other Protostomes. On reflection, therefore, it is not too surprising that these methods had problems with the animal mitochondrial genomes. We may remain optimistic that gene-order phylogeny methods may prove to be more useful in future, as new, larger, and hopefully less badly behaved data sets emerge. Such methods are also applicable to bacterial genomes.

SUMMARY

There are now many complete bacterial genomes available, and comparisons can be made across a large number of species. Bacterial genomes range in size from around half a million to over seven million bases. The number of genes per genome varies almost in proportion to the length, with an average of just over 1000 bases between gene start points. There is relatively little non-coding DNA between genes in bacteria. This suggests that selection for efficiency of genome replication is strong enough to prevent the widespread accumulation of repetitive sequences and transposable elements, in contrast to the situation in many eukaryotes.

Genome size and content can vary quite rapidly between related bacterial species. The smallest genomes are in bacteria that are parasites or symbionts inside other cells. These species manage with a greatly reduced set of genes, presumably because they can absorb useful chemicals from their host cell that would have to be synthesized by their free-living relatives. There are several independent groups of parasitic bacteria where dramatic reduction of genome size has occurred in this way. This suggests that there is a tendency for genes that are no longer necessary for an organism to be deleted from genomes in a relatively short period.

In cases where more than one strain of a bacterial species has been completely sequenced, there can be a surprising degree of variability in gene content between the genomes. A notable case is *E. coli*, where the available pathogenic and non-pathogenic strains contain substantially different sets of genes. In pathogenic strains, it is often possible to identify specific regions of the genome, called pathogenicity islands, that contain genes responsible for their pathogenic lifestyle, and that appear to have been inserted into the genome from elsewhere. Horizontal transfer of genes between unrelated bacterial species appears to occur quite frequently on an evolutionary time scale. It is possible to identify recent insertions by looking for regions with unusual base frequencies and codon usage that does not follow the usual codon bias of the organism. Horizontal transfer also leads to inconsistencies in phylogenetic trees derived from different genes.

There is good evidence that both mitochondria and chloroplasts are descended from bacteria that were taken up by ancient eukaryotic cells, became endosymbionts, and eventually became integral parts of the eukaryote. Analysis of genes from organelle genomes shows that mitochondria are related to α-Proteobacteria, such as the present-day *Rickettsia*, and chloroplasts are related to cyanobacteria. Organelle genomes encode greatly reduced numbers of genes by comparison with even the smallest bacterial genomes. Many of the genes lost from the organelles have been transferred to the nuclear genome. Proteins required in the organelles that are the product of nuclear genes have acquired a signal peptide, which tells the cell to transport these proteins to the organelle. In animal mitochondrial genomes, the process of transfer of genes to the nucleus appears to have come to an end, leaving a small set of essential genes on the mitochondrial genome that has been stable throughout the evolution of the metazoa. In chloroplasts and plant mitochondria, cases of evolutionarily recent gene transfers to the nucleus are known.

There have been attempts to use both gene content and gene order of complete genomes to deduce evolutionary relationships between them. It is possible that phylogenies based on whole-genome characteristics may give a more reliable picture of the evolution of the whole organism than could be obtained from sequence-based phylogeny with any single gene. With bacteria, measures of distance between genomes have been defined, based on shared gene content, and these have been used in distance-matrix phylogenetic methods. The results have considerable overlap with the trees derived from rRNA sequences. In the case of gene order, distance between two genomes can be defined either in terms of the number of breakpoints, or in terms of the number of editing operations (inversions and translocations) needed to convert one gene order into the other. In some cases, shared gene order has given strong information about phylogenetic relationships at specific nodes in a tree. It is not yet clear whether complete phylogenetic trees will be able to be determined reliably from gene order alone. Phylogenetic methods using gene content and gene order are much less developed than sequence-based methods.

REFERENCES

Adams, K.L., Daley, D.O., Qiu, Y.L., Whelan, J., and Palmer, J.D. 2000. Repeated, recent and diverse transfers of a mitochondrial gene to the nucleus in flowering plants. *Nature*, **408**: 354–7.

Andersson, S.G.E., Zomorodipour, A., Andersson, J.O., Sicheritz-Ponten, T., Alsmark, U.C.M., Podowski, R.M., Naslund, A.K., Eriksson, A.S., Winkler, H.H., and Kurland, C.G. 1998. The genome sequence of *Rickettsia prowazekii* and the origin of mitochondria. *Nature*, **396**: 133–43.

Berg, O.G. and Kurland, C.G. 2000. Why mitochondrial genes are most often found in nuclei. *Molecular Biology and Evolution*, **17**: 951–61.

Blanchard, J.L. and Lynch, M. 2000. Organellar genes: Why do they end up in the nucleus? *Trends in Genetics*, **16**: 315–20.

Blanchard, J.L. and Schmidt, G.W. 1996. Mitochondrial DNA migration event in yeast and humans. *Molecular Biology and Evolution*, **13**: 537–48.

Blanchette, M., Kunisawa, T., and Sankoff, D. 1996. Parametric genome rearrangement. *Gene*, **172**: GC 11–17.

Blanchette, M., Kunisawa, T., and Sankoff, D. 1999. Gene order breakpoint evidence in animal mitochondrial phylogeny. *Journal of Molecular Evolution*, **49**: 193–203.

Boore, J.L. and Brown, W.M. 1998. Big trees from little genomes: Mitochondrial gene order as a phylogenetic tool. *Current Opinion in Genetics and Development*, **8**: 668–74.

Boore, J.L., Lavrov, D.V., and Brown, W.M. 1998. Gene translocation links insects and crustaceans. *Nature*, **392**: 667–8.

Boucher, Y., Nesbo, C.L., and Doolittle, W.F. 2001. Microbial genomes: Dealing with diversity. *Current Opinion in Microbiology*, **4**: 285–9.

Bourque, G. and Pevzner, P. 2002. Genome-scale evolution: Reconstructing gene orders in ancestral species. *Genome Research*, **12**: 26–36.

Cedergren, R., Gray, M.W., Abel, Y., and Sankoff, D. 1988. The evolutionary relationships among known life forms. *Journal of Molecular Evolution*, **28**: 98–122.

Chambaud, I., Heilig, R., Ferris, S., Barbe, V., Samson, D., Galisson, F., Moszer, I., Dybvig, K., Wroblewski, H., Viari, A., Rocha, E.P.C., and Blanchard, A. 2001. The complete genome sequence of the murine respiratory pathogen *Mycoplasma pulmonis*. *Nucleic Acids Research*, **29**: 2145–53.

Cole, J.R., Chai, B., Marsh, T.L., Farris, R.J., Wang, Q., Kulam, S.A., Chandra, S., McGarrell, D.M., Schmidt, T.M., Garrity, G.M., and Tiedje, J.M. 2003. The ribosomal database project (RDP-II): Previewing a new autoaligner that allows regular updates and the new prokaryotic taxonomy. Homepage: http://rdp.cme.msu.edu/html/. Prokaryotic backbone tree: http://rdp.cme.msu.edu/pubs/NAR/Backbone_tree.pdf

Cosner, M.E., Jansen, R.K., Moret, B.M.E., Raubeson, L.A., Wang, L.S., Warnow, T., and Wyman, S. 2000. An empirical comparison of phylogenetic methods on chloroplast gene order data in *Campanulaceae*. In D. Sankoff and J.H. Nadeau (eds.), *Comparative Genomics*, pp. 99–121. Amsterdam: Kluwer Academic.

Doolittle, W.F. 2000. Uprooting the tree of life. *Scientific American*, **282**: 90–5.

Doolittle, R.F. and Handy, J. 1998. Evolutionary anomalies among the aminoacyl-tRNA synthetases. *Current Opinion in Genetics and Development*, **8**: 630–6.

Douglas, S.E. 1998. Plastid evolution: Origins, diversity, trends. *Current Opinion in Genetics and Development*, **8**: 655–61.

Embley, T.M. and Hirt, R.P. 1998. Early branching eukaryotes. *Current Opinion in Genetics and Development*, **8**: 624–9.

Fitz-Gibbon, S.T. and House, C.H. 1999. Whole genome-based phylogenetic analysis of free-living microorganisms. *Nucleic Acids Research*, **27**: 4218–22.

Gogarten, J.P., Kibak, H., Dittrich, P., *et al.* 1989. Evolution of the vacuolar H$^+$ATPase: Implications for the origin of eukaryotes. *Proceedings of the National Academy of Sciences USA*, **86**: 6661–5.

Golding, G.B. and Gupta, R.S. 1995. Protein-based phylogenies support a chimeric origin for the eukaryotic genome. *Molecular Biology and Evolution*, **12**: 1–6.

Gray, M.W. 1992. The endosymbiot hypothesis revisited. *International Review of Cytology*, **141**: 233–357.

Gray, M.W. and Doolittle, W.F. 1982. Has the endosymbiont hypothesis been proven? *Microbiology Review*, **6**: 1–42.

Gray, M.W., Lang, B.F., Cedergren, R., Golding, G.B., Lemieux, C., Sankoff, D., Turmel, M., Brossard, N., Delage, E., Littlejohn, T.G., Plante, I., Rioux, P., Saint-Louis, D., Zhu, Y., and Burger, G. 1998. Genome structure and gene content in protist mitochondrial DNAs. *Nucleic Acids Research*, **26**: 865–78.

Iwabe, N., Kuma, K.I., Hasegawa, S., Osawa, S., and Miyata, T. 1989. Evolutionary relationship between archaebacteria, eubacteria, and eukaryotes inferred from phylogenetic trees of duplicated genes. *Proceedings of the National Academy of Sciences USA*, **86**: 9355–9.

Jameson, D., Gibson, A.P., Hudelot, C., and Higgs, P.G. 2003. OGRe: A relational database for comparative analysis of mitochondrial genomes. *Nucleic Acids Research*, **31**: 202–6. http://ogre.mcmaster.ca

Koonin, E.V., Makarova, K.S., and Aravind, L. 2001. Horizontal gene transfer in prokaryotes: Quantification and classification. *Annual Review of Microbiology*, **55**: 709–42.

Korbel, J.O., Snel, B., Huynen, M.A., and Bork, P. 2002. SHOT: A web server for the construction of genome phylogenies. *Trends in Genetics*, **18**: 158–62. http://www.Bork.EMBL-Heidelberg.de/SHOT

Lang, B.F., Burger, G., O'Kelly, C.J., Cedergren, R., Golding, G.B., Lemieux, C., Sankoff, D., Turmel, M., and Gray, M.W. 1997. An ancestral mitochondrial DNA resembling a eubacterial genome in miniature. *Nature*, **387**: 493–7.

Larget, B., Simon, D.L., and Kadane, J.B. 2002. Bayesian phylogenetic inference from animal mitochondrial genome arrangements. *Journal of the Royal Statistical Society B*, **64**: 681–93.

Lawrence, J.G. and Ochman, H. 1998. Molecular archaeology of the *Escherichia coli* genome. *Proceedings of the National Academy of Sciences USA*, **95**(16): 9413–17.

Lobry, J.R. 1996. Asymmetric substitution patterns in the two DNA strands of bacteria. *Molecular Biology and Evolution*, **13**: 660–5.

Ludwig, W., Strunk, O., Klugbauer, S., Klugbauer, N., Weizenegger, M., Neumaier, J., Bachleitner, M., and Schleifer, K.H. 1998. Bacterial phylogeny based on comparative sequence analysis. *Electrophoresis*, **19**: 554–68.

Margulis, L. 1981. *Symbiosis in Cell Evolution*. San Francisco: W.H. Freeman.

Marienfeld, J., Unseld, M., and Brennicke, A. 1999. The mitochondrial genome of *Arabidopsis* is composed of both native and immigrant information. *Trends in Plant Science*, **4**: 495–502.

Martin, W., Stoebe, B., Goremykin, V., Hansmann, S., Hasegawa, M., and Kowallik, K.V. 1998. Gene transfer to the nucleus and the evolution of chloroplasts. *Nature*, **393**: 162–5.

Millen, R.S., Olmstead, R.G., Adams, K.L., Palmer, J.D., Lao, N.T., Heggie, L., Kavanagh, T.A., Hibberd, J.M., Gray, J.C., Morden, C.W., Calie, P.J., Jermiin, L.S., and Wolfe, K.H. 2001. Many parallel losses of *infA* from chloroplast DNA during angiosperm evolution with multiple independent transfers to the nucleus. *The Plant Cell*, **13**: 645–58.

Ochman, H., Lawrence, J.G.R., and Groisman, E.A. 2000. Lateral gene transfer and the nature of bacterial innovation. *Nature*, **405**: 299–304.

Ogata, H., Audic, S., Renesto-Audiffren, P., Fournier, P.E., Barbe, V., Samson, D., Roux, V., Cossart, P.,

Weissenbach, J., Claverie, J.M., and Raoult, D. 2001. Mechanisms of evolution in *Rickettsia conorii* and *R. prowazekii*. *Science*, **293**: 2093–8.

Olsen, G.J., Woese, C.R., and Overbeek, R. 1994. The winds of (evolutionary) change: Breathing new life into microbiology. *Journal of Bacteriology*, **176**: 1–6.

Pevzner, P. 2000. *Computational Molecular Biology: An Algorithmic Approach*. Cambridge, MA: MIT Press.

Philippe, H. and Forterre, P. 1999. The rooting of the universal tree of life is not reliable. *Journal of Molecular Evolution*, **49**: 509–23.

Roehrdanz, R.L., Degrugillier, M.E., and Black, W.C. 2002. Novel rearrangements of arthropod mitochondrial DNA detected with long-PCR: applications to arthropod phylogeny and evolution. *Molecular Biology and Evolution*, **19**: 841–9.

Roger, A.J. 1999. Reconstructing early events in eukaryotic evolution. *American Nature*, **154**: S146–63.

Sankoff, D., Leduc, G., Antoine, N., Paquin, B., Lang, B.F., and Cedergren, R. 1992. Gene order comparisons for phylogenetic inference: Evolution of the mitochondrial genomes. *Proceedings of the National Academy of Sciences USA*, **89**: 6575–9.

Sankoff, D., Deneault, M., Bryant, D., Lemieux, C., and Turmel, M. 2000. Chloroplast gene order and the divergence of plants and algae, from the normalized number of induced breakpoints. In D. Sankoff and J.H. Nadeau (eds.), *Comparative Genomics*, pp. 89–98. Amsterdam: Kluwer Academic.

Scouras, A. and Smith, M.J. 2001. A novel mitochondrial gene order in the crinoid echinoderm *Florometra serratissima*. *Molecular Biology and Evolution*, **18**: 61–73.

Sicheritz-Ponten, T., Kurland, C.G., and Andersson, S.G.E. 1998. A phylogenetic analysis of the cytochrome *b* and cytochrome *c* oxidase I genes supports an origin of mitochondria from within the *Rickettsiae*. *Biochimica et Biophysica Acta (Bioenergetics)*, **1365**: 545–51.

Snel, B., Bork, P., and Huynen, M.A. 1999. Genome phylogeny based on gene content. *Nature Genetics*, **21**: 108–10.

Snel, B., Bork, P., and Huynen, M.A. 2002. Genomes in flux: The evolution of archaeal and proteobacterial gene content. *Genome Research*, **12**: 17–25.

Suyama, M. and Bork, P. 2001. Evolution of prokaryotic gene order: Genome rearrangements in closely related species. *Trends in Genetics*, **17**(1): 10–13.

Tatusov, R.L., Koonin, E.V., and Lipman, D.J. 1997. A genomic perspective of protein families. *Science*, **278**: 631–7.

Tatusov, R.L., Natale, D.A., Garkavtsev, I.V., Tatusova, T.A., Shankavaram, U.T., Rao, B.S., Kiryutin, B., Galperin, M.Y., Fedorova, N.D., and Koonin, E.V. 2001. The COG database: New developments in phylogenetic classification of proteins from complete genomes. *Nucleic Acids Research,* **29**: 22–8. http://www.ncbi.nlm.nih.gov/COG/

Van de Peer, Y., Baldauf, S.L., Doolittle, W.F., and Meyer, A. 2000. An updated and comprehensive rRNA phylogeny of crown eukaryotes based on rate-calibrated evolutionary distances. *Journal of Molecular Evolution,* **51**: 565–76.

Van Ham, R.C.H.J., Kamerbeek, J., Palacios, C., Rausell, C., Abascal, F., Bastolla, U., Fernandez, J.M., Jimenez, L., Postigo, M., Silva, F.J., Tamames, J., Viguera, E., Latorre, A., Valencia, A., Moran, F.Y., and Moya, A. 2003. Reductive genome evolution in *Buchnera aphidicola.* *Proceedings of the National Academy of Sciences USA,* **100**: 581–6.

Viale, A.M. and Arakaki, A.K. 1994. The chaperone connection to the origins of the eukaryotic organelles. *FEBS Letters,* **341**: 146–51.

Welch, R.A. and 18 others. 2002. Extensive mosaic structure revealed by the complete genome sequence of uropathogenic *E. coli. Proceedings of the National Academy of Sciences USA,* **99**: 17020–4.

Wilson, R.J.M., Penny, P.W., Preiser, P.R., Rangachari, K., Roberts, K., Roy, A., Whyte, A., Strath, M., Moore, D.J., Moore, P.W., and Williamson, D.H. 1996. Complete gene map of the plastid-like DNA of the malaria parasite *Plasmodium falciparum. Journal of Molecular Biology,* **261**: 155–72.

Woese, C.R. 1987. Bacterial evolution. *Microbiology Review,* **51**: 221–71.

Woese, C.R. 2000. Interpreting the universal phylogenetic tree. *Proceedings of the National Academy of Sciences USA,* **97**: 8392–6.

Woese, C.R. and Fox, G.E. 1977. Phylogenetic structure of the prokaryotic domain: The primary kingdoms. *Proceedings of the National Academy of Sciences USA,* **74**: 5088–90.

Wolfe, K.H., Li, W.H., and Sharp, P.M. 1987. Rates of nucleotide substitution vary greatly among plant mitochondrial, chloroplast and nuclear DNAs. *Proceedings of the National Academy of Sciences USA,* **84**: 9054–8.

Wood, D.W. and 49 others. 2001. The genome of natural genetic engineer *Agrobacterium tumefaciens* C58. *Science,* **294**: 2317–23.

Yap, W.H., Zhang, Z., and Wang, Y. 1999. Distinct types of rRNA operons exist in the genome of the actinomycete *Thermomonospora chromogena* and evidence for the horizontal transfer of an entire rRNA operon. *Journal of Bacteriology,* **181**: 5201–9.

DNA Microarrays and the 'omes

CHAPTER

13

CHAPTER PREVIEW

Here, we discuss the rationale behind high-throughput experiments to study the transcriptome and the proteome. We describe the manufacture and use of DNA microarrays to measure expression levels of genes on a genome-wide scale. We discuss techniques for normalization of array data, and associated methods for statistical analysis and visualization. We describe the most important experimental techniques used in proteomics, and go on to describe studies of protein–protein interaction networks. We discuss information management issues relating to the storage and effective comparison of complex biological information, emphasizing the importance of standards, such as MIAME (for microarray data). We introduce the concept of a schema, which logically describes the information in a database, and the idea of an ontology, which defines the terms used in a field of knowledge and the relationships between them.

13.1 'OMES AND 'OMICS

Everyone is now familiar with the word "genome" – the standard term describing the complete sequence of heritable DNA possessed by an organism. "Genomics" is the study of genomes, and as there are many approaches to studying genomes, genomics is rather a broad term. In Chapter 12, looking at the structure of genomes, we considered questions about the complete set of genes possessed by an organism, and compared genomes of different organisms. The subject of Chapter 12 can be classed as genomics, because it considers questions relating to whole genomes rather than single genes. However, genomics does not just cover the structure and evolution of genomes; it also encompasses many additional issues relating to how genomes function. As most things that go on inside cells are closely linked to the genome, it seems we might risk including the whole of molecular biology and genetics under the genomics banner. However, this is not really the case. What distinguishes genomics studies from more conventional studies in genetics and molecular biology is that they focus on the broad scale – whole sets of genes or proteins – rather than the details of single genes or proteins.

Genomics is a new way of thinking about biology that has only become possible as a result of technological advances in the last few years. Genome-sequencing projects are, of course, the primary example, but alongside complete genomes have come several techniques for studying the sets of genes in those genomes. It is these so-called "high-throughput" techniques that are the focus of this chapter. High-throughput means that the same experiment is performed on many different genes or proteins very rapidly in an automated or partially automated way. This generates large amounts of data that must be carefully analyzed to look for trends and statistically significant features. High-throughput experiments are closely linked to bioinformatics because they raise questions both of data interpretation and modeling, and also of data management and storage.

The success of genomics, and its holistic way of thinking, has led to the coining of other 'ome and 'omic terms. For example, the complete set of mRNAs that are transcribed in a cell has been dubbed the transcriptome; the complete set of proteins present in a cell is termed the proteome; and the complete set of chemicals involved in metabolic reactions in a cell is the metabolome. Each of these words can be changed into 'omics, meaning the study of the corresponding 'ome. The term proteomics is now used very commonly because it has become associated with a particular set of experimental techniques for studying the proteome: two-dimensional gel electrophoresis and mass spectrometry. The chief tools of transcriptomics are based on DNA microarrays. Microarrays have become very popular recently and the majority of this chapter is devoted to them.

13.2 HOW DO MICROARRAYS WORK?

A DNA microarray is a slide onto which a regular pattern (i.e., an array) of spots has been deposited. Each spot contains many copies of a specified DNA sequence that have been chemically bonded to the surface of the slide, with a different DNA sequence on each spot. Spots are rather small (typically a few hundred nanometers across) and it is possible to deposit thousands of spots on a single slide a few centimeters across, like a glass microscope slide. The DNA sequences in the spots act as probes that hybridize with complementary sequences that may be present in a sample. Fluorescence techniques are used to visualize the spots where hybridization has taken place, as described below. Microarrays are able to detect the presence of the sequences corresponding to all the spots simultaneously in one sample. Two good reviews on the technology of microarrays are Schulze and Downward (2001) and Lemieux, Aharoni, and Schena (1998). There are many steps involved in microarray experiments, as summarized in Fig. 13.1.

There are two types of microarray, which differ with regard to the preparation of both the array and the sample. The first type are called oligonucleotide arrays, because the DNA attached to the array is in the form of short oligonucleotides, usually 25 bases long. Arrays of this type are manufactured commercially by the Affymetrix company. The sequences used in each spot are carefully chosen in advance with reference to the genome of the organism under study. Each oligonucleotide should hybridize to a specific gene sequence of the organism. However, cross-hybridization is possible between genes with related sequences, especially if the probe sequence is only short. For this reason, several separate oligonucleotide sequences are chosen for each gene, and the expression of a gene in a sample is only inferred if hybridization occurs with almost all of them.

Having chosen the oligonucleotide sequences, the slide is prepared by synthesizing the sequences, one base at a time, *in situ* in the appropriate place on the slide. This type of array is sometimes called a "chip", because the surface is made of silicon (or, to be more precise, it is a wafer of quartz – silicon dioxide). The technique used is photolithography. Chemical linkers are attached to the slide that will anchor the sequences. These linkers are protected by a light-sensitive chemical group. The slide is then covered with a solution containing one particular type of nucleotide – say A. Light is then shone at precisely those spots on the slide where an A is required in the sequence. This activates the binding of A to the linker in those positions, but not elsewhere. The A solution is then washed away and replaced by each of the other nucleotides, one after the other. After doing this four times, the first base of each sequence will have been synthesized. The process is then repeated for each base in the sequence. Once the manufacturing process has been set up for a given chip, many copies can be synthesized, containing exactly the same oligomers, in a reproducible way. More information on the manufacture of gene chips is available on the Affymetrix Web site (http://www.affymetrix.com/index.affx).

The aim of the microarray experiment is to measure the level of expression of the genes in a cell, i.e., to measure the concentrations of mRNAs for those genes. Solutions extracted from tissue samples contain large numbers of mRNAs of many different types that happen to be present in the cells at that time. These cannot be used directly on the chip

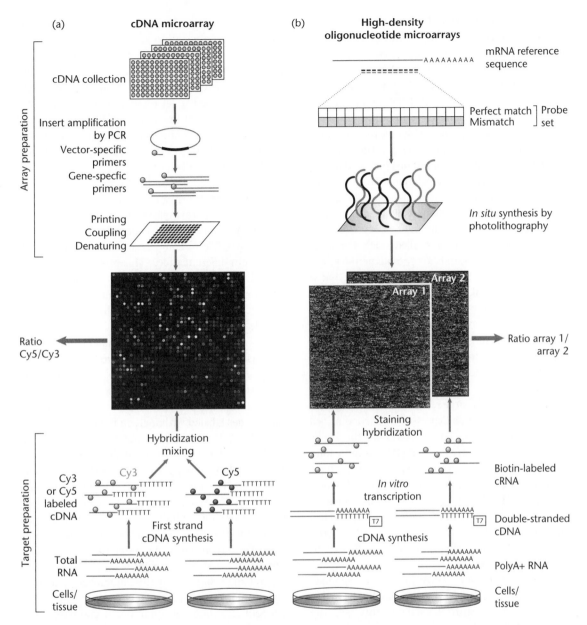

Fig. 13.1 Steps involved in DNA microarray experiments for (a) cDNA arrays and (b) oligonucleotide arrays. The top illustrates preparation of the array and the bottom illustrates preparation of the sample (reproduced from Schulze and Downward, 2001, with permission of Nature Publishing Group).

because they are not labeled. By a series of steps, biotin-labeled RNAs are created with the same sequences as the population of mRNAs extracted from the tissue sample. The chip is immersed in the solution of biotin-labeled RNAs for several hours, during which hybridization occurs with the DNA oligomers on the chip. The remaining unhybridized sequences are then washed away. Biotin is used because of its very strong binding to streptavidin. The final step is to add a streptavidin-linked fluorescent molecule (fluorophore) that will bind to the biotin wherever there is hybridized RNA on the chip. The level of fluorescence can then be measured in each of the spots using optical microscopy. Hopefully, this fluorescence level is proportional to the amount of mRNA in the original extract. Usually, we want to compare the mRNA levels of genes under different experimental conditions, or at different times, or in different cell populations. To do this, the absolute intensities of equivalent spots on different chips treated with the different biological samples are compared with one another.

The second type of array is called a cDNA array. A cDNA (where the c stands for complementary) is a DNA strand synthesized using a reverse transcriptase enzyme, which makes a DNA sequence that is complementary to an RNA template. This is the reverse of what happens in transcription, where an RNA strand is made using a DNA template. It is possible to synthesize cDNAs from the mRNAs present in cells. The cDNAs can then be amplified to high concentration using PCR. Biologists have built up cDNA libraries containing large sets of sequences of different genes known to be expressed in particular cell types. These libraries can be used as the probe sequences on microarrays. Each cDNA is quite long (500–2,000 bases), and contains a significant fraction of a gene sequence, but not necessarily the complete gene. Hybridization to these long sequences is much more specific than with oligonucleotides; therefore, usually only one spot on a cDNA array corresponds to one gene and this is sufficient to distinguish between genes.

Another name for cDNA arrays is "spotted" arrays, because of the spotting process that is used to deposit the cDNAs onto the slide. In this case, the slide is usually made of glass. It is pre-coated with a surface chemical that binds DNA. Gridding robots with sets of pins are then used to transfer small quantities of cDNAs onto the slide. This technology is cheaper than that for preparation of oligonucleotide arrays and is therefore more accessible to many research groups. Preparation of a cDNA array nevertheless requires a large amount of prior labor-

atory work in order to create the cDNAs. Whereas with an oligonucleotide array, knowledge of the gene sequences is necessary, preferably for a whole genome, with a cDNA array, we do not need to know the whole genome sequence, as long as we have already identified a set of suitable cDNAs experimentally. Note that the *in situ* synthesis techniques used for oligonucleotide arrays do not work for sequences as long as cDNAs, although sometimes spotted arrays can be prepared with short oligonucleotide probes instead of cDNAs.

The process of array manufacture is less reproducible for spotted arrays, and the amount of DNA in each spot is not so easy to control. Hence, it is not usually possible to compare absolute intensities of spots from different slides. This problem is circumvented using an ingenious two-color fluorescent labeling system that allows two samples to be compared on one slide. Usually, we have a reference sample to which we want to compare a test sample. We want to know which genes have increased or decreased their level of expression in the test sample with respect to the reference. RNA extracts from these two samples are prepared separately. cDNA is then made from the reference sample using nucleotides labeled with a green fluorophore (Cy5), and cDNA is made from the test sample using nucleotides labeled with a red fluorophore (Cy3). These two labeled populations of sequences are then mixed, and the mixture is allowed to hybridize with the array. The red- and green-labeled cDNAs from the samples should bind to the spot in proportion to their concentrations. The intensity of both red and green fluorescence from each spot is measured. The ratio of the red to green intensity should not be dependent on spot size, so this eliminates an important source of error. In cases where there are many different experimental conditions to compare, each one can be compared to the same reference sample.

13.3 NORMALIZATION OF MICROARRAY DATA

The raw data arising from a cDNA microarray experiment consist of lists of red and green intensities for each spot on the slide. There can be thousands of spots per slide and, for an organism like yeast, with approximately 6,000 genes, it is possible to have a spot for every gene in the organism in a single array. Handling this amount of data from a single experiment requires careful statistical analysis.

Let R_i and G_i be the red and green fluorescence intensities from spot i. These intensities can vary by orders of magnitude from one spot to another, because of real variation in the amount of mRNA for each gene. However, because of the complex nature of sample and array preparation, there are many stages in which biases can creep into the results, and there can be significant spot-to-spot intensity variation on the array that has nothing to do with the biology of the sample. The data analysis is usually done in terms of the ratio, R_i/G_i, of intensities for each spot, rather than in terms of the absolute intensities, because this eliminates a lot of spot-to-spot variability. If the ratio is greater than 1, the gene is up-regulated (i.e., turned on) in the test sample with respect to the reference. If the ratio is less than 1, the gene is down-regulated (i.e., turned off or repressed) with respect to the reference. It is useful to deal with the ratio on a logarithmic scale:

$$M_i = \log_2(R_i/G_i) \tag{13.1}$$

As we are using base 2 logs, a doubling of expression level corresponds to $M_i = 1$, and a halving corresponds to $M_i = -1$. The intensity ratios are often plotted against a measure of the average intensity of the spot. The geometric average of the red and green intensities is $\sqrt{R_iG_i}$, and it is therefore useful to define an average intensity of a spot on a log scale as:

$$A_i = \log_2\left(\sqrt{R_iG_i}\right) = \frac{1}{2}\log_2(R_iG_i) \tag{13.2}$$

An example of real data from over 27,000 spots in a mouse cDNA array (Quackenbush 2002) is shown in Fig. 13.2(a). The scale of average intensity used here is $\log_{10}(R_iG_i)$, which is directly proportional to A_i, as defined above. The figure shows a broad horizontal band containing points for many genes, and a few outlier points, with significantly higher

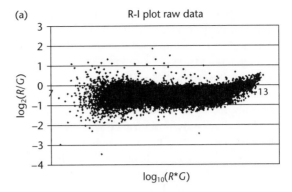

(a)

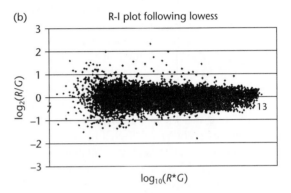

(b)

Fig. 13.2 Intensity-ratio data from a mouse cDNA array (reproduced from Quackenbush 2002, with permission of Nature Publishing Group): (a) raw data before normalization, and (b) after normalization with the LOWESS method. R-I stands for Ratio-Intensity, which we have called M and A.

or lower ratios than average. It is the outliers that we are interested in, because these represent genes whose expression has been significantly up- or down-regulated with respect to the reference. However, before we can interpret the data, it is necessary to correct the positions of the points to account for various biases. This process is called normalization.

The labeled sequences from the test and reference samples are combined in such a way that the total amounts of cDNA are equal. For a gene whose expression level has not changed, we would expect that $R_i = G_i$, and therefore $M_i = 0$. However, in Fig. 13.2(a), the vast majority of points have negative M_i values. This suggests a systematic bias in the experiment, not that the majority of genes have been down-regulated. One possibility, suggested in early experiments, is to normalize the data with respect to "housekeeping" genes, which are assumed to be expressed at a constant level that does not change between test and reference samples. The data would then be shifted so that $M_i = 0$ for the housekeeping genes. However, it is difficult to be sure which genes are really unchanged, and the housekeeping method also fails to account for several other biases. A simple, alternative way of data normalization involves simply adding a constant to all the M_i, to shift them so the mean is at zero. These methods are global normalizations, where all the points are shifted by the same factor, so that the shape of the cloud of points in Fig. 13.2(a) would remain unchanged.

One reason for the systematic downshifting of points apparent in Fig. 13.2(a) is dye bias. The green and red cDNA populations are labeled with different fluorescent dyes. Bias could be introduced if the efficiency of labeling of the two DNA populations were different, or if the binding between the labeled DNA and the probe were affected by the dye in a systematic way, or if the efficiency of detecting the fluorescent signal from the two dyes were different. Global normalization would eliminate these dye biases, if we assume that they affect all spots in the same way.

Another way of eliminating dye bias is to perform dye-flip experiments. Here, a second experiment is performed in which the red and green labeling of the samples is done in reverse. The M_i values from the two arrays are then subtracted, and the result should be twice the unbiased M_i value, because the dye bias will cancel out. This is called self-normalization (Fang *et al.* 2003). The normalized value for each spot only depends on the measured intensity ratios for that spot, and not on any other spot intensities. Self-normalization will eliminate any type of bias that is assumed to be dependent on the particular spot but reproducible between arrays. Unfortunately, biases can be more subtle than this – real biases may neither be equal for all spots (as in global normalization) or independent for all spots but reproducible between arrays (as in self-normalization).

One type of error that is neither global nor completely independent for different spots is when the bias depends systematically on the average intensity of the spot, A_i. In Fig. 13.2(a), it is apparent that there is an upward trend in the M_i values at high A_i. Note that there is no reason to suppose that the cell should up-regulate all the genes that are highly expressed. Whether a gene is up- or down-regulated should, to a first approximation, be independent of its average expression level. The upward curve in Fig. 13.2(a) involves many spots, and hence has the hallmark of a systematic experimental bias. A possible cause for this would be if the fluorescence detector were saturated at high intensity. The detectors must be sensitive enough to detect a signal from the lowest intensity spots, and it may be that they respond in a non-linear way at very high intensity. There could also be steps in the sample preparation process that depend in a non-linear way on the initial mRNA concentration, or the hybridization of the labeled DNA to the spot could be non-linear.

One way of removing this type of bias is called LOWESS (LOcally WEighted Scatterplot Smoothing) normalization. In this method, a smooth curved function $m(A)$ is fitted through the data points (e.g., by least-squares fitting of the curve to the points). Then the corrected values of the intensity ratios are calculated as $M_i - m(A_i)$. Note that the shift depends on A and is therefore not global, but the value of $m(A)$ depends on the data points for many spots at the same intensity, and hence this is not self-normalization either. The results of applying the LOWESS method to the mouse cDNA data are shown in Fig. 13.2(b). The cloud of points is now centered on $M = 0$ for all values of A, and the upward curve in the data is no longer present.

Another good example of LOWESS normalization is given by Leung and Cavalieri (2003), whose data similarly show a systematic trend with A that is successfully removed. This example also shows print-tip bias that arises during array manufacture. Spotting robots typically have a 4×4 grid of print tips. The resulting array is divided into 16 regions, such that spots in different regions are produced by different tips, and spots in the same region are produced sequentially by the same tip. Variation in spot properties from different tips may lead to systematic bias between the different regions of the slide. Leung and Cavalieri (2003) therefore applied

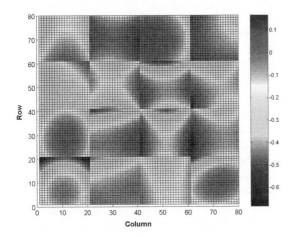

Fig. 13.3 Spot-position dependent bias in a yeast microarray study calculated by fitting a smooth quadratic function of spot coordinates to the intensity ratios of spots in each of the 16 print-tip regions (reproduced from Fang *et al.*, 2003, with permission of Oxford University Press).

LOWESS normalization independently to each of the 16 print-tip groups.

Careful analysis may reveal even more subtle biases that depend on the position of a spot on the slide. Fang *et al.* (2003) fitted a smooth quadratic function of the row and column positions of the spots to each of the 16 print-tip regions in a yeast microarray. The mean values depend on the print tip, but the biases also vary across the region covered by each tip, as shown in Fig. 13.3.

Another issue in microarray experiments is the location of spots during the image analysis. Software is used to distinguish the foreground pixels (i.e., the spot) from background pixels. This works well when the spots are clearly defined, but can be problematic if the slide is contaminated with dust, or if the background intensity is high with respect to the spot. Examples of good and bad spot images are shown in figure 2 of Leung and Cavalieri (2003). Sometimes an additional normalization is applied to the data, whereby the intensity of the background is estimated and then subtracted from the spot intensities. However, Fang *et al.* (2003) show an example where including such a background correction actually makes the data noisier, and therefore advise against making the correction.

There are clearly many issues involved in normalization, and there is not yet a universally accepted, standard way to deal with all the problems. Wolkenhauer, Möller-Levet, and Sanchez-Cabo (2002) have discussed some of these issues, and point out that while normalization is necessary, it can also lead to loss of information from the data. They comment that normalization of microarray data is "an art enjoyed by statisticians and dreaded by the rest".

13.4 PATTERNS IN MICROARRAY DATA

13.4.1 *Looking for significant changes in expression level*

The most fundamental question to ask of microarray data is which genes have been significantly up- or down-regulated in the test sample relative to the reference sample. There is clearly a large amount of scatter in the points on the normalized plot of M against A; much of this may be statistical error in the measurement. A simple way to deal with this is to calculate the mean, μ, and standard deviation, σ, of the M values (note that the mean will be zero for some types of normalization), and then to calculate the z score: $z = (M - \mu)/\sigma$. If the M values are normally distributed, then the probability that a point lies outside the range $-1.96 < z < 1.96$ is 5% (see Section M.10). Genes with z scores outside this range can be said to have significantly changed their expression level.

Sometimes, the statistical error may depend on the intensity – in particular, it may be difficult to accurately measure very low-intensity spots. This effect is apparent to some extent in Fig. 13.2(b), where the scatter of the normalized M values appears slightly larger at the low-intensity end of the range. If this occurs, it is possible to estimate a function $\sigma(A)$ by considering the standard deviations of points in a sliding window of A values. Then the z score for each point can be calculated with the σ that is appropriate for its particular A – see Quackenbush (2002) for an example of this.

Note that z scores are only applicable when the data points have a normal distribution. There is some empirical evidence that the M values (logs of the intensity ratios) are approximately normally distributed (Hoyle *et al.* 2002). Thus, the use of the z score may be reasonable. The intensity ratios themselves, R_i/G_i, are certainly not normally distributed. In fact, we say that the ratios have a log-normal distribution, meaning that logs of these ratios have a normal distribution.

Some factors contributing to statistical errors in array data are truly biological (e.g., variation in gene expression between different individuals), some arise from problems with sample preparation (e.g., samples from multicellular organisms may contain different cell types that may be difficult to separate), and some are just technical (e.g., lack of reproducibility between identical experiments on different arrays). The only way to distinguish between these types of error is to repeat each experiment several times, trying both the same sample on different arrays, and separate, but supposedly equivalent, mRNA extracts from different samples. Despite the associated costs, it is accepted that experiment repetitions are essential for reliable interpretation of the data. Rigorous statistical techniques for determining which genes are differentially expressed are being developed. These need to account for the fact that there are usually only small numbers of replicates, and that multiple tests are being made on the same data (and only a few genes may have changed out of thousands on the array), which has a large effect on significance levels. For more information on statistical testing of array data, see Slonim (2002) and references therein, Tusher, Tibshirani, and Chu (2001), and Li and Wong (2001). Also, Finkelstein *et al.* (2002) give a useful account of experimental design and statistical analysis of arrays used by the Arabidopsis Functional Genomics Consortium.

13.4.2 Clustering

In a typical microarray experiment, we have a set of arrays representing different samples, each with the same set of genes, with intensity ratios measured relative to the same reference. The normalized log intensity ratios form a data matrix whose columns are the different arrays and whose rows are the different genes. A matrix like this can be represented visually, as in the central part of Plate 13.1, using red to indicate up-regulated genes, green to indicate down-regulated genes, and black to indicate no change. In this example, the scale runs from $M = -2$ to 2, i.e., from four-fold down-regulated to four-fold up-regulated.

In this study, Alizadeh *et al.* (2000) were interested in a type of tumor called diffuse large B-cell lymphoma (DLBCL). This is classified as a single disease, although it was suspected that there may be distinguishable subtypes that might differ in their behavior. Identification of these subtypes might then be important in assessing a patient's prognosis, and possibly in deciding suitable treatments. The study compared 96 samples from lymphoid cells, including patients with DLBCL, B-cell chronic lymphocytic leukaemia (CLL), follicular lymphoma (FL), and a variety of normal lymphocytes. The microarrays used contained cDNAs selected from genes thought to be relevant to the function of lymphocytes and other related cells. The reference state was a pool of mRNAs isolated from nine different lymphoma cell lines.

In Section 2.6, we discussed clustering algorithms that can be applied to matrices of data such as this. Alizadeh *et al.* (2000) applied a hierarchical clustering to the columns of their data, to investigate the relationship between the expression patterns in the different samples. The resulting tree is shown at the top of Plate 13.1; the columns of the matrix have been ordered according to this tree. A larger version of the tree is illustrated on the left. It can be seen that the expression patterns differ significantly between the different cell types, and that cells of the same type from different patients almost always cluster together. This means that the expression pattern is able to distinguish between the different medical conditions and is potentially useful in diagnosis. Within the DLBCL samples, two distinct clusters were found that were characterized by high expression levels in two different sets of genes. It was later found that DLBCL patients falling into these two classes had significantly different survival prospects. Thus, the array data permitted the identification of two subtypes of the disease that had been indistinguishable morphologically.

The data matrix can also be clustered horizontally. This allows identification of sets of genes that have similar expression profiles across the samples. In Plate 13.1, some of the horizontal gene clusters have been labeled (Pan B cell, Germinal Center B cell, etc.), to indicate clusters of genes that are highly expressed in particular cell types and are thus characteristic of those cells. One type of study where horizontal clustering is particularly important involves time series data: e.g., in yeast, the response of cells to several types of stimuli has been studied, as has the periodic variation of gene expression that occurs during the cycle of cell growth and division (Eisen *et al.* 1998). Vertical clustering of the samples would not make sense for time series.

As emphasized in Section 2.6, there are many variants of clustering algorithms, and details of the clusters formed will depend on the algorithm. Leung and Cavalieri (2003) give a simple example with artificial data to illustrate this (see Fig. 13.4). This represents the log expression ratios for five genes at seven different time points. Several different measures of similarity/difference between the expression profiles are used. Let M_{ij} be the expression ratio of the *i*th gene at the *j*th time point. This matrix has *N* rows (genes) and *P* columns (arrays). Let μ_i be the mean of the *M* values for the *i*th gene. The "correlation coefficient with centering" is used in this example, defined as:

$$R_{ij} = \frac{\sum\limits_{k=1}^{P} (M_{ik} - \mu_i)(M_{jk} - \mu_j)}{\left(\sum\limits_{k=1}^{P} (M_{ik} - \mu_i)^2\right)^{1/2} \left(\sum\limits_{k=1}^{P} (M_{jk} - \mu_j)^2\right)^{1/2}}$$

The "correlation coefficient without centering" is

$$R_{ij} = \frac{\sum\limits_{k=1}^{P} M_{ik} M_{jk}}{\left(\sum\limits_{k=1}^{P} M_{ik}^2\right)^{1/2} \left(\sum\limits_{k=1}^{P} M_{jk}^2\right)^{1/2}}$$

The Euclidean distance between the gene profiles is

$$d_{ij} = \left(\sum\limits_{k=1}^{P} (M_{ik} - M_{jk})^2\right)^{1/2}$$

and the Manhattan distance is

$$d_{ij} = \sum\limits_{k=1}^{P} |M_{ik} - M_{jk}|$$

where $|\ldots|$ indicates the absolute value.

Each of the tree diagrams in Fig. 13.4 is generated by hierarchical clustering using the UPGMA algorithm, which is described in the context of distances between sequences in Section 8.3 and corresponds to the "group average" rule for similarities described in Section 2.6.2. The difference between the trees arises from the definition of similarity/distance used. In Fig. 13.4(a), the correlation coefficient without centering ranks D and E as the most similar, and also ranks A and B as similar. This measure is sensitive to the magnitude and sign of the data points. The Euclidean distance measure in (e) is fairly similar to (a), and the Manhattan distance in (f) is also fairly similar, except that the order of clustering of C, D, and E is different. The correlation coefficient with centering in (b) is noticeably different from any of these. This measure spots that the profiles for A and C are identical in shape although they differ in absolute level, and that the same is true for D and E. These two pairs have correlation coefficient 1 (distance 0) when this measure is used. If this pattern arose in real data, it might indicate that the genes were regulated by a common transcription factor that up- and down-regulated the genes in a similar way.

The use of the absolute values of the correlation coefficients in (c) and (d) clusters profiles that have mirror-image shapes, as well as those that have the same shape. For example, in (d), genes A, C, D, and E are all perfectly correlated. This might arise in a real example if genes were regulated by the same transcriptional control system, but one was repressed when the other was promoted. Different similarity measures therefore measure different things, and more than one measure might be meaningful

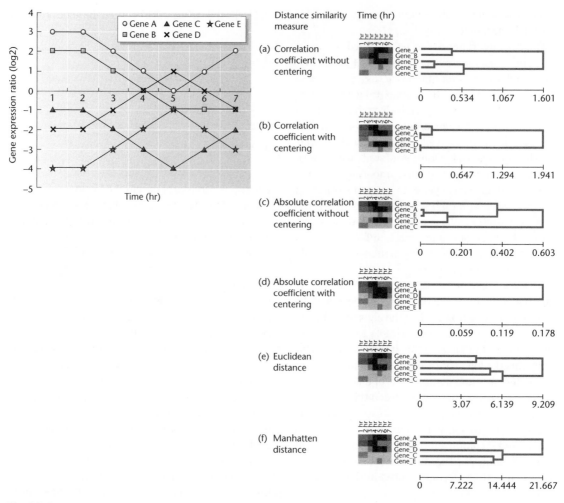

Fig. 13.4 Hierarchical clustering of gene-expression profile data for a series of seven time points. The data are artificial and intended to illustrate the sensitivity of the clustering procedure to the details of the measure of similarity between gene profiles (reproduced from Leung and Cavalieri 2003. Copyright 2003, with permission from Elsevier).

in a real case. It should also be remembered that all these examples use UPGMA. In Section 2.6, we already discussed alternative hierarchical and non-hierarchical clustering schemes, any of which could potentially be used with microarray data, and which could well give different clustering patterns, even if the same similarity measure were used. Although this example is artificial and designed to be sensitive to the similarity measure, it does provide a strong warning that we should not rely too much on any one method of clustering when analyzing real data sets.

13.4.3 Principal component analysis and singular value decomposition

We presented the PCA method in Section 2.5 using the example of amino acid properties. We considered each amino acid as a point in multidimensional space, and we aimed to select a small number of orthogonal directions in that space, known as the principal components, that would capture the greatest variance in position of those points. Plotting the point representing each amino acid in the space defined by the first two principal components showed clearly which amino acids were similar to one another, and suggested groups of amino acids that might be thought of as clusters with similar properties.

This method is also useful for visualizing patterns in microarray data. Plate 13.2 shows an example where PCA has been used to visualize the similarities and differences in the expression patterns of 60 different cancer cell lines (Slonim 2002). A certain amount of clustering of samples of different types is apparent. For example, the leukemia samples form a tight cluster (red dots almost superimposed), and the CNS samples form a cluster quite isolated from the rest (black dots). These clusters also appear in the hierarchical cluster analysis (underlined in Plate 13.2(b)). Other types of cancer appear to be much more variable in this example: e.g., the breast cancer samples (dark blue dots) do not appear close together in the PCA or in the hierarchical clustering.

Clustering algorithms will always create clusters, even if the data do not have well-defined groups. PCA makes it clear when there is variation between gene-expression profiles, but there is no clustering of genes. Raychaudhuri, Stuart, and Altman (2000) show an example where expression profiles of genes during a time-series experiment on sporulation in yeast have a smooth continuous distribution in the space defined by the first two principal components, and argue that clustering would be inappropriate in this case. Other good examples of visualization of structure within microarray data using PCA are given by Crescenzi and Giuliani (2001) and Misra et al. (2002).

The method referred to as singular value decomposition (SVD) by Alter, Brown, and Botstein (2000) is essentially the same as the PCA method described in Chapter 2, but the details of the mathematical presentation differ. An important point emphasized by Alter, Brown, and Botstein (2000) is that we often want to visualize patterns in both the genes (rows) and the arrays (columns) of the expression data matrix. The term eigenarray is used to describe a linear combination of columns of data from different arrays, while the term eigengene is used to describe

a linear combination of gene profiles from the rows. SVD identifies a set of orthogonal eigenarrays and eigengenes that allows the structure in the data to be visualized in a small number of dimensions. Alter, Brown, and Botstein (2000) studied gene expression during the yeast cell cycle, in which many genes have expression levels that rise and fall in a periodic manner. The first two eigengenes in this example are shaped like sine and cosine waves as functions of time, i.e., they are 90° out of phase. Plotting the data using the first two eigengenes/arrays clearly reveals the cyclic nature of these expression patterns (see Plate 13.3). See also Landgrebe, Wurst, and Welzl (2002) for another PCA analysis of cell-cycle data.

13.4.4 Machine-learning techniques

Microarray experiments provide complex multi-dimensional data sets that are interesting to study using a wide variety of techniques. Two techniques from the machine-learning field are self-organizing maps (SOMs) and support-vector machines (SVMs). We will give only the briefest of explanations of these methods here.

The SOM method has some similarity to a non-hierarchical clustering method, like the K-means algorithm discussed in Section 2.6.4. The object is to group the expression profiles of a large set of genes into a prespecified number of clusters. The additional feature of SOMs is that some type of geometric relationship between the positions of the clusters in the multidimensional gene-expression space is specified. This is often a rectangular grid. Tamayo et al. (1999) draw an analogy with an entomologist's specimen drawer, in which neighboring drawers, moving horizontally or vertically, contain insects of similar shape and size. During the calculation of the SOM, the grid points defining the centers of the clusters move as close as possible to the data points, and the grid is deformed but retains its rectangular topology. Each gene is then placed in the cluster corresponding to the nearest grid point at the end of the calculation.

Figure 13.5 shows an interesting application of an SOM to gene-expression data on metamorphosis in *Drosophila*. At the end of their larval period,

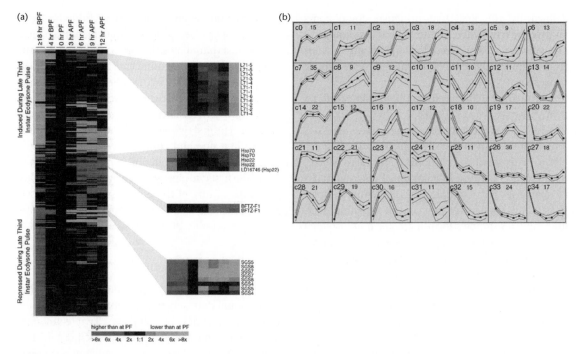

Fig. 13.5 Analysis of expression profiles of 534 genes that vary during the metamorphosis of *Drosophila* using (a) hierarchical clustering and (b) a self-organizing map (reproduced from White *et al.* 1999. Copyright 1999, AAAS). PF stands for puparium formation. Time points are labeled by the number of hours before or after this point (BPF and APF). The PF stage is used as the reference for the other samples; hence the solid black stripe in the third column of data. The SOMs figure shows the mean expression profile of genes in each cluster, together with lines indicating one standard deviation above and below the mean. The number of genes contributing to each cluster is shown above each profile. Cluster c15 corresponds to the co-regulated set of genes that is the uppermost block of expanded gene profiles illustrated in (a).

insects pupate, and emerge from the pupa as adults some time later. The coordinated change of expression of many genes is required to control this major change in the organism's morphology. The shapes of the gene profiles identified by the SOM show different sets of genes that are turned on and off at different times before or after formation of the pupa.

SVMs are used for "yes" or "no" classification problems, e.g., we want to ask whether a gene is or is not a member of a particular class of genes. The input data would be the expression profile for each gene across a series of arrays. We expect that the genes in the class being studied have some similar features in the shapes of their expression profiles, and the SVM is intended to recognize these features. The principles involved are similar to those in neural

networks (NNs) (see Section 10.5). For any given input, the program gives a yes or no output. The value of the output depends on the internal variables of the program. We begin with a training set composed of expression profiles for genes belonging to the class of interest (positive examples) and genes not belonging to this class (negative examples). As with NNs, the SVM is trained by optimizing the internal variables of the program such that it gives the correct answer for as many as possible of the examples in the training set. When the SVM is then used on further data, it will predict that genes with similar profiles to the positive examples are also members of the class.

In Section 10.5.4, we discussed the perceptron, the simplest NN, and showed that it could solve only linearly separable problems, i.e., those in which a hyperplane can be found that separates all the positive and negative points from one another. With NNs, the way to solve more complex problems is to make a network composed of many neurons. SVMs are another way of solving problems that are not linearly separable. They work by mapping each point in the original vector space onto a point in a higher-dimensional space in which the problem becomes linearly separable. The book by Cristianini and Shawe-Taylor (2000) gives a mathematical explanation of the method.

Brown *et al.* (2000) applied SVMs to expression array data from yeast. Many genes in yeast have known functions and have already been classified into functional categories in the MIPS database (Mewes *et al.* 2002). They trained SVMs to recognize several of the gene categories, and showed that this method outperformed several other types of classification algorithm. The results were also interesting from the biological point of view. In several cases, genes were "misclassified" by the SVM according to the expectations from the MIPS categorization; yet, it can be understood why there should have been similarities in the expression profile. For example, the gene for the translation elongation factor EFB1 was misclassified as a member of the ribosomal protein category by the SVM. Although the elongation factor is not actually part of the ribosome, and therefore was not included in this category by MIPS, it nevertheless plays a key role in translation, and is likely to be expressed at the same time as the ribosomal proteins. The SVM works with gene-expression data, and uses different criteria from MIPS for assigning genes to categories – so, in this case, both can be right. SVMs are relatively new to bioinformatics and have not yet been widely used. However, they are very general and have the potential to be applied to many different types of data.

13.5 PROTEOMICS

13.5.1 Separation and identification of proteins

The primary aim of proteomics experiments is to determine which proteins are present in a sample and how much of each one there is. To do this, it is necessary to separate proteins from one another. A standard technique for doing this is SDS-PAGE (sodium dodecyl sulfate – polyacrylamide gel electrophoresis). The proteins in a mixture extracted from a sample are denatured using SDS, a negatively charged molecule that binds at regular intervals along the protein chain: this means that they will be pulled in the direction of an electric field. The proteins are added to one side of a polyacrylamide gel, and an electric field is applied across it. Smaller chains are able to move faster through the pores of the gel, whereas longer ones get entangled and move more slowly. Hence, the proteins are separated according to their molecular weight. A stain is then applied, which reveals each protein as a band at a particular position on the gel.

In proteomics studies, we can have samples containing thousands of different proteins. SDS-PAGE does not have sufficient resolving power here. The most commonly used technique in this case is two-dimensional polyacrylamide gel electrophoresis (2D-PAGE). This involves a combination of standard one-dimensional PAGE, which separates proteins according to their molecular weight, and isoelectric focusing (IEF), which separates them according to their charge.

As some amino acids are acidic and some are basic, and as the numbers of amino acids of each type differ between proteins, the charge on proteins at neutral pH will differ. If the pH of the surrounding solution is varied, the equilibrium of ionization of the acidic and basic groups on the protein will be shifted. At some particular pH, called the isoelectric point, or pI, there will be no net charge on the protein. IEF is able to separate proteins with different pI values. The technique uses a gel in the form of a long narrow strip. Acidic and basic chemicals are bound to the gel in varying quantities along its length, so that an immobilized pH gradient is created from one end to the other (Görg *et al.* 2000). The protein sample is placed on this gel and a potential difference is applied. Charged proteins move through the gel in the direction of the electric field. As they move through the pH gradient, their net charge changes; each will then come to a stop at a position corresponding to its isoelectric point, where there is no net charge, and hence no force from the electric field.

The gel strip is then connected along one edge of a two-dimensional gel sheet that is used to separate the proteins in the second dimension using SDS-PAGE. After interaction with SDS, all the proteins have the same sign charge, so they will all move in the same direction into the 2D gel when a potential difference is applied, but they will be separated according to molecular weight. A stain is then applied, which reveals each protein as a spot at a particular position on the gel. To identify the protein present in a spot, the spot is cut out of the gel and the protein is digested with a protease enzyme (like trypsin), which breaks it down into peptide fragments. These fragments are analyzed with a mass spectrometry technique called matrix-assisted laser desorption/ionization (MALDI), which determines the mass of each fragment. See Pandey and Mann (2000) for more details.

The masses of the fragments provide a characteristic fingerprint for the protein sequence; therefore, the protein can be identified by comparing the measured masses with the masses of fragments that can be generated from sequences in a database. This is known as peptide mass fingerprinting (PMF). Henzel, Watanabe, and Stults (2003) discuss the history of techniques used for protein identification, and consider the improvements that have occurred since the early 1990s in mass spectrometry, sample preparation, and programs for identifying the mass fingerprint. The programs take a sequence from a protein sequence database, calculate the fragments that would be formed if it were digested enzymatically (e.g., trypsin cuts the chain after every lysine and arginine), and compute the masses of these fragments. When several measured fragment masses are available, this can often be matched uniquely to a single database protein. Sometimes, it is even possible to identify individual proteins from the same spot, in cases where these have not been fully resolved on the gel.

PMF may be complicated by several factors: there may be experimental error in the measured fragment masses; the database sequence may contain errors that affect the theoretical fragment masses; and there may be post-translational modifications of the protein, such as phosphorylation or glycosylation of certain residues, that will affect the experimental mass (Mann and Jensen 2003). PMF programs account for these errors. One frequently used program is ProFound (Zhang and Chait 2000), which uses a Bayesian probabilistic model to calculate the posterior probabilities that the set of observed fragment masses was generated from each database protein, and returns a ranked list of hits against the database.

Another obvious point is that PMF can only identify a protein if its sequence is already in the database. The method therefore works well for organisms with complete genomes, where, in principle, we have all the protein sequences. However, in practice, we are not exactly sure which of the ORFs are real genes, and we do not have complete information about the alternative splicing that might occur in a gene, and the many possible post-translational modifications. For organisms without complete genomes, the method is much more limited. It is sometimes possible to identify an unknown protein in one organism by comparing it to the theoretical fragment masses of a known sequence from a related organism. Lester and Hubbard (2002) have compared proteins from many different genomes and investigated how the proportion of conserved fragment masses drops off as the percentage identity between the proteins decreases. In fact, it drops off rather rapidly, which means that cross-species identification is only likely to work for very closely related species. However, there is still a considerable signal for more distantly related species, which may be useful if the fragment mass information is combined with other types of data.

13.5.2 A few examples of proteomics studies

Figure 13.6 shows two examples of 2D gels from Bae *et al.* (2003). Nuclei were extracted from cells of *Arabidopsis* by cell homogenization and density gradient centrifugation. Extracts of the nuclear proteins were then studied. On the gels shown, 405 and 246 spots were detected for the pH gradients 4–7 and 6–9, respectively, and a total of 184 of these were successfully characterized using PMF. Bae *et al.*

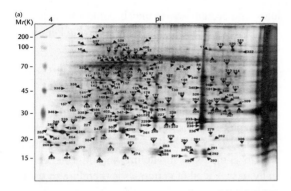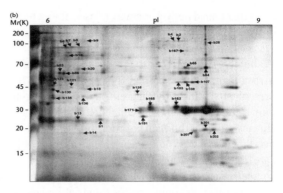

Fig. 13.6 2D gel electrophoresis map of nuclear proteins from *Arabidopsis* (reproduced from Bae *et al.* 2003). The vertical scale is protein molecular weight (in kDa). The horizontal scale is pI. Two gels were prepared from the same sample using different ranges of pH in the isoelectric focusing stage. Numbers indicate protein spots that were successfully identified from their mass fingerprint. Copyright 2003 Blackwell Publishing Ltd.

(2003) were particularly interested in the response of the proteome to cold stress, i.e., in which proteins were up- or down-regulated when the *Arabidopsis* seedlings were grown at unusually low temperatures. They used image analysis software to measure the sizes of spots and to determine which ones had changed under cold stress. Of the 184 identified proteins, 40 were induced and 14 were repressed by more than a factor of two. Proteomics thus allows the identification of genes that are interesting in the context of a particular biological question, and suggests where further biochemical experiments might be done.

Another recent large-scale study concerns the proteome of the human fetal brain. Fountoulakis *et al.* (2002) analyzed protein extracts from the brains of aborted fetuses. They were motivated by an interest in diseases of the central nervous system, including Alzheimer's disease and Down's syndrome. They analyzed 3,000 spots excised from 2D gels and were able to characterize 1,700 of these. Somewhat surprisingly, the 1,700 proteins were the products of only 437 different genes: there were typically 3–5 spots for each gene product; there were only 50 proteins that had a single spot; and there were five proteins with more than 100 spots. Although some of these multiple spots could be the result of artifacts in the 2D electrophoresis, Fountoulakis *et al.* conclude that most of the spots represent post-translationally

modified proteins. We are still far from understanding what the functions of all the protein variants might be.

More than 200 samples were analyzed, and it was found that relatively few proteins were detected in all the samples: ~12% of the proteins were detected in 10 or fewer samples, and most of the proteins were detected in fewer than half the samples. Thus, there is considerable variability in the sets of proteins detected. The authors argue that the proteins detected in only a few samples are likely to be the interesting ones, as they are more likely to be involved in disease-related changes. Nevertheless, it seems very likely that much of the apparent variability between samples is the result of problems with the experimental detection technique. Quantities of different proteins differ by orders of magnitude, resulting in widely different spot sizes (see Fig. 13.6). Faint spots are difficult to detect, and difficult to resolve from nearby larger ones. Fountoulakis *et al.* (2002) also give the distribution of lengths and numbers of fragments in the identified proteins. Longer proteins have more lysine and arginine residues and therefore produce more tryptic fragments. This makes them easier to identify, i.e., the probability of a false identification by chance drops rapidly with the number of fragments. In this study, the proteins have between four and 16 fragments per sequence, with an average around six.

As mentioned above, one way to ascertain which proteins have changed in expression level is to look at variation in spot size in the 2D gel images. Henzel, Watanabe, and Stults (2003) give an example of a myosin light chain 2 protein that was shown to be present at significantly different levels in two cell lines by repeated gel measurements (their figure 10). However, this method can be problematic for faint or overlapping spots, and needs to be performed with care (Moritz and Meyer 2003). In order to avoid comparisons between gels, it is possible to combine extracts from two samples on the same gel, and measure ratios of concentrations between them. This is the same principle as the Cy3/Cy5 labeling system for microarrays. One way of doing this with 2D gels is to use stable isotope labeling. Proteins from one sample are labeled by incorporation of a heavy isotope, like ^{13}C or ^{15}N, and proteins from the other sample are left unlabeled. The labeling can be done *in vivo*, by growing the cells in a medium containing the labeled isotope, or *in vitro*, by chemical modification after extraction of the proteins from the cells (Moritz and Meyer 2003). The labeled and unlabeled samples are combined before the 2D electrophoresis. Both sets of proteins should behave in the same way during the separation process. Hence, each spot should contain both a labeled and an unlabeled protein. The ratio of the two proteins can then be obtained by quantitative comparison of the sizes of the two peaks seen for each fragment in the mass spectrum.

Study of the proteome is not yet as automated or on such a large scale as microarray studies of the transcriptome. There is considerable interest in trying to develop an array-based technology for identifying proteins that would get around some of the limitations of 2D electrophoresis. With DNA arrays, the pairwise interactions between complementary strands make it easy to develop probes that are specific to one nucleic acid sequence. For protein arrays, it would be necessary to develop probe molecules to attach to the array that would bind specifically to one type of protein. One possibility would be to use antibodies, and another would be to select for RNA aptamers that recognize particular proteins. Some progress has been made along these lines, and protein arrays may become an important technique in the near future (Jenkins and Pennington 2001, Cutler 2003).

13.5.3 Protein–protein interactions

A large part of what we think of as "function" in a cell involves proteins interacting with one another – binding to one another, assembling into multiprotein complexes, modifying one another chemically, etc. A goal of genomics is therefore to try to establish which protein–protein interactions occur. Many proteins have unknown functions, but if we can establish that an uncharacterized protein interacts with one of known function, this is an important clue for understanding the function of the unknown protein.

The yeast two-hybrid (Y2H) system is an important experimental technique that has been used to investigate protein–protein interactions on a genomic scale (Vidal and Legrain 1999, Kumar and Snyder 2001). The technique makes use of a transcription factor that has a binding domain and an activation domain, such as Gal4. In nature, these domains are part of one protein. The binding domain recognizes and binds to the DNA sequence in the promoter region of genes that are regulated by the transcription factor. The activation domain interacts with the proteins that carry out transcription, and recruits them to the appropriate site on the DNA, which leads to the initiation of transcription of the regulated gene. In the Y2H experiment, the two domains are split apart and connected to two different proteins, known as the bait and the prey. The object is to determine if there is an interaction between the bait and the prey. The experiment is carried out using two haploid strains of yeast of opposite mating type. One strain contains a plasmid with a gene for the bait protein, to the end of which the DNA sequence for the binding domain has been added; this strain will synthesize a bait protein with an added binding domain covalently attached. The other strain has a different plasmid, with the gene for the prey protein and an added activation domain; this synthesizes a prey protein with a covalently attached activation domain.

Mating occurs between the two strains, producing diploid cells containing both plasmids. The diploid cells also contain a reporter gene, i.e., a gene whose expression causes an easily observable phenotype (Vidal and Legrain 1999). The reporter gene has a promoter that is normally regulated by the transcription factor. In order to express the reporter gene, both the binding and activation domains of the transcription factor are required. These domains are now on separate proteins and will only be brought together if there is an interaction between the bait and the prey. The experiment can be carried out in a high-throughput fashion by creating many different prey strains (i.e., each containing a different protein as prey), and allowing each to mate with the same bait strain. It is possible to tell which of the many prey proteins interact with the bait by selecting the diploid strains that express the reporter gene.

Y2H is rather indirect, and is prone to both false-positive and false-negative results. Nevertheless, it has produced a considerable amount of useful information (Ito *et al.* 2002), and with it we can, in principle, determine the complete set of interactions between proteins in a fully sequenced organism. A new 'ome has been coined for this set of interactions – the "interactome".

Using the available information on protein–protein interactions in yeast, Jeong *et al.* (2001) constructed a network where each node represents a protein and each link represents an interaction between a pair of proteins. This contained 1,870 proteins and 2,240 links. The largest connected cluster of proteins in this network is shown in Plate 13.4. Jeong *et al.* were interested in the distribution of links between nodes, and therefore calculated $P(k)$, the probability that a node has k links. The simplest model to which one can compare this network is a random network in which links are added at random between pairs of nodes. In such a network, each node would have the same number of connections on average, and $P(k)$ would have a fairly narrow distribution about this average. However, this is not the case in the yeast protein network: there are some proteins that have very many interactions, and others that only have one or two. A model that has this feature is a scale-free network

– for such networks, the link distribution is a power law $P(k) \approx k^{-\alpha}$. The observed distribution, shown in Plate 13.4, is approximately scale-free, with an exponential cut-off at large k and small corrections at low k. The equation $P(k) \approx (k + k_0)^{-\alpha} \exp(-k/k_c)$ fits the data quite well.

Biologically, we would expect that highly connected proteins are very important to the cell. Since the completion of the yeast genome, there have been systematic studies of yeast mutants in which one gene has been deleted. Some of these deletions are lethal, i.e., the cell cannot grow without the gene. Others have a quantitative effect on the growth rate, while a fairly large number have little or no observable effect. The coloring system for the nodes of the graph in Plate 13.4 shows this information. Plate 13.4 also shows the percentage of proteins that are essential to growth as a function of the number of links in the network. There is a strong positive correlation, indicating that the more highly connected proteins tend to be more important to the cell, because they carry out many different roles, or because they carry out a role that cannot be substituted by another protein. A similar point has been made by Fraser *et al.* (2002), who show that proteins with a larger number of interactions have a slower rate of evolution, as measured by the evolutionary distance between pairs of orthologs from *S. cerevisiae* and *C. elegans*. This is because for proteins with many interactions, a greater part of the protein is essential to its function, and there are fewer sites at which neutral substitutions can occur.

Another interesting study with yeast protein interaction networks is that of Bu *et al.* (2003), who wished to identify sets of highly connected proteins. A clique is a set of nodes in which every node is connected to every other; a quasi-clique is a set of nodes in which almost all nodes are connected to each other. They identified 48 quasi-cliques containing between 10 and 109 proteins. In most cases, the quasi-cliques contained mostly proteins of a single functional category. Where a protein of unknown function is a member of a quasi-clique, this is then a useful indication that the probable function of the protein corresponds to that of the majority of the other members.

Many proteins assemble into complexes containing many different molecules that together perform a specific function. Whereas Y2H is able to detect pairwise interactions between proteins, another technique, known as tandem affinity purification (TAP), is specifically designed to isolate complexes involving many proteins. The method works by inserting a sequence called the TAP tag onto the end of the gene for a target protein (Rigaut *et al.* 1999). The target protein is then synthesized with the extra section coded by the TAP tag. This contains an immunoglobulin G (IgG)-binding domain. Proteins with this domain can be separated from a mixture using an affinity column containing IgG beads, because they stick to the column. If the target protein is part of a complex, then other proteins associated with the target are also retained on the column. The mixture of proteins obtained in this way is then washed off the column, the proteins are separated from each other with SDS-PAGE, and mass spectrometry is used to identify them. One way of verifying that the mixture of proteins was really part of a complex is to use other proteins that are thought to be from the complex as the target – this should allow isolation of the same set of proteins (see Fig. 13.7).

Using this technique, Gavin *et al.* (2002) tagged 1,739 different yeast proteins, and identified 232 complexes. Of these, 98 were already known, and 134 were not previously known from other techniques. The number of proteins per complex varied from two to 83, with a mean of 12. The functions of many of the proteins found in the complexes were not previously known; therefore, the technique reveals important information about these molecules. Some proteins were found in more than one complex. Gavin *et al.* (2002) show a network diagram where each node represents a complex, and a link between nodes represents a protein shared by the complexes. The information from TAP is complementary to the Y2H studies: when the members of each complex were examined, pairwise interactions between proteins within complexes had already been identified by Y2H in a fairly small fraction of cases.

Dezso, Oltvai, and Barabasi (2003) carried out a computational analysis of the protein complexes identified in the study of Gavin *et al.* (2002). This combined information from the proteomics study with gene-expression data from microarray experiments, and with information on deletion phenotypes (i.e., on whether a gene is essential or non-essential for survival in gene-deletion experiments). They measured correlation coefficients between the mRNA expression profiles for all pairs of genes in each complex, and found that many complexes contained a core of proteins that were strongly correlated with one another according to expression profiles. These were usually of the same deletion phenotype; hence, they were able to classify complexes as either essential or non-essential according to the phenotype of the majority of the core proteins. This suggests that the core proteins are all required for functioning of the complexes, so that deletion of any one of them disrupts the function. The complexes also contained non-core proteins with low correlation in mRNA expression, and often these proteins had different deletion phenotypes from the core. Some proteins that were essential but not part of the core of one complex were also members of the core in another essential complex. Therefore, it appeared that the reason they were essential was because of their role in the second complex, not the first. It may be that the non-core proteins are only temporarily attached to the complexes and do not play a full role in their function, or are spurious interactions arising in the experiment. The study of Dezso, Oltvai, and Barabasi (2003) is a good example of what bioinformatics techniques can achieve in the post-genome age by integrating several types of data in a coherent way.

13.6 INFORMATION MANAGEMENT FOR THE 'OMES

Having read to the end of this book, you will be familiar with biological sequence databases and the type of information they contain. Submitting sequences to the primary databases ensures that everyone will have access to them and that information can be shared. Specifying a format in the primary sequence databases ensures that programs can be written to

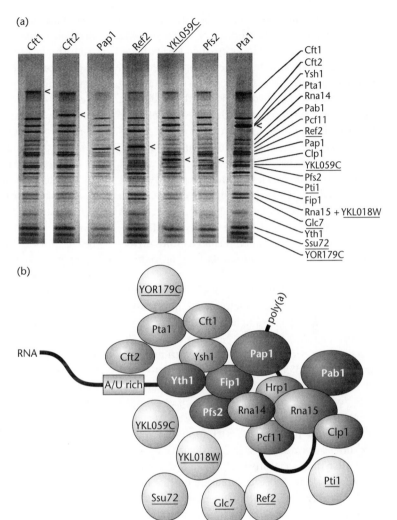

(a)

(b)

Fig. 13.7 An example of the TAP technique applied to the proteins in the polyadenylation machinery complex (Gavin *et al.* 2002, with permission of Nature Publishing Group), which adds a poly-A tail onto mRNAs. A schematic diagram of the complex is shown in (b). In (a), band patterns are shown from SDS-PAGE, with bands labeled according to the corresponding protein. The proteins labeled at the top of each lane were used as the target. The band from the target protein is marked with an arrowhead. This shows that the same bands are observed in each case, confirming that these proteins really do interact in a complex.

read and analyze the information contained in them. Having stable primary sequence databases allows those with expertise in specific areas to build on this information and create more specialized resources with in-depth information in narrower fields. Thus, a network of biological databases is developing that link to one another and that are mutually supportive.

However, sequences are only one type of biological information that we might wish to make available in databases. This chapter has discussed high-throughput experiments that produce huge

amounts of data that are not always easy to analyze. These results can become much more meaningful if they can be compared with similar results from other laboratories, or if they can be combined with entirely different sorts of information, or if they can be re-analyzed at a later date using newer algorithms. To make this possible, efficient means of storage and retrieval of high-throughput data are necessary.

For microarray data, a database has been established at Stanford, USA, where many of the early microarray experiments were performed (Gollub *et al.* 2003). This now contains publicly available data from more than 5,000 experiments, corresponding to information included in more than 140 publications. Experimental data files can be downloaded, and images of the arrays can be viewed, including zoom-in images of individual spots at the pixel level. Links are also available to a wide variety of software for array-data analysis. More recently, the Array-Express database has been established at the European Bioinformatics Institute (Brazma *et al.* 2003). At the end of 2003, this contained 96 experiments (i.e., related groups of arrays designed for a single purpose), 132 arrays (i.e., complete measurements from a single array), and 631 protocols (i.e., descriptions of how samples were prepared, how arrays were designed or how normalization procedures were applied).

In developing ArrayExpress, considerable thought was put into deciding exactly what information should be stored. For example, most data analysis is done on the ratios of red to green intensities, as discussed above. However, if only these ratios were stored, a lot of information necessary for determining errors and checking reproducibility between arrays would be lost. Therefore, it is necessary to include raw data in an unprocessed form. Also, it is much easier for database users if information on the experimental procedures is included with each data set, rather than separately in printed publications. Experimental protocols are separate entries in the database, so that when several different experiments use the same protocol, they can each link to that protocol entry without having to input the same information several times. The ArrayExpress developers have defined a set of standards known as Minimal Information About a Microarray Experiment (MIAME) (Brazma *et al.* 2001). MIAME specifies minimal required information in six categories: (i) experimental design, describing the type of experiment and which samples are used on which arrays; (ii) array design, specifying which genes are on the array and their physical layout; (iii) samples, describing the source of the samples, treatments applied to them, and methods of extracting the nucleic acids; (iv) hybridizations, describing the laboratory conditions under which the hybridizations

were carried out; (v) measurements, including the original image scans of the array, the quantified data from the image analysis, and the normalized gene expression matrix; and (vi) normalization controls, describing the method of normalization used. These standards are likely to be adopted by the Stanford database too, and compliance with MIAME is becoming a requirement for publication in certain journals (Oliver 2003).

The situation in proteomics is less advanced with respect to databases than with microarrays. The need for a central repository clearly exists, and for this reason, Taylor *et al.* (2003) have proposed a design for a Proteomics Experiment Data Repository, or PEDRo. Database design and database management systems are topics regrettably absent from this book, although there are many computer science books covering this field: e.g., Connolly, Begg, and Strachan (1999). Here, we briefly introduce the idea of schema diagrams to represent database designs. A key step in database design is to specify exactly which pieces of information will be in the database and how they are related to each other. Such a design plan is called a schema – the schema diagram for PEDRo, drawn in unified modeling language (UML), is shown in Fig. 13.8. UML is a "language", or a set of conventions, for drawing diagrams representing the relationships between types of information stored in a database. Each box, or "class", represents a type of object, and each line of text in the box represents an attribute of that type of object: e.g., the SampleOrigin class (top left) has attributes that describe the tissue and cell type used and the methods for generating the sample; the Gel class (right) has attributes that include the gel image, the stain used to visualize the proteins, the pixel size in the image, etc.; and the Gel2D class has attributes for the range of pI and molecular mass covered by the gel. When a database is implemented with such a schema, it will store many objects of each type. So, for example, every 2D gel entered into the database must have pI and molecular mass ranges associated with it (the values of these attributes will, of course, be different for every gel). The point is, the schema does not describe the particular entries in the database, but rather the properties

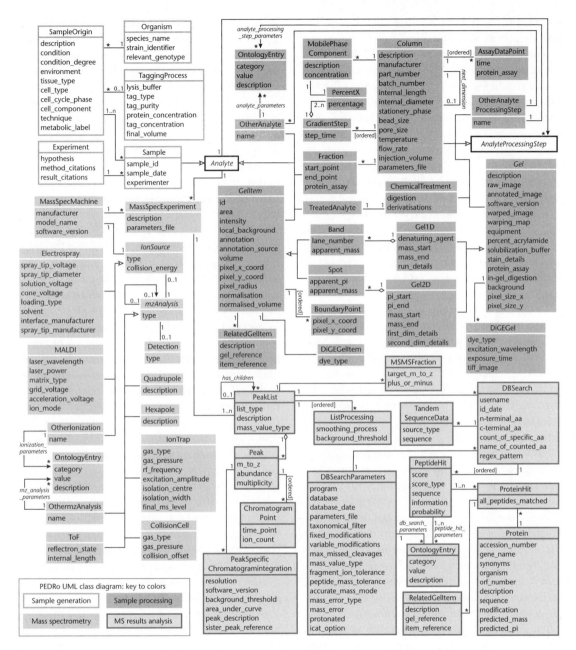

Fig. 13.8 Schema diagram for PEDRo (Taylor *et al.* 2003, with permission of Nature Publishing Group), a proposed Proteomics Experiment Data Repository.

possessed by each entry and the relationships between the objects described.

Relationships are represented in UML by lines connecting the boxes. The most common type of relationship in Fig. 13.8 is a one-to-many relationship, indicated by a line with a 1 at one end and a * at the other. Each SampleOrigin is associated with an Organism, but there may be many different SampleOrigins (e.g., different tissue types) from the same Organism. The Sample table contains brief information needed to identify a single sample, like identification number and date. Many samples may be associated with the same SampleOrigin (e.g., replicates extracted from the same tissue type). One reason for splitting information about samples into two classes is so that it is not necessary to repeat all the information in the SampleOrigin table for every sample.

We cannot cover details of UML here. An introduction to the use of UML in bioinformatics is given by Bornberg-Bauer and Paton (2001), and computer science textbooks are also available. The main point is that careful design of databases is essential if we are to store complex information in a useful way, and that expertise from both computer scientists and biologists is needed to make such ventures work. For further examples, take a look at the schema diagrams for ArrayExpress and the Stanford microarray database, which are available on their respective Web sites. Also, Bader and Hogue (2000) give an instructive account of the data specification of the Biomolecular Interaction Network Database (BIND), which includes information on many different types of interaction between many different types of molecule. The PEDRo schema we used as an example has been proposed to stimulate comments and potential improvements to the design. Let's hope they eventually get round to implementing the database!

In information management systems, we want to know exactly what information we have and where it is in the database – this is what the schema tells us. But we also want to know, in a more abstract way, what the data **mean** – this is where so-called ontologies come in. Ontologies attempt to describe all the terms and concepts used in some domain of knowledge, such as all the background knowledge we need to possess to understand the information in a particular database. The ontology should define the terms in an unambiguous way, so that when people use a term, they mean the same thing. The ontology should also tell us about the relationship between concepts, e.g., when one thing is part of something else, or when one thing is a specific example of a category of things. It should also make it clear when alternative terms are used to mean the same thing.

Unfortunately, scientists are not always as precise in their use of technical terms as we might like them to be. People use the same word in slightly different ways, or they describe the same thing with different words. Experts in a field gradually become familiar with the terminology of their subject, but this takes time. In the genomic age, bioinformaticians need to be generalists who can understand terms used in a wide range of fields and integrate information from different types of study. We cannot all be experts in everything. A biological ontology can help by defining things in a precise way, and encouraging people to use the same terminology in different subdisciplines. It can thus help people to communicate with each other and, at the same time, it can help **computers** to communicate with each other. Clearly, we can only crosslink databases if they deal with the same kinds of object (i.e., sequences, structures, species names, or whatever) and if they use the same name for the same thing. Gene names are a nightmare. There are committees doing important work in standardizing gene names for particular organisms, such as the HUGO Genome Nomenclature Committee, which maintains a database of approved names for human genes and publishes guidelines on gene nomenclature (Wain *et al.* 2002). However, when we compare genes between species, we immediately hit the snag that orthologous genes in different species often have completely different names, even though they do the same thing. This makes it difficult for a person to find all the relevant sequences, and even more difficult to write a computer program to do it. How nice it would be if there were a program that could handle queries like, "Find all the genes that have function X", or "Find all the proteins that interact with protein Y", or "Find all the genes whose expression profiles are correlated with gene Z". We are still far from being able to integrate all our available knowledge in a sufficiently coherent way to allow such complex queries to be performed for all organisms, but a step in the right direction is the establishment of the Gene Ontology (GO).

GO has been developed by a consortium of curators of species-specific genome databases, beginning with those for yeast, mouse, and *Drosophila*, and now extended to include *C. elegans*, *Arabidopsis*, and others (Ashburner *et al.* 2001). GO defines terms that are useful for the annotation of these genomes. The curators of the different species databases then agree to use these terms in the same way. In particular, specific genes in different organisms are associated with each of the relevant GO terms. In fact, GO consists of three ontologies. The first, "molecular function", describes what a gene product does at the biochemical level; the second, "biological process",

describes biological objectives to which usually more than one gene product contributes; and the third, "cellular component", describes places in cells where gene products are found. Examples of GO terms in each of these categories are given in Box 13.1.

The GO consortium (Ashburner *et al.* 2001) insists that GO is not itself a way of integrating biological databases, but that sharing nomenclature is an important step in this process. They emphasize that GO is not a dictated standard, but hope that groups will see the benefit of participating in the system and helping to develop it. They also warn that the association of genes from different organisms to the same GO molecular function is not sufficient to define them as homologs, although clearly this is a useful pointer. There are many things that GO does not include: it is not a complete ontology of biology. Nevertheless, it is rapidly expanding, and promises to be useful for a wide range of different organisms.

BOX 13.1
Examples from the Gene Ontology

Several browsers that allow searching of terms in GO are available from http://www.geneontology.org/. The examples here use the mouse genome informatics database browser at http://www.informatics.jax.org/searches/ GO_form.shtml. We searched GO with the phrase "transcription factor". This gave 11 hits to terms in "molecular function", nine to "biological process" and nine to "cellular component". One hit in the "molecular function" ontology is:
GO term: transcription factor activity
Definition: Any protein required to initiate or regulate transcription; includes both gene regulatory proteins as well as the general transcription factors.
A pathway of terms leading to "transcription factor activity" is shown opposite.

In this notation, ⓘ denotes an "is-a" relationship, and ⓟ denotes a "part-of" relationship. The above pathway is to be read backwards – transcription factor activity is a type of DNA binding function, which is a type of nucleic acid binding, etc., and the molecular function ontology is part of GO.

```
Gene_Ontology
    ⓟmolecular_function
        ⓘbinding
            ⓘnucleic acid binding
                ⓘDNA binding
                    ⓘtranscription factor activity [GO:0003700] (742 genes, 827 annotations)
```

Note that this ontology describes functions and not molecules, and hence avoids saying "a transcription factor is a type of DNA binding protein, which is a type of nucleic acid binding protein". The description in terms of function may seem rather convoluted grammatically, but it makes sense logically. The whole point of ontologies is to be precise about descriptions of concepts.

This above entry also tells us that there are 742 genes in mouse annotated as having transcription factor activity. A link to these genes is available when the ontology browser is used on the Web. GO is not a simple hierarchy, however: there is a second path from "molecular function" to "transcription factor activity":

```
Gene_Ontology
    ℗molecular_function
        ⓘtranscription regulator activity
            ⓘtranscription factor activity [GO:0003700] (742 genes, 827 annotations)
```

This classifies the function of transcription factors in terms of their action as regulators, rather than their DNA binding actions. Both descriptions make sense, and including both of them makes the ontology more flexible than it would be if it were just a hierarchy. The structure relating the terms in GO is called a directed acyclic graph: "directed" means that each link goes from a parent concept to a child concept, and "acyclic" means that a concept cannot be its own great grandfather.

A term from the "cellular component" ontology that matches the query "transcription factor" is as follows:

GO term: transcription factor TFIIA complex

Definition: A component of the transcription machinery of RNA Polymerase II. In humans, TFIIA is a heterotrimer composed of alpha (P35), beta (P19) and gamma subunits (P12).

One of the pathways leading to this term is:

```
Gene_Ontology
    ℗cellular_component
        ℗cell
            ⓘintracellular
                ℗nucleus
                    ℗nucleoplasm
                        ℗DNA-directed RNA polymerase II, holoenzyme
                            ℗transcription factor TFIIA complex [GO:0005672]
```

This says that the transcription factor TFIIA complex is part of the DNA-directed RNA polymerase II holoenzyme, which is part of the nucleoplasm, etc. Note that nucleoplasm is defined as "that part of the nuclear contents other than the chromosomes or the nucleolus".

Finally, an example of a term related to transcription factors from the "biological process" ontology is:

GO term: transcription factor-nucleus import

Definition: The directed movement of a transcription factor from the cytoplasm to the nucleus.

There are seven pathways leading to this term, one of which is:

```
Gene_Ontology
    ℗biological_process
        ⓘcellular process
            ⓘcell growth and/or maintenance
                ⓘtransport
                    ⓘintracellular transport
                        ⓘintracellular protein transport
                            ⓘprotein targeting
                                ⓘprotein-nucleus import
                                    ⓘtranscription factor-nucleus import [GO:0042991] (3 genes, 3 annotations)
```

These examples should serve to illustrate the nature of the three parts of GO, and give some idea of what is involved in developing an ontology.

SUMMARY

High-throughput experimental techniques are now available that allow studies of very large numbers of genes or proteins at the same time. These experiments aim to determine general patterns in the data or relationships that involve large numbers of genes, rather than to determine specific biological details about a single gene. Thinking big in this way is exciting, but it comes at the cost of relatively large experimental errors. The data generated are complex, and require specific computational and statistical techniques to analyze them and to extract the relevant information.

DNA microarrays are now widely used to study the transcriptome, i.e., to measure the concentrations of large numbers of mRNAs for different genes in a cell at the same time. The most frequently used type of arrays use cDNAs as probes. These are "spotted" onto the arrays robotically. Probes for thousands of genes can be attached to the same slide, so that, for organisms with small genomes, like yeast, it is possible to study the complete set of genes on one array. Experiments of this type compare expression levels from two samples on the same array. The reference sample is labeled with a green fluorescent dye, Cy5. The test sample is labeled with a red dye, Cy3. Labeled sequences from both populations competitively hybridize to the same spots. The relative concentrations of the two types are deduced from the relative fluorescence intensities of the two dyes.

Many types of bias arise in these experiments, and normalization procedures are necessary to remove the biases as far as possible. For analysis, data are usually expressed as log intensity ratios, M, for the red and green labels. Positive M indicates the gene has been up-regulated in the test sample relative to the reference, while negative M indicates down-regulation. The expression profile of a gene is the set of M values obtained for the gene in a series of arrays in a given experiment. The arrays might represent different cell types, different patients in a medical study, or different points in a time series. Clustering techniques are used to group genes with similar expression profiles. When genes have similar profiles, this suggests that they are regulated by the same process, and that they may share a common function. Clustering can also be done between arrays. This has revealed, for example, that different tumor cell types have characteristic patterns of up- and down-regulated genes that can be used as diagnostic features. Another important technique is PCA, which reveals structure in multidimensional data sets by projecting the points into a small number of dimensions where the variation between the points is most visible.

Proteomics aims to study the concentrations of large numbers of different proteins in a cell at the same time. Proteins can be separated from one another by 2D gel electrophoresis, and the proteins present in each spot on the gel can be identified by mass spectrometry. Other experimental techniques focus on the identification of interactions between sets of proteins. These may be pairwise protein–protein interactions, as in the Y2H system, or complexes of many proteins, as with TAP experiments.

High-throughput experiments produce huge amounts of data that need to be stored in a way that will be useful to other researchers and allow comparison between different data sets. For microarrays, the MIAME standard is now being adopted; this specifies the minimum amount of information necessary for the experiment to be reproduced by others and compared with other experiments. Databases for complex types of biological information, like results of microarray or proteomics experiments, need to have well-considered designs specifying the relationship between the types of data they contain. This requires input from both computer scientists and biologists.

One challenge for bioinformatics in the post-genome age is to integrate data from many different sources. This requires biologists in different areas of expertise to use a common language so that terms used in different databases really mean the same thing, and that entries in different databases can be systematically cross-linked. A step towards doing this is to define ontologies, which are specifications of the meanings of concepts used in a given field and the relationships between those concepts. The Gene Ontology has been developed as a tool to help unify the annotation of genome databases from a range of different species. It describes concepts related to molecular functions, cellular components, and biological processes.

REFERENCES

Alizadeh, A.A., Eisen, M.B., Davis, R.E., and 27 others. 2000. Distinct types of diffuse large B-cell lymphoma identified by gene expression profiling. *Nature*, **403**: 503–11.

Alter, O., Brown, P.O., and Botstein, D. 2000. Singular value decomposition for genome-wide expression data processing and modeling. *Proceedings of the National Academy of Sciences USA*, **97**: 10101–6.

Ashburner, M. and 18 others. 2001. Creating the Gene Ontology resource: Design and implementation. *Genome Research*, **11**: 1425–33. http://www.geneontology.org/.

Bader, G.D. and Hogue, C.W.V. 2000. BIND – A data specification for storing and describing biomolecular interaction, molecular complexes and pathways. *Bioinformatics*, **16**: 465–77.

Bae, M.S., Cho, E.J., Choi, E.Y., and Park, O.K. 2003. Analysis of the *Arabidopsis* nuclear proteome and its response to cold stress. *The Plant Journal*, **36**: 652–63.

Bornberg-Bauer, E. and Paton, N.W. 2001. Conceptual data models for bioinformatics. *Briefings in Bioinformatics*, **3**: 166–80.

Bu, D., Zhao, Y., Cai, L., Xue, H., Zhu, X., Lu, H., Zhang, J., Sun, S., Ling, L., Zhang, N., Li, G., and Chen, R. 2003. Topological structure analysis of the protein–protein interaction network in budding yeast. *Nucleic Acids Research*, **31**: 2443–50.

Brazma, A. and 23 others. 2001. Minimum information about a microarray experiment (MIAME) – Toward standards for microarray data. *Nature Genetics*, **29**: 365–71.

Brazma, A., Parkinson, H., Sarkans, U., Shojatalab, M., Vilo, J., Abeygunawardena, N., Holloway, E., Kapushesky, M., Kemmeren, P., Lara, G.G., Oezcimen, A., Rocca-Serra, P., and Sansone, S.A. 2003. ArrayExpress – A public repository for microarray gene expression data at the EBI. *Nucleic Acids Research*, **31**: 68–71. http://www.ebi.ac.uk/arrayexpress/.

Brown, M.P.S., Grundy, W.N., Lin, D., Cristianini, N., Sugnet, C.W., Furey, T.S., Ares, M. Jr., and Haussler, D. 2000. Knowledge-based analysis of microarray gene expression data by using support vector machines. *Proceedings of the National Academy of Sciences USA*, **97**: 262–7.

Connolly, T., Begg, C., and Strachan, A. 1999. *Database Systems: A Practical Approach to Design, Implementation and Management*. Harlow, England: Addison-Wesley.

Crescenzi, M. and Giulani, A. 2001. The main biological determinants of tumor line taxonomy elucidated by a principal component analysis of microarray data. *FEBS Letters*, **507**(1): 114–18.

Cristianini, N. and Shawe-Taylor, J. 2000. *An Introduction to Support Vector Machines*. Cambridge, UK: Cambridge University Press.

Cutler, P. 2003. Protein arrays: The current state-of-the-art. *Proteomics*, **3**: 3–18.

Dezso, Z., Oltvai, Z.N., and Barabasi, A.L. 2003. Bioinformatics analysis of experimentally determined protein complexes in the yeast *Saccharomyces cerevisiae*. *Genome Research*, **13**: 2450–4.

Eisen, M.B., Spellman, P.T., Brown, P.O., and Botstein, D. 1998. Cluster analysis and display of genome-wide expression patterns. *Proceedings of the National Academy of Sciences USA*, **95**: 14863–8.

Fang, Y., Brass, A., Hoyle, D.C., Hayes, A., Bashein, A., Oliver, S.G., Waddington, D., and Rattray, M. 2003. A model-based analysis of microarray experimental error and normalization. *Nucleic Acids Research*, **31**: 96.

Finkelstein, D., Ewing, R., Gollub, J., Sterky, F., Cherry, J.M., and Somerville, S. 2002. Microarray data quality analysis: Lessons from the AFGC project. *Plant Molecular Biology*, **48**: 119–31.

Fountoulakis, M., Juranville, J.F., Dierssen, M., and Lubec, G. 2002. Proteomic analysis of the fetal brain. *Proteomics*, **2**: 1547–76.

Fraser, H.B., Hirsch, A.E., Steinmetz, L.M., Scharfe, C., and Feldman, M.W. 2002. Evolutionary rate in the protein interaction network. *Science*, **296**: 750–2.

Gavin, A.C. and 37 others. 2002. Functional organization of the yeast proteome by systematic analysis of protein complexes. *Nature*, **415**: 141–7.

Gollub, J., Ball, C.A., Binkley, G., Demeter, J., Finkelstein, D.B., Hebert, H.M., Hernandez-Boussard, T., Jin, H., Kaloper, M., Matese, J.C., Schroeder, M., Brown, P.O., Botstein, D., and Sherlock, G. 2003. The Stanford Microarray Database: Data access and quality assessment tools. *Nucleic Acids Research*, **31**: 94–6. http://genome-www5.stanford.edu/.

Görg, A., Obermaier, C., Boguth, G., Harder, A., Scheibe, B., Wildgruber, R., and Weiss, W. 2000. The current state of two-dimensional electrophoresis with immobilized pH gradients. *Electrophoresis*, **21**: 1037–53.

Henzel, W.J., Watanabe, C., and Stults, J.T. 2003. Protein identification: The origins of peptide mass fingerprinting. *Journal of American Society for Mass Spectrometry*, **14**: 931–42.

Hoyle, D.C., Rattray, M., Jupp, R., and Brass, A. 2002. Making sense of microarray data distributions. *Bioinformatics*, **18**: 567–84.

Ito, T., Ota, K., Kubota, H., Yamaguchi, Y., Chiba, T., Sakuraba, K., and Yoshida, M. 2002. Roles for the two-hybrid system in exploration of the yeast protein interactome. *Molecular and Cellular Proteomics*, **18**: 561–6.

Jenkins, R.E. and Pennington, S.R. 2001. Arrays for protein expression profiling: Towards a viable alternative to two-dimensional gel electrophoresis? *Proteomics*, **1**: 13–29.

Jeong, H., Mason, S.P., Barabasi, A.L., and Oltvai, Z.N. 2001. Lethality and centrality in protein networks. *Nature*, **411**: 41–2.

Kumar, A. and Snyder, M. 2001. Emerging technologies in yeast genomics. *Nature Reviews Genetics*, **2**: 302–12.

Landgrebe, J., Wurst, W., and Welzl, G. 2000. Permutation-validated principal component analysis of microarray data. *Genome Biology*, **3**(4): research/0019.

Lemieux, B., Aharoni, A., and Schena, M. 1998. Overview of DNA chip technology. *Molecular Breeding*, **4**: 277–89.

Lester, P.J. and Hubbard, S.J. 2002. Comparative bioinformatic analysis of complete proteomes and protein parameters for cross-species identification in proteomics. *Proteomics*, **2**: 1392–405.

Leung, Y.F. and Cavalieri, D. 2003. Fundamentals of cDNA microarray data analysis. *Trends in Genetics*, **19**: 649–59.

Li, C. and Wong, W.H. 2001. Model-based analysis of oligonucleotide arrays: Model validation, design issues and standard error application. *Genome Biology*, **2**(8): research/0032.

Mann, M. and Jensen, O.N. 2003. Proteomic analysis of ost-translational modifications. *Nature Biotechnology*, **21**: 255–61.

Mewes, H.W., Frishman, D., Güldener, U., Mannhaupt, G., Mayer, K., Mokrejs, M., Morgenstern, B., Münsterkoetter, M., Rudd, S., and Weil, B. 2002. MIPS: A database for genomes and protein sequences. *Nucleic Acids Research*, **30**: 31–4. http://mips.gsf.de/genre/proj/yeast/index.jsp.

Misra, J., Schmitt, W., Hwang, D., Hsiao, L.L., Gullans, S., Stephanopoulos, G., and Stephanopoulos, G. 2000. Interactive exploration of microarray gene expression patterns in a reduced dimensional space. *Genome Research*, **12**: 1112–20.

Moritz, B. and Meyer, H.E. 2003. Approaches for the quantification of protein concentration ratios. *Proteomics*, **3**: 2208–20.

Oliver, S. 2003. On the MIAME standards and central repositories of microarray data. *Comparative and Functional Genomics*, **4**: 1.

Pandey, A. and Mann, M. 2000. Proteomics to study genes and genomes. *Nature*, **405**: 837–46.

Quackenbush, J. 2002. Microarray data normalization and transformation. *Nature Genetics* supplement, **32**: 496–501.

Raychaudhuri, S., Stuart, J.M., and Altman, R.B. 2000. Principal component analysis to summarize microarray experiments: Application to sporulation time series. Citeseer Scientific Literature Digital Library (http://citeseer.nj.nec.com/287365.html).

Rigaut, G., Shevchenko, A., Rutz, B., Wilm, M., Mann, M., and Seraphim, B. 1999. A generic protein purification method for protein complex characterization and proteome exploration. *Nature Biotechnology*, **17**: 1030–2.

Schulze, A. and Downward, J. 2001. Navigating gene expression using microarrays – A technology review. *Nature Cell Biology*, **3**: E190–5.

Slonim, D. 2002. From patterns to pathways: Gene expression data analysis comes of age. *Nature Genetics* supplement, **32**: 502–7.

Tamayo, P., Slonim, D., Mesirov, J., Zhu, Q., Kitareewan, S., Dmitrovsky, E., Lander, E.S., and Golub, T.R. 1999. Interpreting patterns of gene expression with self-organizing maps: Methods and application to hematopoietic differentiation. *Proceedings of the National Academy of Sciences USA*, **96**: 2907–12.

Taylor, C.F. and 24 others. 2003. A systematic approach to modelling, capturing and disseminating proteomics experimental data. *Nature Biotechnology*, **21**: 247–54. http://pedro.man.ac.uk/schemata.shtml.

Tusher, V.G., Tibshirani, R., and Chu, G. 2001. Significance analysis of microarrays applied to the ionizing radiation response. *Proceedings of the National Academy of Sciences USA*, **98**: 5116–21.

Vidal, M. and Legrain, P. 1999. Yeast forward and reverse "n"-hybrid systems. *Nucleic Acids Research*, **27**: 919–29.

Wain, H.M., Lush, M., Ducluzeau, F., and Povery, S. 2002. Genew: The Human Gene Nomenclature Database. *Nucleic Acids Research*, **30**: 169–71. http://www.gene.ucl.ac.uk/cgi-bin/nomenclature/searchgenes.pl.

White, K.P., Rifkin, S.A., Hurban, P., and Hogness, D.S. 1999. Microarray analysis of *Drosophila* development during metamorphosis. *Science*, **286**: 2179–84.

Wolkenhauer, O., Möller-Levet, C., and Sanchez-Cabo, F. 2002. The curse of normalization. *Comparative and Functional Genomics*, **3**: 375–9.

Zhang, W. and Chait, B.T. 2000. ProFound: An expert system for protein identification using mass spectrometric peptide mapping information. *Analytical Chemistry*, **72**: 2482–9.

SELF-TEST

This test covers material in Chapters 11, 12, and 13.

1 The comparative method for the deduction of secondary structures of RNA molecules relies on the fact that:

A: Substitutions cannot occur in the paired regions of RNA structures.
B: There are always three hairpin loops in the secondary structure of tRNAs.
C: Stems in RNA structures do not change their length.
D: Substitutions in sites that are paired with one another occur in a correlated way.

2 The alignment below is thought to be part of an RNA gene with conserved secondary structure. Which of the four conclusions would you draw? (The species names are abbreviations and are not relevant to the question.)

A: There is evidence for two hairpin loop structures in this region.
B: This region probably contains the anticodon loop of a tRNA.
C: This region probably contains a pseudoknot structure.
D: There is no evidence for secondary structure in this region.

```
LAMFLU : TAAACAAGAA-TAGTTTGTCTTGTTAAAGCGTACCGGAAGGTGCGCTTA :
PETMAR : TAAACAAGAA-TAGTTTGTCTTGTTAAAGCGTACCGGAAGGTGCGCTTA :
RAJRAD : TAAGAAAAACTTAGGATATTTTTCTAAAGTGTACTGGAAAGTGTACTTA :
SCYCAN : TAAGAGAAAGTCTGAGTATTTCTCTAAAGTGTACTGGAAAGTGCACTTA :
MUSMAN : TAAGAGAAGATCTGAGTACTTCCCTAAAGTGTACTGGAAAGTGCACTTA :
SQUACA : TAAGGAAAAGTCAGAGTATTTTCCTAAAGTGTACTGGAAAGTGTACTTA :
CHIMON : TAAGATAAAAATAGTATATTAATCTAAAGCGTACCGGAAAGTGCGCTTA :
LEPOCU : TAAAAAGGACATAGAGAGTCCTTTTAAAGTGTACCGGAAGGTGCACTTA :
AMICAL : TAAAAAGGACGCAGAGAGTCCTTTTAAAGTGTACCGGAAGGTGCACTTA :
ENGJAP : TAAGAGGGGAATGGAGTGCCCCTCTAAAGTGTACCGGAAGGTGCACTTA :
SARMEL : TAAAGGGGAAAGAGAGCGTCCCCTTAAAGTGTACCGGAAGGTGCACTTA :
CARAUR : TAAAAGGGAAAGAGAGTGTCCCTTTAAAGTGTACCGGAAGGTGCACTTA :
POLLOW : TAAGTAAGAAATAGAGCGTCTTACTAAAGCGTACCGGAAGGTGTGCTTA :
POLJAP : TAAGTAAGAAATAGAGCGTCTTACTAAAGCGTACCGGAAGGTGTGCTTA :
GADMOR : TAAGTAGGAACTAGAGTGTCCTGCTAAAGCGTACCGGAAGGTGCGCTTA :
ARCJAP : TAAGCAGGAAATAGAGCGTCCTGCTAAAGTGTACCGGAAGGTGCGCTTA :
PAROLI : TAAGCAGGGAATAGAGTGTCCTGCTAAAGTGTACCGGAAGGTGCACTTA :
```

3 Which of the following statements about pseudoknots is true?
A: Pseudoknots are helices formed between two separate RNA strands rather than between different parts of the same strand.
B: Pseudoknots do not occur in rRNA.

C: The interaction energies in pseudoknots are too strong to be accounted for within the dynamic programming algorithm for minimum free energy structure prediction.
D: Pseudoknots are usually ignored in the dynamic programming method because this greatly

simplifies the algorithm needed for structure prediction.

4 The likelihood of a set of sequences evolving on a specified tree is calculated according to the JC model (L_{JC}) and the HKY model (L_{HKY}). The two likelihoods are compared with a likelihood ratio test. Which of the following is true?

A: Since L_{HKY} is always greater than L_{JC}, the likelihood ratio test cannot tell us any useful information in this case.

B: If the two models do not have significantly different likelihoods we cannot reject the JC model.

C: These two models are nested, therefore the two likelihoods are bound to be equal.

D: If L_{HKY} is significantly greater than L_{JC} we can conclude that the sequences evolved according to the HKY model.

5 The number of synonymous substitutions per synonymous site, d_S, and the number of non-synonymous substitutions per non-synonymous site, d_N, are calculated for a pair of gene sequences.

A: We would expect d_N to equal d_S in mitochondrial genes because they evolve faster.

B: If the ratio d_N/d_S is unusually large, this may be an indication that the gene has been horizontally transferred.

C: The ratio d_N/d_S can vary between genes because of the varying strength of stabilizing selection acting on the sequences and because of the possibility of positive directional selection acting on some genes.

D: If the ratio d_N/d_S is unusually small, this may be an indication that selection on this gene is less stringent than on most genes.

6 Which of the following sets of species is listed in order of relatedness to humans from closest to most distantly related?

A: Baboon, Marmoset, Elephant, Tyrannosaurus, Shark, Amphioxus, Sea urchin.

B: Chimpanzee, Echidna, Armadillo, Sparrow, Xenopus, Hagfish, Locust.

C: Haddock, Amphioxus, Locust, Sea urchin, *Saccharomyces cerevisiae*, *Plasmodium falciparum*, *Yersinia pestis*.

D: Armadillo, Cobra, Haddock, Earthworm, *Arabidopsis thaliana*, *Escherichia coli*, *Saccharomyces cerevisiae*.

7 Phylogenetic studies using small sub-unit rRNA showed that . . .

A: there are three principal domains of life known as animals, plants, and bacteria.

B: the root of the tree of life is thought to be on the branch separating the bacteria from the other two domains of life.

C: there was an RNA world in the early stages of evolution of life on earth in which RNA carried out the roles played by DNA and proteins in today's organisms.

D: the rRNA genes in chloroplast genomes have a recognizable similarity to those in Cyanobacteria.

8 Which of these statements about the origin of mitochondria is true?

A: An organism in which functional copies of genes have been transferred from the mitochondrion to the nucleus is said to have undergone secondary endosymbiosis.

B: The mitochondrial genomes of all known eukaryotes are thought to share a common origin in a single occurrence of endosymbiosis.

C: One piece of evidence that mitochondria arose from endosymbiotic bacteria is that the mitochondrial genome of *Trypanosoma brucei* has many organizational features in common with bacterial genomes.

D: Eukaryotes may be defined as cells possessing mitochondria.

9 Which of the following statements about bacterial genomes is true?

A: The smallest known bacterial genome has approximately 500 genes.

B: The smallest known bacterial genome is approximately 65 kb long.

C: A single-celled eukaryote like *S. cerevisiae* has roughly 100 times as many genes as a bacterium like *E. coli*.

D: The genomes of bacteria living as intracellular parasites tend to be larger than those of free-living bacteria because the parasites require complex genetic systems to combat the immune system of their hosts.

10 Comparison of bacterial genomes between species indicates that . . .

A: over 95% of genes on a typical genome have orthologs on every other genome.

B: the more closely related the species, the larger the number of shared orthologous genes they possess.

C: it is very hard to change the order of genes on a genome without disrupting the function.

D: variation in length of bacterial genomes is due principally to variation in length of non-coding regions rather than variation in number of genes.

11 Two species are found to share a cluster of eight genes, but the genes are in different orders in the two species. The orders are represented by signed permutations.

species X 1,2,3,4,5,6,7,8
species Y 1,2,−5,−4,−3,8,6,7
The transformation between the two gene orders. . . .

A: cannot be achieved by inversions alone.
B: can be achieved by one translocation and one inversion.
C: can be achieved by three inversions.
D: requires six separate genome rearrangement events.

12 Which of the following biases in a microarray experiment could be eliminated by a dye flip experiment?
A: Errors arising from varying numbers of cells of different types in a tissue sample.
B: Systematic variation of spot size across an array.
C: Non-specific hybridization of RNA to the probe sequence.
D: None of these.

13 Which of the following statements about clustering algorithms is true?
A: Hierarchical clustering algorithms work by repeatedly dividing a set of points into two until there is only one point remaining in each cluster.
B: Hierarchical clustering algorithms require the user to pre-specify the desired number of clusters.
C: *K*-means clustering minimizes the distances between the points and the centers of the clusters to which they belong.
D: *K*-means clustering maximizes the mean distance between points in different clusters.

14 Which of the following statements is true?
A: Cy3 and Cy5 are fluorescent dyes used to visualize the protein spots in polyacrilamide gel electrophoresis.

B: In a two-dimensional gel, proteins are separated according to size and hydrophobicity.
C: Affymetrix chips require probe sequences several hundred nucleotides long.
D: The probe sequences on microarray chips are usually made of DNA.

15 The yeast two-hybrid system . . .
A: makes use of a transcription factor domain that recruits RNA polymerase to transcribe a reporter gene.
B: is a way of locating transcription factor binding sites in the 3′ regions of genes.
C: makes use of a DNA binding protein from the thermophilic bacterium *Thermus aquaticus*.
D: is a way of locating duplicated genes in the yeast genome.

Mathematical appendix

ENCOURAGEMENT TO "MATHEMOPHOBICS"

As a reader of this book, you will probably be studying for a science degree or you may already have one. This means that you will have been exposed to mathematics courses at some time in your past. However, we recognize that many biological scientists do their best to forget everything they were taught in mathematics lessons. We hope that most of the concepts in this book can be understood qualitatively without worrying too much about the mathematics. However, if you want to understand how some of the methods in bioinformatics work, then you will need to brush up on your mathematics a bit. The book *Basic Mathematics for Biochemists* by Cornish-Bowden (1999) is a good basic-level textbook that we recommend; however, there are many others, so find one you like.

Sections M.1–M.8 of this chapter briefly cover some basic mathematical definitions that we hope you know already (they are all in the Cornish-Bowden book, for example). If so, these sections will just be a reminder and a confidence builder; if not, they will summarize some important results that we make use of in this book, and they will tell you what you need to read up in more detail elsewhere. Sections M.9–M.13 cover the ideas of probability distributions and statistical tests. These distributions will be familiar to you if you have done any statistics previously. Examples are given of problems in sequence analysis where these probability distributions crop up.

If you think you already know this stuff, check the self-test and the problem questions at the end of this Appendix.

M.1 EXPONENTIALS AND LOGARITHMS

We know that $1,000 = 10^3$. Another way of saying this is that $\log(1,000) = 3$. Here, "log" means "logarithm to the base 10". In general, if $y = 10^x$, then $\log(y) = (x)$.

We can take logarithms to any base; for example, base 2. So, if $16 = 2^4$, then $\log_2(16) = 4$. The subscript 2 means that we are using base 2 logarithms. If the subscript is omitted, it is usually assumed that we are using base 10, i.e., $\log(x)$ means $\log_{10}(x)$.

As powers of 2 come up a lot, it is worthwhile remembering them. A couple of easy ones to remember are $2^6 = 64$ (because it begins with 6), and $2^{10} = 1,024$ (because it begins with 10). If we know 2^{10} is roughly a thousand, then 2^{20} must be roughly a thousand times a thousand = a million, and 2^{30} must be roughly 10^9, etc.

The most frequent type of logarithms used are **natural logarithms**, or logarithms to the base e. The constant e is approximately $2.718 \ldots$ – in fact, e is an irrational number, which means that its value cannot be written down precisely in a finite number of decimal places. As logs to the base e are important, there is a special symbol for them: $\ln(x)$ means $\log_e(x)$. Following the usual rules of logs, if $y = e^x$, then $\ln(y) = (x)$.

As an exercise, plot graphs of the functions $y = e^x$ and $y = \ln(x)$ over a range of values of x. Compare these graphs to the functions $y = x^2$ and $y = x^{1/2}$. (As you will remember, $x^{1/2}$ is another way of writing $\sqrt{x}$).

Some general rules of logarithms that are worth remembering are:

$$\ln(ab) = \ln(a) + \ln(b)$$

$$\ln(a/b) = \ln(a) - \ln(b) \qquad (1)$$

$$\ln(a^n) = n\ln(a)$$

These formulae are true for any values of a, b, and n. Try them out on a calculator.

Another way of writing e^x is $\exp(x)$. This is usually used when we want to write the exponential of a more complex function. For example, rather than write $e^{-(x-a)^2}$, it is easier to read if we write $\exp(-(x - a)^2)$. The exp and ln functions are inverses of one another and it is useful to think of them as canceling each other out, so that

$$\exp(\ln(x)) = x$$
$$\qquad (2)$$
$$\exp(n\ln(x)) = \exp(\ln(x^n)) = x^n$$

M.2 FACTORIALS

The factorial function, denoted by an exclamation mark, is defined as the product of the integers between n and 1:

$$n! = n \times (n-1) \times \ldots \times 3 \times 2 \times 1 \qquad (3)$$

Factorials are big numbers. For example 10! is more than three million, and 70! is greater than 10^{100} (my pocket calculator is unable to work it out). There is another special property of the factorial function: $0! = 1$. (Don't worry about this – it just **is**).

M.3 SUMMATIONS

The sign Σ is a Greek S and means that we are to take a **sum** of whatever follows it. For example,

$$\sum_{i=1}^{n} i = 1 + 2 + \ldots + n \qquad (4)$$

and for any function f:

$$\sum_{i=1}^{n} f(i) = f(1) + f(2) + \ldots + f(n) \qquad (5)$$

In the first case, we are to sum all the integers, starting with $i = 1$ and ending with $i = n$. There is a simple formula for the value of this sum:

$$\sum_{i=1}^{n} i = \frac{n(n+1)}{2} \qquad (6)$$

This works for any value of n. If $n = 3$, then $1 + 2 + 3 = 6$, and $3 \times 4/2 = 6$. Try it for a higher value of n.

The sum of the squares of the integers can also be written down:

$$\sum_{i=1}^{n} i^2 = \frac{n(n+1)(2n+1)}{6} \qquad (7)$$

For example, $1^2 + 2^2 + 3^2 = 3 \times 4 \times 7/6 = 14$.

Another important sum is known as a geometric series:

$$\sum_{i=0}^{n} x^i = 1 + x + x^2 + \ldots + x^n = \frac{(1 - x^{n+1})}{(1 - x)} \qquad (8)$$

For example, $1 + 1/2 + 1/4 = (1 - (1/2)^3)/(1 - 1/2) = 7/4$. As long as $x < 1$, then this series can be summed to an infinite number of terms:

$$\sum_{i=0}^{\infty} x^i = \frac{1}{(1 - x)} \qquad (9)$$

Many functions can also be written as series. Three that will come up in this book are:

$$e^x = \sum_{i=0}^{\infty} \frac{x^i}{i!} = 1 + x + \frac{x^2}{2} + \frac{x^3}{6} \ldots \qquad (10)$$

$$\ln(1-x) = -\sum_{i=1}^{\infty} \frac{x^i}{i} = -x - \frac{x^2}{2} - \frac{x^3}{3} \dots \qquad (11)$$

$$(1+x)^n = 1 + nx + \frac{n(n-1)}{2}x^2 + \dots \qquad (12)$$

The first of these works for any value of x. Try summing the terms on the right using a calculator for some small value of x (say 0.1 or 0.2) and show that they give the same figure as calculating e^x. The sum for $\ln(1-x)$ only makes sense when $x < 1$, otherwise you would have the log of a negative number. Again, pick a small value of x and show using a calculator that this series gives the right answer.

M.4 PRODUCTS

The sign $\prod$ is a Greek P and means that we are to take a **product** of whatever follows it.

$$\prod_{i=1}^{n} f(i) = f(1) \times f(2) \times \dots \times f(n) \qquad (13)$$

If we take the logarithm of a product, we get a sum, i.e.,

$$\ln\left(\prod_{i=1}^{n} f(i)\right) = \sum_{i=1}^{n} \ln(f(i)) \qquad (14)$$

M.5 PERMUTATIONS AND COMBINATIONS

Suppose a man has 50 CDs and a rack to put them in that has 50 slots. How many ways can he arrange his CDs in the rack? Well, he has 50 choices of CD for the first slot, 49 choices for the second slot, 48 for the third, and so on. The number of ways of doing it is therefore $50 \times 49 \times 48 \times \dots 3 \times 2 \times 1 = 50!$ An ordering of distinct objects is called a permutation. The number of permutations of n objects is $n!$, as in this example. In fact, this is the same problem as the number of possible routes for a traveling salesman who has to visit n cities (as in Section 6.1).

Now suppose the man wants to play three CDs in an evening. How many different possible selections does he have for his evening's entertainment? Simple – he has 50 choices for the first CD to be played, 49 for the second and 48 for the third. Thus, there are $50 \times 49 \times 48 = 117,600$ selections, which is more than enough for him to listen to a different selection of three CDs every day of his life. In making this calculation, we have assumed that the man is bothered about the order in which he listens to the CDs (Madonna followed by Bach followed by the Sex Pistols is different from Bach followed by the Sex Pistols followed by Madonna). The three CDs are a permutation of three objects selected from 50. The number of permutations of r objects selected from n has the symbol P_r^n. Following the same argument as for the CDs, we have

$$P_r^n = n \times (n-1) \times (n-2) \times \dots \times (n-r+1)$$

$$= \frac{n!}{(n-r)!} \qquad (15)$$

Note that when we take the ratio of the two factorials in the above equation, all the numbers from $(n-r)$ downwards cancel out on the top and bottom.

Our friend now decides to lend five CDs to his sister. How many different ways can he choose five CDs from his collection? As we now know, the number of ways of drawing five CDs one after the other from the rack is $P_5^{50} = 50!/45!$ However, the sister is not bothered about the order in which the CDs were drawn from the rack. She only cares about which five CDs she gets. This unordered set of five CDs is called a combination. The number of combinations of five CDs drawn from 50 is equal to the number of permutations of five CDs drawn from 50, divided by the number of different ways in which those same five CDs could have been drawn (= 5!). The number of combinations of r objects drawn from n is written C_r^n. In this case

$$C_5^{50} = \frac{50 \times 49 \times 48 \times 47 \times 46}{5 \times 4 \times 3 \times 2 \times 1} = 2,118,760$$

In the general case,

$$C_r^n = \frac{n!}{r!(n-r)!} \qquad (16)$$

Note that $C_r^n = C_{n-r}^n$ because the above formula is unchanged when we swap r for $n-r$. In other words, the number of ways in which the man can select five CDs to give to his sister is the same as the number of ways he can select 45 CDs to keep for himself.

There are some problems of permutations and combinations for you to try at the end of this Appendix.

M.6 DIFFERENTIATION

Suppose we plot a graph of a quantity y that is a function of a variable x (Fig. M1). Consider any point (x,y) on this curve, and another point $(x+\delta x, y+\delta y)$ slightly further along the curve. The gradient of the dotted line that connects these two points is $\delta y/\delta x$. In the diagram, the dotted line lies close to the original curve $y(x)$. As the distance, δx, between the points gets smaller and smaller, the line gets closer and closer to the curve. In the limit, where δx tends to zero, the slope of the line is equal to the slope of the

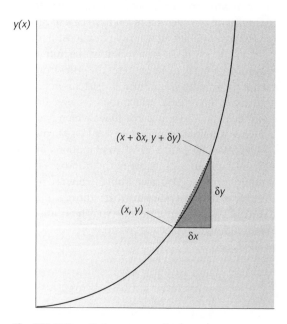

y(x)

$(x + \delta x, y + \delta y)$

δy

(x, y)

δx

Fig. M1 Differentiation measures the slope of a curve.

curve, and this is written as $\dfrac{dy}{dx}$. The gradient $\dfrac{dy}{dx}$ is referred to as the derivative of the function y, or the rate of change of y with respect to x. The process of taking the derivative of a function is known as differentiation. Here are two derivatives that you should definitely remember:

$$\frac{d}{dx}(x^n) = nx^{n-1} \qquad (17)$$

$$\frac{d}{dx}(e^{kx}) = ke^{kx} \qquad (18)$$

If the constant k is equal to 1 in Eq. (18), this just says that the derivative of e^x is equal to itself. One way of proving this is to take the series expansion for e^x in Eq. (10) and differentiate each term separately. The derivative of each term is equal to the previous term in the series (try this!) and so the series is unchanged by differentiation. Occasionally, we will require the derivative of the exponential of a more complicated function of x:

$$\frac{d}{dx}(\exp(f(x))) = \exp(f(x))\frac{df}{dx} \qquad (19)$$

This is an example of the "function of a function" rule.

The second derivative of a function y is written $\dfrac{d^2y}{dx^2}$. It is the derivative of the derivative, i.e.,

$$\frac{d^2y}{dx^2} = \frac{d}{dx}\left(\frac{dy}{dx}\right)$$

Returning to Fig. M1, the second derivative may be thought of as the curvature of the function $y(x)$. If we want to estimate the value of the function y at the point $x+\delta x$, then we can use a rule known as a Taylor expansion:

$$y(x + \delta x) = y(x) + \delta x\frac{dy}{dx} + \frac{1}{2}(\delta x)^2\frac{d^2y}{dx^2} + \ldots \qquad (20)$$

where each of the derivatives is to be evaluated at the point x, and the dots indicate that higher-order terms have been neglected. If δx is small enough, then the second-order term can also be neglected, and we are left with just the first-order term, as in the definition of differentiation above.

M.7 INTEGRATION

Integration is the inverse of differentiation. If y and z are two functions of x, and z is the derivative of y, then y is the integral of z:

$$z = \frac{dy}{dx} \Leftrightarrow y = \int z \, dx$$

A definite integral is one where the limits of integration are specified. If we have a graph of a curve $y(x)$ and two specified limits of integration x_1 and x_2 (as in Fig. M2), then the definite integral

$$\int_{x_1}^{x_2} y(x) \, dx$$

is the area of the shaded region.

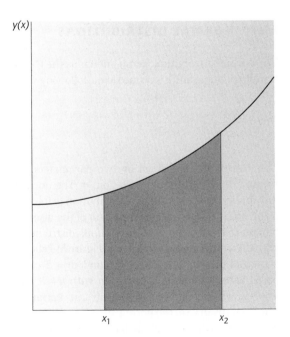

Fig. M2 Integration measures the area under a curve.

M.8 DIFFERENTIAL EQUATIONS

Suppose we know that the function y satisfies the equation

$$\frac{dy}{dx} = ky + c \qquad (21)$$

where k and c are known constants. What is y? This is one of the most common forms of differential equation. Whenever we see an equation of this form, we can use our prior experience to tell us that the solution must be of the form

$$y = Ae^{kx} + B \qquad (22)$$

where A and B are constants whose values are yet to be determined. To prove that this solution satisfies Eq. (21), we need to calculate the left- (LHS) and right-hand sides (RHS) of Eq. (21) and show that they are equal.

$$LHS = \frac{d}{dx}(Ae^{kx} + B) = Ake^{kx}$$

$$RHS = k(Ae^{kx} + B) + c = Ake^{kx} + Bk + c$$

We see that the term Ake^{kx} appears on both the left- and right-hand sides, so these terms are certainly equal. We end up with $Bk + c$ on the right, and if both sides are to be equal, then this must be zero. Thus, we have determined that B must equal $-c/k$, if the solution (22) is to satisfy Eq. (21). We have not yet determined A, because the solution will satisfy the equation for any value of A. In order to determine A, we need what is known as a boundary condition, or an initial condition. This is the value of the function specified at a particular point (usually at $x = 0$). Suppose that additional information that we have about the problem tells us that $y = y_0$ when $x = 0$. Putting these values into our trial solution (22) gives us $y_0 = A + B$, because $e^0 = 1$, and so

$$A = y_0 - B = y_0 + c/k$$

Thus, finally, we know that the solution to the equation is

$$y = (y_0 + c/k)e^{kx} - c/k \qquad (23)$$

This type of differential equation crops up several times in this book and the method described here is a general way of solving it.

M.9 BINOMIAL DISTRIBUTIONS

For any two constants a and b, we know that

$$\begin{aligned} (a+b)^2 &= b^2 + 2ab + a^2 \\ (a+b)^3 &= b^3 + 3ab^2 + 3a^2b + a^3 \end{aligned} \qquad (24)$$

We can calculate these things by multiplying out the brackets. For larger powers, n, we cannot do this by hand, so we need to remember the general formula

$$(a+b)^n = \sum_{r=0}^{n} C_r^n a^r b^{n-r} \qquad (25)$$

where C_r^n is the number of combinations of r objects selected from n, as in Eq. (16).

The binomial expansion (25) is related to the binomial probability distribution in the following way. Let a and b be the probabilities of two opposite outcomes of an event, e.g., if a is the probability of rolling a six on a die, then b is the probability of not rolling a six. As one or other of these opposing outcomes must happen, the sum of the two probabilities must be 1, so $b = 1 - a$. In the die-rolling example, $a = 1/6$ and $b = 5/6$ if the die is a fair one. Suppose the event occurs n times, then the probability that the outcome we are interested in occurs r times is given by the appropriate term in Eq. (25):

$$P(r) = C_r^n a^r b^{n-r} \qquad (26)$$

So this formula gives us the probability of rolling a six r times out of n die-rolls. As another example, let us return to the man with 50 CDs. Every day for 100 days running he selects three CDs at random to play. The probability that the Madonna CD is selected on any one day is therefore $3/50$. The probability that he listens to Madonna r times during this period

is given by Eq. (26), with $n = 100$, $a = 3/50$, and $b = 47/50$.

The binomial probability distribution is normalized so that summing up the terms over r gives 1:

$$\sum_{r=0}^{n} P(r) = 1 \qquad (27)$$

In fact, we know this already from Eq. (25), because the left-hand side is 1 owing to the fact that $a + b = 1$. The mean value of r and of r^2 and the variance of r (which we will call σ^2) can be calculated fairly easily, but as they crop up so often, it is worth remembering them:

$$\bar{r} = \sum_{r=0}^{n} rP(r) = na \qquad (28)$$

$$\bar{r}^2 = \sum_{r=0}^{n} r^2 P(r) = nab + n^2 a^2$$

$$\sigma^2 = \bar{r}^2 - (\bar{r})^2 = nab \qquad (29)$$

M.10 NORMAL DISTRIBUTIONS

The normal distribution is also known as the Gaussian distribution, and it is defined in the following way:

$$P(r) = \frac{1}{\sqrt{2\pi\sigma^2}} \exp\left(-\frac{(r - r^2)}{2\sigma^2} \right) \qquad (30)$$

The distribution is defined by two parameters: its mean, $\bar{r}$, and its standard deviation, σ. The normal distribution is a very common distribution in its own right, but it is also an approximation of the discrete binomial distribution by a continuous distribution with the same mean and variance. Figure M3 shows a normal distribution with $\bar{r} = 4$ and $\sigma^2 = 2$ compared with a binomial distribution with $n = 8$ and $a = 0.5$, which has the same mean and variance. The smooth curve of the normal distribution is quite a good approximation of the binomial distribution, although the binomial is only defined for integer

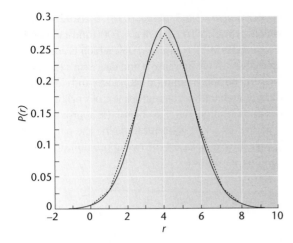

Fig. M3 A binomial distribution can sometimes be approximated by a normal distribution with the same mean and variance. The dashed line is a binomial distribution with mean 4 and variance 2, and the solid line is the normal distribution approximation.

values of r in the range $0 \leq r \leq 8$, whereas the normal distribution is defined for all values of r, and has some small probability of giving $r > 8$ and $r < 0$. The larger the value of n, the closer the binomial distribution becomes to the normal.

As the normal distribution is a continuous function, the normalization condition is an integral rather than a sum, as in Eq. (27).

$$\int_{-\infty}^{+\infty} P(r)dr = 1 \qquad (31)$$

The integral means that the area under the curve is equal to 1, which must be true for any function that is a probability distribution.

All normal distributions have the same bell-shaped curve. This means that they can be related to the distribution of a standard normal variable z, defined by:

$$z = \frac{r - \bar{r}}{\sigma} \qquad (32)$$

The distribution of z is a normal distribution, with mean equal to zero and standard deviation equal to 1:

$$P(z) = \frac{1}{\sqrt{2\pi}} \exp\left(-\frac{z^2}{2}\right) \qquad (33)$$

The quantity z is very useful for statistical tests. If we know that a quantity r has a normal distribution, and we know its mean and variance, then z tells us the number of standard deviations away from the mean. Suppose that we have a genome that we believe has a frequency of 25% of each of the four bases. We sequence a stretch of $n = 100$ bases from this genome and observe that the number of G+C bases is $r = 70$. The expected value of r is $\bar{r} = 100 \times 0.5 = 50$, and the variance is $100 \times 0.5 \times 0.5 = 25$, from Eq. (27). Therefore,

$$z = \frac{70 - 50}{\sqrt{25}} = 4$$

Now let us ask what is the probability of observing a stretch of length $n = 100$ with $\geq 70\%$ GC content? We could do it by using the binomial distribution, but we would have to calculate $P(70) + P(71) + P(72) + \ldots P(100)$. Life is too short for this! As n is quite large, the normal approximation should be good enough. The probability of observing a value of z greater than or equal to the observed value z_{obs} is

$$p = \int_{z_{obs}}^{\infty} P(z)dz \qquad (34)$$

Tables of integrals of the standard normal distribution are available in statistical textbooks. However, the value $z_{obs} = 4$ is very large, and is off the end of my statistical table (Daniel 1995). This means that p must be very small. Using Microsoft Excel, I found that $p = 3.17 \times 10^{-5}$. This confirms what we might have expected: it is extremely unlikely to observe a stretch of DNA with 70 out of 100 G and C bases if the frequencies of the bases are equal.

We can now formulate a statistical test to check whether an observed value of r is consistent with the normal distribution. We make the hypothesis that r has a normal distribution, with known mean and variance. We calculate z using Eq. (32), and we know that z should have a standard normal distribution if the hypothesis is true. The statistical tables tell us that there is a probability of 2.5% that $z > 1.96$, and by symmetry, there is a probability of 2.5% that $z < -1.96$. Thus, there is a probability of 95% that z lies in the interval $-1.96 \leq z \leq 1.96$. We say that the hypothesis is rejected at the 5% level if the observed value of z lies outside this range. This is a two-tailed test that rejects the hypothesis if z is either very high or very low. 5% is the probability that z is in one or other of the two tails of the distribution (see Daniel 1995, if you need more details on this).

M.11 POISSON DISTRIBUTIONS

The Poisson distribution is a limiting case of the binomial distribution that occurs when the number of events n is extremely large, but the probability of the outcome of interest a is very small. The archetypal example of this is radioactive decay. We have a sample of a material containing $n = 10^{23}$ radioactive nuclei. Each nucleus has a very small probability $a = 5 \times 10^{-23}$ of decaying per minute, and this is detectable via a click on a Geiger counter. Let r be the number of clicks we count in one minute. The probability distribution of r is given by Eq. (26),

$$P(r) = \frac{n!}{r!(n-r)!}a^r(1-a)^{n-r}$$

However, it is possible to derive a simplified formula for $P(r)$ that applies in this case. The mean number of clicks per minute is $\bar{r} = na$, which is 5 here. Typical values of r will be close to this value, whereas n is very much larger than this. As we know $n >> r$, we can make the approximation

$$\frac{n!}{(n-r)!} = n(n-1)(n-2)\ldots(n-r+1) \approx n^r$$

We can also use the approximation that

$$(1-a)^n \approx e^{-na} \qquad (35)$$

This approximation applies whenever n is large and a is small. In our case, $e^{-5} = 6.74 \times 10^{-3}$. If we set

$n = 100$ and $a = 0.05$, we get $0.95^{100} = 5.92 \times 10^{-3}$, which is roughly 10% out. If $n = 1000$ and $a = 0.005$, we get $0.995^{1000} = 6.65 \times 10^{-3}$, which is considerably closer. If we increase n and decrease a (keeping $na = 5$), the approximation gets closer and closer (try this yourself with a calculator). For our original values in the radioactivity problem, the approximation is almost exact. It is useful to define the parameter $\lambda = na$. If we now insert the two approximations into Eq. (26), we obtain

$$P(r) \approx \frac{n^r}{r!}a^r e^{-na}(1-a)^{-r}$$

$$\text{i.e., } P(r) = \frac{\lambda^r}{r!}e^{-\lambda} \qquad (36)$$

This result is the Poisson distribution. (Note that, in the final step, we forgot about the $(1-a)^{-r}$ as this is very close to 1 in the limit we are interested in). The Poisson distribution depends on a single parameter, λ, whereas the binomial distribution depends on two parameters, n and a. It is worth remembering that the variance of a Poisson distribution is equal to its mean λ.

Figure M4 shows a Poisson distribution with $\lambda = 5$, a binomial distribution with $n = 50$ and $a = 0.1$

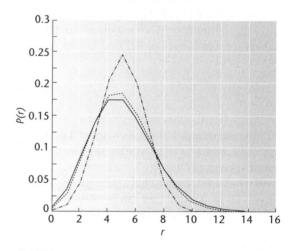

Fig. M4 A binomial distribution tends to a Poisson distribution when $n >> 1$ but the mean value remains constant. Solid line: Poisson distribution with mean $\lambda = 5$. Dashed line: a binomial distribution with $n = 50$ and $a = 0.1$. Dot-dashed line: a binomial distribution with $n = 10$ and $a = 0.5$.

(which is quite close to the Poisson distribution), and a binomial distribution with $n = 10$ and $a = 0.5$ (which also has a mean of 5, but is significantly different in shape from the Poisson distribution).

The Poisson distribution is relevant in the context of genome sequencing. Suppose we have a genome of total length L. We are sequencing many short fragments, each of length l (which is very much less than L), and we will suppose these to be positioned randomly within the genome. Consider one particular nucleotide in the genome. The probability that this falls within the range sequenced in any one fragment is $a = l/L$. After n fragments have been sequenced, the probability that the particular nucleotide has been sequenced r times is given by a binomial distribution, defined by n and a. However, we are in the limit where we can approximate it by a Poisson distribution with $\lambda = nl/L$. Therefore,

$$P(r) = \left(\frac{nl}{L} \right)^r \frac{\exp(-nl/L)}{r!} \tag{37}$$

M.12 CHI-SQUARED DISTRIBUTIONS

Let $z_1 \ldots z_k$ be k random numbers, each of which is drawn from a standard normal distribution $P(z)$ (Eq. (33)). The quantity

$$y = \sum_{i=1}^{k} z_i^2 \tag{38}$$

has a probability distribution called a χ^2 distribution, with k degrees of freedom. The shape of this distribution is:

$$P(y) = \frac{e^{-y/2} y^{\frac{k}{2}-1}}{2^{k/2} \Gamma(k/2)} \tag{39}$$

In the above formula, $\Gamma(k/2)$ is the gamma function discussed in Section M.13 below. This is just a constant that normalizes the distribution in Eq. (39).

The importance of this distribution comes from its use in statistical tests. Here we will give two brief

examples of the use of chi-squared tests on frequencies of bases in DNA (see Daniel 1995, if you need refreshing on this).

In the small subunit ribosomal RNA gene from *E. coli* (EMBL accession number J01859), the observed numbers of bases of each type are given below. The expected number in brackets is calculated under the hypothesis that all four bases have equal frequency.

i	$n_i^{obs}(n_i^{exp})$
A	389 (385.25)
C	352 (385.25)
G	487 (385.25)
T	313 (385.25)
Total	1,541

We calculate the quantity X^2 defined by

$$X^2 = \sum_i \frac{(n_i^{obs} - n_i^{exp})^2}{n_i^{exp}} \tag{40}$$

where the sum runs over the four types of base. It can be shown that X^2 has a distribution that is very close to the χ^2 distribution if the hypothesis is true. The number of degrees of freedom is 3 in this case (it is always one less than the number of categories). From statistical tables, the values of χ^2 corresponding to p values of 5% and 1% are

	0.05	0.01
1 d.o.f.	3.841	6.635
2 d.o.f.	5.991	9.210
3 d.o.f.	7.815	11.345

In this case, $X^2 = 43.3$. This is very much larger than 11.345; therefore, the probability of observing this distribution of bases in the gene if the bases have equal frequency is very much less than 1%. Microsoft Excel comes in handy again here: the probability of observing a value greater than or equal to 43.3 is 2.1×10^{-9}. Thus, there is a significant deviation from equal base frequencies.

Now let us compare the large subunit rRNA gene from *E. coli* (EMBL accession number V00331) with the small subunit gene. The observed numbers of bases for the two genes are given below (m_i for the large subunit and n_i for the small subunit). The expected numbers are calculated under the hypothesis that the frequencies of the bases are the same in both genes. For example, the average frequency of A is $1,151/4,445$; therefore, the expected number of A's in the first gene is $1,541 \times 1,151/4,445 = 399.0$.

i	$n_i^{obs}(n_i^{exp})$	$m_i^{obs}(m_i^{exp})$	Row totals
A	389 (399.0)	762 (752.0)	1,151
C	352 (343.6)	639 (647.4)	991
G	487 (485.0)	912 (914.0)	1,399
T	313 (313.4)	591 (590.6)	904
Total	1,541	2,904	4,445

X^2 is the sum of eight terms:

$$X^2 = \sum_i \frac{(n_i^{obs} - n_i^{exp})^2}{n_i^{exp}} + \sum_i \frac{(m_i^{obs} - m_i^{exp})^2}{m_i^{exp}} \quad (41)$$

which adds up to 0.71. The number of degrees of freedom, when dealing with tables like this, is equal to (number of rows -1) $\times$ (number of cols -1), which is $3 \times 1 = 3$ in this case. From the table of χ^2 values, 0.71 is much less than 7.81, hence the probability of this distribution of bases occurring is greater than 5% (in fact $p = 87\%$ according to Excel). Therefore, we have no reason to reject the hypothesis that the bases have the same frequency in the two genes. We can say that the processes of mutation and selection acting on these genes cause a significant bias away from equal base frequency, but that the same types of process seem to be working on both genes.

M.13 GAMMA FUNCTIONS AND GAMMA DISTRIBUTIONS

The gamma function $\Gamma(a)$ is defined by the integral

$$\Gamma(a) = \int_0^\infty x^{a-1} e^{-x} dx \quad (42)$$

We can show directly from this that $\Gamma(1) = 1$, and $\Gamma(2) = 1$. Also, for any value of a

$$\Gamma(1 + a) = a\Gamma(a) \quad (43)$$

Therefore, it follows that for an integer n

$$\Gamma(n) = (n - 1)! \quad (44)$$

Another handy fact is that $\Gamma(1/2) = \pi^{1/2}$. This, together with Eq. (43), allows us to calculate all

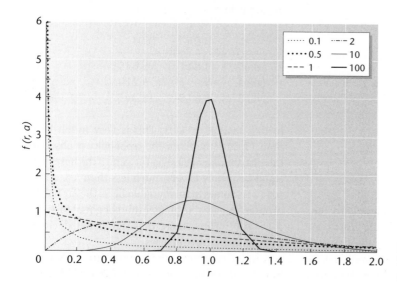

Fig. M5 Gamma distributions $f(r,a)$. Labels indicate the value of a on each curve.

the values $\Gamma(3/2)$, $\Gamma(5/2)$, etc. that occur in the chi-squared distribution (Eq. (39)).

The gamma distribution is a probability distribution for a variable x that can take any positive value between 0 and infinity. It is defined as

$$f(x,a) = \frac{x^{a-1} e^{-x}}{\Gamma(a)} \qquad (45)$$

From the definition of $\Gamma(a)$ in Eq. (42), we can see that Eq. (45) is a properly normalized probability distribution. In fact, Eq. (45) describes a whole family of distributions whose shapes are controlled by the value of a. The mean value of x is equal to a, with the definition above. It is useful to rescale the distribution so that the mean is at 1. Defining a new variable $r = x/a$, and changing variables, we obtain

$$f(r,a) = \frac{a^a r^{a-1} e^{-ar}}{\Gamma(a)} \qquad (46)$$

The shapes of the distributions $f(r,a)$ for several values of a are shown in Fig. M5. When $a < 1$, the weight of the distribution is close to $r = 0$ and there is a long tail extending to high r. When $a = 1$, the curve is a simple exponential. When $a > 1$, the distribution becomes peaked around $r = 1$. For very large a, the curve becomes a delta function. The mean value of r is 1 for all values of a.

REFERENCES

Cornish-Bowden, A. 1999. *Basic Mathematics for Biochemists*, second edition. Oxford, UK: Oxford University Press.

Daniel, W.W. 1995. *Biostatistics: A Foundation for Analysis in the Health Sciences*, sixth edition. New York: Wiley.

PROBLEMS

M1 In a card game, a player is dealt a hand of 13 cards from the pack of 52. How many possible hands of 13 cards are there? How many hands of cards are there that contain no aces? Hence, what is the probability that a player is dealt a hand with no aces?

In the same game, there are four players and each receives 13 cards. How many distinct ways are there of distributing the cards between the players?

M2 A genome contains 25% A, G, C, and T. What is the probability that a 10-base stretch of DNA will have a GC content $\geq 70\%$?

M3 In a stretch of DNA of length n bases, we observe 70% GC content. How long must the sequence be before we can reject the hypothesis that all bases have equal frequency? Use an argument based on the z score.

M4 The *Drosophila* genome is of length 180 Mb, and is to be sequenced in many short randomly positioned fragments of length 500 bases. How many fragments must be sequenced so that the fraction of the genome that has been missed by the random-sequencing process is less than 1%? What is the average number of times each base has been sequenced when this level of coverage is achieved?

SELF-TEST
Mathematical background

For each of questions 1–4, which of the mathematical statements is correct?

1

A: $\dfrac{1}{3} + \dfrac{1}{4} = \dfrac{1}{12}$

B: $\dfrac{1}{3} + \dfrac{1}{4} = \dfrac{1}{7}$

C: $\dfrac{1}{3} - \dfrac{1}{4} = \dfrac{1}{12}$

D: $\dfrac{1}{3} + \dfrac{1}{4} = \dfrac{5}{12}$

2
A: $\log(2) + \log(3) = \log(5)$
B: $\log(8) = 3\log(2)$
C: $\log(8) = 8\log(1)$
D: $\log(2) + \log(3) = \log(8)$

3
A: $2^{10} + 4^{10} = 6^{10}$
B: $2^{10} + 2^6 = 2^{16}$
C: $3 \times 2^{10} = 2^{30}$
D: $8 \times 2^{10} = 2^{13}$

4
A: $\sum_{n=4}^{n=8} n^2 = \sum_{n=1}^{n=8} n^2 - \sum_{n=3}^{n=8} n^2$

B: $\sum_{n=4}^{n=8} n^2 = \sum_{n=1}^{n=8} n^2 - \sum_{n=1}^{n=3} n^2$

C: $\sum_{n=1}^{n=8} n^2 = \frac{n(n+1)(2n+1)}{6}$

D: $\sum_{n=1}^{n=8} n^2 = 64!$

5 In a class of 20 students, the teacher asks for two volunteers. How many different ways can the two volunteers be chosen?
A: 2^{20}
B: 20×19
C: $20 \times 19/2$
D: $20!$

6 The class of 20 students is having a games lesson. How many different ways can the class be divided into two teams of 10?
A: $20!$
B: $10! + 10!$
C: $20! / (10! \times 10!)$
D: $20! / (2 \times 10! \times 10!)$

7 The value of the definite integral $\int_4^5 2x\,dx$ is

A: 9
B: 5
C: 2
D: 1

8 Which of the functions $f(x)$ is a solution of the following differential equation?

$$\frac{df}{dx} = 2f - 1$$

A: $f(x) = \frac{1}{2}(e^{2x} + 1)$

B: $f(x) = \frac{1}{2}e^{2x} + 1$

C: $f(x) = e^2 + 1/2$
D: $f(x) = e^x + 1/2$

9 A solution contains many RNA oligomers of length six bases. All four bases are present with equal frequency and are present in random orders in the oligomers. The proportion of oligomers containing exactly two guanines should be approximately:
A: 0.2966
B: 0.0625
C: 0.0198
D: 0.000244

10 A DNA sequence of length 100 nucleotides is generated randomly such that each of the four bases A, C, G, and T is chosen with equal frequency. The base at each position is chosen independently of the others. Let $P(n)$ be the probability that there are n Gs in the sequence. Which of the following statements is correct?
A: $P(n)$ can be approximated by a normal distribution with mean 20 and variance 75.
B: $P(n)$ can be approximated by a normal distribution with mean 25 and variance 18.75.
C: $P(n)$ can be approximated by a Poisson distribution with mean 25 and variance 25.

D: $P(n)$ is a bell-shaped curve with a peak at $n = 50$.

11 A DNA sequence of length 100 is obtained from a real sequence database. The numbers of the four different bases are:

A = 20; C = 27; G = 34; T = 19

The 5% significance values in the χ^2 table are

No. of degrees of freedom	5% significance value
1	3.84
2	5.99
3	7.82
4	9.49

A: The χ^2 test shows that the content of G+C bases differs significantly from 50%.
B: The χ^2 test shows that the content of G+C bases does not differ significantly from 50%.
C: More information is needed to carry out a χ^2 test.
D: A χ^2 test is not valid in this case.

12 A student answers a multiple choice exam paper with 12 questions on it. He guesses randomly from the four possible answers on each question. Which of the following is true?
A: The probability of guessing all 12 questions correctly is slightly greater than 3%.
B: The probability of getting exactly three correct answers is 0.0156.
C: The probability of getting exactly three correct answers is 1.17×10^{-3}.
D: The probability of guessing all answers wrong is greater than the probability of guessing all answers right.

List of Web addresses

NUCLEOTIDE SEQUENCE

EMBL	http://www.ebi.ac.uk/embl/
GenBank	http://www.ncbi.nlm.nih.gov/GenBank/
DDBJ	http://www.ddbj.nig.ac.jp/
INSD	http://www.ncbi.nlm.nih.gov/projects/collab/
dbEST	http://www.ncbi.nlm.nih.gov/dbEST/
UniGene	http://www.ncbi.nlm.nih.gov/UniGene/

ORGANISM

FlyBase	http://flybase.bio.indiana.edu/
SGD	http://yeastgenome.org/
WormBase	http://www.wormbase.org/
MGD	http://www.informatics.jax.org/
TAIR	http://www.arabidopsis.org/
Ensembl	http://www.ensembl.org/
TDB	http://www.tigr.org/tdb/

PROTEIN SEQUENCE

PIR	http://pir.georgetown.edu
MIPS	http://mips.gsf.de/
Swiss-Prot	http://www.expasy.org/sprot
NRL-3D	http://www-nbrf.georgetown.edu/pirwww/dbinfo/nrl3d.html
UniProt	http://www.uniprot.org/

PROTEIN FAMILY

PROSITE	http://www.expasy.org/prosite/
PRINTS	http://www.bioinf.man.ac.uk/dbbrowser/PRINTS/
Blocks	http://blocks.fhcrc.org/
Profiles	http://hits.isb-sib.ch/cgi-bin/PFSCAN/
Pfam	http://www.sanger.ac.uk/Software/Pfam/
eMOTIF	http://fold.stanford.edu/emotif/
InterPro	http://www.ebi.ac.uk/interpro/

PROTEIN STRUCTURE

PDB	http://www.rscb.org/pdb/
SCOP	http://scop.mrc-lmb.cam.ac.uk/scop/
CATH	http://www.biochem.ucl.ac.uk/bsm/cath/
PDBsum	http://www.biochem.ucl.ac.uk/bsm/pdbsum/
CDD	http://www.ncbi.nlm.nih.gov/Structure/cdd/cdd.shtml

METABOLIC/PATHWAY

KEGG	http://www.genome.ad.jp/
WIT	http://www-wit.mcs.anl.gov/Gongxin/
EcoCyc	http://ecocyc.org/
UMBBD	http://umbbd.ahc.umn.edu/

OTHER

OMIM	http://www3.ncbi.nlm.nih.gov/ entrez/query.fcgi?db=OMIM
BIND	http://www.blueprint.org/bind/ bind.php
GO	http://www.geneontology.org/
NAR	http://nar.oupjournals.org/
DBcat	http://www.infobiogen.fr/services/ dbcat

Glossary

accession number a unique identifier for a database entry (e.g., a protein sequence, a protein family signature) that is assigned when the entry is originally added to a database and should not change thereafter.

advantageous mutation a mutation that causes the fitness of a gene to be increased with respect to the original sequence.

algorithm a logical description of the way to solve a given problem that could be used as a specification for how to write a computer program.

alignment an arrangement of two or more molecular sequences, one below the other, such that regions that are identical or similar are placed immediately below one another.

allele one of several alternative sequences for a particular gene present in a population.

annotation (i) text in a database entry (e.g., for a gene sequence) that describes what the entry is (including relevant background information, literature references, and cross-references to other databases) to a human reader; (ii) the process of assigning readable explanations to all the entries in a database (e.g., the complete set of gene sequences in a genome).

archaea one of three principal domains of life. Archaeal cells can be distinguished from bacterial cells based on gene sequence and gene content. Many archaea are specialized to conditions of extreme heat or salinity.

bacteria (singular bacterium) one of three principal domains of life. Bacterial cells contain no nucleus and usually possess a cell wall of petidoglycan.

base chemical group that is part of a nucleotide. May be either a purine or a pyrimidine. The sequence of bases in a nucleic acid stores the information content.

BLAST (Basic Local Alignment Search Tool) a widely used program for searching sequence databases for entries that are similar to a specified query sequence.

BLOSUM (BLOcks SUbstitution Matrix) type of amino acid scoring matrix (derived from blocks in the Blocks database) used in aligning protein sequences.

bootstrapping a statistical technique used to estimate the reliability of a result (usually a phylogenetic tree) that involves sampling data with replacement from the original data set.

cDNA DNA strand that is complementary to an RNA strand and synthesized from it by a reverse transcriptase.

chloroplast (also called plastid) an organelle in eukaryotes such as plants and algae that possesses its own genome and is responsible for photosynthesis. Thought to have arisen by the incorporation of a cyanobacterium into an ancient eukaryotic cell (cf. endosymbiosis).

clade a group of species including all the species descending from an internal node of a tree and no others.

coalescence the process by which the lines of descent of individuals or sequences in a present-day population converge towards a single common ancestor at some point in the past.

codon triplet of three nucleotides that specifies

one type of amino acid during the translation process.

codon usage   the frequency with which each codon is used in a gene or genome, particularly the relative frequencies with which synonymous codons are used.

compensatory substitutions   a pair of substitutions at different sites in a sequence such that each one alone would be deleterious, but are neutral when both occur together (e.g., in RNA helices).

complementary sequences   two nucleic acid sequences that can form an exactly matching double strand as a result of A-T and C-G pairing. Complementary sequences run in opposite directions: ACCAGTG is complementary to CACTGGT.

deleterious mutation   a mutation that causes the fitness of a gene to be reduced with respect to the original sequence.

dynamic programming   a type of algorithm in which a problem is solved by building up from smaller subproblems using a recursion relation (e.g., in sequence alignment and RNA folding).

electrophoresis   method of separating charged molecules according to their rate of motion through a gel under the influence of an applied electric field.

endosymbiosis   process by which a bacterial cell was enclosed within a eukaryotic cell and eventually became an integral part of the cell. Thought to have given rise to chloroplasts and mitochondria.

eukaryotes   one of three principal domains of life. Eukaryotic cells possess a nucleus, and usually contain other organelles like, e.g., mitochondria, endoplasmic reticulum, Golgi apparatus, a cytoskelton of actin and tubulin.

exon   part of a gene sequence that is transcribed and translated to give rise to a protein (cf. intron).

fixation   the spread of an initially rare mutant allele through a population until it becomes the most frequent allele for that gene.

gap   a space in a sequence alignment indicating that a deletion has occurred from the sequence containing the gap, or that an insertion has occurred in another sequence.

gap penalty (or gap cost)   part of the scoring system used for sequence alignments intended to penalize the insertion of unnecessary gaps.

generalize   the ability of a supervised learning algorithm (e.g., a neural network) to perform well on data other than that on which it was specifically trained.

genetic code   set of assignments of the 64 codons to the 20 amino acids.

genome   the complete sequence of heritable DNA of an organism.

genomics   the study of genomes. Usually applies to studies that deal with very large sets of genes using high-throughput experimental techniques.

global alignment   an alignment of the whole of one sequence with the whole of another sequence.

heuristic algorithm   an algorithm that usually produces fairly good answers most of the time, but cannot be proven to give the best answer all the time.

hidden Markov model   a probabilistic model of a protein sequence alignment in which the probability of a given amino acid occurring at a given site depends on the value of hidden variables in the model.

high-throughput experiments   experiments that allow large numbers of genes or gene products to be studied at the same time using partially automated methods.

homologs   sequences that are evolutionarily related by descent from a common ancestor (cf. orthologs and paralogs).

horizontal gene transfer   the acquisition of a gene from an unrelated species by incorporation of "foreign" DNA into a genome.

hybridization   the binding of a nucleic acid strand to its complementary sequence.

hydrophobic   an amino acid is said to be hydrophobic if its free energy would be higher when in contact with water than when buried in the interior of a protein, i.e., it would "prefer" not to be in contact with water. In contrast, a hydrophilic amino acid would "prefer" to be in contact with water.

IEF (isoelectric focusing)   technique for separating proteins from one another according to the value of their pI (isoelectric point). Used as the first dimension in 2D gel electrophoresis.

indel   an insertion or deletion occurring in a protein or nucleic acid sequence. The term denotes the fact that we do not know whether a gap in a sequence is actually a deletion in that sequence or an insertion in a related sequence.

interactome the complete set of protein–protein interactions that occurs in a cell.

intron part of the DNA sequence of a gene that is transcribed, but is cut out of the mRNA (i.e., spliced) prior to translation. Introns do not code for protein sequences.

knowledge-based an approach that uses information from previously known examples in order to make predictions about new examples. Used in contrast to an *ab initio* approach, in which a prediction would be made based on a fundamental theory or principle.

local alignment an alignment of a part of one sequence with a closely matching part of another sequence.

locus a position on a chromosome corresponding to a gene or a molecular marker used in population genetics studies.

log-odds a quantity that is calculated as the logarithm of the relative likelihood of an event to its likelihood under a null model. A positive log-odds score indicates that the event is more likely than it would be under the null model.

machine learning an approach in which a computer program learns to solve a problem by progressive optimization of the internal parameters of the algorithm.

MALDI (Matrix-Assisted Laser Desorption/ Ionization) mass spectrometry technique used to measure the masses of peptide fragments.

metabolome the complete set of chemicals in a cell that are involved in metabolic reactions.

microarray a glass slide or silicon chip onto which spots of many different DNA probes have been deposited. Used for the simultaneous measurement of gene-expression levels of many genes in the same tissue sample.

mitochondrial Eve hypothetical female from whom all present-day human mitochondrial DNA is descended.

mitochondrion (plural mitochondria) an organelle in eukaryotic cells that possesses its own genome and is the site of aerobic respiration. Thought to have arisen as a result of the incorporation of an α-proteobacterium into an ancient eukaryote (cf. endosymbiosis).

monophyletic adjective describing a group of species on a phylogenetic tree that share a common ancestor that is not shared by species outside the group. A clade is a monophyletic group.

motif characteristic sequence of conserved or partly conserved amino acids appearing in a protein sequence alignment that can be used as a diagnostic feature of a protein family.

Muller's ratchet the accumulation of deleterious mutations in asexual genomes via a stochastic process that is virtually irreversible.

nested models two probabilistic models such that the simpler one is a special case of the more complex one.

neural network a type of machine-learning program used for pattern recognition, composed of a network of connected neurons.

neuron an element of a neural network having several inputs and one output. The output is a simple function of the combined inputs.

neutral mutation a mutation whose fitness is equal to the fitness of the original sequence, or whose fitness is sufficiently close to that of the original sequence that the fate of the mutation is determined by random drift rather than natural selection.

normalize for a probability distribution, to divide all probabilities by a constant factor so that the sum of probabilities over all possible alternatives adds up to 1. For microarray data, to adjust the measured intensities in order to correct for biases in the experiment.

nucleic acid a polymeric molecule composed of nucleotides. May be either DNA (deoxyribonucleic acid) or RNA (ribonucleic acid).

nucleotide chemical unit that forms the building block for nucleic acids. Composed of a nitrogenous base, a ribose or deoxyribose sugar, and a phosphate group (cf. nucleic acid, purines, pyrimidines).

null model a simple model used for calculating the probabilities of events that ignores factors thought to be important. If the observed data differ significantly from expectations under the null model, it can be concluded that the additional factors are necessary to explain the observation.

oligonucleotide   a nucleic acid fragment composed of a small number of nucleotides.

ontology   specification of the set of terms and concepts used in a domain of knowledge giving definitions of the terms and relationships between them.

ORF (open reading frame)   a continuous sequence of DNA in a genome beginning with a start codon and ending with a stop codon in the same triplet reading frame that could potentially be a protein-coding gene.

organelle   structural component of a eukaryotic cell.

orthologs   sequences from different species that are evolutionarily related by descent from a common ancestral sequence and that diverged from one another as a result of speciation.

outgroup   a species (or group of species) that is known to be the earliest-diverging species in a phylogenetic analysis. The outgroup is added in order to determine the position of the root.

PAGE (polyacrylamide gel electrophoresis)   technique for separating proteins from one another according to their molecular weight.

PAM (point accepted mutation) **matrix**   a matrix describing the rate of substitution of one type of amino acid by another during protein evolution.

paralogs   sequences from the same organism that have arisen by duplication of one original sequence.

parsimony   a principle that states that the simplest solution to a problem (or the solution using the fewest arbitrary assumptions) is to be preferred. In molecular phylogenetics, the tree that requires the fewest mutations is preferred.

pathogenicity island   region of a bacterial genome containing genes responsible for the pathogenic behavior of some strains of bacteria. The island is usually not present in all strains of the species and is thought to have been acquired by horizontal transfer.

PCA (principal component analysis)   technique for visualizing the relationship between points in a multi-dimensional data set in a small number of dimensions (usually two or three).

PCR (polymerase chain reaction)   experimental technique by which a specified sequence of DNA is multiplied exponentially by repeated copying and synthesis of both complementary strands.

perceptron   simplest type of neural network, having only a single layer of neurons.

peptide   a sequence of amino acids connected in a chain via peptide bonds. Usually refers to a sequence of a few tens of amino acids, in contrast to a protein, which usually contains a few hundred amino acids.

phylogeny   an evolutionary tree showing the relationship between sequences or species.

polymorphism   the presence of more than one allele for a given gene at an appreciable frequency in a population.

polynomial time   for a problem with input data of size N, a polynomial time algorithm will run in a time of order N^α, for some constant α. If α is small, the algorithm is efficient, and will work for large sized problems.

polyphyletic   adjective describing a group of species on a phylogenetic tree for which there is no common ancestor that is not also shared by species outside the group. A polyphyletic group is evolutionarily ill-defined.

posterior   in Bayesian statistical methods, the probability of an event estimated after taking account of the information in the data (cf. prior).

primary structure   refers to the chemical structure of a protein or nucleic acid. As far as bioinformatics is concerned, this is just the sequence of the amino acids or bases in the molecule.

primer   short nucleic acid sequence that is used to initiate the process of DNA synthesis from a specified position, e.g., during PCR.

prior   in Bayesian statistical methods, the probability of an event estimated from expectations based on previous experience, but before taking account of the information in the data to be analyzed (cf. posterior).

prokaryotes   cells lacking a nucleus. Thought to be divided into two principal groups: archaea and bacteria.

promoter   region of DNA upstream of a gene that acts as a binding site for a transcription factor and ensures that the gene is transcribed.

proteome the complete set of proteins present in a cell.

proteomics the study of the proteome, usually using two-dimensional gel electrophoresis and mass spectrometry to separate and identify the proteins.

purines the bases A (adenine) and G (guanine) that are present in nucleic acid sequences.

pyrimidines the bases C (cytosine), T (thymine), and U (uracil) that are present in nucleic acid sequences.

random drift change in gene frequency owing to chance effects in a finite sized population rather than to natural selection.

redundant gene a gene that is not necessary to the organism owing to the presence of another copy of the same gene or of another gene performing the same function.

regular expression a way of representing a sequence of conserved or partly conserved amino acids in a protein motif.

ribosome the organelle that carries out protein synthesis. Composed of small and large subunit RNAs and a large number of ribosomal proteins.

ribozyme an RNA sequence that functions as a catalyst (by analogy with an enzyme, which is a protein that functions as a catalyst).

root the earliest point on an evolutionary tree. The common ancestor from which all species on the tree have evolved.

rooted tree an evolutionary tree in which the root is marked, and in which time proceeds from the root towards the tips of the tree. In contrast, an unrooted tree does not specify the direction in which time is proceeding.

schema specification of the types of object in a database, the attributes they possess, and the relationship between the objects. A diagram representing this specification.

secondary structure elements of local structure from which the complete tertiary structure of a molecule is built. In proteins, the α helices and β strands. In RNA, the pattern of helical regions and loops.

selectivity the proportion of objects identified by a prediction algorithm that are correct predictions. Selectivity = True Positives/(True Positives + False Positives).

sensitivity the proportion of the true members of a set that are correctly identified by a prediction algorithm. Sensitivity = True Positives/(True Positives + False Negatives).

stabilizing selection when natural selection acts to retain ancestral sequences or characters and hence slows down the rate of evolutionary change.

stacking interaction between one base pair and the next in a double-stranded helix of RNA or DNA.

substitution the replacement of one type of nucleotide in a DNA sequence by another type owing to the occurrence of a point mutation and the subsequent fixation of this mutation in the population. Also the replacement of one type of amino acid in a protein by another type.

supervised learning a learning procedure where a program is given a set of right and wrong examples and is trained to give optimum performance on this set. In contrast, an unsupervised learning algorithm uses an optimization procedure without being given examples.

synapomorphy a shared derived character in parsimony analysis that marks a group of species as evolutionarily related.

synonymous substitution a substitution at the nucleic acid level that does not lead to change of the amino acid sequence of the protein (because of redundancy in the genetic code). A non-synonymous substitution is one that **does** change the amino acid.

symbiont an organism that lives with another organism of a different species in a mutually beneficial way.

TAP (tandem affinity purification) technique for simultaneously isolating sets of proteins that interact in the form of complexes.

taxon (plural taxa) name for a group of species defined in a hierarchical classification.

tertiary structure the complete three-dimensional structure of a molecule, such as a protein or RNA, built from the packing of its secondary structures.

Test set a set of known examples used for verifying the performance of an algorithm (e.g., a neural network). The test set must be distinct from the training set.

training set a set of known examples on which an algorithm (e.g., a neural network) is trained. The algorithm is optimized to give the best possible results on the training set.

transcription the synthesis of an RNA strand using a complementary DNA strand as a template.

transcription factor a protein that binds to a regulatory region of DNA, usually upstream of the coding region, and influences the rate of transcription of the gene.

transcriptome the complete set of mRNAs transcribed in a cell.

transition in the strict sense, the substitution of a purine by another purine, or of a pyrimidine by another pyrimidine. In the more general sense, any substitution in a DNA sequence, and sometimes also used for substitutions of amino acids.

translation the process by which the ribosome decodes the gene sequence specified on a mRNA and synthesizes the corresponding protein.

transversion the substitution of a purine by a pyrimidine or vice versa (cf. transition).

yeast two-hybrid experimental technique for determining whether a physical interaction occurs between two proteins.

Index

adaptation/adaptationist 49–53
additive trees 166, 171
algorithms 119–21
alignment
    global 123
    local 125
    multiple 130–6
    pairwise 121–30
    progressive 130–5
    statistics 147–55
alleles 7, 39, 51, 56
allozymes 51
amino acids
    physico-chemical properties
        22–32, 72–4, 245
    relative mutability 68–9, 72
    structure 14–15
aminoacyl-tRNA synthetase 20, 293
anticodon 20, 56–7
archaerhodopsin 208
Archezoa 301
Atlas of Protein Sequence and Structure
    89

Back-propagation 249
Baum–Welch algorithm 237
Bayesian methods 180–5, 229–33
bacteriorhodopsin 208–9
BIND 114
bioinformatics (definition) 6
BLAST 143–7, 153–5, 195
Blocks (database) 75, 95–6, 102–4
    (motifs) 95–6, 205–7
BLOSUM matrices 74–5, 147, 196
bootstrapping 169–71
BRCA1 40–3

breakpoint distance 306–8
Buchnera 288–9, 295

Captain Kirk 143–4
CASP 252
CATH 113
CDD 114
central dogma 16–22
chloroplasts 298–300, 303
clade 161
ClustalX/ClustalW 128, 132–5
clustering
    direct 33
    hierarchical 28–33
    K-means 33
    microarray data 320–2
    phylogenetics 162–9
coalescence 44–6
codon bias 53–4, 292
codons 19–20, 42, 53–4
COGs 294–6
coiled coils 237–8
color blindness 210
compensatory substitutions 258–9,
    265
cones 209
consensus trees 170, 184–5
correlation coefficient 28, 30
database
    accession numbers 82, 92, 97, 98,
        99, 102, 104, 110
    divisions 84
    ID codes 82, 92, 97, 99, 101, 102,
        105
data explosion 1–7, 11

dbCAT 81, 114
dbEST 87
DDBJ 1, 88
Dirichlet distribution 232–3
distance matrix 30, 132
DNA
    replication 21–2
    structure 12–13
dynamic programming 123–7, 238,
    260–4

EBI 84
EcoCyc 114
electrophoresis 5, 51, 325–6
EMBL database 1, 84–5
eMOTIF 95–6, 108
endosymbiosis 298–300
Ensembl 114
Entrez 85
ESTs 84, 85, 87
Escherichia coli 291–4
eukaryotes (evolution of) 277–8,
    298–301
E values 139–41, 145–7, 153–4
evolution (definition) 37–8
ExPASy 93, 105
Expectation-maximization algorithm
    237
extreme value distribution 149–53

false positives/false negatives 146
FASTA (file format) 83
    (program) 143–7
feature table 85, 92
file formats 82–111
fingerprints 95–6, 200–5, 213–14

Fitch–Margoliash method 171
fitness 38, 48–9
fixation 46–50, 56
flatfiles 83
FlyBase 114
forward and backward algorithms 239

G protein-coupled receptors 208–16
G protein coupling 215–16
gamma distribution 64, 175, 352
gap penalties 122, 126–7, 134
G–C skew 290
GenBank 1–4, 85–7, 154
gene content phylogenies 296–8
gene deletion 288, 304–5
gene duplication 131, 290–1, 304–5
gene ontology 110–11, 335–6
gene order phylogenies 305–9
general reversible model 63
generalization (in neural networks) 245, 248
genetic code 19–20, 73
genomics 5–6, 313–14
gradient descent/ascent 248
GSSs 84, 85

halorhodopsin 208
Hasegawa, Kishino, and Yano (HKY) model 63, 175
hexokinase 70, 128–34, 294
hidden Markov models 95–6, 234–43
high-throughput techniques 5, 313–30
hill-climbing algorithm 173, 181
homologs/homology 8, 161, 209
homoplasy 177
horizontal gene transfer 290–4
HTCs 84, 85
HTGs 84, 85
hydrophobicity 24–5

indels 40
informative sites 45, 178, 187
INSD 88
InterPro 108–11, 141–2
intron/exon 41
inversions 288, 304

ISI Science Citation Index 5–6
ISREC 96, 105

JIPID 88, 89
Jukes–Cantor model 60–2, 162

KEGG 114
Kimura two-parameter model 63

life (definition) 37
ligand binding (of GPCRs) 215–16
likelihood ratio test 267–70
log-odds scores 71–5, 229
long-branch attraction 179, 277, 301

machine learning 227–8, 323–5
mammals 42–3, 162–9, 175–85, 272–6
Markov Chain Monte Carlo 181–5, 232, 307
Markov model 234
mass spectrometry 5, 326
maximum likelihood 173–6, 178–81
MEDLINE 85
metazoa 276–7, 307–8
Metropolis algorithm 181
MGD 114
MIAME standard 332
microarrays 5–6, 314–25
microsatellites 40
MIPS 90
mitochondrial DNA/mitochondrial Eve 44–5
mitochondrial genomes 298–309
model selection 266–70
molecular clock 165, 275–6
Moore's law 2–4
motifs 96, 99–102, 197, 204
MPsrch 139–42
mutations 39–43

NCBI 85
nearest neighbor interchange 171–2
Needleman–Wunsch algorithm 123–5
neighbor joining 132, 166–9, 187
neural networks 244–53
neutral evolution 38, 41–7, 49–54

normalization 316–19
NRL3D 94–5
nucleic acid (see DNA and RNA)

'omes (definitions) 313–14
OMIM 90, 114
ontologies 334–6
open reading frames 286
opsins 210–11
orthologs 209–10, 291, 294
outgroup 160

PAGE (see electrophoresis)
PAM model
  derivation 65–72
  distances 69–72
  scoring matrices 71–4, 139–42, 147, 196, 202, 270
paralogs 209–10, 294
parsimony 65–6, 177–9, 187, 306
pathogenicity islands 291–2
pattern recognition 227–8
PCR 22–3
PDB 3, 111–12
PDBsum 113
peptide mass fingerprinting 326
perceptron 246–7
Pfam 95–6, 107–8, 243
pI 24–5, 325–7
PIR 89, 94–5
PIR SUPERFAMILY 109
polymorphism 39, 45
position-specific scoring matrix 144–5, 196, 205–8
principal component analysis 25–9, 322–3
PRINTS 95–6, 98–102, 213–14
prion proteins (in databases) 96–111
prior/posterior probabilities 180, 229–33
ProDom 107, 109
profiles 95–6, 105–7, 196, 205–8, 231
profile HMM 241–3
prokaryotic genomes 283–98
PROSITE 95–8, 197–200, 219–21
protein
  folding 14, 20–5
  structure 14–16
  structure prediction 250–3

protein family databases 95–111, 197–208
protein–protein interaction network 330
protein structure databases 111–13
protein tyrosine phosphatase 139–42, 145, 154
proteobacteria 284–94
proteomics 5–6, 325–30
pseudocounts 231
PSI–BLAST 144–5
PubMed 85

quartet puzzling 179–80

random drift 38, 47–50
Reclinomonas americana 299–300
recombination 46
regular expressions 95–7, 198–200
replicator 38
restriction fragment length polymorphism 51
Rhizobia 289–90
rhodopsin 209–11
ribosome 21
ribozyme 21
Rickettsia 284–5, 288–9, 300
rods 209
rooted/unrooted trees 158–61, 166
RNA
    free energy parameters 263–4
    processing/splicing 18
    rRNA 21, 162, 257–8
    structure 13–14, 257–66
    tRNA 14, 54, 56–7, 307–8
rules (regular expression) 198–9

SCOP 112
selection 48–54

selective sweep 49, 53
self-organizing map 323–4
sensitivity and selectivity 146
sensory rhodopsin 208
sequence alignment (*see* alignment)
serpentine receptors 209
SGD 114
Shakespeare 124–7, 131–2
SIB 90
signal peptides 250, 302
single nucleotide polymorphism 10, 52
SMART 109, 111
Smith–Waterman algorithm 126, 139–42, 153
sorting 119–20
sparse matrix 201, 202
splicing 18, 41
SRS 85
STSs 84, 85
substitution rate matrices 58–64, 264–7
SUPERFAMILY 109
supervised/unsupervised learning 236, 244
support vector machines 324–5
Swiss–Prot 85, 90–3, 139–42, 145, 154, 221–5
synapomorphy 177
synonymous/non-synonymous substitutions 40, 53, 56–7, 271–2

TAIR 114
tandem affinity purification 330–1
Taq polymerase 23
T–Coffee 135–6
TDB 114
TIGRFAMs 109

time reversibility 63
top 500 supercomputers 2–4
training set/test set 228, 244
transcription 16–18
transfer of genes from organelles to the nucleus 301–4
transition/transversion 39, 42, 63
translocations 288, 304
transmembrane helices 208–10, 215, 225, 229, 240–1
traveling salesman problem 120–1
tree optimization criteria 171
tree space 171–3
TrEMBL 85, 93, 223–4

ultrametricity 169
UM–BBD 114
unified modeling language 332–4
UniGene 114
UniProt 95
untranslated regions 18
UPGMA 164–6, 187

variability of rates between sites 64, 175
visual pigments 209–10
Viterbi algorithm 238, 242

WD-40 repeats 105–6
WIT 114
wobble 20
WormBase 114
Wright–Fisher model 47–50

Y chromosome 44–5
yeast two-hybrid 328–9

z score 148, 349